THE WORLD ALMANAC BOOK OF

INVENTIONS

THE WORLD ALMANAC BOOK OF

INVENTIONS

Valérie-Anne Giscard d'Estaing

WORLD ALMANAC PUBLICATIONS
New York, New York

A Note to the Reader

The subject of inventions is a controversial one. In numerous cases, many individuals and, often, many nations claim credit for an invention. We have done our best to present the facts as our research has shown them, but recognize that errors may appear in this book. We welcome readers to write to us with helpful comments and criticism in order that we may correct any erroneous information in future editions.

We thank the following for their meticulous work in translating the original French text: Guy Dess, Marc Chanliau, Ana Ciechanowska, David L. Clark, Suzanne Clonan, Bernard Cohen, Kevin Fenwick, Noah Hardy, Alexandra Harnden, Cole Harrop, Simone Hilling, George and Madeleine King, Philip Le Mire, William Mahder, Jessica Newell, Grant A. Poth, Arnold Rosin, Ashley Rountree, Marjorie Schwartz, Scribo, Nancy Swenton, Peter Skarimbas, Peter Weisman, and Serge Winkler.

Photo and illustrations were provided courtesy of Inter Twelve (France) with the exception of the following: P. 50, U.S. Department of Defense; P. 62, Massey-Ferguson; P. 100, Polaroid Corp.; P. 121(2) and page 122, Bell Laboratories; P. 176, American Express; P. 279 (all three), Seiko; P. 308, Renault; and P. 351, Boeing Aerospace Company.

First published in 1985.

Distributed in Canada by Macmillan of Canada

10 9 8 7 6 5 4 3 2 1
Newspaper Enterprise Association ISBN 0-911818-96-0

Printed in the United States of America

World Almanac Publications
Newspaper Enterprise Association
A division of United Media Enterprises
A Scripps Howard company
200 Park Avenue New York, New York 10166

contents

1. TRANSPORTATION

Railroads   4
Public Transport   6
Two-Wheelers   9
Automobiles   13
Balloons   17
Airplanes   19
Helicopters   22
Boats   25
Navigation Aids   28
Diving Equipment   31

2. ARMAMENTS

Small Arms   35
Artillery   39
Armor   42
Explosives   44
Chemical, Electronic and Nuclear
Weapons   45
Naval Vessels   48
Aviation   53

3. AGRICULTURE

Agriculture Machines   59
Phytobiology   65
Gardening   67
Insecticides   70
Animal Husbandry   71
Foods   74
Beverages   76
Preserving   77

4. THE ARTS

Writing   81
Writing Materials   83
Painting and Drawing   84
Printing   84
Publishing   86
Office Equipment   87
Instruments   89
Musical Composition and Performance   93
Sound Reproduction   93
Still Photography   96
Motion Pictures   101

5. MEDIA AND COMMUNICATIONS

Radio   109
Televsion   111
Video   115
Post and Telecommunications   117

6. GAMES, TOYS AND SPORTS

Games and Toys   125
Sports   135

7. EVERYDAY LIFE

Cooking   151
Clothing and Accessories   155
Beauty and Cosmetics   159
Dieting   164
Home and Leisure   164
Business   174
Crime   177

8. THE BIZARRE AND THE FUTURE

The Bizarre   181
The Future   190

9. MEDICINE

Assistance to the Sick   197
Investigations   197
Vaccinations   201
Respiration and Blood   203
Surgery   205
Gynecology   211
Drugs and Treatments   215
Optics and Acoustics   220
Dentistry   222
Parallel Medicine   225
Addendum   227

10. SCIENCE

Arithmetic   231
Geometry   231
Algebra   234
Physics   237
Statics—Mechanics   238
Changes of State—Thermodynamics   242
Electricity   247
Atomic Physics   251
Mineral Chemistry   254
Organic Chemistry   257
Genetics   260
Oceanography   264

11. INDUSTRY

Petroleum   269
Building Materials   271
Glass   273
Clocks and Watchmaking   275
Security Devices   279
Spinning   280
Looms   281
Knitting   282
Refrigerating   283
Machine Tools   285
Printing   287
High Technology   289

12. INFORMATION SYSTEMS

Data Processing Machines   293
Implementation of Machines   298
Robotics   307

13. ENERGY

Hydraulics   313
Steam Engines   315
Generation of Electricity   317
Electro Technology   320
Engines   321
Turbomachines   325
Rocket and Aircraft Engines   327
Nuclear Energy   328
Alternate Energy Sources   330

14. SPACE

Astronomy   337
Meteorology   340
The Conquest of Space   344

Index   353

transportation

Evolution of the Wheel

The wheel rolls. But it does much more than that: Think of the potter's wheel, the gear wheel, the driving wheel of a motor, etc.

The first wheel to appear in human history was the potter's wheel, circa 3500 B.C. It was followed by the chariot wheel (Wheel of Ur), circa 3200 B.C. The hydraulic wheel, which was actually the first motor, appeared in the Middle East, circa 400 B.C. About the same time, the cog-wheel ("cogged wheel"), which forms the basis of all gear systems, became known in Greece. The Greeks also invented the pulley.

For a long time, however, certain peoples remained ignorant of this important tool of progress. The Indians of North America probably never saw it until the arrival of the white man.

The wheel is also the motor of industry and is found everywhere: in watches, in motors, in pulleys—in virtually everything mechanical that we use today.

The Wheel of Ur (3000 B.C.) reconstituted from the bas-relief it figures upon.

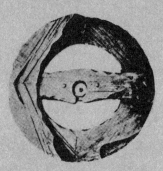

1400 B.C.: The Wheel of Mercurago, the first European wheel.

500 B.C.: A Persian chariot wheel of very modern design. It is even notched.

Assyrian wheel (700 B.C.) The appearance of spokes dates from 2000 B.C.

Wheel of a Gallic chariot (50 B.C.) The warriors of Vercingetorix fought Julius Caesar from these chariots.

The wheel with slanting spokes already existed under the Romans. But it was Leonardo da Vinci who laid down the theoretical principle.

The first wheels for mounting tires were still thought of in terms of their distant ancestors.

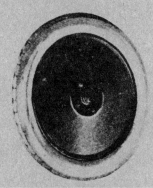

In 1914 Michelin invented the metal wheel, which led to the appearance of demountable tires.

The wheel is, above all, the motor of industry and is found everywhere.

GROUND TRANSPORTATION

1. Railroads

Locomotives

Trevithick locomotive (1802)

In 1802—1803 the English mechanic, **Richard Trevithick** (1771—1833), built the first steam locomotive at the Coalbrookdale ironworks. Soon afterwards, urged on by Samuel Monfrey, an iron master near Cardiff, he built a second one.

Conclusive tests began on February 2, 1804. Trevithick's locomotive, with a 6-ton convoy, ran on the Pen-y-Darren-Abercynon line (24 miles). A few passenger chariots were added. Empty, its speed was 32 mph; loaded, 13 mph. Unfortunately, the boiler was poorly suspended and a few rails were broken. Moreover, the use of horses remained more economical.

Trevithick invented high-pressure steam machines and also designed several prototypes of steam automobiles.

Blenkinsop locomotive (1812)

The first steam locomotive to be mass produced was built by an Englishman named **Blenkinsop**, beginning in 1812. An engine with an ordinary, nontubular steam boiler, Blenkinsop's locomotive was designed to carry goods and traveled at a very low speed. Its special characteristic was that it ran on a toothed railroad track.

Beginning in 1812, Blenkinsop's locomotives were used in England to equip the railway line running between Leeds and Middleton. They were used on this line until 1853, when they were replaced by more advanced models.

Seguin locomotive (1829)

In February 1828, **Marc Seguin** imported two early **Stephenson** loco-

motives from England for use on the railway line he was building between Lyons and St. Etienne. But he was not quite satisfied with them so, in 1829, began building his own locomotive, one with a tubular boiler. This locomotive could pull much heavier loads than Stephenson's but its speed hardly went beyond 6 mph. For this reason, Seguin stopped building them after trying about ten models. In turn, Stephenson adopted Seguin's boiler (see below).

Stephenson's Rocket (1829)

In 1829, the Englishman **Robert Stephenson** (1781—1848), developed the first high-speed steam locomotive. As early as 1813, he had built a steam locomotive equipped with driving wheels joined together by connecting rods. These wheels were smooth and rolled on rails.

His *Rocket* won the locomotive competition organized at Rainhill in 1829 by the contractors for the Liverpool—Manchester railroad line. Under the competition conditions (pulling a 40-ton train), the *Rocket* attained a speed of 16 mph, but with no train to pull it could reach up to 35 mph. This exploit marked the birth of the railroad. In comparison, the ordinary speed of a horse at full gallop is 28 mph. (The record speed of 36½ mph set by the English racehorse Consul in 1869 remained exceptional for that period.)

Patented locomotive (1833)

Perfected by George Stephenson in 1833, it differed from a previously built model, *Planet*, by the presence of an additional axle under the boiler's cylindrical body. This locomotive was called the *Patented* because its builder filed a patent to protect his invention (October 7, 1833).

Crampton locomotive (1846)

In 1846, an Englishman, **Thomas Russel Crampton,** built the first two locomotives embodying a principle he had conceived three years earlier. Crampton's idea was to build a high-speed locomotive modeled on Stephenson's *Long Boiler* but without the latter's major disadvantage; namely, a lack of stability resulting from the overhanging position of the boiler furnace in relation to the wheel axles. Crampton shifted one driving axle to the rear of the firebox, leaving only two axles under the cylindrical body of the boiler. His two locomotives, put into service in

The first steam locomotive, the Trevithick.

Upper left: George and Robert Stephenson's Rocket. Above, Crampton's locomotive. Left: the Pacific locomotive.

Belgium on the line running between Liege and Namur, easily reached a speed of 60 mph.

In 1848, Crampton built the *Liverpool,* a locomotive of huge size intended for use on the London-Wolverton line of the London and Northwestern Railway. It reached a speed of nearly 80 mph. It was quickly taken out of service because it put too much strain on the railroad tracks.

Pacific *locomotive (1892)*

The famous *Pacific* was put into service for the first time in the United States on the Missouri—Pacific Railroad in 1892. The *Pacific* developed 2200 horsepower and a steam pressure of 16 kilograms. It weighed 93 tons.

Electric *locomotive (1895)*

In 1895, the first electric locomotive was inaugurated in the United States, on the Baltimore—Ohio line. It weighed 96 tons, had four all-engine axles, and used a direct current of 550 volts.

In 1902, a 323-mph speed record was set by an electric locomotive between Zossen and Marienfelde, Germany.

Atlantic *locomotive (1900)*

The *Atlantic* locomotive appeared about 1900 on the Philadelphia-Atlantic City line, hence its name.

The use of the *Atlantic* spread to Europe in 1901. It then had 1500 horsepower, weighed 64 tons, and developed steam pressure of 16 kilograms.

Diesel *locomotive (1912)*

The first diesel locomotive was made in 1912, in Winterthur, Switzerland, by the firm of Sulzer. It weighed 85 tons and developed 1200 horsepower, weak when compared to the power of steam locomotives at the time.

Mountain *locomotive (1912)*

The *Mountain* appeared on the American railway network in 1912. It was a rapid steam locomotive that developed 2700 horsepower, weighed 112 tons, and had a maximum steam pressure of 17 kilograms.

Rails (1785)

The first metal rails for railroad use were designed by the Englishman **Jessop** in 1785.

These were the descendants of a long series of guiding systems. Paved roads with ruts for chariot wheels appear to have been used in antiquity, but this system disappeared in the Middle Ages. A document of 1550 mentions the existence of rails made of wood in an Alsatian mine at Leberthal. Similar rails were widely used in English coal pits in the 17th century. Wooden rails reinforced with nailed castiron strips appeared at Whitehaven in 1738.

The first wholly castiron rails, laid from 1763 to 1768 between Horsebay and Coalbrookdale, seem to have been the work of **Abraham Darby's** son-in-law, **Richard Reynolds.** These were not railways but tram roads for ordinary flangeless wheels.

The castiron projecting rail invented by Jessop underwent a certain number of improvements. In 1820 **John Birkinshaw** made wroughtiron rails. Castiron rails were replaced by steel rails when steam traction came into general use.

Rack railroads (1868)

In 1868, **Sylvester Marsch** built the Mount Washington line in New Hampshire, ascending inclines that reached gradients of up to 30 degrees. In 1870, the first rack railway in Europe was built at Righi, Switzerland. In 1885, a Swiss engineer from Lucerne, **Abt,** invented a system of triple gears with staggered cogs for the Harz Mountain railway in Germany. With this gear system, backlashes could be avoided.

Railroad signals (1849)

The running frequency of trains was originally based on the time interval that separated them. After numerous accidents caused by one train overtaking another, a block-signaling system was adopted by the New York & Erie in 1849. This first manual block system was improved upon by **Vignier**, a railroad worker with the Compagnie de l'Ouest who, in 1855, had the idea of combining the signals with the points, thus making dangerous maneuvers impossible. One year later, **Tyer** invented an electric signaling device that was adopted for use in the lower Blaisy tunnel on the Paris-Dijon line.

The automatic block system appeared in the United States in 1867. In 1871, **Franklin Pope** installed on the Boston & Lowell Railroad the first signaling system to be centrally controlled. In 1898, the first electric points system was developed in Paris to control the lines at the Gare de Lyon.

Railroad Cars

Freight car (1801)

Public freight traffic appeared on the Surrey Iron Railway in England, in 1801. The tariff varied according to the nature of the goods.

Passenger car (1832)

The oldest railroad passenger cars in the world were those introduced in 1832 on the Linz—Budweiss line in Austria. Designed by **Franz Anton von Gerstner**, the passenger cars, mounted on four wheels, consisted of a case in the form of a post chaise. There were two seats for the coachman, one in front and the other in the rear, so that the car could be harnessed in both directions. Motorized passenger trains did not yet exist; these first passenger cars to run on a railway were drawn by horses.

Sleeping cars (1836)

As early as 1836, cars with couchettes were in general use on certain American railroads. On the Atlanta—Augusta line, the six couchettes with which the cars were furnished were used as sofas during the day. In 1864, **George M. Pullman** invented the sleeping car that bears his name, which is famous for its great comfort. Soon afterwards, in 1871, the Belgian engineer **Georges Nagelmackers** created the International Sleeping Car Company.

Dining car (1840)

Dining cars were first built in the United States in 1840. They were designed for trains on which one could walk from one car to another. The Imperial Russian Railway line that serviced Odessa and Kiev in 1860 was well equipped with a dining car that had a table for 16 guests; its kitchen had a coal stove and ice compartments.

Air brakes (1869)

Originally, the brakes used on railroad cars consisted of a simple brake shoe that pushed against the wheel. They were maneuvered by hand. In 1869, an American, **George Westinghouse** (1846-1914), invented automatic air brakes. Compressed air was distributed to each car by a central air tank that fed auxiliary air tanks through a triple valve. When the pressure was lowered, a supply of air was released to the brake cylinder. This system was tried out on a passenger train for the first time in 1872. It contributed to the improvement of rail transport and the same principle has remained in use up to the present day.

George Westinghouse established the Westinghouse Electric Corporation, one of the largest corporations in the United States.

2. Public Transport

Taxi (1640)

Nicolas Sauvage, a French coachman, opened the first taxi business in 1640 on the rue Saint-Antoine in Paris. He started with a fleet of 20 coaches. In 1703, the police laid down laws for their circulation and gave each an easily readable number, thus

Heilman's electric locomotive.

Left: An early Pullman sleeping car. From left to right: the beds, the bathroom, and the compartment during the day. Below: A dining room in a carriage, around 1890.

introducing the first form of vehicle registration.

The first real use of the automobile as an individual means of public transport with a meter registering both speed and distance is attributed to **Louis Renault** who, in 1904, launched small, specially designed two-cylinder cars. On September 7, 1914, 700 Paris taxis requisitioned by **General Gallieni**—the famous red Renault two-cylinder models— helped **General Maunoury** to pile some 3000 reservists on the flank of the German army, thus contributing to its being routed.

First public transport system (1662)

Blaise Pascal (1623—1662), the French philosopher, was the inventor of the public transport system. In 1661, he proposed a system of

coaches that would "circulate along predetermined routes in Paris at regular intervals regardless of the number of people, and for the modest price of five sols." On January 19, 1662, the King's Council authorized the project's financiers, the **Marquis de Sourches** and the **Marquis de Crenan**, to begin running coaches in the city of Paris and its suburbs. The first coach went into service March 16, 1662; it ran between Porte Saint-Martin and Porte du Luxembourg. The company had four vehicles which covered the route, in both directions, every 8 minutes.

This public transport system attracted a good deal of curiosity, but its success was short-lived, for the coaches, ill-adapted to the tortuous, crowded medieval streets, were far too slow. Lacking patrons, the price rose to 6 sols, and the company went bankrupt 15 years later.

Tramway (1755)

The tramway was invented by the Englishman **John Outran** in 1775. This public transport vehicle ran on castiron rails and was drawn by two horses. It was not used in the city.

In 1832, **John Stephenson** built the first urban tramway between Upper Manhattan and Harlem. It ran for only three years. In 1852, a Frenchman, **Emile Loubat**, thought of embedding the rails in the pavement. That same year, he used his idea to build the Sixth Avenue line in New York. The cars were horse-drawn and open at both ends. In 1853, Loubat built the first Paris line between the Place de la Concorde and Saint-Cloud.

Electric tramway (1888)

The first operational electric tramway line was built in 1888 by the

Right: The omnibus was a great success. Below: Edison driving the first electric trolly.

American **Frank J. Sprague**. He obtained a concession for a 17-mile line to Richmond, Virginia. Ten years later 40,000 tramways were in use in the United States. Sprague's "premiere" had been preceded by a few prototypes: that of **Siemens and Malske** in Berlin in 1879, and that of **Edison** in Menlo Park in 1880.

Omnibus (1825)

In 1825, 150 years after the first public transport system was abandoned, a former soldier in the French Imperial Army, **Colonel Stanislas Baudry**, thought of using vehicles derived from the diligence (which could hold some fifteen passengers, including a conductor) to provide public transportation. Baudry made his vehicles available to Parisian customers of his bath house, which was in the suburbs. But, after noticing that many people who lived in the suburbs also used his coaches, he expanded the service. His terminus in the city was located at the Place du Commerce, in front of the shop owned by a certain **Monsieur Omnes**, whose sign included the words *Omnes omnibus*. Baudry found the word *omnibus* (which means "for

everybody") appealing and decided to use it for his transport line.

In Great Britain, the first omnibus line was inaugurated on July 4, 1819, by **George Shillibeer**. The buses had 22 seats and were drawn by three horses, and the conductors wore midshipmen's uniforms.

Motor bus (1831)

Walter Hancock, an Englishman, provided his country with the first motor bus in 1831. This 10-seat bus, powered by a steam engine, could carry 10 passengers. It was experimentally put into service between Stratford and the London that year; it was named the *Infant*.

Steam-powered buses were replaced by the gas-engine bus, first built in 1895 by the German firm Benz. The Benz bus was put into service on March 18, 1895, on a 15-kilometer route in the northern Rhineland. There was room for six to eight passengers and two conductors, who remained outside.

Subway (1863)

The first underground urban railway was built in London in 1863. It was

6.4 kilometers long and used steam traction. By 1890, the famous "Underground" had become the world's first underground urban rail network.

The first elevated railway was built in New York City in 1868. A viaduct supported the line, called the "New York Elevated Railway" (soon shortened to "the El").

Subway science fiction

Currently under study in the United States is the *Planetran*, a super-subway circulating in a vacuum-sealed tunnel. Circa 3,000 A.D., it will link New York to Los Angeles (3950 km.) in thirty-five minutes at a speed of 22,500 km/h.

Trolleybus (1911)

The first trolleybuses date from 1911. As an invention, they are of more recent vintage than motorized monibuses (now conventionally known simply as buses). The trolleybus has an electric motor fed by an overhead line, two wires distributing a constant voltage. After 1945, they were rapidly replaced in Western countries by omnibuses. The Soviet Union, however, continues to increase its lines of trolleybuses and today has several tens of thousands of them in service.

3. Two Wheelers

Origins

Although no actual example is known to have existed prior to the 18th century, it is likely that the ancients had already envisaged two-wheel locomotion. Drawings of two-wheeled vehicles have been discovered in China. The Egyptian obelisk taken from the temple of Luxor in 1831 to ornament the Place de la Concorde in Paris has among its hieroglyphs a representation of a man astride a horizontal bar which is mounted on two wheels. The obelisk dates from the reign of Rameses II, sometime between the 13th and 15th centuries B.C.

Closer to us historically, one can cite a drawing by **Leonardo da Vinci** (1452—1519). And in the Stoke Page Church in England, one can see, depicted on a 17th-century stained-glass window, a sort of two-wheeler straddled by a cherub.

Celerifer and the velocifer (1790)

Two-wheelers made their first appearance in France in 1790, in the form of **Count de Sivrac's** celerifer (the combination of two Latin words means "carrier of speed"). This two-wheeled vehicle had neither a steering mechanism nor any means of propulsion other than the vigorous use of the feet against the ground.

The celerifer soon took on the name of *velocifer*; people decided to adorn it, sometimes making it look like a lion, a horse, or even a dragon. It quickly became identified in the mind of the public as an oversized toy rather than as a means of locomotion.

Dandy horse (1817)

The famous dandy horse was presented for the first time in 1817 in Paris's Luxembourg Gardens by a German, **Baron Carl von Drais von Sauerbronn,** who brought the two wheeler back into fashion. It had fallen into neglect following the quick passing of the celerifer and the velocifer. There was still no means of propusion other than pushing the feet against the ground, but this time the machine was equipped with a pivoting steering mechanism that operated a kind of rudder. This was the ancestor of the modern handlebars.

Nicephore Niepce, the inventor of photography, conceived a similar vehicle, also in 1817.

The dandy horse developed rapidly in England from 1819 on; there it was called the hobby horse. It was given an adjustable seat. A chest support in front made it possible to push more vigorously, and the metallic frame, in the form of an open cradle, made it possible for fashionable people to mount it gracefully. Nevertheless, the dandy horse remained uncomfortable (although in 1830 a Burgundian sportsman, using the pivoting steering mechanism, raced from Beaune to Dijon in 2 hours, 30 minutes—that is to say, 23 miles (37 kilometers) at an average speed of 3 mph.

Velocipede (1855)

In Paris, in 1855, the blacksmith **Pierre Michaux** and his son **Ernest** had a brilliant idea. While repairing a dandy horse, they decided to attach what was subsequently called a pedal-and-gear mechanism to the front-wheel axle. The innovation worked and by 1865 the firm of Michaux & Co. had sold more than 400 velocipedes. (The name comes from a combination of two Latin words and means "fast feet.") Meanwhile, Michaux's partner, **Pierre Lallemant,** emigrated to the United States, where he founded the American bicycle industry with another pioneer, **James Carrol.**

Safety or chain transmission (1869)

As early as 1869, the Frenchman **Guilmet** had constructed an entirely

new velocipede, one with pedals no longer situated directly on the front hub. Placed where we find them today, at the center of the vehicle, these pedals were linked to a chain that transmitted movement to the rear wheel. Unfortunately, Guilmet was killed in 1870 in the Franco-Prussian War, and his invention remained unutilized in an attic. It was not until 1879 that an Englishman, **Harry Lawson,** revived the design, which was later put to practical application by the manufacturer **Starley.**

From that moment the bicycle took on its definitive form. The wheels, which were very disproportionately sized on some 19th-century two-wheelers, regained a sensible and equal diameter, with a corresponding increase in stability. It was on account of this stability that Lawson's bicycle came to be nicknamed "the safety."

Cycling

See Chapter 6, Sports

Vertical pedal bicycle (1978)

This bicycle was patented in 1978 by its two inventors, the Korean **Man Te Seol** and the American **Marione Clark.** The vertical pedal bicycle enables the rider to attain remarkable speeds without getting tired—thanks to its 29 gears and the unusual movement of its pedals, which puts much less strain on the legs.

Motorcycle with a four-stroke engine (1885)

Historians still argue over the question of who the father of the first

The dandy horse.

In 1869, New York cyclists had to have a license. Here is a driving school.

motorcycle really is. Keeping strictly to the concept of a two-wheeled vehicle, this honor falls to two Germans, **Wilhelm Maybach and Gottlieb Daimler** (1834—1900), the father of the automobile. In 1885, he built a motorcycle with a wooden frame and wooden wheels, powered by a four-stroke internal combustion engine.

The English, on the other hand, claim that **Edward Butler** invented the motorcycle a year earlier. But this claim derives only from the patent for a tricycle with a gasoline engine, which was not built until three years later.

Neither one was developed any further in its country of origin—for economic reasons in Germany and because of overly strict regulations in England.

"Self-propelling" bicycle (1888)

In France, on December 22, 1888, **Felix Millet** took out a patent on a rotary engine composed of five cylinders in the shape of a star. This engine was placed on the front wheel of a tricycle in 1889. In 1895, with this same star-shaped engine mounted on the rear wheel, Millet introduced his "self-propelling bicycle." It was considered too complicated, and only a few models were actually built.

First mass-produced motorcycle (1894)

In 1894, two Germans, Hildebrand and Wolfmuller, built a motorcycle that was mass-produced in over a thousand models. Because of the anti-German sentiment rampant in

France during that period, this motorcycle was marketed there under the trade name of **Duncan & Suberbie**, who manufactured it under franchise at their factory at Croissy, near Paris.

Motorbike (1897)

The motorized bicycle, baptized the motorcyclette (motorbike) by its French inventors, **Eugene** and **Michel Werner**, was exhibited for the first time at the Paris Salon of 1897. After having constructed a gramophone, a cinematographic apparatus, and a typewriter in 1896, these two journalists of Russian origin concentrated their efforts on finding a way to put a small engine (already designed by **H. Labitte**) on a bicycle. Initially they placed it in a horizontal position above the rear wheel, subsequently in front of the handlebars, with direct front-wheel drive by means of a leather belt. Exhibited to the public, the motorbike was an immediate hit. From 1898 on, millions of the front-wheel drive model were produced, as many in France, where it sold for 1000 francs, as in England, where it was built under license by the Motor Manufacturing Company of Coventry. In Germany it was sold by the firm of Eisenbach.

The definitive model (1900)

The earliest motorbikes (with the engine located over the front wheel) remained fairly unstable. A new Werner's model, put on the market in 1900, can be seen as an important breakthrough, for it gave the motorbike its definitive form. From then

on the motorbike was equipped with electric ignition and a small 217-cc Werner engine located in the lower loop of the frame in place of the pedal and gear mechanism, which was now moved to the back. This new layout, which was much more logical, gave the motorbike a degree of safety that has not yet been superceded, and it was subsequently adopted by nearly all manufacturers.

The fortune of the Werner brothers should have been assured, but unfortunately they invested their profits in imperial Russian bonds. This patriotic attachment ruined them, and they died penniless.

Scooter (1902)

The scooter appeared in France in 1902 and was called the *autofauteuil* (auto-easy chair). It was a motorcycle equipped with a protective shield, small wheels, and an open frame that allowed the rider to sit while driving. It was invented by **Georges Gauthier** and was built in Blois, France until 1914. A variety of models succeeded the prototype, the first in 1919. Successor models include the Autoglider and the Skootamota, both English; the German DKW Golem; the French 140-cc. Lumen, Velauto, and Monet Goyon; and the American Ner. The Italian Vespa met with enormous success beginning in 1946.

Motorbike with a two-stroke engine (1900)

The development of the first two-stroke engine used on a motorbike is attributed to a Frenchman, **Cormery**, who patented this invention in Paris on August 20, 1900. In 1901, another French bicycle maker, **Leon Cordonnier**, filed a patent for his Ixion engine, which featured the first rotary distributor. It was to be popularized by the Tour de Monde ("Around the World") model manufactured by the Belgian firm Gillet d'Herstal in 1919.

An Englishman, **Alfred A. Scott**, received British Patent No. 3367 on February 11, 1904 for the first two-stroke, two-cylinder engine. However, it was not built until 1908, and only six models were ever produced.

It was not until 1911, with the appearance of a water-cooled engine, that motorbikes with two-stroke engines became widely used. The two-stroke engine only really matured in the 1930s, thanks to important research carried out in Germany by DKW (**Schnurle's** patent for the expulsion of exhaust fumes in

1932). Separate lubrication became a standard technique with the arrival of Japanese motorcycles at the end of the 1960s, but it had already been featured much earlier on some makes: the Scott in England (in 1914) and the DFR in France (in 1924).

Military Motorcycle (1899)

This type of motorcycle appeared for the first time at the turn of the century. The earliest known example dates back to 1899 when an Englishman, **Simms**, introduced a De Dion tricycle which had front-wheel drive, and was equipped with a Maxim machine gun for use in the Boer war. At the same time, a tricycle that could haul a light cannon was developed in Norway.

It was in 1914 that the motorcycle was discovered to have genuine military potential. The English and the Americans equipped them with sidecars for use as ambulances, or outfitted them with machine guns. Ordinary motorcycles were used by couriers. The first motorcycles specially designed for military use appeared in 1939. These were the motorized sidecars, made by BMW and Zundapp in Germany, FN and Gillett in Belgium, Gnome and Rhone in France, and Norton in England. The most unusual was the Simca-Sevitane, a French model that was entirely waterproof and could be converted into a landing craft for disembarkations.

Sidecar (ca. 1910)

The idea of attaching a sidecar to a bike was discussed as early as April 29, 1894 in the magazine *Le Cycle*. But this addition had to wait until motorcycles became more solid and powerful; sidecars only started to evolve around 1910. An articulated assembly was perfected in America in 1916. It allowed the motorcycle to lean when going into a curve or around a corner. A motorized sidecar wheel was developed in 1939 and adopted by the best-known makes in all industrial countries.

Four-cylinder Motorcycle (1901)

Beginning in 1901, an Englishman, **Colonel Holden**, produced a motorcycle engine with four opposing cylinders. Its driving rods, like the connecting rods on a locomotive, propelled the rear wheel without using a driving belt. A four-cylinder, V-shaped racing engine was built by **Clement** in France in 1902, and in 1904 it enabled the dream of a com-

Top: Daytona Beach and the cult of the motorcycle. Above: A scooter in Berlin, around 1925. Left: An unknown inventor presents his motorcycle.

mercially successful four-cylinder motorcycle to be realized.

The Belgian National War Armaments Factory at Liege-Les Herstal introduced its straight four-cylinder motorcycle with transmission by propeller shaft and ignition by magneto at the 1904 Paris Motor Show. Commercial production of this model, with many improvements, continued until 1926. In Germany, **Laurin-Klement**, one of the pioneers of the motorcycle, introduced a straight four-cylinder motorcycle similar to the Belgian model at the 1904 Leipzig Motor Show. Production stopped when the factory shut down in 1908. It was in the United States however, that the manufacture of

four-cylinder motorcycles really took off: the Pierce appeared in 1909, the Militaire in 1910, the Henderson in 1912, the Ace in 1919, the Cleveland in 1925, the Indian in 1927, etc.

Front suspension (1903)

The first front suspension to be marketed for use on a motorcycle was the extremely complex Truffault fork, which was exhibited at the Paris Motor Show in December 1903 on a Peugeot motorcycle. Many different systems of front suspension had been imagined for the bicycle, but their adaptation to the motorcycle did not follow automatically, because of its much heavier weight.

Rear suspension (1904)

The suspension principle was invented for velocipedes as far back as 1898. For motorcycles, the first development was the French Stimula of 1904, which had rear suspension with a cantilevered rocking arm and a spring under the seat. In 1911, the German firm NSU inaugurated rear suspension with a rocking arm and two lateral shock absorbers with helicoidal springs, as used today. The English ASL of 1912 inaugurated pneumatic suspension, with inflatable front and rear shock absorbers. In addition, the seat contained an inflatable cushion.

All the basic principles of rear suspension currently in use had already been invented at the beginning of the century, but the technology was not far enough advanced to apply them. It was only after World War II that sliding suspensions appeared, then rocker suspensions with greater and greater wheel clearance. Finally, in 1979 and 1980, the first variable geometric suspensions were marketed: Honda's Pro-Link, Kawasaki's Uni-Track, Suzuki's Full-Floater, etc. These suspensions contain a combination spring and shock absorber, activated by an intricate arrangement of articulated levers.

Motorcycle transmissions (ca. 1900)

Belts, chains, and propeller shafts are the three main systems to transmit the energy produced by the engine to the rear driving wheel. All three appeared with the introduction of the motorcycle at the beginning of the 20th century. These devices already existed on agricultural equipment and automobiles. A true transmission system had become widespread on motorcycles by the beginning of World War I. At first, control was manual, using a shifting lever and gear selector attached to the gasoline tank. Foot control appeared in England during the 1920s with the "Velocette" but never became widespread until after World War II.

Electric Starter (1913)

In 1913, an American firm, Indian, introduced the Hendee Special, with a V-shaped, two-cylinder, 998-cc. engine. This was the first motorcycle equipped with an electric starter, the "dynastart," although other manufacturers had in fact already proposed electric ignition by then. Too advanced for its time, the Hendee Special was a complete failure, and all Indian models with an electric starter were recalled to the factory for removal of this accessory, as unperfected as the fragile batteries supplying its current. The electric starter did not really make its appearance commercially until the 1960s, on mass-produced Japanese motorcycles.

Motorcycle racing

See Chapter 6, Sports

Futuristic Bicycles

Micro Bicycle (1981)

The micro bicycle was invented by the British firm, Product Systems Ltd., in 1981. Once folded, the Brompton bicycle is much less cumbersome and thus easier to transport than any of the "folding" bicycles now on the market. Once closed, it can fit into a suitcase. Thanks to many hinges, everything pivots, including the pedals. Yet it can be unfolded in 10 or 15 seconds. The Brompton is a completely developed adult bicycle with two brakes, two speeds in the hub, and rear suspension.

The bicycle that never rusts (1982)

Introduced in May 1982 by the French Micmo Corporation and made of polyester, it could completely change the methods traditionally used in the manufacture of bicycles.

The pneumatic shock-absorber, with springs, was invented in 1912 by the British company, ASL.

A wide choice of geometric forms is possible, which gives leeway to many innovations in form.

The bicycle of tomorrow

Now being built in the United States, it will be manufactured according to specifications being worked out by NASA technicians. It will weigh less than 11 pounds but will still offer sufficient rigidity.

4. Automobiles

Cugnot's steam-car, built in 1771.

Cugnot steam car (1771)

Under orders from the **Duc de Choseul**, Minister of War, the French engineer, **Nicolas Joseph Cugnot** (1725—1804), set out to build the ancestor of all automobiles in 1771. This, the first car with mechanical traction, was known as the "fardier." It had a strong wooden chassis supported by three wheels. The front wheel, turning on a vertical axle, provided locomotion; the driver steered the vehicle by turning a two-handled crank. The engine was a vertical, simple-effect, two-cylinder steam machine. But it was the first high-pressure steam engine used in a car. Power was transmitted from the engine to the front driving wheel by two piston rods. Cugnot's car had a speed of roughly 2.3 mph.

Bollée steam car (1873)

This 12-passenger steam car was created in 1873 by the Frenchman **Amedee Bollée**. It was the first mechanical traction vehicle capable of transporting a large number of people. The car was equipped with two steam engines.

In 1875, the Bollée made the Le Mans—Paris run (154 miles) in 18 hours, including stops for water and rest. Its maximum speed was 13 mph with a load of 4800 kg. Because this car was easy to drive, it was called "the Obedient."

Daimler-Maybach automobile (1889)

The Daimler-Maybach, built in 1889 by the Germans **Gottliev Daimler** (1834—1900) and **Wilhelm Maybach** (1847—1929) was introduced at the Paris Exposition that same year as the "Quadricycle." It was the first gasoline automobile. It had four wheels and a chassis made of steel pipes; a V-shaped, two-cylinder gasoline engine that produced 1.5 horsepower at 600 rpm; a four-speed transmission for a speed of between 3 and 10 mph; a tapered friction clutch; steering controlled by two front wheels slightly smaller than the rear driving wheels; and a mechanical braking system set on the rear wheels. The engine was placed at the rear, on the chassis.

Model T Ford (1896)

Built at Detroit in 1896 by the American **Henry Ford** (1863—1947), the Model T Ford was put into mass production on the assembly line in 1908. It was a far cry from the hand-crafted automobile built by Henry Ford at Detroit in 1896, with its 4-stroke engine and Kane Pennington cylinders. The assembly-line production, a Ford innovation, was an application of the principles of **Frederick W. Taylor**, the creator of Scientific Management, who stressed strict specialization, elimination of all superfluous motion, and maximum utilization of plant and equipment. It was this system that made it possible for 18 million Model T's to roll out of the Ford factories between 1908 and 1927.

The traditional Model T was equipped with a straight four-cylinder, four-stroke engine and a high-tension ignition system. The engine was cooled by water circulating through a thermo-siphon. A system of multiple clutchplates along with a planet-gear transmission, a planet differential, and a bevel-wheeled rear axle transmitted drive power to the rear wheels. Steering was controlled through a planet housing.

Gas turbine automobile (1950)

The first automobile powered by a gas turbine was built in 1950 by Rover, a British manufacturer. On June 26, 1952, one of these vehicles attained a speed of 150 mph.

The gas turbine that powers this racing car has an open circuit design. It is composed of a compressor, one or several compression chambers, and the actual turbine itself. The turbine consists of two barrels: one that activates the compressor and keeps it at a constant speed, and a second that transmits the power to the wheels by means of a speed reducer, which moderates the compresssor's constant energy output.

To improve the efficiency of the gas turbine, the exhaust fumes are reused to heat the compressed air inside a heat exchanger before this compressed air enters the combustion chamber. Other manufacturers have taken up the process invented by Rover: General Motors, Chrysler, Boeing, Fiat, and Renault, whose Etoile Filante ("Shooting Star") attained a speed of 192 mph in 1956.

The answering car (1981)

Thanks to a system invented by **M. Kempf**, a Strasbourg industrialist, a Renault 5 appeared in 1981, then a Renault 11 in 1983, both of which are specially equipped for the handicapped and able to be controlled by the voice rather than the hands, the steering alone being done by the left foot. For example, the word "bugle" triggers off the horn, "to the right" sets the right blinker in motion, etc.

This model is obviously only produced in a very limited series. Renault and the Japanese firm, Nissan, are continuing their research in this still relatively new field of voice recognition by machine.

MECHANICS

Ball Bearings (Middle Ages)

Ball bearings have been used since the Middle Ages, especially in mills. But it was only with the first developments of the velocipede that the

This 1950 Rover, equipped with a turbo-propeller, was the first jet car.

French mechanic, **Jules Pierre Suriray**, actually registered the design and description of a device for applying ball bearings to a vehicle's wheel axles.

Cardan, or universal joint (1545)

In 1545, the famous Italian mathematician, **Geronimo Cardano** (1501—1576), described the joint he had invented, and which now bears his name, in a treatise on physics entitled *De subtilitate rerum*. The cardan allows for the relative angular movement of two shafts whose geometric axes converge at a single point. It is used in an automobile to couple two turning shafts whose positions can vary in relation to each other.

Engine ignition devices

Induction coil (1841)

The induction coil was invented by the French physicists, **Antoine Masson**, and **Louis Breguet** (1804—1884), in 1841. This is an apparatus that uses the phenomenon of electromagnetic induction to create a high tension alternating current. It fits onto all ignition systems used in internal combustion engines. As early as 1836, Antoine Masson had been able to produce high tension currents by inducing rapid interruptions in the normal current produced by a battery. The coil that he and Breguet built in 1841 produced electrical discharges in rarefied gases. The induction coil was perfected in 1851 by the German physicist, **Heinrich Daniel Ruhmkorff**, and it bears his name (Ruhmkorff coil).

Ignition Magneto (1880)

The ignition magneto was invented in 1880 by a German named **Giesen-**berg. The Frenchman **F. Forest** and the German **N. Otto** perfected it in 1883 and 1884 respectively. The magneto is a small generator of electric current that transforms the mechanical energy produced by the turning of the wheels of the automobile into electrical energy. This produces a spark that fires the fuel-air mixture in the combustion chamber of the engine. The first magnetos were of low tension.

Electric ignition by storage battery and induction coil (1883)

The system of electric ignition by storage battery and induction coil is by far the most widely used in internal combustion engines. In 1883, it was used for the first time, simultaneously in two countries: in France, by a Frenchman who originally came from Luxembourg, **Etienne Lenoir** (1822—1900), on a four-cylinder engine; and in Germany, by the engineer, **Karl Benz**, on a two-cylinder engine. In 1893, Karl Benz improved the system and got a much slower discharge from the battery. In 1895, the Frenchmen **Albert de Dion** and **Bouton** invented the cam distributor, which allowed for much greater precision in the appearance of the spark.

Spark plug (1885)

In 1885, the Frenchman **Etienne Lenoir** invented an electric spark plug quite similar to those still used today. Fuel ignition by electric spark had been foreseen by the Italian scientist, **Volta**, already in 1777. Isaac de Rivaz mentioned it for an internal combustion engine he had described in 1807.

Delco distributor (1900)

With the improvement of storage batteries, automobile manufacturers returned to high voltage distribution about 1900. The American firm Dayton Engineering Laboratories Co., of Ohio, developed a distribution system whose distributor became familiar through using the firm's initials: the Delco.

High tension magneto (1902)

The high tension magneto was invented in 1902 by the German engineer, **Gottlieb Honold**. Two systems of ignition, by high tension magneto and by storage battery, were combined in airplane engines to take advantage of both types of generators. A fierce competition ensued between the two systems, but after 1925 the use of a storage battery in automobiles won out.

Transmission

Direct transmission (1899)

In 1899, **Louis Renault** (1877—1944) equipped his first automobile, completed in 1898, with a transmission coupled directly to the engine and gear-shifting by selector rod. The transmission on the 1899 Renault had three speeds and a reverse gear. The fastest speed, third gear, was reached directly, the primary and secondary propeller shafts turning at the same speed.

Automatic transmission (1910)

The automatic transmission invented by the German **Fottinger** in 1910 was nothing more than a torque converter. The torque converter is characterized by the unlimited number of possible gear ratios that it can establish between the rotation speed of the main drive shaft and the mainshaft. It acts in such a way that, when the resistance torque goes beyond the maximum speed of the driving torque, the driving torque is progressively multiplied.

Electronic control transmission (1965)

In 1965, the Renault Company launched the first automatic transmission with electronic control, on its Renault 16. This allowed faster shifting from one speed to another, with greater reliability.

Carburetor (1893)

The carburetor was invented by the German engineer, **Wilhelm Maybach**, in 1893. The carburetor is indispensable to any gasoline engine since it is in the carburetor that the fuel mixture, air and gasoline vapor is prepared before being sucked into the engine's cylinder. It was used for the first time on a parallel two-cylinder engine called the "Phenix," built by Maybach and Daimler. This was one of the first operational gas-powered cars, and it had great success.

Butterfly valve (1893)

In the same year that the carburetor was invented (1893), it was perfected by **Karl Benz**, who added the butterfly valve. This valve regulates the quantity of the fuel-air mixture drawn in by the engine and, consequently, its power and its turning speed.

Straight six-cylinder engine (1888)

The first straight six-cylinder engine was conceived by a Frenchman, **Fernand Forest**, in 1888. In this type of engine, the rods of the six cylinders are parallel and situated (most often vertically) at the same level. This construction gives a balance and regularity of torque that are superior to all other straight multicylinder engines. For this reason, it has continually been used in the construction of automobiles. In the United States, where the manufacture of heavy automobile engines continues, this engine is still very popular.

Straight four-cylinder engine (1889)

The straight four-cylinder engine conceived by **Fernand Forest** in 1889 (a year after the straight six-cylinder engine) is characterized by a four-stroke cycle. The four cylinders, situated at the same level, are cast into the engine block and covered by a quadruple detachable cylinder head. This is the type of engine most commonly used today in European-built automobiles.

Engine cooling

Water-cooled engine (1823)

In 1823, an Englishman, **Samuel Brown**, invented a system of engine cooling by water. In Brown's engine, the water circulated around the cylinders, surrounded by a casing or cylinder liner. The water was kept constantly in motion by a pump and recooled by contact with the surrounding air. In 1825, Samuel Brown founded a company that built several models of his engine. One of them

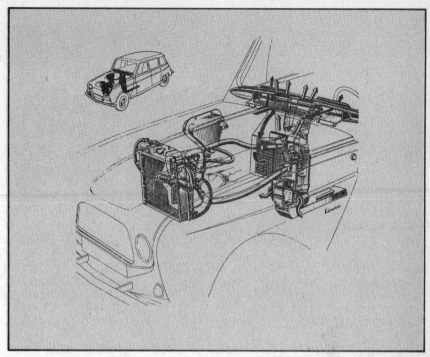

The sealed cooling system invented by Renault in 1960.

Maybach invented the radiator. The most famous grill belongs to Rolls Royce.

Left: The Chinese added sails to their wheelbarrows. Below: A California version of the VW Beetle. Bottom: Brooks Walter invented the fifth wheel—for parking in town.

blown in by a fan placed in front of it could circulate. This radiator formed a major part of the water-cooling circuit. The first radiators were built by the German automotive corporation, Daimler Motoren Gesellschaft, where Maybach worked.

BODY AND ACCESSORIES

Tires (1888)

An Irishman from Belfast, **John Boyd Dunlop** (1840-1921), invented the tire in 1888. This was one of the most significant leaps forward as far as locomotion by wheels was concerned.

Dunlop, a horse vet by profession, discovered the tire by accident when he tried attaching air-inflated rubber tires to the wheels of his son's bicycle. The same idea had already been put forward by a Belgian, **Dietz**, in 1836, and by an Englishman, **Robert W. Thomson**, in 1845, but it was not put into practice. Abandoning his old profession, Dunlop patented his invention and founded the first tire factory, where he utilized Goodyear's vulcanization process. Through the mediation of a German subsidiary, Dunlop tires were put on the first mass-produced motorcycles, the Hildebrand and the Wolfmuller in 1894. The Dunlop firm immediately received complete support from most manufacturers. On the other hand, it was the French brothers **Michelin**, who, in 1895, first used tires on an automobile.

Removable tire (1891)

Invented in 1891 by the French firm, Michelin, which was directed by two brothers, **Andre** (1853—1931) and **Edouard Michelin** (1859—1940), the removable tire proved to be revolutionary. A blow out, which formerly meant calling a specialized repairman, could now be fixed by the driver in less than 15 minutes. This invention had immediate success. In 1895, the Michelin brothers created a sensation by being the first to enter an automobile with tires in the Paris—Bordeaux—Paris race. It was called the *Lightning*.

Dual tires (1908)

In 1908, the French company, Michelin, found a solution to the problem posed by heavy vehicles (trucks and buses) riding on pneumatic tires. Too heavy to be supported on four wheels, the load was balanced out on four combination sets, consisting of several wheels placed side by side.

was installed on a carriage and another on a small boat. The first usable internal combustion engine, invented in 1860 by the Frenchman **Etienne Lenoir** (see Engine Without Preliminary Compression), used water-cooling.

Air-cooled engine (1875)

In 1875, the Frenchman **Alexis de Bischop** used air to cool an engine for the first time. His engine, of a mixed type and without preliminary compression, contained a cylinder surrounded by vertical blades. Air-cooling is used particularly on motorcycles and on airplanes engines.

Radiator (1897)

The radiator, which efficiently cools an engine, was invented in 1897 by the German engineer, **Wilhelm Maybach**. After numerous attempts, Maybach finally perfected the type of radiator called a "honeycomb." It was composed of a network of short straight ducts through which the air

This system was generally known under the name "dual tires" or "twin tires." The city buses of Champeix, in the south of France, were the first vehicles to run on dual tires.

Radial-ply tires (1937)

Michelin, the French tire manufacturer, brought out the "Metallic," the first steel-ply tire, in 1937. Fine steel wire, supple but resistant, reinforced the rubber of the tire. Up until that time, the only method used for protecting the canvas or cotton-fiber framework of the tire had been to add more layers of rubber.

Because of constant research carried out in this area during the 1930s, a rough model was finally worked out for a steel-belted tire baptized the cage a mouches ("fly net"). The tire tread, reinforced with strands of crossed wire, was held in place by separate strands of steel wire placed perpendicular to the wire belt of the tire. The invention, remarkable for its improved road-holding ability, was also the first tire to take into account the two different functions served by the flange and the crown of the tire. The idea of the radial tire was born, and the patent was registered on June 4, 1946.

Tire chains (1904)

As early as August 23, 1904, **Harry D. Weed** of Canastota, New York, invented chains that fit on automobile tires for use when it snows.

Bumpers (1905)

Rubber bumpers were patented by the Englishman **F. R. Simms**, in 1905. The Simms Manufacturing Company of Kilburn put bumpers on the Simms-Werbeck car in the summer of 1905. The first bumper had appeared on a Czech Prasident, built in 1897, but during its first test run it fell off after 10 miles and was never put back on.

Windshield wiper (1916)

The first mechanical windshield wipers appeared in the United States in 1916. In 1921, the Englishman **W. M. Folberth** invented windshield wipers that worked automatically, using compressed air supplied by the engine. The first electric windshield wipers were manufactured in the United States by Berkshire.

Automatic lock (1955)

It was 1955 that Ford introduced an automatic security lock on the market that was capable of closing all the vehicle's doors automatically while it was in motion.

Safety Belt (1959)

The idea of equipping private cars with safety belts was derived from aeronautical experiments. It was the Swedish firm, Volvo, that, in 1959, was first to equip its vehicles with safety belts.

Fifth door (1961)

The fifth door, which allowed one to have access to the rear of the car by lifting up rear panel, appeared for the first time on the Fiat 600 on January 1, 1961.

Plastic Safety Shield (1971)

In 1971, Renault became the first manufacturer to protect the front of an automobile by using a polyester safety shield.

AIR TRANSPORTATION

First human flight (1783)

Jean-Francois Pilatre de Rozier and the **Marquis d'Aalandes** made the first human flight on November 21, 1783. Ascending in a basket supported by a hot-air balloon, the first two aeronauts left the gardens of the Chateau de la Muette, near the gates of Paris, and landed on the Butte aux Cailles, close to the center of the city. This first piloted flight lasted 25 minutes, during which time they rose to an altitude of about 3000 feet (1000 meters) and covered a distance of about 5 miles (8.3 kilometers).

Pilatre de Rozier, the first pilot, was also to become aviation's first victim. He was killed on June 15, 1785, while attempting to cross the English Channel. Soon after departure, the balloon caught fire and crashed 3 miles from Boulogne.

1. Balloons

Hot-air, or Montgolfier balloon (1783)

The first flight by an aerostat took place at Annonay, near Lyon, France, on June 4, 1783. This hot-air balloon was constructed by two brothers, the paper manufacturers **Joseph** (1740—1810) and **Etienne** (1745—1799) **Montgolfier**. The Mont-

golfier was made of pack-cloth covered with paper. The balloon carried a portable stove in which wool and straw were burnt to produce hot air, that is, a gas lighter than air. This first balloon attained an altitude of 1500 feet (500 meters) within 10 minutes. When the news of the flight reached Paris, it caused a sensation. The famous physicist, **Jacques Charles** (1746—1828), decided to build his own balloon. It was constructed in the workshop of the Robert brothers. Charles had the balloon filled with hydrogen, a gas that is lighter than air, isolated in 1766 by the British chemist, **Henry Cavendish**. This first hydrogen-filled balloon flew on August 17, 1783, reaching an altitude of 3000 feet (1000 meters) in less than 2 minutes.

On September 19, 1783, the Montgolfier brothers repeated their first experiment in front of Louis XVI and his court at Versailles. This time the balloon carried a suspended cage containing the first air passengers: a rooster, a duck, and a lamb. This flight was also witnessed by Benjamin Franklin, then United States minister to France.

The modern Montgolfiere (1950s)

The revival of interest in Montgolfieres is due to research done by the U.S. navy in the early 1950s, and it was in 1963 that touring balloons

Ballooning has become a recreational activity.

first appeared. Balloon coverings are now made of very light, synthetic material; the basket is usually wicker, and the pilot heats the air with the help of a flame from a propane gas burner.

Basketless balloon (1980)

The British firm, Coll Balloons, brought back the Montgolfier in 1980 by designing a balloon without a basket. The pilot is suspended in a harness. On his back he carries a 44-pound (20-kilogram) propane reservoir. There is enough fuel for him to stay aloft for 60 to 90 minutes, and directional control is possible.

Gas balloons (1783)

It was French physicist, **Jacques Alexandre Charles** (1743-1823), who at one go invented all the rules gov-

erning modern ballooning. The balloon he designed, inflated with hydrogen, went up in Jardin de Tuileries in Paris on December 1, 1783, with Charles and the Robert brothers on board. It was equipped with a valve connected to the ballast, allowing the balloonist to control the balloon's ascent and descent. The cargo-passenger wicker basket was attached to the balloon by a net. The American balloonist, **Wise**, perfected Charles' balloon by introducing the use of a rip cord. Modern sporting balloons differ from that of Charles only in the use of a noninflammable gas, helium, and in the quality of the covering used.

Dirigible (1852)

The dirigible was invented in 1852 by the Frenchman, **Henry Giffard**, who perfected a balloon furnished with a

means of propulsion. This dirigible used a steam engine, and it flew for the first time on September 24, 1852, covering a distance of 17.3 miles (28 kilometers) at a speed of 4.3 mph.

In 1872, the Frenchman **Dupuy de Lome** designed an elongated dirigible powered by a handcrank. That same year, another dirigible with a steam engine was tried out by the Englishman, **C. E. Ritchell**. In 1883, the Frenchman, **Tissendier**, utilized an electric engine.

The first closed-circuit (with return to the takeoff point) dirigible flight was accomplished on August 9, 1884 by the French army captains, **Charles Renard** and **Arthur Krebs**. Their dirigible, *France,* was powered by an electric engine. It departed from and returned to Chalais-Meudon, near Paris.

The Zeppelin (1890)

Count Ferdinand von Zeppelin undertook experiments as early as 1890 and took out a series of patents for a streamlined dirigible that would not lose its shape but remain rigid. A movable hangar enabled it to be taken out without risk, no matter what the wind direction was. The first ascension of the L-21 took place on July 2, 1900. The odyssey of the enormous rigid forms came to an end with the outbreak of World War II.

Heli-Stat (1980s)

The Heli-Stat is neither dirigible nor helicopter nor airplane, but a compromise between the three. Invented by a sixty-two-year-old American engineer, **Frank N. Piasecki**, this machine has an original design. The prototype of the Heli-Stat is presently under construction in former U.S. navy hangars at Lakehurst, New Jersey. It consists of a dirigible envelope, approximately 328 feet long (100 meters) with a diameter of about 82 feet (25 meters), to which is attached an aluminum framework supporting four U.S. navy surplus Sikorsky H-43 helicopters.

The Heli-Stat is designed to carry very heavy payloads at much lower cost than standard helicopters. If it proves successful, it could be used for passenger transportation over medium distances, with the added advantage of its being able to land or take off from almost any surface.

Cyclo-Crane (1980s)

This balloon was designed by the American firm, Aerolift, whose director is **Mr. Crimmins**. It is a rather large balloon whose lift during takeoff is reinforced by the motion of kelvar ailerons, which are retracted afterward. The Cyclo-Crane is still a prototype. Like the Heli-Stat, it has been designed to carry very heavy cargo loads, and possibly passengers at a later date. Plans are already being made for a scheduled route service between Las Vegas and Los Angeles.

2. Airplanes

Origins

From the beginning of time, man has dreamed of imitating the birds. During the 15th century **Leonardo da Vinci** scientifically observed birds in flight and drew plans for a mechanical wing. His analysis was later taken up by **Borelli**, in a study made in 1680. Thereafter, research was aimed at finding a way to use fixed, rather than movable wings, with pro-

Leonardo da Vinci.

pulsion provided for by some sort of propeller.

Keeping this principle in mind, the Englishman **Sir George Cayley** perfected an engineless aircraft in 1809. Later he added a steam engine, but it made the plane too heavy. It was not until August 18, 1871 that a French engineer, **Alphonse Penaud**, was able

to fly a reduced model airplane based on Sir George Cayley's 1809 aircraft design.

Glider (1809)

The Englishman **Sir George Cayley** (1773—1857), deserves credit for having built a glider in 1809, and therefore the first airplane. However, the Frenchman **Jean-Marie Le Bris** is credited with having made the first glider flight, in 1856.

Airplane (1890)

The first engine-powered, piloted takeoff in a plane took place on October 9, 1890. The Frenchman **Clement Ader** (1841—1925) took off in the steam-powered Eole. The plane's wheels lifted from the ground, a few inches at most, for a distance of approximately 160 feet (50 meters). The Eole is generally credited with having made the first takeoff. During the same period the German **Otto Lilienthal** (1848—1896) constructed several gliders that could remain aloft for a considerable distance when launched from the top of a hill. However, Lilienthal was unable to resolve control problems and was killed in 1896 during a trial flight. Lilienthal made hundreds of flights, and his glider designs were carefully studied by the Wright brothers. In the year that Lilienthal died, the Englishman **J. Challis** was the first to use the word airplane to describe flying machines.

First flights (1903)

Two Americans, the Wright brothers, **Wilbur** (1867—1912) and **Orville** (1871—1948), invented the first airplane capable of sustained and controllable flight. These two Ohio bicycle manufacturers mounted a gasoline engine of their own design on a wood and a fabric biplane they had previously used as a glider. The airplane's first flight took place on December 17, 1903 at Kitty Hawk, North Carolina. The plane flew 131 feet (40 meters) in 12 seconds, at an average speed of 7.5 mph (12 km/h). It was a modest flight, but it heralded great things to come.

On January 13, 1908, the Frenchman **Henri Farman** (1874—1958), flew a Voisin (manufactured by **Gabriel Voisin**, 1880—1973) to complete the first closed-circuit kilometer course at Issey-les-Moulineaux, near Paris. That same year, the Wright brothers came to France, and Wilbur demonstrated the superiority of the Wright Flier by breaking the

A Zeppelin dirigible over Manhattan, around 1930.

Dirigibles overlook the landing of troops on D-Day, 1944.

The first aviators needed a great deal of courage.

100-kilometer (62-mile) mark with a flight of 120 kilometer (74.5 miles) in one hour and 54 minutes. On July 25, 1909, a second historic mark in aviation occurred: the crossing of the English Channel by **Louis Bleriot** (1872—1936).

Seaplane (1907)

The Frenchman Henri Fabre (1882—), was the first to replace airplane wheels with floats. His seaplane was equipped with flat-bottom, swivel floats. It made its first water takeoff at Martigues, France, on March 28, 1910.

The idea for a seaplane originated in 1907 when **Gabriel Voisin** (builder of the Bleriot monoplane, see above) made plans to mount a glider fuselage on floats. The glider was to be towed on the Seine by a speed boat until it got up enough speed to take off.

The invention of the hull seaplane in 1912 by **Donnet Leveque** led to the eventual design of large seaplanes such as the huge Pan American Airways Sikorsky and Boeing Clippers, which were to inaugurate the first regular transatlantic and transpacific passenger air routes.

Following the first seaplane flight, numerous other seaplanes were constructed: Farman's seaplane with catamaran floats, the single-float Curtiss seaplane, the 1912 Donnet-Leveque hulled seaplane, etc.

Boeing 247: the first modern airliner (1933)

The Boeing 247 was put into service by United Airlines in the United States in March 1933. It carried 10 passengers. For the first time, travelers could cross the United States in less than 20 hours. This low-wing, twin-engine aircraft had retractable gear; wing de-icers; constant-speed, full-feathering propellers (which permitted the automatic pilot mechanism to be used and also guaranteed maximum engine efficiency under all conditions); and the ability to maintain flight on a single engine.

Douglas DC-3: the first successful commercial airliner (1935)

Equipped with the same advanced technology as the Boeing 247 but able to carry more passengers, the Douglas DC-3 was the first transport airplane to fly over the Himalayas, between India and China. During World War II, it was frequently used as a military transport aircraft. Over 10,000 of these commercially profitable airliners were built, and some are still in service.

Viscount V-630: first turbopropeller airliner (1948)

The Viscount V-630, constructed by the British firm, Vickers-Armstron, was a low-wing monoplane powered by four Rolls-Royce turboprop engines. Its cruising speed was only 273 mph (440 km). This airliner made its first flight on July 29, 1948; it was put into service for a period of only two weeks (July 29—August 14, 1950) on the London—Paris route. It was the first turboprop to be used for commercial service. However, it was almost immediately dropped in favor of the larger and more powerful Viscount V-700.

Havilland Comet 1: the first turbojet airliner (1949)

The Comet 1, built by the British firm, Havilland, made its first flight on July 27, 1949. Powered by four turbojets, each with 4450 pounds of thrust, this aircraft could cruise at 490 mph (788 km), at an altitude of 39,000 feet (12,000 meters). It carried 36 passengers. Entering service with BOAC on May 2, 1952, this airliner could fly from London to Johannesburg (6686 miles, or 10,760 kilometers) in less than 24 hours. The Comet was withdrawn from airline service following a series of accidents.

Boeing 707: first modern jetliner (1954)

The first trials of the B-367-80 (prototype of the Boeing 707) were made in Seattle, Washington on July 15, 1954. The Boeing 707-120 aircraft made its first commercial flight with Pan American Airways on December 20, 1957. This aircraft, whose wings were placed at a 35 degree sweep, was powered by four turbojets. It had a wing span of 131 feet (39.9 meters) and could carry 179 passengers at a cruising speed of 566 mph (912 km).

Left: 280 Caravelles were built between 1958 and 1980. Top right: A Boeing 747. The dimensions of this model are 70 meters long with a wingspan of 60 meters. Bottom right: The world's fastest commercial plane, the Concorde, is a technological success and a financial failure.

More than 3000 Boeing 707s were built, and many are still in use. Larger and more powerful than any airliner of the period, the Boeing 707 became the standard long-range airliner.

Caravelle: first Airliner with fuselage-mounted engines (1955)

The revolutionary idea of mounting jet engines at the rear of the fuselage, rather than on the wings originated with the French Caravelle, which made its first flight on May 27, 1955. Rear mounting of the jet engines allowed undisturbed airflow over the wings, as in gliders, and brought increased aerodynamic efficiency, better stability, and decreased cabin noise. The Caravelle was also one of the first airplanes to utilize a rear, integrated stairway.

This efficient, rear-mounted engine system was soon adopted by almost all aircraft manufacturers, and the Caravelle did not meet with the commercial success that was expected of it. Entering service in 1958, it would carry 130 passengers. Production of this aircraft ended in 1970; only 280 airliners were built.

Yet the Caravelle did have the lowest accident rate of any airliner.

C-5A or Galaxy: the largest aircraft (1968)

Construction of the C-5A prototype by the Lockheed Aircraft Co. began in August 1966; the first flight took place on June 30, 1968. The Galaxy is the largest logistic transport aircraft in the world. This giant U.S. air force transport plane weighs 380 tons, is 251 feet long (76.67 meters) and has a wing span of 222 feet (67.73 meters). It can carry a payload of 57 tons (without in-flight refueling) a distance of 8000 miles (12,900 kilometers).

Jumbo jet (1969)

The first flight of a Boeing 747 took place on February 9, 1969. The aircraft was put into service by Pan American Airways on January 21, 1970. With this aircraft, the Boeing Company launched a new generation of large-capacity planes called "jumbo jets." This aircraft has a wing span of 196.8 feet (60 meters) and it can cruise (with a full payload) over

4600 miles (7400 kilometers) at Mach 0.89. Empty, it can fly over 6134 miles (9,200 kilometers). It can carry 490 passengers. Its cabin, 19.6 feet wide, allows ten passenger seats per row.

Concorde: first supersonic passenger plane (1969)

The British-French (BAC and Aerospatiale) Concorde made its first flight on March 12, 1969. It is the fastest airliner in the world. With a wing span of 83.8 feet (25.56 meters), a length of 204 feet (62.17 meters), it can carry from 100 to 139 passengers. Its maximum takeoff weight is 407,848 pounds (185,000 kilograms); its maximum range is 3,852 miles (6,200 kilometers). The Concorde flew at Mach 1.05 on October 30, 1969, and it flew over Mach 2.0 for the first time on November 4, 1970. Today this aircraft is able to attain a speed of 1448 mph or Mach 2.2.

Concorde entered airline service on January 21, 1976. Air France and British Airways started carrying passengers on the same date. It was only on November 27, 1977 that Concorde was given authorization to land in New York. Up until that time the Federal Aviation Administration had feared the aircraft would be too noisy.

Conceived prior to the 1973 oil crisis, the Concorde's high fuel consumption has doomed the plane to financial failure.

Tupolev 144: civil supersonic transport (1968)

The Soviets' supersonic Tupolev 144 has a wing span of 94.4 feet (28.8 meters), a length of 211.4 feet (64.45 meters) and can carry up to 126 passengers. It made its maiden flight on December 31, 1968.

Some authorities consider this aircraft a copy of the Concorde, while others regard it as an original aircraft. The latter opinion is based on the plane's wing shape, profile, and engine design, which are said to make it a very different aircraft. The real point in common is that the Concorde and Tupolev are the only supersonic planes flying in the world today. Following flight trials in October 1974, the Tupolev entered service on December 26, 1975, flying between Moscow and Alma-Ata (capital of Kazakhstan). However, on initial flights, only mail and light cargo were carried—and a few nonpaying passengers. It is thought that the aircraft did not yet perform satisfacto-

rily, vibrating excessively and producing too much engine noise—things that would make a trip uncomfortable. Since November 1, 1977, the Moscow—Alma Ata route has been open for regular passenger service. Average speed during the flight is 1245 mph, or Mach 1.9.

Solar aircraft (1980)

American **Paul McCready** invented a solar aircraft, the Gossamer Penguin, which made its first flight on June 28, 1980 at 8:25 A.M. The plane took off from a runway at the Sheafter Airport, in southern California. It was piloted by a 32-year-old schoolteacher who weighed only 97 pounds. (The aircraft's maximum payload is 152 pounds.) It flew 2634 feet (800 meters) at an altitude of 11.8 feet (3.6 meters) and at a speed of 15 mph (24 km/h).

The Gossamer Penguin carries solar panels on top of the control cabin. Those panels consist of 3640 photoelectric cells (see p. 318), which transform solar energy into direct electric current that energizes a small 450 watt electric motor. The Gossamer Penguin's total weight is only 55.1 pounds. To keep the weight down, ultralight materials were used: The cabin is made of a specially modified bicycle frame; wings and cockpit are covered with a strong plastic film, mylar; wing structure and struts are made of expanded polystyrene; and control cables are made of kevlar, a material five times stronger than steel. The photoelectric cells in the solar panels are U.S. army surplus; they were previously used on a surveillance aircraft that crashed (September 1975) without damaging its solar panels. These salvaged panels are presently on the Gossamer Penguin, baptized Solar Challenger.

On July 7, 1981, the Solar Challenger, piloted by American Stephen Ptacek, crossed the English Channel. It covered the 160.3 miles between Cormeilles-en-Vexin, near Paris, and Canterbury, England, in 5 hours and 23 minutes at an average speed of 29.8 mph (48 km/h), maintaining an average altitude of 4921 feet (1500 meters).

3. Helicopters

Origins

The helicopter appeared at the end of the 19th century. Its origins date to the 15th century. **Leonardo da Vinci** imagined an "air-screw," which,

The Russian-built Tupolev looks very much like the Concorde.

"rapidly rotating, and which by digging itself into the air, must rise by itself." Strangely enough, the first helicopter to fly was a toy designed by the Frenchmen, **Lannoy** and **Bienvenu.** This toy consisted of two contrarotating propellers fixed at both ends of a shaft, and it was powered by a piece of whale bone that had been bent into an arc.

First flights (1877)

A Frenchman, Ponton d'Amecourt, realized the first aluminum helicopter, powered by a steam engine in 1863. In 1876, **Sir George Cayley** drew plans for a helicopter with four rotors and eight blades, but it was never built.

On April 15, 1877, a helicopter, constructed by the Italian **Enrico Forlanini,** lifted off the ground for the first time and rose to an altitude of 42.6 feet (13 meters) for approximately 20 seconds. In November 1907, the helicopter of the French mechanic, **Paul Cornu,** powered by a gasoline engine, rose to an altitude of 6.5 feet (2 meters) with its pilot on board. In March 1908, this machine flew for 20 minutes; it had two rotors and a blade span of 19.6 feet (6 meters).

Rotor (1908)

The history of the rotor is inseparable from that of the helicopter. In 1908—that is, shortly after Paul Cornu's first takeoff in 1907— **Igor Sikorski,** a Russian emigrant engineer in the United States, tackled the problem posed by the mechanism of the blades and the rotor that together comprised the lifting force and the propulsion of the machine. Numerous solutions were devised during the evolution of the helicopter, among which must be mentioned the invention of the variable pace cycle, which enabled the position of the blades to

be modified, in relation to their rotation thus remedying the dissymmetry caused by the machine's going forward. In 1939, Sikorski's work led to the fitting of a monorotor, the VS-300. In 1941, a Sikorski helicopter beat the world record with an independent flight of one and a half hours. (In 1937, a two-rotor system was fitted on a German Focke-Aghelis FA 61 birotor helicopter.)

Autogyro, or autogiro (1923)

The autogyro was conceived by the Spaniard **Juan de la Cierva,** and made its first flight, with its inventor on board, on January 9, 1923. The autogyro has a free-wheeling rotor as well as an engine-driven propeller furnishing horizontal motion. The autogyro should not be confused with the helicopter, as it is incapable of lifting off vertically and cannot hover or stop its forward motion in the air. It is, however, capable of landing almost vertically.

Tail rotor or anti-torque rotor (1939)

With a mechanically driven, vertical rotor, the helicopter's fuselage has the tendency to turn in the opposite direction of rotor rotation. To counteract this major inconvenience, **Igor Sikorsky,** an American of Russian origins, built (in 1939) a small, side-mounted propeller known as an anti-torque rotor or, more commonly, a tail rotor, since that is where it was located. This propeller, which counterbalanced the torque of the single rotor, is no longer necessary today, in either dual-rotor machines or in helicopters having a single pylon with two contrarotating rotors.

Giant helicopter (1968)

The Mil Mi-12 Homer, built by the Soviet firm, Mikhail Mil, made its first flight in 1968. Also known as the V-12, it is the world's largest helicopter. It weighs 105 tons, has a rotor span of 219.8 feet (67 meters), and is 121.3 feet long (37 meters). On August 6, 1969, this machine established a world record by lifting a 40.2-ton payload to an altitude of 7398 feet (2,255 meters).

4. Miscellaneous

Rocket belt (1961)

The rocket belt was perfected by American's Bell Aerosystem Company. It was shown to the public in 1961. The device was invented by **Wendell F. Moore**. To date, experimental flights with the belt have not exceeded 650 feet (200 meters) horizontally and 65 feet (20 meters) vertically. The rocket belt consists of two vertical exhaust tubes and a fuel reservoir strapped onto the user's back. Two motorcycle-type handles are used to control it. This belt allows vertical takeoffs and stationary flight, and it can turn 360 degrees.

Air hostesses (1930)

On May 15, 1930, the first air hostesses, or stewardesses, began service on airliners. They were eight in number and were assigned to the San Francisco—Chicago route. "Invented" by the American S. A. Stimson, their role was to attract train passengers to airline travel. However, it was not until 1946 that stewardesses were to enter the airlines in large number. In that year, Air France and TWA decided to hire 26 applicants.

Parachute (1802)

The first true parachute was patented on October 11, 1802 by the Frenchman **Jacques Garnerin** (1769—1823) just after he had invented it. Since antiquity, Chinese acrobats had used paper and bamboo parachutes to amuse their public. Several centuries later, **Leonardo da Vinci** (1452—1519) drew a sketch of a parachute: a 36-foot (11 meter) cloth fixed to a wooden framework, supported by ropes.

In 1783, the French physicist, **Sebastien Lenormand**, jumped from a second floor, holding a parasol in each hand. However, **Garnerin** made the first true parachute jump on October 22, 1797. He ascended over the Monceau Gardens in Paris in a balloon; when he reached approximately 2600 feet (800 meters), at 5:35 P.M., he cut the cord that held the bas-

ket to the balloon. Supported by a parachute, the basket made a safe descent and landing.

The first parachute jump from an airplane was made by U.S. Army Captain **Berry** on March 1, 1912 over St. Louis, Missouri. He jumped from a biplane piloted by Jannus. The country first to use paratroops (in 1935, during peacetime maneuvers) was the Soviet Union. The Soviet Union was also the first to practice parachuting as a sport, in 1946. The first World Parachuting Championships, in 1951, were won by a Frenchman and a Frenchwoman.

Automatic pilot (1914)

From 1910 on, numerous mechanical systems were conceived and utilized to replace the manual control of an aircraft. The first efficient system was designed by an American, **Elmer Sperry**, and test flown in 1912 by his son, Lawrence, on a Curtiss seaplane.

In 1914, **Lawrence Sperry** presented this system at an air safety competition held in Paris. Sperry demonstrated the system's dependability by flying with his hands in the air while his passenger stood on the wing to show the stability of the Curtiss. Sperry was awarded second prize; no first prize was awarded.

In-flight refueling (1921)

An American, **Wesley May**, should be credited for having performed the very first in-flight refueling. On November 12, 1921, May jumped from the wing of a Lincoln Standard to the wing of a JN-4 with a fuel tank strapped to his back. He then climbed on top of the engine and poured the fuel into the upper wing tank. The first in-flight refueling using a pipe to connect the two airplanes took place over San Diego, California on June 26, 1923. An airplane piloted by U.S. Army Lieuten-

An American Farrington gyroplane, very convenient for short flights.

The airline stewardess is an American "invention".

ant **Seifert** refueled an airplane piloted by Captain **Smith** and Lieutenant **Richter**.

February 28—March 2, 1949: After flying exactly 94 hours 1 minute, U.S. Air Force Captain **James Gallagher** completed the first nonstop, around-the-world flight. His B-50 Lucky Lady II was refueled in the air several times during the flight.

Ejection seat (1941)

In 1941, the German firm, Heinkel, built the first operational ejection seat; although an Austrian inventor, **Odolek**, had experimented with an ejectable seat at Issy-les-Moulineaux, France as early as December 12, 1912. The ejection seat was designed to avoid the risks of a parachute jump from an aircraft in distress. The seat designed by Heinkel was ejected by compressed air. It was first mounted on the Heinkel HE 280 fighter prototype, which made its

first flight on April 2, 1941 from Rostock-Marienehe. The seat was soon put to the test, since on January 12, 1942, the HE 280 crashed at Rechlin, Germany. Unable to stay aloft because the plane's wings had iced over, the pilot, Major Schenk, was forced to eject at an altitude of 7775 feet (2370 meters). The parachute opened normally and the pilot landed safely.

The Swedes developed a seat ejected by an explosion. Initial tests were carried out in a Saab 21 on July 30, 1943. Today this superior ejection system is still being used, and it has rendered obsolete the Germans' compressed air system.

Ejectable capsule and ejectable cabin (1961)

The U.S. firm, Stanley Aviation Corporation, first experimented with an ejectable capsule (carrying a

dummy) on March 6, 1961. Use of an ejection seat at supersonic speeds is highly dangerous. Thus, a more suitable means of safely ejecting a pilot was sought, such as a highly resistant, waterproof capsule that would make high-speed ejections less dangerous. The March 1961 trial was performed on the ground, at a standstill. An initial in-flight test was conducted from a piloted aircraft at Mach 0.8 at an latitude of 20,000 feet (6,000 meters). But the ejectable cabin was rather complex and prone to mechanical problems in its separation from the aircraft.

In spite of problems, the ejection capsule was put in service with the U.S. Air Force's B-58s during the 1960's. The first ejection cabin to enter squadron service was the General Dynamics F-111 ejectable cabin, in which the ejection of the whole crew compartment is effected by rocket engines. The cabin is brought to the ground under a large parachute and is capable of staying afloat for extended periods. Crew comfort is increased by not wearing parachutes, and the cabin is able to carry a substantial amount of survival equipment.

Retractable landing gear (1911)

Plans for retractable landing gear had been drawn as early as 1876 by the Frenchmen **Alphonse Penaud** and **Paul Gauchot**, who wanted to reduce the aerodynamic drag on their small models. In 1911, a German named **Wiencziers** designed retractable landing gear for his monoplane, but his system needed improvements. During the Gordon-Bennet race in 1920, another system was seen. An American entry, the Dayton-Wright, was equipped with mechanically retracting landing gear. In 1929, the Frenchman **Georges Messier** designed the first hydraulically operated workable retractable landing gear.

Pressurization (1920)

With the advent of jet transports and the consequent increase in flight altitudes, machines had to be pressurized so that passengers could breathe normally. Although in reality a plane can be flown at any height from 10 to 12,000 meters, a passenger cannot be flown higher than 2450 meters without air pressure maintenance. The first airliner to be pressurized was the Boeing 307 Stratoliner, dating back to before World War II, but the first trials with pressurized cabins go back to 1920.

Modern parachutes permit great precision in direction and have an air of fantasy about them.

Left: The first ejector seats were tested using dummies. Right: Modern retractable landing gear, invented in the 20s.

MARITIME TRANSPORTATION

1. Boats

Steamboats

A boat with a steam boiler (1730)

As early as 1690, a Frenchman, **Denis Papin**, wrote about the possibility of developing a mechanism that could transmit the driving power of his steam engine to two wheels placed on each side of a building. He carried this project out in 1707, but did not follow up on it.

On the other hand, when Newcomen's steam pump (*See* Large Volume Steam Boilers, p. 315) had spread into the coal-mining regions of England during the 1730's, a mechanic named Jonathan Hulls used it to equip a tugboat. Placing a crank at the end of the beam of Newcomen's machine, he transformed the back-and-forth movement of the piston into a rotating movement which was transmitted to the paddle wheel of the boat. But the mechanical irregularity of this atmospheric engine and

the large quantity of coal that it consumed made Hull's project impractical and it was forgotten.

Steamboat with a paddle wheel (1793)

Watt's invention of the single-acting steam engine, which permitted an increase in the driving power of the engine and used less combustible fuel, led to a breakthrough in the progress of steam-powered navigation. Using Watt's machine, which the Perier brothers had imported from Birmingham, the **Marquis de Jouffroy** succeeded in launching a boat on the Seine in 1773. Unfortunately for the inventor, the huge counterweight on the fire pump (it weighed nearly 145 pounds) fell through the bottom of the boat and the boat sank. After Jouffroy's failure the ball passed into the camp of the English. Development proceeded rapidly now, because Watt had developed his more advanced double-acting expansive engine in 1783 (see

Watt's steam engine, p. 326). But it was in America that steam naviagation was to undergo its greatest improvements.

Steamboat with oars (1787)

A curious demonstration, witnessed by George Washington and Benjamin Franklin, took place on the Delaware River during the summer of 1787. Two American builders, **John Fitch** and **James Rumsey**, introduced a boat with oars fixed to a horizontal wooden rod, operating in the same way as ordinary oars but powered not by men but by a steam engine. The boat worked so well (it sailed up river against the tide at 5½ miles per hour) that a company was established and Fitch obtained a license from the U.S. government for exclusive development of all steam navigation in five states during the next 14 years. But the practical results did not measure up to general expectations and, after having worked hard to improve his apparatus, Fitch was abandoned by all his supporters and wound up going to France in 1792, where he hoped to meet with a better reception. France was at war with the rest of Europe, so Fitch returned to the United States. He killed himself by jumping into the Delaware River.

The Clermont, the first river steamboat (1807)

The *Clermont* (named for Robert Livingston's home on the banks of the

The first American paddle-steamboat, on the Delaware River.

The Boeing jetfoil, the most modern type of hydrofoil.

rules. In spite of this inauspicious beginning, Fulton convinced his contemporaries. When he died in 1815, there were numerous testimonials of mourning and respect throughout the United States.

The Savannah, *the first transatlantic steamer* (1818)

In 1818, Captain **Moses Rogers** of Savannah, Georgia, planned to build a steamboat intended for regular service between America and Europe. A corporation launched the operation, acquiring a handsome sailboat and installing a steam engine and paddlewheels. These paddlewheels could be dismounted and folded on deck. The *Savannah* left port on May 26, 1819 and arrived in Liverpool in 25 days. Its engine had functioned only 18 days of that time, since the captain wanted to take advantage of a favorable wind to economize on coal.

After this successful run, the *Savannah* made its way into the Baltic Sea to Kronstadt and St. Petersburg, where it was visited by Tsar Alexander I. But the *Savannah* ended up in obscurity. After returning to the United States, it was reconverted into a passenger sailboat and ended its adventurous career on the Long Island coast. It wound up sinking in the harbor during its last voyage.

An English steamer, *Enterprise,* traveled to India in 1825.

Hydrofoil (1869)

The principle for the first hydrofoil was laid down by a Frenchman, **Farcot,** in 1869. The hydrofoil is a boat with a standard hull, on the underside of which are attached fins that, under hydrodynamic pressure, create buoyancy proportional to the speed of the craft. When the hydrofoil reaches a certain speed, its hull is completely above water. Thus the force of friction against the hull is nearly eliminated.

Many technicians have studied the hydrofoil and improved it: an Italian, **Forlanini,** in 1900 (his device was tested by the Wright brothers; **Alexander Graham Bell,** inventor of the telephone; a Russian, **Alexeyev;** and a German, **Hans von Shertel,** after World War I. In 1962, the United States put into service the *Plain View,* a 320-ton antisubmarine boat that can navigate in troughs of 4.5 meters and whose fins weigh 5 tons.

At the present time, two main types of hydrofoils exist: hydrofoils

Hudson River) was built by the American **Robert Fulton** in New York in 1807. It was the first successful commercial steamboat. It measured 164 feet long by 16 feet wide, had a capacity of 150 tons, and its paddle-wheels were 16 feet wide. Few people, however, believed that this powerful riverboat would succeed, and even though its trial runs had taken place with no major problems, no passengers showed up for its maiden voyage. Only Fulton and his crew made that first run up the Hudson River from New York to Albany. During its night-time voyage, the *Clermont* spread terror; Fulton fired the boiler with pine boughs, which produced a lot of smoke and sparks. This column of flame, along with the noise of the engine and the paddle-wheels crashing on the water, terrified the people living along the banks of the river. One passenger showed up for the return trip, a Frenchman named Andrieux, who lived in New York; he paid his $6 passage, according to the

The Savannah, the first steamboat to cross the Atlantic, May 26 to June 21, 1819.

with fixed fins—generally ferries such as the Canadian *Bras d'Or* which weighs 200 tons and attains speeds of 60 knots (73 mph); and hydrofoils with swiveling fins whose surfaces are constantly under water.

Oil tanker (1886)

The first ship to be specially equipped for the transport of oil was a German ship, the *Gluckauf,* launched in 1886. In 1861, a shipment of casks containing oil extracted from wells in Titusville, Pennsylvania was transported by the sailing ship, *Elizabeth Witts.*

Ice breaker (1898)

The *Ermak* was the first ice breaker; it was built in 1898 by the English Armstrong shipyards, following a design of Russia's Admiral **Makarov.**

Aluminum boats (1890)

In 1885, **Heroult** in France and **Hall** in the United States invented a process for the industrial production of aluminum. In 1890, the Swiss shipbuilder, Escher-Wyss, built a 19-foot boat in aluminum, then a 39-foot steamer yacht. In 1893, the Frenchman **Godinet** designed the first aluminum sailboat, the *Vendenesse,* measuring 57 feet. In 1895, **Her-**

reshof built the *Defender* for the America's Cup race. This 124-foot sailboat had a copper hull (which glides better in water than wood or steel) and an aluminum deck. Unfortunately, the two metals are incompatible; the aluminum, disintegrated by electrolytic coupling, formed a deposit on the copper—the *Defender* became a sort of electric battery and self-destructed. It was necessary to wait until 1944 and the development of alloys combining aluminum and manganese for aluminum boat construction to become widespread. The use of this alloy in building the ocean liner *United States,* in 1952, allowed this vessel to be lighter by several thousand tons.

Atomic-powered ships (1958)

Laid on the stocks on May 22, 1958, in the United States, the *Savannah* was the first commercial atomic-powered ship. Launched in 1959, it was given the name of its glorious ancestor, the first steamer ever to cross the Atlantic (see above). It measured 181.5 meters by 23.8 meters, displaced 21,840 tons, and could transport 9500 tons of cargo transport and 60 passengers. Attended by a crew of 100, it could attain speeds of 20.5 knots (25 mph) and could develop 20,000 horsepower.

Theoretically the *Savannah's*

endurance range was 300,000 miles (550,000 kilometers), the equivalent of three years' uninterrupted navigation. In truth, although the *Savannah* was undoubtedly a technical success, she was a commercial failure. The operating company had to take her out of service in 1967—the ship was running an annual deficit of $1.5 million.

Hovercraft (1959)

This type of boat was invented in 1959 by the Englishman **Christopher Cockerell.** His Hovercraft is a boat which floats on a cushion of air. It does not touch the water, but remains above it, thanks to a cushion of air. It can also land on flat shores to unload passengers. A French company, Bertin, developed the same technology at the same time as the English.

Modern wind-powered cargo vessels (1980)

The last great commercial sailing vessel disappeared from service in 1936. But the rising cost of fuel has again made it necessary to utilize the wind as an auxiliary energy source. In 1980, the Japanese shipbuilding corporation, Nippon Kokan, launched the cargo ship, *Shin Aitoko Maru,* which was equipped with two rigid sails, each opening like a book, with a total surface of 2100 square feet. This auxiliary surface allows for a 30% reduction in fuel consumption. A second ship, the *Aitoku Maru*, was equipped with a single rigid sail like those used by the *Shin Aitoko Maru*. At the end of these trials, Nippon Kokan built two cargo vessels of 2100 tons with a forward sail of 1483 square feet and a rear sail of 1033 square feet. The sails are controlled by computer. The same shipbuilder is making plans for an ore carrier of 3500 tons with a sail of 36,000 square feet.

In the United States, the coasting vessel, *Mini-Lace*, 3100 tons, was equipped in 1980 with a fabric sail (the Japanese sails are metal) that can be rolled up and down the length of a pivoting mast. Fuel consumption has been reduced by 20%.

Fire-resistant boat (1982)

In 1982, the Dutch shipbuilders, **Mulder** and **Rijke**, developed an entirely closed survival boat capable of resisting temperatures up to 2200 degrees F (1200 degrees C) for 10 minutes. Oxygen tanks furnish the air necessary for the passengers and the operation of the motor. Depending upon the size of the model, this lifeboat can transport between 16 and 63 persons, as well as all the equipment necessary for their survival.

2. Navigation Aids

Anchors (3000 B.C.)

The first anchors, used by Chinese and Egyptian sailors during the 3rd millenium B.C. and later by the Greeks and the Romans, were stones or bags containing sand or pebbles that could simply be thrown overboard. Pliny, Strabo, and other Roman authors attribute the invention of the metal anchor to several different seafaring peoples. The first approach tried was the single-palm grapnel used around 600 B.C. An important improvement was made in the 18th century when better quality, less brittle iron was used, and when the arms were given a new camber. Around 1770, iron-stock anchors completely supplanted wood-stock anchors. (The stock is a bar perpendicular to the shank of an anchor; its purpose is to make the anchor swing so that one of the flukes grips the bottom.)

In 1821, the Englishman **Hawkins** engineered the mooring hawse-pipe anchor with palms. With this system the stock is no longer necessary, the arms being mounted in such a way that they automatically lean to the same side to grip the bottom. This type of anchor was modified between 1872 and 1887 by the Englishmen C. and **A. Martin, S. Baxter,** and **W. Q. Byers**.

The CQR anchor, or "plow anchor," which the Englishman **G. I. Taylor** patented in 1933, is of an extremely original design. At the extremity of the shank, a double plowshare is mounted, and the gripping power this provides is twice as great as that of a standard anchor.

Lighthouse (3rd century B.C.)

In 285 B.C., one of the seven wonders of the world was built on Pharos (formerly an island, now part of the city of Alexandria, Egypt). This was the lighthouse built according to the instructions of the Egypt's King **Ptolemy II**. This lighthouse is said to have measured over 130 meters in height. A wood fire was kept burning all night at its top. It was destroyed by an earthquake in 1302.

Many lighthouses were built under the Roman Empire, but this practice fell into disuse at the beginning of the Middle Ages. Lighthouses reappeared in the 11th century with the rebirth of trade. The decisive advance in lighthouse construction was the invention, by the Englishman **John Smeaton** in 1759, of a cement that could set in water.

As to the light, it was provided by wood fires until the 18th century. In 1780, a Swiss, **Argand**, designed the flat-wick oil lamp called the "Argand burner" (see p. 169). In 1901, a new development appeared; this was the petroleum burner, invented by **Arthur Kitsen**. This mechanism, improved further in 1921 by **David Hood**, is still used today in nonelectrified lighthouses.

The progress made in optics found lighthouses to be a perfectly suited field of application. As of 1752, the parabolic reflector designed by the Englishman **William Hutchinson** increased the power of the light signal. The most decisive step was an advance in the use of lenses, conceived in 1821 by the French engineer **Augustin Fresnel** (1788—1827). Fresnel replaced ordinary lenses by plano-convex lenses whose external face was cut into plane strips. In this way, the rays were made parallel and the light condensed. The use of electric lighting (beginning in 1859 in Great Britain) greatly improved the effectiveness of lighthouses.

Weighted Keels (200 A.D.)

In order to prevent sailing ships from capsizing under the lateral force of their sails, it was always necessary to concentrate the weight of the ship as low as possible. When commercial ships had no cargo, they had to be filled with stone ballast that acted as counterweight to the sails. As war ships had no cargo to transport, they were ballasted with something heavier than stones. At the end of the 17th century, the king of England, Charles II, had the bottom of one of his yachts covered with lead plates, in much the same way that these had been attached to Roman vessels discovered at Nenni (200 A.D.). In 1796, the Royal Navy had two frigates built, the *Redbridge* and the *Eling;* both had long lead ingots fixed to the outside of the hull. In 1844, the American sloop, *Maria,* designed by **Robert L. Stevens,** had lead-lined wooden railings. The first sailing ship to be completely lead-ballasted on the outside of the hull was the *Peg Woffington,* owned by the Scot **George L. Watson,** in 1871.

Keel of the Australia *(1983)*

Thanks to a particularly original keel, the America's Cup was won by the yacht, *Australia,* in 1983, after 132 years of American domination.

A keel functions like the wing of a plane, with a decrease in pressure on the one side and increase on the other. Unfortunately, at the extremity of the keel, the depression sucks pressure from the other side and reduces efficiency. The designer of the *Australia,* **Ben Lexcen,** envisaged fins under the keel to prevent this loss of pressure. Thanks to this, it was possible to design an even more efficient keel, broader above than below. This keel is beginning to be fitted on cruise ships, allowing them to reduce their draught by 20% to 30% with no loss in efficiency.

Compass

Tradition has it that the compass was invented by the Chinese. They passed it on to the Arab sailors who plied the northern part of the Indian Ocean. The Venetians are said to have brought the compass back from their trading expeditions in the Orient. Navigation contracts published in the 13th century indicate that the needle was then fixed onto a wooden disk divided into 32 quarters. And legend

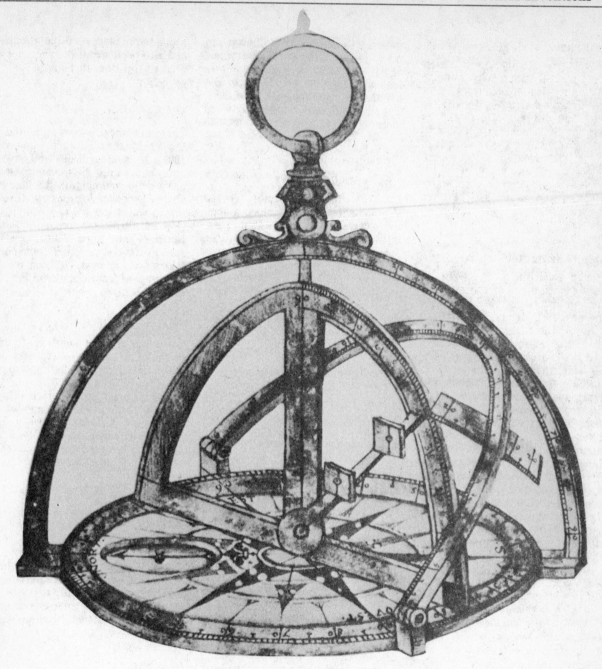

A compass from the middle ages, truly a work of art.

has it that the first "wind rose" (the first marine compass) was created by **Flavio Giova**, an Amalfi craftsman, south of Naples, Italy.

After centuries of searching and designing, the model that surpassed all others was the Thompson rose of 1876. The bars of this "dry compass" are cylindrical and tapered, mounted on silk thread and fastened to a compass cord made of very fine paper, whose total weight is no more than 20 grams.

Use of the "liquid compass," resulting from the work by the Englishmen, **Dent** (1833) and **Ritchie** (1855), became widespread around 1880. It faced fierce competition after World War I from the gyroscopic compass.

Tide tables (13th century)

The first known tide tables were drawn up in the 13th century by the monks of Saint Albans to give the height of the Thames at London Bridge. Their method for calculating the hours of the high sea was based on the phase of the moon within its cycle of 29½ days revolution around the earth.

In the 16th century, the hour of the passing of the moon's meridian was included in the computation. During the second half of the 17th century, a diagram (table) was designed for each harbor so as to avoid calculations: By a simple reading, one obtained the hour of the tides from the phase of the moon. The first printed tables were published in the Nautical Almanack by the Breton **Brouscon** in 1546. One of the most famous tables is that of **Richard** and **George Holden**, published in Liverpool in 1773. For a long time, it was used to determine the tides of that harbor, although these are very irregular. In 1858, the Nautical Almanack, published by the British Admiralty, gave a tide table for all of the harbors in the world. As of 1910, it gave tables for only 26 of them.

Canal lock (14th century)

The double, or chamber, canal lock is an invention attributed by some to 14th-century Dutch engineers (later to be improved upon by the Italian genius **Leonardo da Vinci**) and by others to da Vinci himself around 1480. By regulating the conditions of passage, the invention of the chamber lock solved the problem of getting vessels from one reach of water to another and simplified the often dangerous lowering and lifting of boats.

Marine barometer

Although **Evangelista Torricelli** invented the ancestor of the mercury barometer in 1643 (*see* Measuring the Atmospheric Pressure, p. 244), at the end of the 17th century, sailors still used a rudimentary instrument called the "barometer of Liege"; the receptacle was half-filled with water that could flow out by a spout. The variations in the water level expressed the variations in atmospheric pressure. The process was very inaccurate, as differences in the temperature of the water greatly influenced results.

It was in the 18th century that the first marine barometers with mercury appeared. But their use did not become widespread because sailors were reluctant to give up traditional methods of predicting the weather.

On land, in 1858, England's Admiral Fitzroy thought of equipping all fishing ports with a barometer in order to warn sailors of probable bad weather. A terrible storm had just laid waste to the English coast, causing the loss of many ships.

Ship model tests (1770)

The first person to test model ships in order to deduce the performance of the real vessels was the Dane **Frederick H. af Chapman** in 1770. Several attempts at using models were undertaken by England's Colonel **Beaufoy** (in one of the reservoirs of the family brewery), then by the American **Robert L. Stevens** (son of John Stevens, the first to use the propeller); and the Scot **John Scott Russell**.

It was an English professor, **William Foude**, who discovered the law of similarity allowing an extrapolation from the test models to full-size boats. This made it possible to use models scientifically (*see* Fluid mechanics, p. 240). In 1874, he built the first test tank (90 x 11 x 3

meters), which was followed by many others. Between World War I and II, test tanks almost a kilometer long were built to study the hydrodynamics of seaplanes. Nowadays, computer simulation tends more and more to replace tank tests.

Ship's propeller

First propellers (1785)

In 1785, the Englishman **Joseph Bramah** patented a 16-bladed propeller to drive boats. But the first experiment was made by an American, **John Stevens**, and a Frenchman, **Marc Brunel**, who coupled two four-bladed propellers to their one-cylinder steam boiler to sail up the Passaic River in New Jersey. (The machine and its propellers are now on display at the London Science Museum.)

Propulsive propellers (1837)

After the passable, but unfruitful, tries of the Frenchman **Dallery** and the Austrian **Russel** (to whom the Viennese nevertheless erected a monument "for having invented the propeller applied to steam navigation"), the attempts of the Swedish-American **John Ericsson** were at last crowned with success in 1837. His propulsive system, composed of two propellers, was applied to a tugboat, the Francis Ogden. At the same time, two English builders, **Smith** and **Rennie**, developed a comparable system and founded the Company for Steam Propulsion. They built a large steamship, the Archimedes, which definitively placed the propeller in the field of navigation. In 1843, **Isambard K. Brunel** built the first propeller-driven transatlantic steamship, the Great Britain, which was 312 feet long. Its six-bladed propeller was 15½ feet in diameter. (This boat is being restored at Bristol, England.) Two years later, the British Admiralty attached two steamships by their sterns: the Alecto (800 tons, 200 horsepower) had paddlewheels; the Battler (888 tons, 200 horsepower) was equipped with a propeller. All the power of the Alecto's paddlewheels could not stop its being towed by the Battler. The superiority of the propeller had been proven.

Super-cavity propellers

During the 1970s, in order to obtain greater speeds, the American navy developed super-cavity propellers, which take advantage of a depression

around the blades of the propeller and also profit from the vaporpocket created by the turbulence of the propeller.

Mechanical log (1801)

The mechanical log was created by the Englishman **Edward Massey** in 1801. It standardized an already ancient, but imprecise, system. Originally quite rudimentary, the log was simply a wooden log that was thrown in the water toward the bow of the ship and then recovered when it had reached the stern. The apparent speed of the vessel was deduced by taking into account the time the log took to pass from one end of the ship to the other.

Beaufort scale (1806)

This scale, used to note wind intensity (indexed from 0 to 12), was proposed in 1806 by an English admiral, **Sir Francis Beaufort** (1774—1857). It has been used by all of the Royal Navy's vessels since 1834 and makes it possible to standardize sailing records in logbooks. The Beaufort scale was adopted by all navies in 1854, then by international meteorology as of 1874. It is still very important to all navigators.

Outboard engines (1905)

Two models appeared in 1905: an American engine designed by **Ole Evinrude**, who called it "outboard" because it had the characteristic of being screwed vertically to the boat's outer hull; and a German engine built by **Fritz Ziegenspeck**, who named it "Elf-Zett."

Bow bulbs (1929)

Attempts have been made since the beginning of this century to reduce resistance to the forward movement of sea-going vessels by designing hulls that make fewer waves and expend less energy. In 1929, the German ocean liner Bremen, winner of the transatlantic Blue Ribbon, had a protuberance at the foot of its bow designed to diffuse the pressure of resistance. In 1962, after twenty years of research, the Japanese **Takeo Inui** proposed equipping vessels with huge bow bulbs. The penetration of this volume under the water creates a wave on the surface. At the hollow of this wave, another wave is created by the bow of the vessel itself, and the volume of this second wave is thereby limited. The amount of energy saved can reach as high as 30%, but such results can only be achieved by perfectly synchronizing the two waves, either by keeping the vessel at a constant speed or by controlling each forward movement of the ship.

Radio beacons

The French physicist, **Andre Blondel** (1863—1938), created the first radio beacons. Radio beacons emit signals that are intercepted by ships equipped with a radiogoniometer. The radiogonimeter indicates the position of the transmitter in relation to the ship so that the vessel, which has identified the source by noting the frequency of its signal, can determine its own position. The same principle applies to airplanes. Since 1960, radio navigation has profited from the important progress made in space technology.

3. Diving Equipment

Diving Bell (ca. 400 B.C.)

The diving bell was already known in antiquity. Aristotle (384—322 B.C.) described it in these words: "Divers are given the possibility of breathing underwater when a kettle or tank made of bronze is submerged. It does not fill up with water but keeps the air if it is forced to go down perpendicularly."

The diving bell disappeared in the Middle Ages. It appeared again in the 16th century in Spain and Italy. In 1538, before an amazed audience of several thousands, including Emperor Charles V, two Greeks dove on a bell to the bottom of the Tagus, in Toledo, and came up dry with their lamp still lit. In 1552, fishermen of the Adriatic also experimented, in the presence of the Doge and senators of Venice. Their tank measured 5 meters in height and had a diameter of 3 meters. The same Venetians designed, at the same time, a diving hood called a "bagpipe." It was a big upturned tank over the divers' heads; to the top of which were connected flexible tubes through which assistants on shore sent air by using huge bellows.

Halley's Bell and "Cap" (1721)

It was the English astronomer, **Edmund Halley**, who engineered the first diving bell worthy of the name, in 1721. He invented a way of con-

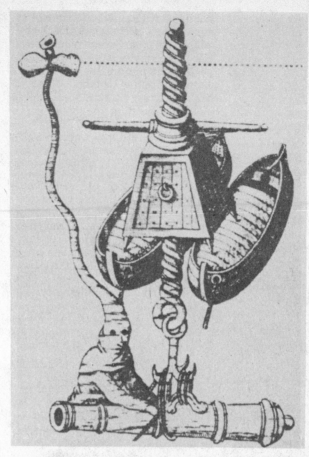

Left: A fantastic conception of an early diver. Below: An ancestor of the diving suit. The diver certainly has a strange encounter.

stantly renewing the air within the device and of sufficiently compressing the air to prevent water from penetrating, whatever the depth might be. His lead-covered, wooden bell was connected to a cask of compressed air that was lowered by its side. As the air exhaled by the divers was hotter (thus lighter) than the air coming from the cask, it could be expelled by a tap simply placed at the top of the bell.

To enable a diver to move away from the bell, Halley designed a "cap of maintenance." This was connected to the inside of the big bell by a hose.

The arrangement was not convenient, since the wearer of the "cap" had to keep his head perfectly horizontal all the time lest water enter and he drown. In 1786, another Englishman, the engineer **Smeaton**, greatly improved Halley's bell by replacing the compressed air casks with a lifting pump that sent a constant air supply.

Primitive diving suit (1796)

The first real diving suit was invented by a German of Breslau, named **Klingert**, in 1796. It consisted of a thick cylinder made of tin and rounded off into a dome. It completely covered the head and chest of the diver, leaving the arms free. A short-sleeved jacket and short pants protected the limbs against water pressure. All of this was perfectly airtight. Two eye-level holes fitted with glass enabled the diver to see. A tube coming from above water ended in a hole made at nose-level while another, placed close by, was supposed to evacuate the air the diver exhaled. Two lead weights were attached to the waist to weigh the diver down. Outfitted in this very uncomfortable attire, **Frederic-Guillaume Joachim** nevertheless succeeded in sawing a tree trunk on the bed of the River Oder, on June 23, 1797.

Diving suit with pump (1829)

The first truly effective diving gear is attributed to an Englishman, **Siebe**, who designed it in 1829 and was entrusted with equipping the French navy until 1857. The model was adapted by the Frenchman **Cabriol**. His diving suit included a tinned copper helmet pierced with three windows that were protected against shock by a strong metal lattice. The air pipe ended in the back of the helmet. A valve that could be maneuvered by the diver enabled him to evacuate the air he exhaled, as well as that provided in excess by the pump. The suit was of strong linen, lined with a one-piece layer of rubber and fastened to a metal pelerine screwed to the helmet. The suit also included lead-soled half-boots and a leather belt to which were attached tools, a dagger, and the rope with which the diver communicated with the surface.

Independent diving suit (1867)

It appeared in France during the World's Fair of 1867, where it was presented by the Submarine Company of New York. The outfit, like Cabriol's, consisted of a metal helmet and a waterproof garment, but the diver also carried on his back a tank containing compressed air at 17 atmospheres, an amount sufficient for a man to breathe normally for 3 hours at a depth of 20 meters. The tank connected with the helmet via a tube equipped with a valve. For great depths, the outfit was complemented by a "protector" made of wooden rings, which formed a kind of armor around the diver's body, attenuating the harmful effects of water pressure.

Diving suit with automatic tank regulator (1865)

The drawback of the diving suit whose air supply depended either on a standard tank carried by the diver or on an air pump placed outside resided in the impossibility of making the air pressure vary according to the diver's needs. The pressure of the air inhaled must counterbalance the water pressure, and this latter varies according to whether the diver ascends or descends.

It is thanks to the collaboration of a French industrial mining engineer, **Rouquayrol**, and a naval lieutenant, **Denayrousse**, that this device, essential to the improvement of the diving suit, was invented in 1865. The tank regulator played the role of a true artificial lung insofar as the diver's breathing regulated the intake of air by acting directly on a distribution valve.

Bathyscaphe (1948)

This diving device was invented and tested by the Swiss physicist, **Auguste Piccard** (1884-1962), in 1948. The bathyscaphe makes it possible to explore underwater depths. The first model went to a depth of 138 meters off the coast of Dakar, without a passenger.

The American bathyscaphe, the Trieste, built by Professor Piccard in 1953, holds the record for depth. On January 23, 1960, it reached the ocean bed at 10,916 meters in the Challenger trench, the deepest in the world, off the Mariana Islands in the North Pacific.

armaments

ARMAMENTS

1. Small Arms

Longbow (12th century)

The British longbow dominated battlefields for over a hundred years, in fact retarding the adoption of portable firearms in England until the 17th century. The bow had been used since the earliest recorded warfare, but the longbow, averaging 5 feet in length, became the decisive British weapon during the Hundred Years War. Requiring a 60- to 90-pound pull and necessitating long training for proper use, its yard-long shaft could be fired almost 500 yards and was deadly when used in mass fire at 200 yards. Flights of arrows decimated French knights at Crecy (1346) before they could reach the English lines, and the victory in this battle established the supremacy of the weapon when used against slow-firing crossbows, or the even slower early firearms, for more than a century.

Matchlock (15th century)

Until the middle of the 15th century, guns were held in the left hand while the explosive charge was ignited by the right hand, which touched a lighted wick or "match" to the weapon's touchhole or priming powder pan. The advent of a mechanically activated S-shaped match holder during this period liberated the user's right arm, allowing a better grip on the firearm. This was an enormous step forward.

Wheel lock (15th century)

By producing its own sparks, the wheel lock liberated the gun user of the problems associated with a lighted match or wick, which had to be kept constantly aglow in order to utilize the firearm. Probably invented by **Leonardo da Vinci** (1452—1519), who made drawings of it in his *Codice Atlantico,* the wheel lock was apparently perfected by German artisans from Nuremberg and marketed from 1525 on.

Its good qualities, in particular its sureness of fire, were counterbalanced by its fragility and its exorbitant cost. The articulated chain found in the wheel lock is the direct ancestor of the present bicycle chain. The wheel lock's operating principle was applied to clock and watch-making toward 1630, replacing catgut, which had previously served to transmit the movement of the expanding spring to the indicating needles. Chain watches were produced until the invention of the cylindrical movement, around 1825—1830.

Bayonet (16th century)

According to legend, the bayonet was invented, or at least manufactured for the first time, in 1640 in Bayonne, France, the city from which it takes its name. In fact, this weapon appeared as early as the 16th century and was occasionally mounted on the barrel of hunting arms.

Flintlock (16th century)

The classic flintlock is the synthesis of two earlier primitive systems, and it uses the flint and strikeplate as spark-producing elements. Appearing toward the end of the 16th century, the *snaphaunce* and the *miquelet* were two early types of flintlock, the former used in Holland and the latter in Spain and southern Italy. Further development of these weapons resulted in a simpler and less expensive flintlock. Perfected definitively by the British toward the end of the 17th century, the flintlock remained in service until it was replaced by the percussion cap at the start of the 19th century.

William Tell with a longbow.

Cartridges (17th century)

The first cartridges consisted of a paper cylinder containing the ball and powder charge. The cartridge was torn open with the soldier's teeth and a small quantity of powder was placed in the priming pan, which was then closed. The rest of the powder charge was put in the barrel along with the ball and rammed down the barrel with a rod. To fire the weapon, the piece was cocked and the trigger pulled. The flint-carrying hammer, on striking the pan cover, produced a shower of sparks and ignited the powder in the pan, which then detonated the main powder charge.

In 1807, the invention of the priming cap by the Scot, **Alexander Forsyth,** eliminated the flint and its common wet weather misfires. A brass detonating cap placed over the "nipple" was struck by the hammer. The shock caused the cap to spark, and that in turn fired the main powder charge. However, the soldier still had to tear the cartridge open with his teeth and place its contents in the barrel with the help of the ramrod.

Starting in 1850, progress in armament technology made it possible to design breech-loading arms, secured by locking action. The next step to further simplify loading was to attach

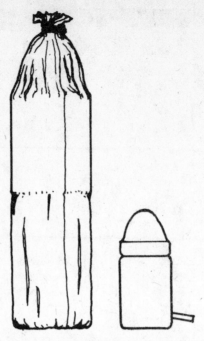

Left: The first combustible cartridge by the German Dreyse. Right: The pin cartridge of Lefaucheux.

the priming cap directly to the cartridge. Ignition was effected by a "needle" that pierced the base of the cartridge to which the primer cap was attached. With the arrival of the breech-loading gun came the idea of incorporating the detonating cap in the cartridge. The principal inventors were **Casimir Lefaucheux,** with his pin cartridge (1836), the Prussian **Dreyse** (1841), and the gunsmiths **Beringer** and **Flobert** (annular percussion, 1845).

The Franco-Prussian war of 1870 demonstrated that paper cartridges were too fragile for wartime conditions. Such cartridges were therefore abandoned after the war and replaced by copper cartridges and "centerfire" percussion caps, perfected in the United States immediately after the Civil War (1865). This was considerable progress, since it made loading and preserving ammunition from deterioration easier; however, ballistics as such remained unchanged.

Revolver (Modern, 1836)

An old European invention dating from the 16th century, the revolver did not really develop until the appearance of the percussion cap at the start of the 19th century. An American, **Samuel Colt** (1814—1862), is given credit for perfecting the revolver from 1836 on.

The principle of the revolver is based on the utilization of a cylindrical block pierced by several cham-

The S.S.G. Steyr carbine.

bers which rotate around a fixed axis. The rotation system allows successive alignment of the chambers with the single barrel through which they are fired.

Browning semiautomatic pistol (1900)

John Moses Browning (1855—1906), an American, worked in the Winchester firm toward the end of the 1870s. Around 1897, he left the firm and went to Europe, where he joined the National Belgium Arms Factory (FN), bringing with him plans for his pistol and hunting shotgun. The first .32 caliber (7.65 mm.) Browning pistol was marketed in 1900. It met with immediate success: 250,000 units were sold in the first six years and a total of 1 million pistols were sold in 12 years. The Browning pistol provided the basic design for Colt pistols, particularly the famous .45 caliber Model 1911 automatic. The largest version of this weapon system is the HP (High Power) Browning pistol dating from 1935.

Rifles

Winchester carbines *(1860s)*

American **B. Tyler Henry** (1821—1898), completed the design of the first repeating rifle in 1860 at the request of his first employers, Smith and Wesson, directors of the Volcanic company.

The first weapons produced by Volcanic were mechanically sound but used very weak cartridges; this weakness resulted in the failure of the company in 1857. The firm was then bought by **Oliver Winchester,** who changed its name to the New Haven Repeating Arms Company.

In 1860, Winchester started production of an improved version of the Volcanic rifle. The new versions, sold as the Henry rifle or Henry carbine, met with tremendous success. The name of the company was changed to Winchester in 1866, at which time the rifle's design was further improved by the addition of a spring trap mounted on the right side of the magazine, which allowed rapid loading. The new design was the first of the famous Winchester Carbines.

Springfield service rifles *(mid-19th century)*

These rifles, produced by the U.S. army ordnance factory in Springfield, Massachusetts, were the standard infantry weapon of the U.S. army during the American Civil War, World War I, and the beginning of World War II. The best known model is the

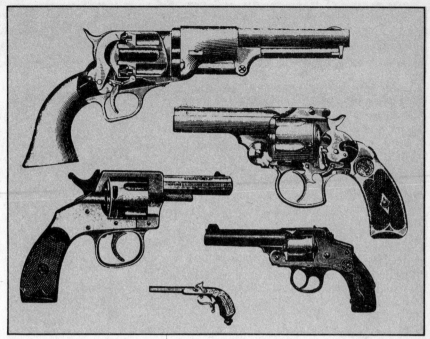

Some examples of American revolvers.

1903 Springfield, with an effective range of 2000 yards, and a maximum range of 5800 yards. Replaced as the standard U.S. infantry rifle by the Garand, it remained in limited service during World War II and the Korean and Vietnam Wars as a sniper rifle due to its extreme accuracy and range.

Automatic rifle (or light machine gun) *(1902)*

The automatic rifle was invented by the German **Madsen** in 1902. He gave it the name of light machine gun, a name later adopted by most countries.

The idea of the automatic rifle attracted other inventors such as Hotchkiss, and later Berthier, who produced guns used in small quantities by the British army just prior to 1914. However, at the time, this type of weapon met with only limited acceptance.

The hastily concieved French Chauchat, a rather crude automatic rifle, was introduced in 1915. Its ease of manufacture allowed widespread utilization by French troops and, later on, by American troops when they entered the war in 1917.

Weighing 15 to 26 pounds (7 to 12 kilograms), the automatic rifle can be carried and fired by a single infantryman, who is usually accompanied by a "loader" carrying extra ammunition. Today the automatic rifle has been replaced by all-purpose machine guns derived from the German MG43 and MG42 guns of World War II.

Garand semiautomatic rifle *(1937)*

Designed in 1937 by the ordnance department of the U.S. army, the Garand semiautomatic rifle was the first semiautomatic rifle to be used as a standard infantry weapon. This gas-operated, .30 caliber (7.62 mm.) rifle utilized an eight-rounds clip and had an effective range of about 2000 yards. Previous semiautomatic rifles had been issued in limited quantities during World War I (St. Etienne and Mauser), but were experimental rather than standard equipment.

Replacing the 1903 Springfield utilized during World War I and the Japanese invasion of the Philippines, the Garand became the standard U.S. infantryman's weapon during World War II, the Korean War, and the early part of the Vietnam War. It was also used in large numbers by the French army in Indochina and Algeria.

Machine guns (1850)

Maximum infantry rate of fire has always been a major military goal, as a barrage of projectiles can compensate for the mediocre accuracy of average infantry marksmanship. Machine guns were first used during the American Civil War (1861—1865) and the Franco-Prussian War (1870—1871).

Mechanical machine guns *(1850)*

Appearing towards 1850, they were composed of either a single barrel re-

Modern Machine Guns

Recoil energy from the first round automatically initiates succeeding operations. The first automatic machine gun was invented by the American Maxim, in 1884. It operated on the "short barrel recoil" principle.

SHORT BARREL RECOIL PRINCIPLE

A. The barrel is locked to the breech, the round ready to be fired.

B. Breech-block barrel recoil. Start of return spring compression.

C. Unlocking of breech from barrel. Recoil of breech-block only. Further compression of return spring and ejection of empty cartridge.

D. Return to firing position. Return spring decompression introduces a new cartridge and locks breech and barrel.

GAS DIVERTING PRINCIPLE (HOTCHKISS)

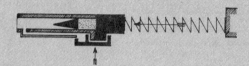

A. Round is ready to be fired. The barrel is locked to the breech-block.

B. The bullet has passed the gas port. A small quantity of gas enters the gas tube and provokes the unlocking and recoil of breech-block.

C. Decompression of the return spring. Return of the breech-block; new cartridge introduced; breech-block locks to the barrel.

Guns have always been a part of the movies, from the earliest silent westerns to today's thrillers. Above: Ingrid Bergman and Gary Cooper with a 1917 Colt. Right: Raquel Welch holding an 1875 Remington.

quiring the manual maneuver of a starting handle that induced the passage of a cartridge and its firing, or of a group of barrels charged and fired one after the other in rapid succession by turning a crank. The rate of fire was directly proportional to the cranking speed. The most popular model was that of the American **Richard J. Gatling** (1818—1903).

Early cranking machine gun
(19th century)

Cranking brings up a cartridge, introduces it into the firing chamber, cocks the firing pin, and detonates the explosive charge by percussion.

Both the Gatling gun and the similar French Montigny *mitrailleuse* were very efficient guns. The Vulkan cannon, to which an electric motor has been added, allowing firing of 6000 35 mm. shells per minute, utilizes the same principle as these earlier weapons.

Automatic machine guns *(1884)*

The first automatic machine gun with nonstop firing was invented in 1884 by the American **Hiram Maxim** (1840—1916). It functioned with a kicking effect.

Recoil energy produced by firing the first round was utilized to: 1) eject the fired cartridge; 2) bring up to the firing chamber a new cartridge from a belt or magazine; 3) fire the second round. The cycle repeated itself as long as the trigger was depressed and continued until the ammunition was exhausted. Approximately 600 rounds per minute could be fired.

In 1892, the American **John Moses Browning** (1855—1926) produced the first machine gun working off gas produced in firing. In 1893, the Czech Obdolek protected a similar invention under a patent taken out by the French company Hotchkiss. Today the gas-operating principle is still widely used in light automatic weapons, e.g., automatic and assault rifles. Maxim's recoil principle has been retained in the majority of current machine guns.

Submachine gun, or machinepistol *(1915)*

The first model of a submachine gun is probably the one made by the Italian **Villa-Perosa** in 1915. The Bergman was utilized in 1917 by frontline German troops. This weapon, firing standard pistol ammunition at a rate of 400 rounds per minute, was effective in spite of a weakly constructed magazine.

The machinepistol, more commonly known as a submachine gun, is a light automatic weapon using standard pistol ammunition. This weapon is being relegated to police, rather than military use. The submachine gun, along with semiautomatic rifles and pistols, is being replaced in most armies by a multipurpose "assault rifle" capable of firing either "bursts" or single rounds.

2. Artillery

An artillery piece is fundamentally a large rifle capable of throwing a heavy projectile at long range. Artillery actually predates portable or small firearms.

The first barrels were made of wood, circled with iron hoops; later they were made of wrought iron or brass. Finally, in the second half of the 19th century, they were made of steel.

Loading cannons was always a problem. To facilitate the loading of guns placed in "gun ports," attempts were made at loading them from the rear of the barrel or from the top of the firing chamber. However, the impossiblity of achieving an effective gas seal created dangers to the gun crew. Breech-loading systems were tried in the 15th century with the earliest cannons but were abandoned from the 16th to the 19th century in favor of muzzle-loading artillery. With the appearance of Bessemer's steel manufacturing process in 1856, it was possible to return to the original breech-loading system. Current artillery, with the exception of mortars, utilizes breechloading.

Slow rate-of-fire problems were not resolved until the last years of the

An early artillery piece

A Gatling gun.

A Maxim automatic gun

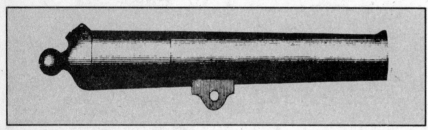

Dalghren 24-pound Howitzer, 1862.

Early cannons

19th century, when three new elements appeared: 1) breech-loading systems; 2) cased ammunition; and 3) recoil absorption through hydraulic braking. The "French 75" of 1896 is the best example of this progress. Prior to the invention of "barrel brakes," a field artillery piece had considerable recoil, the gun and its carriage moving several meters to the rear after firing each round.

A constant factor in artillery evolution has been ballistics: Flat trajectories with initial high velocity shells are fired from long barrels and are associated with direct, long distance fire by field and marine artillery; high trajectories, with initial low velocity and large caliber shells, are fired from heavy howitzers and mortars to reach short-range targets behind obstacles

Carriage (14th century)

The gun carriage was designed to support and aim artillery tubes. Appearing in the 14th century, the first gun carriages were crude, wooden tube supports without wheels. In the 15th century, the **Bureau** brothers perfected a horse-drawn, wheeled gun carriage

Howitzer (17th century)

The howitzer appeared toward the end of the 17th century in Anglo-Dutch armies. The name is derived from the Dutch word *houwitser*. This type of artillery piece fires a high trajectory explosive shell at targets that cannot be reached by flat trajectories. The explosive shell was at first armed with a fuse and lighted prior to firing. The howitzer was the first wheeled artillery piece to fire an explosive shell and remained the sole example of this type until the middle of the 19th century.

Cannon (18th century)

The cannon's ancestors were the "firetubes" that appeared in the 14th century following the invention of gunpowder. They were known under different names, such as bombards, firevase, etc. The word cannon was adopted in the 18th century by the Frenchman Valliere to designate all nonportable firearms.

Antitank cannon (1917)

With the appearance of the assault tank in 1916 came the need for a specialized cannon to use in battle against the new machine. From 1917, the French army equipped its anti-

A sailor cleaning a cannon on a naval ship turns his task into an opportunity to get a new perspective on his surroundings.

tank batteries with small-caliber, high-speed cannons, initially supplied by the navy.

The first antitank cannon worthy of that name was the 37 mm. of the American Bethlehem Steel Company (1918). The progressive increase in the defensive armoring of tanks entailed an increase in the power of the antitank cannons which, at the end of World War II, had reached a caliber of 128 mm.

The hollow charge and missiles supplanted the antitank cannon during the 1960s.

Modern mortar (1917)

Short in length, but large in caliber, the mortar is an artillery piece firing projectiles that have a parabolic trajectory. Its first appearance coincided with that of other early artillery pieces. The early mortar launched large stone balls, which caused extensive destruction since they fell from great heights. The inventor of the

The French competitor to Big Bertha.

The mortar was invented in 1917 by the British.

modern mortar was the Englishman **Stockes**, whose device, an 81 mm. caliber artillery piece, was put into service in the British army in 1917. The Frenchman Edgar Brandt worked with Stockes and improved on his invention.

In the 1930s, Brandt perfected a whole series of mortars going from 45 mm. to 155 mm. caliber. The Stockes-Brandt mortars are the immediate predecessors of today's mortars whose range is up to 10 kms. for the 120 mm. caliber piece.

Recoilless rifle (1940)

The German *Leichtgeschutz*, manufactured in 1940, was the first recoilless rifle. This gun, mounted on a standard gun carriage with rubber tires, weighed only 855 pounds (388 kilograms). Maximum range of its 105 mm. shell was 7270 yards (7950 meters). A smaller 75 mm. version was utilized in the 1941 airborne invasion of Crete.

Lack of recoil allowed even further weight reductions in later models. The U.S. 75 mm. recoilless rifle was mounted on a modified M1917 ma-

chine gun tripod. In spite of its very light weight, this M20 75 mm. rifle had a maximum range of 6400 yards (7000 meters). The major advantage of its light weight was offset by the disadvantage of a powerful gas stream ejected to the rear of the weapon during firing. The recoilless rifle is found less and less since the use of missiles has become widespread.

3. Armor

Armored combat vehicles

The idea of rapid displacement to strike an adversary is as old as the history of warfare. Chariots carrying crews armed with bows and javelins were utilized in already ancient times. The Persian King Cyrus employed war chariots at the battle of Thymbre (ca. 540 B.C.) against the Lydian King Croesus.

Armored Car (1902)

The French firm of Charron, Girardot & Voight marketed the first armored

car in 1902. This was the first car in which modern armored vehicle design concepts appeared. It was a motorized vehicle with armor plating and was equipped with an armament-carrying turret. The armor plating protected the crew against small arms fire (from rifles and machine guns), and the armament in the turret could be fired in all directions.

Armored troop carriers (1918)

The first armored troop carriers were built in 1918 by the British. About 50 vehicles, transporting 20 men each, were constructed. They utilized tracked chassis. Protected only by light armor and armed with one machine gun, these carriers did not participate in combat operations. The vehicles were first named "rolling coffins" by infantrymen who appreciated armor protection but were used to open-air battle conditions.

Tank (1908)

In 1908 the Englishman **Roberts** demonstrated an armored, tracked vehicle near London. In 1912, the Austrian **Gunter Burstyn** perfected a similar vehicle, but it was armed with a cannon. Both models were dismissed as unpractical by the military.

The trench warfare that began in 1914 rekindled the idea of a land battleship and led to feasibility studies by the British navy and the French artillery staff under Colonel Estienne. In both Britain and France the development of the new vehicle was surrounded by secrecy. The British code name for the project, "Tank," provided the vehicle its popular designation.

The tank is an all-terrain, tracked vehicle, covered with armor; it carries cannon and/or machine guns in movable or fixed turrets. The first tanks were utilized to crush barbed wire defenses, to cross trenches, and to engage the enemy with direct machine gun or cannon fire.

The first tank engagement took place on September 15, 1916, in the vicinity of Flers (France). The small number of British tanks engaged (49) gave away the secret without great material advantage, as only 9 tanks reached their objective. The others broke down due to mechanical failures.

Rapid countermeasures by the Germans led to the failure of the April 16, 1917, Berry-au-Bac attack by 132 French tanks. The tanks caused panic among German infantry troops, but proved vulnerable to direct artillery fire from the gun crews that stood their ground.

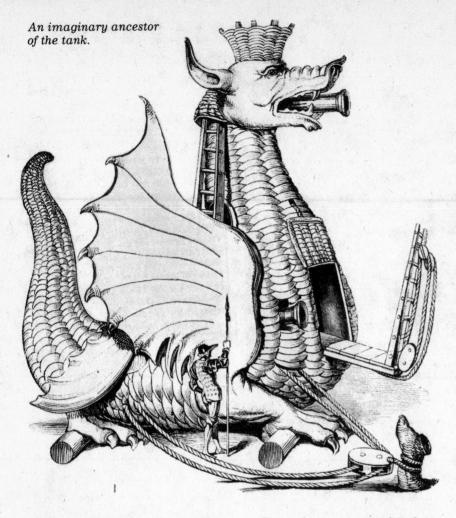

*An imaginary ancestor
of the tank.*

ently, a 40-ton tank is the mainstay of NATO and Warsaw Pact armored divisions.

The most striking progress, made from the end of World War I to the middle of World War II (1918—1943), was in speed increases. From the slow, walking pace (3 mph, or 5 km/h) of the World War I tank, speed increased to the point that only motorized infantry could follow tank attacks (31 mph, or 50 km/h). Tank evolution in the past 30 years has been mainly in the areas of fire control systems, and of very high velocity armament—105 mm. to 125 mm. guns. Advances in speed during the same period have been rather modest—from 15 to 20 km/h increases. The average tank weight today is approximately the same as that during World War II.

Underwater tank (1944)

In 1944, a team of German engineers produced in a matter of weeks a miniature, caterpillar-tracked submarine that was able to function as well under water as on land. A crazy invention? No. The idea was far from being a bad one. A machine traveling on the seabed was difficult to detect. This caterpillar-tracked submarine carried as weapons two torpedoes which were capable of sinking a battleship. An invention which perhaps came too late to be used by the Germans but that wasn't lost to the rest of the world. In fact, in the early 1980s, a mysterious submarine was detected off the coast of Sweden—and caterpillar tracks were found on the seabed.

This table shows, by category, some of the most representative tanks since their origin. The British MK. IV was the first tank to be used in combat (September 15, 1916, Flers-Courcelette, France) and was derived from the MK. I prototype.

Currently, two types of tanks are considered obsolete: the light tank, with insufficient armor and armament, and the very heavy tank (such as the "Maus" prototype) which is unable to maneuver rapidly on the battlefield due to excessive weight. Pres-

	NAME	DATE	COUNTRY	WEIGHT	SPEED	ARMAMENT*
Light Tank	FT-17 Renault	1918	France/USA	7 t	8km/5 mph	1 37 mm. Can., or MG
	MK. IV	1918	Britain	28 t	5 km/3 mph	2 57 mm. Can., 4 MG
	M2A4 Chrysler	1935	USA	14 t	56 km/35 mph	1 37 mm. Can., 3 MG
	B1. Bis	1936	France	32 t	28 km/17 mph	1 75 mm. & 1 47 mm. Can., 2 MG
	AMX-13	1952	France	15 t	60 km/37 mph	1 75 mm. Can., 1 MG
Medium Tank	T-34	1940	USSR	32 t	50 km/31 mph	1 85 mm. Can., 2 MG
	M4 Sherman	1941	USA	30 t	56 km/35 mph	1 75 mm. Can., 2-3 MG
	Panther	1943	Germany	45 t	46 km/28 mph	1 75 mm. Can., 2 MG
	AMX-30	1963	France	36 t	65 km/40 mph	1 105 mm. Can., 2 MG
	T-62	1963	USSR	35 t	50 km/31 mph	1 115 mm. Can., 2 MG
	Leopard	1965	Germany	39 t	65 km/40 mph	1 105 mm. Can., 2 MG
Heavy Tank	FCM 2-C	1919	France	70 t	12 km/7 mph	1 75 mm. Can., 4 MG
	King-Tiger	1944	Germany	69 t	38 km/23 mph	1 88 mm. Can., 2 MG
	Maus (prototype)	1945	Germany	188 t	20 km/12 mph	1 128 mm. Can.

TANKS

*MG +- machine gun; Can. +- cannon.

4. Explosives

Black powder (ca. 1300)

Black powder came about from the continual experimentation with and development of nitrous compounds, utilized since ancient times as incendiary means. Early incendiary weapons appeared in areas where the necessary raw materials—saltpeter, potassium, nitrate, sodium or calcium—were available. They were developed almost simultaneously in the Near East, India, and China. Black powder is a mixture of saltpeter, sulfur, and charcoal. This mixture was perfected around 1300 and made possible the invention of firearms.

Safety (Bickford) fuse (1831)

The safety, or Bickford, fuse was invented in 1831 by the Englishman

The modern armored tank.

A black powder explosion in combat.

Louis Nobel (top) and his brother Alfred (right).

William Bickford (1774—1834). This extremely important invention completely changed blasting techniques in mines due to its ease of operation and unprecedented safety. Wartime employment of the Bickford fuse facilitated the utilization of buried charges, such as mines, to be detonated under the enemy.

Nitroglycerin (1847)

The Italian chemist **Ascanio Solaro** produced nitroclycerin in 1847 by slowly pouring half of a measure of glycerin into a mixture composed of one part nitric acid and two parts sulfuric acid. This experiment resulted in a very powerful and dangerous explosive. The slightest shock could cause an explosion, and the mixture's sensitivity often resulted in terrible accidents.

Two of the Nobel brothers experimented with the dangerous compound in order to reduce its frightening sensitivity. One, Emil, was killed by an explosion, but the second, Alfred (1833—1896), courageously continued his experiments.

Dynamite (1866)

During one of his 1866 experiments, **Alfred Nobel** made a discovery. The nitroglycerin from a broken flask was absorbed by a *kiesel-guhr* (an insulating substance). Nobel found that the absorbed nitroglycerin retained its explosive qualities but was considerably more stable and much easier to handle. Dynamite was born.

Gum dynamite (plastic explosive) (1871)

In 1871, Nobel made gum dynamite, which he patented four years later. This explosive was obtained by dissolving 10 grams of nitrocellulose in 100 grams of nitroglycerin. The gelatine mixture resembled gum and later came to be known as "plastic." It is the proportion of nitroglycerin (12% to 93%) that characterizes various dynamite types.

Smokeless powder (1884)

The Frenchman **Paul Vieille** (1833—1896) invented smokeless powder in 1884. This invention was a tremendous step forward. Detonation of this explosive did not leave any combustion residues and produced little or no smoke. It also had unexpected ballistic advantages. Smokeless powder allowed caliber reduction in service rifles from the then current .44 caliber (11 mm.) to .27 or .32 calibers (7 or 8 mm.), resulting in range increases, better accuracy, and lower trajectories. However, while Paul Vieille was the first to perfect a smokeless powder, he was soon followed by inventors from other industrialized countries. Within two or three years following his invention, every major country had its own smokeless powder, perfected through domestic research.

5. Chemical, Electronic, and Nuclear Weapons

Incendiary devices

Crude incendiary mixtures were used in warfare from ancient times to the end of the Middle Ages. Most of the incendiary agents were derived from animal or vegetable oils and were combined with resins, tars, and other readily flammable compounds. The addition of sulfur and saltpeter to these mixtures took place at an early date. A good example is "Greek fire," made from naphtha (crude petroleum), sulfur, saltpeter, and quicklime. However, in spite of its name, the ancient Greeks did not utilize "Greek fire" or naphtha; it was not until the Middle Ages that it was used by the Byzantine Greeks.

Thermit (ca. 1890)

Thermit is the trademark of a mixture perfected around 1890, consisting of white phosphorous (which combusts spontaneously in air) and aluminum and magnesium powders mixed with a small quantity of an adjuvant, such as an iron metallic oxide.

Incendiary bomb (1910)

In 1910, the appearance of giant, long-range German dirigibles created considerable worry in military circles. As a possible defense against this new threat, it was suggested that airplanes be armed with small incendiary arrows that they could drop on the enemy. In November 1910, the Frenchman **Sazerac de Forges,** an engineer at the Chalais-Meudon Research Center, made an incendiary bomb that was tested from the Eiffel Tower. In May 1913, the first of these devices was dropped by a single aircraft on Moroccan crops.

Flame thrower (1914—1918)

This weapon, first used by the Germans, appeared in various forms during World War I. It is designed to destroy structures and to set armored vehicles on fire by igniting their fuel tanks. Flame throwers are also often mounted on tanks and used against infantry.

Molotov cocktail (1939)

Invented by the Finns in 1939, the Molotov cocktail is simply a bottle filled with a gasoline-base mixture, to which is attached a lighted wick. The bottle explodes when it hits its objective. It is a favorite rioters' weapon.

Napalm bomb

The napalm bomb was invented in 1943. It is a simple reservoir of sheet metal, provided with a fuse and filled with a mixture of gasoline and benzene, thickened with polysterene. The mixture is long-burning and will adhere to almost anything. It causes terrible suffering to its victims and is a cheap, powerful incendiary.

Chemical warfare gases

Chemical warfare gases are divided into incapacitating gases (such as tear gas), which temporarily disable humans, and local- or systemic-action toxic gases capable of causing death (such as nerve gas).

Mustard gas (yperite) (1917)

The first blistering and burning chemical warfare gas, mustard gas was invented in 1917 by the Germans and utilized in their attack near Ypres, Belgium, hence the name yperite. It is commonly known as mustard gas due to its very noticeable smell, recalling the smell of mustard. The first chemical warfare gas attack was made in 1915 by the Germans, utilizing chlorine gas, which had previously been proposed for use by Union Forces during the American Civil war some half century earlier.

Nerve gas (1936)

Discovered in 1936 during German pesticide research, the German government realized its possible military potential and started secret production, even though employment of chemical warfare gases was prohibited by the Geneva Convention in 1925. Continuing German research led to the 1938 discovery of Sarin. This toxic chemical agent belongs to the "G," immediate-action, gas category. Absorption of a single milligram of this colorless and odorless liquid causes immediate death.

The U.S. army may still have supplies of Sarin and of a VX delayed action gas. The Soviet army may have large stocks of the immediate-action Soman G. A less volatile gas than Sarin, Soman can be thickened with a synthetic polymer.

Neurotoxic gases have never been utilized due to the danger of an unpredictable wind shift, which might return the gases to the senders.

Radar (1935—1940)

In January 1935, Sir **Robert Watson-Watt,** Chief of the Radio Department of the British National Physical Laboratory, was asked by military authorities about the possibility of inventing a "death ray." Watson-Watt's answer was negative, but he suggested that it might be possible to invent a system utilizing radio waves to detect enemy aircraft. This suggestion was accepted and secret laboratories were constructed at Bawetsey Manor on the English east coast. In spite of the secrecy surrounding the project, several 245-foot (75-meter) towers could not be hidden from view. Soon after their construction, it became possible to detect an aircraft flying at an altitude of 1500 ft. (500 meters) and at a distance of 75 miles (120 kilometers).

Ten million pounds sterling was spent in 1939 by the British government for the project, which by then was capable of detecting an aircraft flying at 10,000 feet (3000 meters) and at a distance of 100 miles (160 kilometers). In the same year, on the eve of the invasion of France, the ITT's European representative, Maurice Deloraine, was invited to visit the British radar installations to see what could be done to improve this area of the war effort. Returning to Paris, Deloraine requested his office to assemble all nonclassified information on the subject. He then realized that the jealously guarded British secrets had been already published in scientific journals. Any electronic specialist who had gathered all the published material would have been in possession of the secret. However, German

intelligence was still under the impression that the famous towers were designed for some sort of radio guidance for ships or aircraft, not for aircraft detection. From August to September 1940, during the Battle of Britain, the Germans lost over 2300 aircraft and the British only 100 fighters, mainly due to the help of the radar installations.

Rockets and missiles

The essential difference between rockets and missiles is the lack of a guidance system in the former, which simply follow a ballistic course. A missile, on the other hand, is directed from an outside source or has a completely self-contained guidance system and thus is able to modify its ballistic trajectory to reach its target.

Rockets appeared with the earliest artillery and date from the invention of gunpowder. Abandoned due to lack of accuracy, rockets reappeared during World War II. They were used as antiarmor weapons and were fired from the still current "bazooka" type tubes, from the ground or mounted on aircraft, or from multiple tube launchers such as the "Stalin Organs," which were used for saturation fire.

Missiles did not make their appearance until almost the end of World War II, when the first miniaturized electronic components small enough to be contained in the projectile itself began to appear. The Germans were the first to experiment with air-to-ground and ground-to-air missiles. The short-range missile family developed rapidly after the war.

Currently, missiles can be divided into several categories: infantry missiles (such as the British Blowpipe, fired from the shoulder) or missiles fired from light vehicles (such as the British Rapier or American Stinger); short-range ground-to-air missiles (such as the Soviet SAM), ground-to-sea, sea-to-sea, and air-to-sea missiles (such as the French Exocet), categories of missiles that have been tested in Vietnam, the Middle East, and the Falklands; antiarmor missiles, which are generally wire guided and launched either from the ground or helicopters (among these are the well-known Franco-German Milan and Hot and the American Tow).

Intercontinental ballistic missiles, or ICBMs, are deterrence, or "balance of terror" weapons, with enormous destructive power of several megatons each. They can have multiple warheads and tremendous ranges—over 6000 miles (9000 kilometers). Launched from silos, the best known ICBMs are the American Minuteman and the Soviet SS-19. Intermediate range ballistic missiles or IRBMs, are ground-launched missiles (from a silo or the surface), with inertial guidance, usually three separate warheads, and a maximum range of around 3100 miles (5000 kilometers). The best known IRBMs are the Soviet SS-20, the American Pershing and the French S-3. The American and British Polaris, the Soviet SS-N-20, and French M-20 are current missiles classified as SLBMs, submarine-launched missiles, with about the same characteristics as IRBMs. All long-range missiles are directly derived from the German "V-2," or Retaliation Secret Weapon No. 2, used at the end of World War II.

A last family consists of "nonballistic missiles," which are really high-speed, low altitude, pilotless air-craft having an explosive or nuclear warhead. Such craft were only experimental in 1918 but were utilized by the Germans in 1944 in large numbers against London. These were the pulso-jet V-1 "flying bombs." This type of small, pilotless aircraft is currently represented by the still experimental U.S. cruise missile.

Cruise missiles (1982)

These are nonballistic missiles flying at very high speeds and very low altitudes. They derive from the German V-1 "flying bomb." In 1982, Boeing delivered its first ALCM (air-launched cruise missile) to the U.S. Air Force, a small plane with dart-shaped, foldable wings (3.6-meter span, 6.33 meters long, with a weight of 1450 kilograms) propelled by a double-flux, turbojet engine. The submarine cruise missile *Tomahawk,* built by General Dynamics and McDonnell Douglas, is still at an experimental stage.

For the period 1983—1990, it is expected that some 9000 cruise missiles will be manufactured, which represents an expenditure of 20 billion.

Atomic bomb (1945)

The atomic bomb is the result of the work of a large international group of scientists. The major researchers involved in the project were the two Americans, **Arthur H. Compton** and **Robert Oppenheimer**, the Italian, **Enrico Fermi**, and the Hungarian-born **Leo Szilard**. On July 16, 1945, the first experimental explosion of an atomic device took place 217 miles (350 kilometers) south of Los Alamos, New Mexico. The atomic bomb, or more

correctly, the nuclear bomb, is a weapon that uses nuclear reaction as a source of energy.

A letter written by Albert Einstein (1879—1955) to President Roosevelt, indicating the feasibility of constructing an atomic weapon, was the origin of the project. However, this letter, written by Einstein on August 2, 1939, was not read by Roosevelt until October 11, 1939, over a month after the start of World War II (September 1, 1939).

Initial research authorized by the American president led to the first self-generated chain reaction of a uranium-graphite atomic pile on December 2, 1942. It was built by Enrico Fermi. This atomic reaction took place in a squash court at the University of Chicago. In this experiment, uranium was the combustible material whose fission (the splitting of its atomic nucleus into fragments) liberated neutrons with enormous energy; the graphite was utilized as a moderating agent to slow down the neutron-generating fission of additional atomic nuclei and to prevent an explosion. (See under Energy: Nuclear Reactors.) Atomic bombs are defined as nuclear devices exploding by fission action.

Following this test, which confirmed Einstein's theories, the go-ahead was given for the Manhattan Project. A factory to obtain uranium 235 by isotopic separation through gaseous diffusion was built at Oak Ridge, Tennessee. Secret installations for plutonium preparation were constructed at Hanford, Washington, and a planning agency with manufacturing facilities for the bomb itself was located at Los Alamos, New Mexico.

On August 6, 1945, an atomic bomb was dropped on the Japanese city of Hiroshima, killing 80,000 and wounding 50,000. A second was dropped a few days later, August 9, on the city of Nagasaki, destroying it almost completely. The dropping of the bombs resulted in Japan's surrender but created the specter of a nuclear war that could destroy humanity.

H-bomb (1952)

The first experimental H-bomb was detonated by the United States on October 31, 1952. This bomb, a tower-mounted nuclear device, was exploded on the Pacific atoll of Eniwetok. It was the work in great part of **Edward Teller**.

In August of the following year, the USSR exploded a similar nuclear device, followed by Great Britain (1957), China (1967), and France (1968).

An atomic bomb explodes during the early days of the American nuclear testing program.

Instead of the atomic fission principle of the atomic bomb, the H-bomb utilizes the inverse fusion principle of the atomic nucleus. Since ordinary hydrogen possesses no neutrons, deuterium (also known as heavy hydrogen and possessing one neutron) fused with tritium (another heavy hydrogen, possessing two neutrons). The name H-bomb is a reference to those two heavy hydrogens.

Neutron bomb (1958)

Research on the neutron bomb started in the United States around 1958, under the direction of the physicist **Samuel Cohen**. Specialists prefer to call these particular nuclear weapons "reinforced radiation weapons" or "neutron shells." In these devices, extremely high-temperature fusion is obtained by lasers, which excite deuterium and tritium atoms; this causes a neutron flux of hundreds of billions of radioactive particles. These neutron rays are very deadly, but the weapon is described as a "clean bomb" because it is less destructive than an H-bomb and has been specially designed to kill by radiation effect. Thus, it kills all forms of life within a wide radius but does not destroy buildings or material. This means that as soon as the radioactivity level returns to normal, enemy installations can be used by the victors. In view of industrial production difficulties, these weapons are to be considered as tactical weapons, primarily designed to neutralize large tank formations in a specific theater of operations.

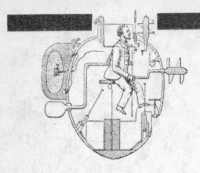

NAVAL

6. Naval Vessels

Galley (3rd millennium B.C.)

Already known by the Cretes (second and third milleniums B.C.), the galley, propelled principally by oars, dominated the Mediterranean for almost 3000 years. The Greeks, who were good navigators and excellent naval architects, thrust the art of building war galleys far forward. Their greatest hour of glory came on September 28, 480 B.C., at Salamis when the 470 or so ships led by Themistocles defeated the Persian fleet of Xerxes, which numbered more than 1100 ships.

Drakkar (8th century A.D.)

Invented by the Vikings, the drakkar, which takes its name from the dragon decorating the prow of the vessel, appeared on the coast of Norway toward the end of the 8th century. The vessel distinguished itself both by the grace of its lines and by its means of propulsion, combining sail and oar. The Vikings ranged the coast of Europe with their drakkars, particularly dominating the North Sea, penetrated the Mediterranean, and even got as far the Black Sea. They may also have reached the shores of North America.

Man-of-war (1514)

The *Great Harry*, constructed in 1514 by the English, closely followed by the *Grand Francois*, prestige naval vessels that weren't lacking in faults, foreshadowed the appearance of ships with modern lines. The three centuries or so that followed saw the successive preeminence of men-of-war of the second (two decks) and third rank (three decks), with sides pierced by portholes. The man-of-war, was perfected as a naval weapon at the end of the 18th century by the French naval engineer Sane.

Fire ship (16th century)

The fire ship, which appeared at the beginning of the 16th century, was a vessel that had been taken out of service, stuffed full of flammable material, and—when the wind was favorable—set afire and launched against enemy vessels. Its main role was to cause a break in the enemy naval lines.

Floating batteries (1782)

During the 1782 siege of Gibraltar, the French utilized floating gun batteries invented by General **Darcon**. These barges, built of heavy oak, were provided with additional inclining wooden panels that acted as armor and deflected or absorbed solid shots.

In 1810, the American, **Robert Fulton**, built the *Demologon* to defend New York harbor. The boat was provided with a steam engine.

Armored ship (1859)

After partially covering with armor the Gloire class of French frigates, **Dupuy de Lome** decided that a fully armored ship was even a better protective solution. In 1859, this naval engineer started construction on the *Magenta* and the *Solferino*, two heavy, armor-clad warships.

The first iron armor-clad ship was the *British Warrior*, completed in 1861.

Armor-clad turrets (1854)

The idea of a pivoting turret allowing artillery pieces to fire in all directions originated with an American, **John Ericsson**, in September 1854. This idea found a practical application in the construction of the famous *Monitor*, which played an important part in the Civil War (1861-1865). The battle of the two iron-clads, the *Virginia* (formerly the *Merrimac*) and the *Monitor* at Hampton Roads, on March 9, 1862, is still famous.

The *Monitor* was very much inspired by the 1855 project for a "steam raft and cupola" of the English Captain Coles. With very little freeboard above water, it was difficult for the enemy to hit it. The turret consisted of an iron cylinder with a

General Darcon's floating gun batteries during the seige of Gibraltar in 1782.

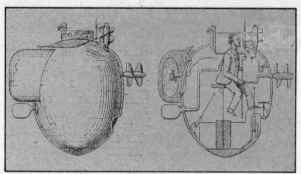

Left: Bushnell's Turtle in 1776. Right: Fulton's Nautilus in 1797.

bombproof roof and was operated by a below-deck engine. Designed to operate on large American rivers, *monitors* were far from seaworthy.

Submarine (1624)

The Dutchman **Cornelis Drebbel** is generally credited with the invention of the submarine. He is reported to have made the first trials with his submarine in 1624 in the Thames River. His submarine remained underwater for several hours, its twelve rowers breathing an artificial gas made by Drebbel. This gas (oxygen) greatly

interested the physicist Robert Boyle (1627—1691), but no one was able to obtain the exact composition of Drebbel's mysterious "chemical liquid."

Bushnell "turtle" (1776)

This submersible single-place boat was built by the American **David Bushnell**. In the shape of two turtle shells set together, the boat was powered by a hand-cranked screw propeller. In 1776, Bushnell personally instructed the boat's pilot, Sergeant Ezra Lee, before sending him out to attack the British warship *Eagle*. The

attack took place in New York Harbor, causing fear aboard but little damage to the target.

Nautilus (1797)

A propeller-driven submarine was invented in 1797 by the American mechanic **Robert Fulton**. This ship was designed to place explosive charges under the hulls of enemy ships.

The *Nautilus* was constructed in 1798 at Le Havre, France, due to lack of U.S. interest. The elongated, oval-shaped submarine had a length of 21.25 feet (6.48 meters) and a width of

An AMX 30 Tank crossing a river. The Schnorkel, a tube allowing fresh air in and bad air out, was invented for an American submarine in 1897. It was adapted to tanks by the Germans in 1940.

The nuclear submarine Nautilus in Manhattan harbor in 1959. It was taken out of service in July, 1985.

6.36 feet (1.94 meters). It was controlled through the use of two movable planes, one for depth control and the other for directional control. Trials took place in the Seine River but failed to impress Napoleon. Fulton then went to England but was not well received there either, as further progress in the submersible field was perceived as a threat to the supremacy of the British navy.

David (1864)

The first feat of arms accomplished by a submarine armed with a torpedo—actually a simple barrel of gunpowder towed by a long rope—goes back to the Civil War. During the night of February 17, 1864, a submarine—the *David*, designed by Captain **Hunley**, and crewed by nine men—sent the Northern frigate *Housatonic*, participating in the Charleston blockade, to the bottom. The event however remained without any follow-up because the David itself was also destroyed with loss of the crew, and the Union forces were able to take protective countermeasures.

Nuclear submarine (1955)

The first nuclear submarine was launched on January 17, 1955, for the U.S. Navy, under the direction of **Admiral Hyman G. Rickover**. According to naval tradition it was given the name *Nautilus* to honor both Robert Fulton and Jules Verne. The *Nautilus*, 311.6 feet (98.5 meters) in length and 27.5 feet (8.4 meters) in width, displaced 3200 tons, and had a speed of 20 knots (23 mph). Its crew consisted of 10 officers and 95 men. Having a range of over 100,000 nautical miles, this submarine could dive to 722 feet (220 meters). On August 3, 1958, it became the first submarine to reach the North Pole by navigating under the polar icecap.

Since 1954 the United States has built several dozen nuclear submarines, with some types reaching speeds of 45 knots. With the advent of nuclear propulsion, submarines, whose original function was to attack ships, have become underwater SLBM bases. A submarine such as the British *Revenge* can launch 16 multiple-warhead atomic missiles with a destructive power equivalent to 700,000 tons of TNT, more than the total amount of explosives used by all World War II combatants.

Modern battleships (1906)

The disappearance of masts from sailing vessels allowed the installa-

USS Shangri-La, the first attack carrier.

The British take a sea plane out of an M2 submarine.

battleships were armored mainly on the top part of the hull, with the submerged part of the ship protected by a double-hull construction. The space between the two hulls was filled with water and acted as anti-torpedo protection. Steel thickness around the waterline could reach 14 inches (355 mm.) and was up to 16 inches (400 mm.) in the main turrets and around the fire-control center. Sixteen-inch armor plate weighs 648 pounds per square foot. In spite of this, battleships proved vulnerable to carrier-based aircraft once these came into use.

The last wartime engagement of a battleship was during the Vietnam War (1968—1969) in which the *New Jersey* fired 5688 16-inch shells; seven times more than the ship had fired during World War II. Withdrawn from service in 1971, the *New Jersey* was recommissioned in 1978. It joined the U.S. Sixth Fleet in the Mediterranean in 1983 and played a role in the 1983 Lebanese conflict. However, this type of ship is now considered obsolete by many naval strategists because of the danger posed by aircraft carriers and the new generation of heavier and more accurate missiles.

Torpedo boats (1860s)

Although they were not called torpedo boats at the time, the first vessels answering to that description appeared between 1860 and 1865. They were small boats carrying long, sloping booms, to which were attached charges designed to explode under enemy ships' waterlines. They ap-

tion of turrets on the center-line of a ship. In 1906, the shape that modern battleships were to take appeared in the design of the British *Dreadnought*. Its appearance revolutionized naval construction. The turrets sheltered three or four large caliber guns, which were often superimposed. From 1915 on, battleships could fire 15-inch shells weighing 1895 lbs (860 kilograms), a distance of 20 miles (32 kilometers). Initial propulsion was

created by exploding a 423-pound (192 kilogram) charge of cordite powder. The shells left the gun muzzle at more than twice the speed of sound.

During World War II, two Japanese battleships, the *Yamato* and the *Musashi*, carried 18-inch (457 mm.) guns, the heaviest ever seen at sea. Salvo fire (all gun batteries firing simultaneously) threw 15 tons of shells into a high parabolic trajectory prior to hitting the target. For this reason,

A sketch of an early submarine.

proached enemy craft silently during the night. Use of these boats prompted the idea of attacking underwater, which was tried during the Civil War.

Many small, fast steamboats were designed in the 1870s to attack moored enemy ships, and they used electrically fired, boom-mounted torpedos.

Automotive torpedo (1866)

The "automotive torpedo" was invented in 1866 by **Robert Whitehead** (1823—1905), a British engineer. This torpedo was a small, self-propelled submarine carrying a powerful explosive in its bow. Later models carried an automatic guidance system and a compressed air motor, since replaced by an electric motor.

The first Whitehead torpedo was 14 feet (4.25 meters) long and had a diameter of 16.5 inches (42 centimeters). It had a range of 656 yards (600 meters) and a speed of 6 knots (7 mph). It carried 68 kilograms of dynamite and was a very advanced weapon for its time. A hydrostatic piston regulated its running depth by acting on a horizontal rudder and another piston regulated a directional control vertical rudder. The torpedo was launched by placing it on the surface of the water and directing it toward the enemy. It was a very expensive device for the period and not always reliable.

A typical torpedo today is 24 feet (7.3 meters) long and 20.4 inches (52 centimeters) in diameter; it carries a 485-pound (270 kilogram) warhead and weighs over a ton. Its maximum range varies with the chosen speed: a speed of 28 knots gives a range of 8 miles (13 kilometers); a speed of 34 knots reduces the range to 5.6 miles (9 kilometers); and a further speed increase, to 46 knots, reduces the range to 3.4 miles (5,500 meters).

Aircraft carrier (1911)

On November 14, 1910, an airplane took off from a warship for the first time. The American airman **Eugene Ely** took off from the U.S. navy cruiser *Birmingham*, in a 50 horsepower Curtiss biplane. The 1907-built cruiser was under the command of Captain Irving Chambers. A platform, 80 feet x 23 feet (24.6 meters by 7 meters), was constructed over the ship's forecastle. This platform, built 32 feet (10 meters) above the ship's deck, was inclined 5° toward the bow.

On January 8, 1911, Ely landed the same aircraft on a platform placed over the quarterdeck of the battleship *Pennsylvania*. The skid landing gear of the Curtiss was provided with a hook designed to catch ropes attached to sandbags. Landing on a measuring platform 105 feet by 31.4 feet (32 by 9.6 meters), the pilot caught the ninth rope and stopped the aircraft in less than 30 feet (9 meters). The heavy fabric barrier that had been placed at the end of the platform to stop the Curtiss as a last resort was unnecessary. Several minutes later, Ely took off from the same deck. The aircraft carrier was born.

Deck landings were, and still are, a problem. The first steps towards a solution were made by the British. In 1919, they furnished the steamship *Argus* and the obsolete battleship *Eagle* with unobstructed flight decks, which were built on top of the boats' hulls. These were the first operational carriers. Essential structures were progressively moved to the side of the ship. The next big step forward was the invention of the angled landing deck, which allowed a go-around after a missed approach and reduced the danger of crashing into aircraft parked at the forward end of the take-off deck. The U.S. navy built 120 carriers during World War II. The carrier largely changed traditional ship-to-ship naval battles into air-to-sea or air-to-air battles. Today, aircraft carriers are able to handle 35-ton aircraft; these are steam-catapulted into the air within seconds, at speeds of over 185 mph (300 khr); the pilots are subjected to 6 G's acceleration—that is, they must bear six times their own weight during the takeoff.

The two largest aircraft carriers are the U.S. navy's *Enterprise* and *Nimitz*, whose deck lengths measure over 1,083 feet (330 meters). The power plants of these ships consist of eight nuclear reactors, delivering 300,000 horsepower, enough to move their 85,000 tons at speeds of over 35 knots (40 mph). With a crew of over 4600 men, each ship can carry over a hundred aircraft.

Mines (17th century)

In spite of its earlier appearance at the start of the 17th century, the naval mine was not really utilized until the Civil War. Large numbers of them were first used during World War I. Most naval mines were exploded either by shock through a long plunger (British Elia type) or by electrochemical action (German Hertz type). During World War II, the German invent-

The New Jersey fired 5688 406mm. shells during the Vietnam War.

ed magnetic mines, which exploded from the magnetic effect of ships' iron hulls.

Antiship missile (1967)

On October 21, 1967, Western sailors brutally entered into the era of the antiship missile. That day, the Israeli destroyer *Eilat* was destroyed by Styx missiles of Russian manufacture launched by Egyptian sentry boats. Since then, this type of weapon has been used regularly in combat, one recent example being the destruction by Argentinian Exocets of the British ships *Atlantic Conveyor* and *Shef-* field during the Falklands campaign.

Two modern types of antiship missiles exist that differ essentially in the means or propulsion: missiles with powder propulsion (the French Exocet, the Israeli Gabriel, the Norwegian Penguin) and turboreactor missiles (the French-Italian Otomat, the American Harpoon, the Swedish RBS 15). The turboreactor requires a delay before being put into action, but it permits greater independence.

Antiship missile defense

In reply to the new menace of antiship missiles, surface units can count on a complete arsenal of both active and passive means. Figuring in the active means are short-range sea-to-air missiles (6-12 kilometers) and very short-range (6 kilometers and less), systems of multi-tube cannons of small caliber (20 to 30 mm.) and a high rhythm of fire (from 2000 to 6000 shots per minute), and electromagnetic and infrared ray decoy launchers. Among the passive means are: the use of shapes and materials less detectable to radar; camouflage; and cooling of hot parts (in particular funnels) of vessels that might otherwise be detected via infrared rays.

AIR POWER

7. Aviation

Military balloon corps (1794)

The first military balloon corps was created on April 3, 1794 by the French revolutionary "Public Safety Committee." The idea was to use balloons held in place by cords as observation posts. *The Entreprenant*, a balloon built by **Nicolas Conte**, was used for the first time in the battle of Fleurus, in 1794. Its commander was Captain Coutelle. However, the balloon's military career was shortlived since Napoleon, as consul, was as suspicious of balloons as he later would be (as emperor) of submarines. He did not think they had any military value.

Fighters (1914)

The first fighter aircraft appeared in 1914 and were used to attack enemy aircraft and balloons. The first fighters were armed with individual weapons such as carbines or pistols, sometimes with small cannons. They were relatively ineffective in spite of the planes' slow speed. A Hotchkiss machine gun, mounted on a French Voisin aircraft, shot down the first enemy aircraft with automatic fire in October 1914. The first machine gun that eliminated the need for a gunner, firing outside the propeller's arc, was mounted soon after, in 1914, on a Morane-Saulneir N.

The problem of firing a machine gun forward without hitting the propeller was resolved by the Dutch engineer **Antony Fokker**. Working for the Germans, he designed the first propeller-synchronized machine gun. The evolution of the fighter aircraft is shown in the table below:

In 1913, military aviation was somewhat fragile.

Bombers (1917)

At the start of World War I, aircraft were only used for reconnaissance. However, it did not take long for pilots to arm their aircraft with bombs, which they dropped on enemy targets. Thus the first bombers were reconnaissance aircraft, often civilian models, carrying a few bombs. It was not until 1917 that the first aircraft specially designed for bombing was built. This was the British Airco DH-4. Major steps in the evolution of the bomber are shown in the table on the opposite page.

First turbojet aircraft

The turbojet engine was first conceived by the British pilot and engi-

Machinegunner Hotchkiss on a two seater mono-plane.

One of the first military aerial photographs.

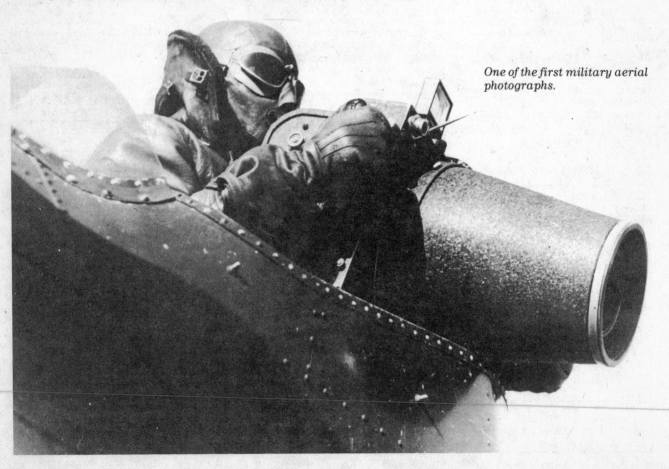

neer **Frank Whittle**. In 1930, at the age of only 23, Whittle patented a gas turbine engine, which he later tried to adapt as a fighter plane power plant. The first bench trials of his aircraft engine took place in 1937.

The race for the turbojet aircraft was, however, won by the Germans. On August 27, 1939, the Heinkel firm made the first test flight of a jet aircraft, piloted by E. Warsitz, in the utmost secrecy. Whittle meanwhile made numerous improvements in his original engine design and these led to the construction of the first Rolls Royce jet engine. This engine was installed in the Gloster Whittle E-28, which, on May 5, 1941, made its first flight at Cranwell, England. The plane remained in the air for 15 minutes.

Rocket engine aircraft (1947)

On October 14, 1947, the American Bell X-1 was the first rocket propelled aircraft to break the sound barrier by flying at 714.5 mph (1150 khr). This type aircraft does not take off from the ground but is carried aloft and launched by a carrier aircraft. By definition, the rocket engine does not need outside air to operate. Its fuel consumption is very high at low altitudes.

Much faster rocket-propelled aircraft are presently in service; some attain speeds of more than six times the speed of sound. The North American X-15, flown by William Knight, reached 4534.3 mph (7297 khr) on October 30, 1967. Its cabin was covered with materials allowing it to support

a temperature of 2998o F (1648o C). Its landing speed was 242 mph (389 km/h).

Vertical takeoff and landing aircraft (VTOL) (1954)

The first machine to take off and land vertically that wasn't a helicopter was the 1954 "lift cage" built by the British firm Rolls Royce. Two vertical jet engines were utilized to lift the machine from the ground. Decreasing engine thrust landed the machine.

Convair XFY-1

The first vertical takeoffs and landings by the Convair XFY-1 took place on June 2, 1954. This aircraft made its first flights, piloted by J. F. Coleman, at the Mofett Naval Air Station, in

FIGHTER PLANES				
AIRCRAFT	DATE	MANUFACTURER	COUNTRY	CHARACTERISTICS
Vickers FB-5	1914	Vickers	Britain	Named "Gunbus," this two-seater biplane was the first in the world to be designed for combat.
Fokker E-1	1915	Antony Fokker	Germany	First monoplane with a propeller-synchronized machine gun.
Polikarov I-16	1936		USSR	First fighter with retractable gear and cantilever wing.
BF-109	1936	Messerschmitt	Germany	Largest production run of any WW II fighter; a formidable opponent. Best fighter at the start of the war, with speed of 334 mph (570 km/hr). Direct injection engine.
Spitfire	1936	R. J. Mitchell	Britain	Greatest rival of the BF-109. 20,334 Spitfires were produced up to 1947. Slightly slower than the Messerschmitt but more maneuverable and easier to fly.
Gloster Meteor	1943	Gloster	Britain	Slower than the ME-262; only operational Allied jet aircraft; utilized against the V-1.
ME-262	1944	Messerschmitt	Germany	First operational jet fighter.
ME-163 Komet	1944	Messerschmitt	Germany	First rocket-powered fighter.
P-80	1944	Lockheed	USA	Utilized during the Korean War, it was the first fighter to be downed by another jet aircraft (MIG-15).
F-102 Delta Dagger	1948	Convair	USA	First delta-wing fighter; electronically very advanced. Modified, 1959, version had a speed of over Mach 2.
F-86	1950	North American	USA	Widely used in Korea; 9500 of those fast and maneuverable sweep-wing fighters were built.
MIG-15	1950	Mikoyan and Gourevitch	USSR	Largest production jet fighter—over 25,000 built. Exceptionally maneuverable and fast-climbing aircraft.
F-100 Super Saber	1953	North American	USA	First supersonic (Mach 1) fighter.
Mirage-III	1956	Dassault	France	Powered by a French Atar 101, this well-known Mach 2 fighter has participated in numerous Middle East conflicts.
MIG-25	1965	Mikoyan	USSR	Code name "Foxbat." Still the fastest combat jet; this aircraft has been clocked at 3.2 Mach (2110 mph).
F-16	1975	General Dynamics	USA	Short-range fighter-bomber; a good example of new fly-by-wire, all-electronic fighters, with thrust superior to clean aircraft weight (24,250 lbs thrust).

California. This short-wing, turbo-propelled aircraft had the peculiarity of landing directly on its tail, with its nose pointed straight up. Its counter-rotating propellers provided sufficient lift for a vertical takeoff. Once in the air, the aircraft pivoted from a vertical to a horizontal position, then accelerated to normal flying speed. A reverse procedure was utilized for landing.

Hawker Siddeley Harrier (first VTOL combat aircraft)

The first flight of the Hawker Siddeley Harrier took place in October 1960. The aircraft was designed by Sir Sidney Camm, for the British firm of Hawker Siddeley. It was to become the first operational VTOL aircraft. It has been used by the Royal Air Force since 1969.

The American F-16 is one of the most versatile fighters flying.

BOMBERS				
AIRCRAFT	DATE	MANUFACTURER	COUNTRY	CHARACTERISTICS
Handley-Page 0/100	1916	Handley-Page	Britain	Strategic bomber; limited WWI role.
DH-4	1917	Airco	Britain	First genuine bomber; could carry a 460-lb bomb load.
Ilya-Mourometz	1917	Sikorsky	Russia	First four-engine bomber, modification of the first prewar multiengine transport.
TU-TB1	1927	Tupolev	USSR	First twin-engine, monoplane bomber.
TU-TB3	1932	Tupolev	USSR	First four-engine bomber.
B-29	1942	Boeing	USA	Not operational until 1944. A B-29 piloted by Col. P. W. Tippets dropped the first atomic bomb on Japan on August 6, 1945. The aircraft was named Enola Gay in honor of the pilot's mother.
AR-134	1943	Arado	Germany	First jet designed as a bomber.
Stuka	1946	Hugo Junkers	Germany	First widely-used dive bomber.
B-36	1949	Convair	USA	Several hundred of these intercontinental bombers were produced. Largest bombers ever produced with four jets and six piston engines.
B-58 Hustler	1956	Convair	USA	First supersonic bomber.
A-5 Vigilante	1958	North American	USA	Largest carrier aircraft (2 tons).
F-111	1964	General Dynamics	USA	First swing-wing bomber (Mach 2.5).

agriculture

1. Agricultural Machines

Swing plow (antiquity)

The swing plow, a primitive tool, is the ancestor of the modern plow and dates back to Neolithic times. It can be found towards 3500 B.C. in Egypt, from 3000 to 2500 B.C. in India and western Europe, but only from about 800 B.C. in China. This rudimentary tool, consisting of stag antler picks and wooden spades, was used to prepare the soil before planting. This was an'example of the change from prehistoric hand gardening to agriculture as such. Later improvements did not concern the tool itself but the mode of pulling and the plowshare. The swing plow was at first formed by a single piece of wood. Little by little it came to be composed of a number of assembled parts.

Swing plow with wheels (1st century B.C.)

The addition of two small wheels to the front of the swing plow came during the 1st century B.C. in a Gallic province in Rhaetia (Switzerland). The wheels gave added weight to the instrument and permitted the plowing depth to be regulated to a greater extent.

Plow (ca. 1st century B.C.)

The modern plow goes back to the Celts of pre-Roman Britain, but it underwent a long eclipse and did not reappear until the 6th century A.D. in central Europe, and it developed in western Europe from the 8th century on. Carpenters, blacksmiths, and plowrights built plows completely by hand until 1653, when the first treatise on plow construction was published by Walter Blith in England.

Aside from having a plowshare slightly different than that of a swing plow, the plow has a coulter—a large vertical cutting blade placed in front of the plowshare—which parts the earth, and a mouldboard, which channels the soil to one side, completely turning it over. From the 18th century on, there were many different models.

Robert Ransome's factory plow (1785)

In 1785, the Englishman **Robert Ransome** patented cast iron plowshares. Then in 1808, he decided to make his plows with a number of detachable pieces.

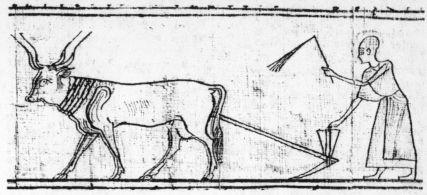

An Egyptian plow-seeder.

The tractor eased the task of the plowman and replaced beasts of burden.

The image of the plowman has inspired many artists.

Steel plow (1833)

In 1833, at a time when only the carpenter's saw was made of steel, **John Lane**, of Chicago, was the first to manufacture plows made of steel. It was John Deere, however, who patented the only steel reservoir plow, in 1837.

Disc plow (1847)

The disc plow appeared in the United States in 1847, but its use did not become widespread at the time. In Australia, on the other hand, the Sovereign plow, built by John Shearer and Sons, was a great success from 1877 on. It had from 10 to 24 discs and a device permitting it to go over s it to go over stumps. The disc plow reappeared in the United States in 1893.

Sulky plow (1864)

The invention of the two-wheeled sulky plow, known as such because it had a seat, is attributed to the American **F. S. Davenport**, who patented it in 1864. The model was developed rapidly before it was replaced by the Flying Dutchman, the three wheeled sulky model produced by the Moline Plough Company in the United States in 1884.

Carried plow (1927)

In conception, the plows of today are identical to those of the 19th century, but between World War I and II, a radical change took place with the coming of tractors that could haul heavy equipment. Plows, until then had been pulled, could now be carried as well.

One of the first plows carried by tractor was the French plow **Huguet-Huard**, patented in 1927. Since then, none of the large agricultural equipment companies has made any innovations in this area, except Ferguson, which in 1941, presented, a hydraulic system for controlling depth of tillage, the pressure being regulated by the tractor. This process was first developed in Great Britain.

Reaping machine (1st century A.D.)

The reaping machine existed during the early Christian era but was forgotten until its reinvention at the beginning of the 19th century. Pliny the Elder (1st century A.D.) of Rome described it as "an enormous case with teeth, carried on two wheels, driven by an ox or donkey pushing it from behind. The wheat collected by the teeth fell little by little into a chest."

Disc reaping machine (1811)

From 1800 to 1810, several Englishmen (Meares, Taylor, Gladstone, Salmon, etc.) improved the reaping machine. In 1811, **Smith** developed a horse-drawn machine that cut the grain with a moving disk turning level with the ground.

Reaping-sheafing machine (1820)

In 1820, the Englishman **Brown** invented a reaping-sheafing machine. It

Smith's disc harvester, 1811.

Patrick Bell's harvester, 1826

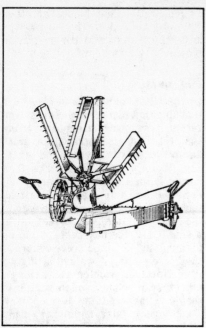

A reaping machine.

In addition to hard work, early farm laborers also had to endure pollution from farm equipment.

Walter Wood's reaper binder, 1871

was the first pulled (rather than pushed) machine of its type, and it cut the grain with a serrated horizontal bar. The wheat, reaped by a sharp blade that moved back and forth laterally, was transported by a rotary disc onto a platform situated behind the cutting bar.

Reaping planer machine (1826)

In 1826, the Englishman **Patrick Bell**, vicar of Angus, Great Britain developed a planer: a horse-drawn machine that gathered wheat by means of a system of rotary blades. The wheat was then cut wit a double-serrated rotary blade. Cut near the root, the wheat was thrown by the blades onto an inclined cloth where sheaves formed. This clever model was hand-made and the machine generally lasted only one season.

Horse-drawn reaping machine (1831)

In 1831, **Cyrus McCormick**, pursuing the efforts of his father, developed a reaping machine with a serrated cutting bar that moved back and forth like a saw. The machine was designed to be drawn by two horses.

Hussey's reaping machine (1833) had a simple lateral cutting bar with coupled blades like those of a scissors; it was also horse-drawn.

These two machines made a sensation at the London Exposition of 1851. There was much disagreement between the partisans of Bell's machine, which was pushed, and the partisans of the American machines. The latter

finally won out, since the devices for pulling were simpler.

Reaper binder (1858)

In 1871, **Walter Wood**, an American, patented a model of a wire binder which was exhibited in England. As early as 1858, however, **John F. Appleby** had invented a string binder. It was not developed until twenty years later, as string was expensive. Nevertheless, this machine remained the standard to which others were compared due to its ingenuity. From 1878 on, binders made by Deering were equipped with Appleby tying devices.

Threshing machine (1732)

The origin of the threshing machine, used to separate grain from the chaff, is obscure. Nevertheless, it is attrib-

uted to the Scot **Michael Menzies**, who built such a machine in 1732. It consisted of a set of flails attached by short chains to a horizontal bar turned by an enormous hydraulic wheel. The wheel did as much work as thirty men using handheld flails. Unfortunately, the flails tended to break and the machine was eventually abandoned.

Meikle threshing machine (1786)

In 1786, the Scot **Andrew Meikle** developed a threshing machine so efficient that it remained the standard model for this type of apparatus. The grain was separated from the chaff by a drum equipped with four vertically placed threshing blades. The sheaves were introduced between two channeled metal wheels situated at

An American farmer driving a Massey-Ferguson Toronto Harvester.

the front of the drum and were repeatedly struck by the threshing blades to separate the kernels of wheat. The sheaves were then removed from between the wheels. Later improvements included sweeping rakes and shakers added to separate the grain from the chaff and to drop it through a sieve.

Transportable threshing machine (1850)

To replace the stationary threshing machines driven by water, wind, or horses, several factories proposed light transportable models. In 1850, the Englishman **Tasker** marketed a threshing machine that could be transported to the fields. This machine was very successful and was exported.

Winnower (18th century)

Andrew Meikle's father, **James Meikle**, had observed handmade grain-cleaning devices in Holland during a trip there. Back in Scotland, he built a mechanical winnower during the latter half of the 18th century. It was a hand-powered machine which had a series of mobile sieves in which the grain was put. Around them were four large pieces of cloth used to create the necessary draft to separate the grain from the chaff and straw dust. This machine replaced the traditional method, which was to spread the grain on a dry flat surface and let the wind blow away the chaff. Another

method was to place the grain in the draft between the two open doors of a barn.

Combined thresher-winnower (1837)

In 1837, **Hiram** and **John Pitt** of Maine were the first to build a steam-powered machine that simultaneously threshed and cleaned grain. The grain was separated by a threshing drum equipped with numerous metal-lic pins, rather than blades, and then fell into a hopper. The straw was swept by a rake and the chaff was blown away by a ventilator.

Haymaker (1820)

In 1820, **Robert Salmon** developed a machine that picked up cut grass and turned it over, so as to expose it to the sun and wind. Up to that time, grass had been cut with scythes and left to dry. Women with pitch-forks then spread it out, turned it over and gathered the hay into stacks. This task took much time and left harvests at the mercy of intemperate weather. In a single day, the haymaker could do as much as 15 women, even more if the horse pulling it walked at a good pace. However, problems with this elementary machine, which scattered the hay, hindered its development. Twenty years later, haymakers using forks were invented. They were used up until the 1930s, when they were replaced by swath turners and side-delivery rakes.

Mowing machine (1822)

The American **Jeremiah Bailey** was the true inventor of the mowing machine. Patented in 1822, his reaper-mower used a cutting disc that turned horizontally a few centimeters above the ground. The first machine especially made for cutting grass was patented in 1812 by the American Peter Gaillard, of Pennsylvania, who tried to reproduce the action of the scythe. His machine, however, proved to be inefficient. The cutting arm, specific

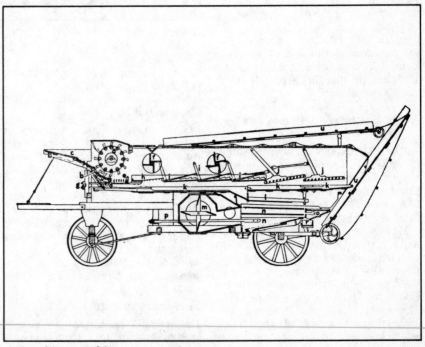

A threshing machine.

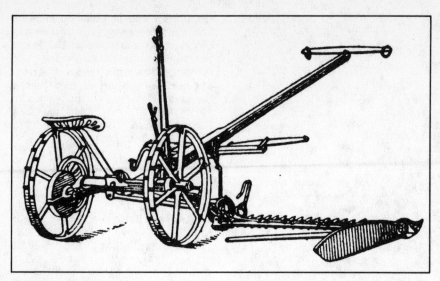

A mowing machine

Hay loader (1875)

The hay loader was developed by **Keystone**, an American, in 1875. The machine was made up of two basic components a cylindrical cage bristling with teeth, centered on an axle between two wheels and a ramp with a conveyor belt. When the apparatus was pulled, the movement of the wheels turned the cage, which picked up the hay from the ground and hoisted it up onto the conveyor belt. The hay was transported up the ramp until it reached the top, where a roller ejected it into a trailer.

Cylindrical straw bale press (1958)

In order to put straw into cylindrical rather than rectangular bales, the American firm Allis-Chalmers marketed the Roto-Baler in 1958. It made cylindrical bales weighing 40 to 100 pounds, 14 to 22 inches wide and 3 feet long. Present models can make bales weighing over 1320 pounds.

Combine harvester (1828)

The first combine harvester was patented by **Lane**, an American, in 1828. The first truly operational machine, however, was developed by Moore and Hascall in 1836. It was built in Michigan and transported to San Francisco via Cape Horn in 1854 to demonstrate its capabilities.

During the same period, the Australian H. V. McKay built an approximately identical model, which was

to hay mowing, is attributed to the American William Manning, who developed it in 1831. However, it was not manufactured industrially, until 1850 by Obed Hussey's firm. "Fingers" grouped the grass stalks into bunches that were then cut by coupled blades working like scissors.

Hay baler (1853)

The hay baler was invented in 1853 by the American **H. L. Emery** of Albany, New York. In America large quantities of hay had to be delivered to the big cities to feed the horses used in industry and transportation, and it was necessary that the hay be in as compact a form as possible. Emery's machine produced five bales per hour, each weighing 250 pounds. Several men were needed to run the machine. The rate of output was accelerated with P. K. Dederick's continuous production machine. That was first manufactured in 1872 and gained considerable success in the United States and Great Britain.

Steam-powered "perpetual production" balers were introduced in 1884, following the ever-growing need to increase yield. When the bale was sufficiently compressed, it was ejected automatically and then tied with wire by hand.

The Axial flow Harvester.

Inside the Axial Flow Harvester.

These hay bales can weigh 1500 pounds.

A modern tractor

later greatly improved by a team of Australian technicians. At the turn of the century, International Harvester and Massey-Harris not only adopted some of their ideas but employed some of the same Australians in their research departments.

Self-propelled combine harvester (1888)

The prototype of the self-propelled combine harvester was introduced by Best, an American, in 1888, but his project was not developed at the time. As early as 1922, Massey-Harris was marketing a horse-drawn combine harvester equipped with an auxiliary motor running the parts of the machine that cut, threshed, cleaned, and put the harvest into sacks. It was only in 1938 that the Australian engineer Tom Carrol developed the use of a wholly self-propelled machine (in Argentina) for the American firm Massey-Harris. Its practical introduction, however, took place in 1944, when Massey-Harris received special authorization from the American government to manufacture and run 500 self-propelled combine harvesters. This "harvest brigade" started working. in May, in Texas and California, and progressed northward. By the end of the season, the brigade had harvested thousands of acres, and the entire world was impressed by the yield of this new harvester.

Combine harvester with axial threshing (1975)

The American firms International Harvester and New Holland started to develop the axial threshing combine harvester in 1962, but the first prototype was not built until 1975. This new machine, baptized Axial Flow, represents the biggest change in harvesting since motorization.

In conventional machines, the wheat was shaken before being threshed. The capacity increase of the machine was linked to the increase in width of the thresher and the increase in length of the shakers. The separation and threshing system is totally new in the Axial Flow. It uses shock, friction, and centrifugal force. The harvest goes around several times inside a cylindrical cage equipped with threshing blades. In the center, a rotor with beaters set in a helix turns and advances the wheat. This system allows more compact machines to be built and it means we will no longer encounter the monsters which totally obstruct the roads.

Tractors

Steam tractors (1829)

Because of their immense fields, American agronomists chose to develop direct traction. They built three wheeled machines. The two big rear wheels supported a boiler, and a small front wheel steered the vehicle. These tractors were built by the Case company as early as 1829. Three years later their use was widespread on large western American farms. The most famous developments were the steam locotractor made by the Frenchman Albaret in 1856, and Obed Hussey's steam plow, made in 1885 (Baltimore, Maryland).

Burger tractor (1889)

This tractor was developed when people began thinking about adapting oil engines to the frames of steam machines. The first one was built by an American, **Burger**, in 1889. The chassis was very similiar to those used to support steam engines, and his engine was built by the Charter Gas Engine Company.

Froelich tractor (1892)

In 1892, the American **John M. Froelich**, in Iowa, asked the Van Duze Gas and Gasoline Engine Company of Cincinnati to build an internal combusion engine that could be used in a tractor. This model, the first gas engine tractor to be really operations, was the precursor of John Deere's tractors.

Ivel tractor

In 1901, the Ivel tractor appeared. It was a three-wheeled tractor invented by the American **Dan Albone**. Its front was too light, so it sometimes turned over when it was pulling heavy loads. Despite this defect, it became rather successful.

Hart and Pan's tractor (1902)

Like Froelich, the Americans **C. W. Hart** and **C. H. Pan** built, in 1902, a four-wheeled model that was heavy but efficient. In 1905, they founded the first American firm specializing in the manufacture of tractors. It was their comercial director, W. H. Williams, who decided to use the word *tractor* instead of *gasoline traction engine*, which had been used until then in advertisements.

Caterpillar tractor (1905)

In 1905, **Benjamin Holt** developed the idea of the caterpillar tractor, which would be able to increase the carrying surface of steam engines. As engine power increased, the machines grew heavier and had more and more of a tendency to get bogged down. Holt, with two of his friends, thought they could solve this problem by providing their tractors with larger and larger wheels. The result was a three-wheeled giant over 7 feet wide. It had a total span of 49 feet and was one of the biggest tractors ever built.

Holt finally found the solution to the problem. He equipped an ordinary steam tractor with a set of wooden caterpillar tracks, and his idea led to the founding of the Caterpillar Company. Only eight steam-driven models were built In 1906, the manufacture of gasoline-powered caterpillar tractors began.

Ford tractor (1907)

Henry Ford (1863—1947) was the son of an American farmer, and even before he began producing cars he used to dream of building tractors. In his first model (1907) the rear wheels were those of a sheaf-binder, while the axle and the steering mechanism for the front wheels were from a K model Ford car.

Equipment-driving tractors (1907)

It was quickly ascertained that certain forces caused problems when pulling equipment by tractor. For example, the plow has a tendency to pull down on the hookup and cause the front of the tractor to lift up. Transporting equipment to the fields also caused problems.

Tourand of France (1907) and International Harvester (1918) were the first to come out with portable equipment systems. The revolutionary invention in this field, however, is attributed to an Irishman, Harry Ferguson, who created the standardized three-point hookup and introduced it in the United States in 1939. Ferguson showed his system to Ford and the two men entered into partnership to mass produce a tractor with a hydraulically driven hookup. In less than 15 years, Ferguson had the leading position in the world tractor market. In 1953, his firm merged with Massey-Harris. The group took the name Massey-Ferguson in 1958.

Wallis Cub (1913)

The year 1913 stands out as an important date in tractor history since this was the year in which smaller, lighter, and faster models, like the Bull Tractor, built in Minneapolis, were introduced. Bull Tractor's first model was named the 1913 Wallis Cub. It was referred to as frameless since it no longer had the heavy frame common to steam engines.

Frame tractor (1917)

This tractor was developed in 1917 by Ford Tractors, and it was undeniably more advanced than any other model available at the time, known as boiler steel, made the chassis heavy. To get

around this problem, Ford decided to build a cast iron chassis, which was lighter. The tractor ran on four steel wheels and was powered by an oil or kerosene engine. Light, considering its strength, and relatively cheap, it was very successful, especially since it came out the year the United States entered World War I, a period when a large number of draft horses were requisitioned and sent to Europe.

Farmall tractor (1924)

The Farmall tractor was introduced in June 1924 by International Harvester. It was the model from which multiple-use tractors originated and represented a revolution in the conception and utilization of tractors. It was suitable for several kinds of work: It could plow, harrow and do a number of other jobs. As new accessories were invented, the number of its functions grew. It had a high rear axle supported by two large, well-spaced wheels. The front was provided with two small wheels placed close together so it could go between rows, and also with a hook to which the necessary accessories could be attached.

Tractors with tires (1931)

In the lemon plantations of Florida, farmers noticed that tractors were

damaging the tree roots, so in 1928, they came up with the idea of covering the steel wheels with rubber. Pneumatic tire manufacturers followed their example and in 1931, B. F. Goodrich invented a solid tire. It consisted of a thick rubber strip, fixed onto a perforated steel frame.

In 1932, the Firestone Tire and Rubber Company put pneumatic tires on an Allis-Chalmers tractor, but when tested the tires proved to be too susceptible to punctures and they did not provide sufficient traction since they were not heavy enough. The firm then came up with water tires, in which water provided the weight needed to insure traction.

The most recent innovation was made by Dunlop, who, in 1965, introduced the Stabilarge case. It performs better—it is more flexible and does a better job of distributing the tire's stress on the ground as well as in the tire itself.

2. Phytobiology

Biodynamic Agriculture (ca. 1920)

It originated at the beginning of the 20th century with the work of a German agronomist, H. Pfeiffer and an

Crosskill invented the roller in 1841; the principle has not changed since.

English agronomist, **A. Howard.** Pfeiffer, in 1924, developed the biodynamic method, which emphasizes the role of compost. Howard did a number of tests on composting, which involves putting fermented organic substances in the ground. He published his results in 1940 in his "Agricultural Testament." Since 1945, these basic methods have been developed and refined by H. Muller and H. P. Rush in Switzerland and Germany, and in 1963 by Lemaire and Boucher in France.

The main objective of biodynamic agriculture is to improve the nutritional quality of food. This is achieved by a long-term increase in soil fertility, resulting from enrichment with humus, correction of mineral deficiencies, and rectification of biological activity. While this is being done, any kind of pollution, not only of food but also of water and the environment, must be avoided. These objectives can be attained thanks to special techniques: (1) the use of natural products to make fertilizers (manure, slaughterhouse wastes, guano, turning over of crops after harvest, minerals which have undergone no chemical treatment; (2) the practice of large-scale crop rotation, giving importance to leguminous plants (clover, lucern); (3) thorough and frequent cultivation of the top layers of the soil; (4) the abolition of weed killers, fungicides, and insecticides of chemical origin.

Compared to the techniques used by modern agriculture, the yield obtained is smaller in quantity but better in quality. Nevertheless, the prof-

Vine growers using an atomizer in fight phylloxera around 1868.

itability of this type of agriculture still depends on the existence of "organic marketing" networks to assure better sales prices for the producers.

Sprayer (1884)

Although the idea of a sprayer was put forth already in 1781 by Father Rosier, the first prototype, developed by the Frenchman **Vermorel**, was built only in 1884. Vermorel was inspired by the "nozzle" presented by the American entomologist Riley to the farmers of Herault, France. Riley proved that it was possible to treat mildew on vines by spraying them with copper sulphate. Vermorel improved Riley's invention by adding a pump to the nozzle. In 1887, he perfected a dorsal sprayer, which proved to be very convenient. Mechanical powder-sprinklers appeared in 1889.

Powder-Sprinkling Planes, or crop dusters (1921)

It was in 1921 that planes (World War I Curtis JN6H's) were used for the first time in agriculture to spread arsenate powder over catalpa forests in order to destroy sphinx grubs that were attacking the trees.

In 1924, the first powder-sprinkling operations were carried out in the Mississippi Delta region by planes built by Huff-Daland Company of Ogdensburg, New York. These planes released calcium arsenate to kill cotton weevils. They carried approximately 1000 pounds of powder and their cruising speed was 60 mph. They were nicknamed "puffers."

The first liquid larvicides were

FERTILIZERS		
DATE	INVENTOR	PRINCIPLE
1550	Bernard Palissy (1510-1590, France)	The first to speak about "vegetable salts" and to introduce the notion of restitution: Since vegetables absorb minerals, minerals must be put back into the earth.
1600	Olivier de Serres (France)	Spoke about crop rotation in his "Theater of Agriculture and Field Upkeep."
1792	Antoine de Lavoisier (1743—1794, France)	Established the fact that vegetables feed themselves with minerals they draw from the earth.
1841	Lawes (Great Britain)	Invented superphosphates, which came from treating natural phophates. Rapid acting fertilizer.
1864	G. Ville (France)	Suggested the use of artificial fertilizers.
1876	Sidney G. Thomas (Great Britain)	Invented phosphatic slag, bottom fertilizers for soil lacking lime.
1906	Norsk Hydro (Norway)	Invented nitric fertilizers (family of introgenous fertilizers
1912	Haber and Bosch (Germany)	Developed ammonia synthesis, upon which nitrogenous fertilizers are based. This revolutionary process was part of the reparations Germany had to pay France at the end of World War I in 1918.

Ultra-light sprinkling plane.

Greenhouses and Glass Bells (Antiquity)

It is thought that the Romans already knew about the use of glassed-in shelters and artificial heat for remedying the production difficulties of some plants during poor seasons.

It was not until 1600 and the Agricultural Theatre of Olivier de Serres, the first French agronomist, that the use of glass bells designed for the cultivation of melons came into existence. The greenhouse is probably a Dutch invention of the 16th century and was not introduced into France until 1660 when the court of France ceased being a tributary of the Netherlands for its supply of fresh fruits and vegetables.

sprayed, in 1930, in New Jersey and California to exterminate mosquito grubs.

Farming without soil (1959)

This process, which could be used in space stations, was developed in 1959 by two Soviet research scientists at the University of Leningrad, **Timiriadzer** and **Prianichkov**. Called hydroponics, it consists of feeding mineral salt solutions to plants placed in concrete containers. The hydroponic yield is twice as high as that of the same species cultivated in soil.

3. Gardening

Fantastic Trees (Antiquity)

The ancient *ars topiaria*, or art of landscaping and gardening, today called topiary art, is a very particular technique. It consists in giving trees and shrubs ornamental shapes which are more or less fantastic by playing with successive sizes and sometimes trailing them over pre-erected metallic frames.

Pliny the Younger described the garden of his Tuscan villa, containing hedges and shrubs shaped in the form of animals. Preserved by some afficiandos across the centuries, topiary art enjoyed its apogee from the 16th to the 17th century.

The Spade (Prehistory)

A hand farm tool, the first appearance of the spade probably goes back to prehistory, when humanity was progressing from a gathering to an agricultural stage.

Greenhouse around 1900.

Mini-greenhouse (1829)

Nathaniel Bagshaw Ward was a naturalist. In 1829, while studying the evolution of a butterfly chrysalis contained in a bottle, his attention was drawn to the survival and even the growth of some herbs and ferns which were growing out of the soil after three years of imprisonment, without any water and without renewal of air.

He discovered that condensation water alone sufficed to rehydrate the earth and allowed the proper functioning of the vegetative cycle.

Filling two narrow-necked bottles with various plants, he sent them by boat to Australia in 1833. They were watered only once during the voyage, which lasted four months. The plants were so beautiful when they arrived in Sydney that it was necessary to place guards near them to keep a constant surveillance over the Wardian Cases, as they were called. The plants returned to London in the same manner.

This thus proved that plants can travel without harm or stunting of their growth.

The Sickle (Prehistory)

The Romans attributed the origin of the sickle to Sylvan, the god of the forests. Historians believe that the sickle goes back to the very origins of agriculture and that, well before the metal age, it was made up of a simple piece of curved wood, equipped with grooved flint.

The first Herbarium (1450 B.C.)

The first botanist-collector of the ancient world was the Pharaoh Tuthmosis III. In 1450 B.C. he invaded Syria and brought back from his campaign 275 different plant species. He had them drawn in bas-relief on the great temple of Karnak, and they can still be seen today.

Pruning Shears (1815)

Exiled during the turmoil of the French Revolution, the Marquis **Bertrand de Moleville**, a former minister to Louis XIV, used his forced leisure to develop an instrument which today has become the very symbol of the gardener's work: pruning shears. In 1815, with the return of the Bourbons to the throne, the tool made its first appearance in France on the occasion of a communication to the Royal Society of Agriculture.

As for so many novelties, pruning shears had a difficult time being accepted. Horticulturists, wary indeed, criticized them for "compressing and tearing wood before it had been cut, thus causing the death of the extremity of the branch that had been pruned and the weakening of the nearest eye".

Pierre Antoine Poiteau, one of the great names in fruit arboculture at the time, promoted the pruning shears which were very quickly taken up by the best professionals for use in their work.

The derivative of the 17th century pruning shears has been widely improved upon by ingenious craftsmen, and since the end of the last century hundreds of models of pruners have been available.

Therapeutic Pruning Shears (1981)

An inventor from Bordeaux (France), **Henri Begout,** developed in 1981 a pair of pruning shears which simultaneously cuts, protects and cares for plants by spraying biological products, disinfectants or fungicides.

Initially designed for the treatment of grape vines, this instrument may be perfectly adapted to shrubs, fruit trees, etc.

Lawnmower in 1931.

Hedge-Trimming Machine
(1880)

In 1880, an Englishman, **R. Hornsby**, invented an entirely metallic machine designed for the trimming of hedges. Drawn by horses, it was supported by two large wheels and had a seat in the front for the driver, and another in the rear for the man who handled the hand cutter, made up of two large perpendicular blades.

Lawn Mower (1831)

The invention of the lawn mower in 1831 was due to two Englishmen, **Budding** for the design, and **Ferrabee** for the manufacture. Horse shoes were literally covered with rubber boots so that they did not damage the lawn while pulling the cylindrical blades.

At the beginning of the 19th century the vapor mower was produced, followed by the motor-driven lawn mower, for which manufacturers improve the qualities every year while reducing the annoying sound so familiar to Sunday gardeners.

Electric Lawn Mower (1958)

The first electric lawn mower with a rotating blade was launched by Wolf in 1958, and one of the last models (1984 silver medal at the Agricultural Show) is the self-propelled "Eurotondor", also developed by the Wolf Company.

Bicycle Lawn Mower (1982)

A marriage of the useful and the pleasureful—this was the goal for this original mower, invented by **A. R. Despland**, of Switzerland. Rather than pedal at home on an "ordinary" home-trainer, the inventor thought up this non-polluting and silent mower.

This pedal-driven bicycle mower may also be used on an incline to mow sloping terrains (very useful in Switzerland!), while the cyclist remains on a vertical plane. This machine had some ancestors. In 1898, an American, William Burnet, also invented a bicycle-mower, but the rotation of the blades took place from a wheel base and not from a pedal.

Garden Hose (1850)

The imitation of rain for providing plants with water is a procedure which was described by a Benedictine monk during the time of Charlemagne (742—814). At that time, water was poured from a ewer into the hand, so that it would slide out from between the fingers.

Towards 1850, the initial gutta-per-

Swiss inventor Raymond Despland on his bicycle-lawnmower.

cha hoses began to take the place of watering cans and the watering cart which was drawn by a donkey.

In 1848, a certain Mr. Combaz thought up an ingenious system which allowed a "rain-like" watering to be obtained; its first application took place in the Bois de Boulogne in Paris.

The Chrysanthemum (1789)

In 1789 a florist from Marseille, Pierre-Louis Blancart, brought the first varieties of the flower back from China; they quickly spread to all the gardens in Provence.

During the 19th century it was the passion of numerous gardeners who sowed it and hybridized it, thus leading to the very diversified forms known today.

Granny Smith and Golden Apples (19th century)

In 1868, Miss Granny Smith managed to obtain this apple in Australia; it has a green skin with white spots and a juicy and acidulated flesh.

The famous Golden apple was born in western Virginia at the end of the 19th century, in the orchards of Anderson H. Mullins. This apple has been known and planted in various regions of the United States since 1914.

Coconut (1674)

Charles Perrault, the famous story teller, the creator of Tom Thumb and Cinderella, must be credited with the discovery of coconut. In a communication to the Academy of Science in 1674, he presented to the members of that impressive assembly the first specimen of the fruit which had just arrived in France.

The world's largest chysanthemum which bears around 680 flowers.

The Tomato (1596)

Of South American origin, the tomato was brought to Europe in 1596. At first cultivated as a vegetal and ornamental curiosity, it was considered to be a violent poison.

Under the lovely name of "apple of love", this vegetable-fruit had to wait more than two centuries until its gustatory and alimentary qualities were recognized. The president of the United States, Thomas Jefferson, cultivated the tomato in his garden, but he didn't eat it.

The first harvest sold in a market was in 1812, in New Orleans (USA).

The Clementine (1900)

The clementine is a truly new fruit since it was obtained in 1900 by Father Clement, a member of a religious order in the region of Oran and an enthusiastic agronomist.

By fertilizing the flowers of the mandarin tree with pollen taken from a Seville orange tree which produces bitter oranges, Father Clement devised a fruit which, in honor to him, bears his name.

The Tulip (1554)

Augier de Busbecq, ambassador of Ferdinand I to Suleiman the Magnificent, was an amateur botanist. In 1554, he discovered a new flower in Persia and he shipped seeds and bulbs from it to Vienna. The botanist, Charles de l'Ecluse, born in Arras, sent the first tulips to a trial garden in Leyden, Holland some years later. The reign of the Dutch tulip was thus begun.

It was not until 1730, however, a year marked by "tulip-mania", a veritable speculation madness which made and lost fortunes, that the flower really spread to gardens in Europe and then throughout the whole world.

It is perhaps its oriental origin which gave birth to the legendary black tulip. However, like the blue rose or the red iris, this mythical flower only exists in the imagination of gardener-poets, or. . .of poet-gardeners.

The Rose of Peace (1945)

On April 29, 1945, in Pasadena, California, a rose was presented at the annual Conference of American rose growers; it would conquer America under the name of "Peace", and, later Europe as well, under the name of "Madame A. Meilland". The Meilland dynasty, represented by Antoine Meilland, the creator of this famous rose, won fame by obtaining, in 1949, the first invention patent ever granted for a flower in France for the "Meilland Red" variety.

4. Insecticides

Pyrethrin (ca. 1st century B.C.)

Two thousand years ago the Chinese were using dried, pounded pyrethrum (chrysanthemum) leaves to kill fleas. The use of pyrethrin as an insecticide took root in Europe in the 18th century. In 1925, the English entomologist Tattersfield and the French entomologist Vassieres tried to cultivate the oriental chrysanthemum from whose flowers the powder was derived. They had to abandon this project, however, because the climatic conditions in Europe were not favorable to the plant.

Some modern aerosol sprays contain natural pyrethrin, which acts almost immediately on household insects.

Arsenic (1681)

The toxicity of arsenic oxides was discovered in the Middle Ages. The famous Borgia family used it against their enemies, while entomologists employed it against insect plagues.

Arsenic compounds were first recommended to protect plants in 1681. Arsenic compounds were used in 1872 to fight the Colorado potato beetle, which devours potato leaves. Until World War II, lead arsenic was effectively employed against ravaging insects (caterpillars and beetles) to protect fruit trees, vines, and potato fields.

Tobacco (1690)

In 1690, the French agronomist **Jean de la Quintinie** (1622—1688) discovered tobacco's insecticidal properties. The pear trees of his garden were being ravaged by aphids. Since he knew that tobacco leaves contained toxic agents, he soaked some in water. He obtained a dark juice with which he sprinkled the fruit trees. It killed the insects immediately.

Cryolite (1806)

In 1806, the Dane **Giesecke** discovered cryolite. Cryolite is a mineral consisting of sodium-aluminum fluoride. The insecticidal properties of this mineral, which is virtually nontoxic as far as humans are concerned, were discovered in 1920. Cryolite lodes are found in Greenland and in the Esterel Mountains in France.

Sulfur (1843)

Sulfur has been used as an insecticide since 1843 to fight red spiders and plant lice. Today it is used in many synthetic products. Mixed with lime, it is commonly used to treat fruit trees.

Biological control (1873)

Biological control is the use of one animal species to provoke the disappearance of another, unwanted species. A predatory species is introduced into a given environment, where it acts as a nonpolluting insecticide and destroys the unwanted species. The first example of biological control dates back to 1873, when the American C. V. Riley sent a French vintner, Jules Planchon, a mite (the *Rhizoglyphyus Phylloxerae*) to destroy the microscopic plant lice (*Phylloxera*) that were attacking his vines.

In 1888, A . Koebek implanted California orchards with coccinella, a genus of beetle native to Australia that feeds on the cochineal (wood louse), which then infested citrus orchards. The cochineal were destroyed and this method of control came to be used throughout the entire world.

Rotenone (1920)

In 1920, entomologists from different countries discovered that rotenone, which is extracted from the roots of tropical plants, acts on the nervous system of fish. Several hundred plants from South America—e.g., the haiari from Guiana, the nikoe from Surinam, and the timbo from Brazil and Ecuador—contain this compound, which paralyzes not only the nervous system of fish but also of insects. After the discovery of its insecticidal properties, the cultivation of plants containing rotenone was considered for crop protection. With the advent of DDT, the project was abandoned, but rotenone may be reconsidered in the near future.

Organophosphoric insecticides (1930s)

Before World War II, the German chemist Schrader invented, at the request of his government, combat gases made from phosphorus compounds. These gases showed themselves to be very efficient against insects and many insecticides were developed from them. The most well known are tetraethyl-pyrophosphate and parathion. After the war, Schrader discovered organophosphorus compounds that could penetrate plant tissues, diffusing themselves in the sap and spreading throughout the plant without losing their insecticidal properties. This was a very important discovery.

The first organophosphorus compounds were very toxic but fortunately very short-lived. Since then, thousands of organophosphorus compounds, such as malathion, have been synthesized. Like DDT, they are nowadays proscribed almost everywhere because of the pollution they have caused.

First synthetic insecticide (1935)

In 1935 American research scientists from the Department of Agriculture obtained the first synthetic insecticide, phenotiazine, which allows for efficient treatment of fruit trees.

DDT (1939)

Dichloro-diphenyl-trichloro-ethane (DDT) was synthesized for the first time in 1874 by a German chemist. In 1939, the chemists P. Muller and Weissmann, working for the Swiss firm Geigy, in Basel, discovered the insecticidal properties of this product. The American military forces used DDT extensively for the first time during World War II in the Pacific, where malaria-carrying mosquitoes were as dangerous as the Japanese. Thanks to this remarkable insecticide, sicknesses like malaria have diminished. Unfortunately, it is a very stable agent and pollutes the environment. Its use is now totally or partially proscribed in many countries.

Pyrethroids (1949)

In 1949, The Swiss chemist **Schachter** while working for the United States Entomology and Quarantine Office synthesized a compound very close to natural pyrethrin. The Interdepartmental Committee on Pest Control named this compound allethrin.

In 1967, an English research scientist, **Elliot**, synthesized remesthrin. This insecticide has a prolonged shock effect on insects so that they can't recover but die. However, it is unstable in light, as is allethrin.

In 1973, at the Rothamsted experimental station. Elliot and his team of research scientists developed permethrin, which stands up to ultra-violet light very well and stays effective for several days after being sprayed onto plant leaves. It is called pyrethrin. The same team later synthesized another pyrethroid, one that is ten times more effective than permethrin, decamethrin.

Pyrethroids are now also made in France and Japan. The Japanese firm Sumitomo has synthesized new one.

5. Animal Husbandry

Poultry farming (1749)

It was the French physician **Rene Antoine Ferchault de Reaumur** (1683–1757) who established the foundations of poultry farming in his book *The Art of Hatching and Raising Domestic Birds of All Kinds Year Round* (1749). He described a hot room, heated by the warmth from manure or by a stove. In this room, he put the hen coop with a wood stove and an incubator, a box equipped with grates containing the feeding troughs for chickens, a butter thermometer, and a basket full of eggs that had been brooded by the heat of the room.

However, poultry farming did not greatly develop until after the discovery of the importance of vitamin D, in 1930. Up to that time, birds raised in great numbers were subject to rachitis, which caused their eggs to break. Cod liver oil, rich in vitamin D, was first fed to poultry in 1930 by the GLF Mill cooperative of New York.

Artificial insemination

Though commonly practiced only for the last few decades, artificial insemination had precursors. Arabs were known to have had the idea as early as 1322. In 1725, Ludovic Jacobi obtained salmon fries by sprinkling male milt on eggs that had been drained from a female's abdomen.

However, it is to the Italian **Lazzaro Spallanzani** (1729–1799) that we owe the first insemination experiment. He meticulously recorded his work. His methods were not far from those used today. In 1780, he injected a dog's sperm (collected after the dog had been masturbated) into the vagina of a female in heat. Sixty-two days later, the female gave birth to three healthy puppies.

In 1933, research by **Milanov**, a Russian, led to the development of an artificial vagina for gathering stallion sperm. Russia had already been the first country to discover the possibilities of artificial insemination in horse breeding, thanks to the work of Ivanov in 1907. A cylinder in metal, ebonite, or thick rubber, lined with a thin rubber sleeve full of lukewarm water, was used as a false vagina to collect the sire's sprerm.

Vaginal sponge for ewes (1965)

In 1965, a process for planned lamb births was thought up by an Australian, **Robinson**. It has been progressively perfected in France, Since 1970, by the National Institute for Agronomic Research. A polyurethane sponge, impregnated with a hormone, is placed in a ewe's vagina and left from 12 to 14 days. The withdrawal of the sponge, sometimes accompanied by the injection of another hormone, provokes the start of a new sexual cycle. By this process, groups of ewes can be synchronized so they can all be inseminated on the same day.

The Chimera; half sheep-half goat.

On the 145th day of gestation, the pregnant ewes get an injection that provokes birth. Thus all the lambs are born on the same day, under the shepherd's watchful eye, and later he can rest easy.

Beekeeping (modern, 1844)

The mobile frame have was developed by **Debeauvoys**, a Frenchman, in 1844 and perfected by **Langstroth**, an American, in 1851. After hundreds of innovations, it is the Langstroth hive that is the most widely used in the world.

The mobile frame is a case used to house the wax cells made by the bees. Removable, it facilitates the apiculturist's work while respecting the bee's "private life." It even save the bees work, as the honeycombs do not have to reconstructed after each collection of honey.

Mythology, technical innovations, and observations of bee life have proceeded jointly since the first recorded studies, by Aristotle. Moreover, it is from the mobile honeycomb hive existing in Greek antiquity that the modern hive is derived.

Milking machines (mid-19th century)

Before the milking machine, invented in the middle of the 19th century, the first known mechanical aid to help with milking were wheat straws, which were introduced into the cow's udder. The Egyptians were employing this system approximately 38 centuries before Christ.

Tubes for milking appeared again in 1836, in England. In the United States, a New England farmer's journal mentions, in 1852, that the bones of robins' feet were used for milking.

Vacuum milking machines (1862)

In 1862 **Colvin**, an American, conceived the first model of a vacuum milking machine. The cows' udders were placed in plastic sleeves, inside of which a vacuum was created. The milk, drawn from the udder ran into the container set under the udder. Later on, in 1889, the Scot William Murchland developed a permanent installation: A pipe branched to two openings in each stall. A vacuum was created by a water-column and when it was time for milking, the sleeves were plugged into a tube and placed on the udders. The milk would flow out into a bucket held under the cow by a strap. Unfortunately, this machine had two disadvantages: First, it frequently inflamed the udders; second, one person had to work the vacuum pump while three others manipulated six to nine buckets that were used.

Milk refrigeration (1850)

In 1850, an American, **Lawrence**, invented the first milk refrigeration system. This invention represented a big step forward. The warm temperature of fresh milk (103° F) is favorable to the proliferation of bacteria. Rapid refrigeration represented a fundamental element in milk conservation and made possible the expansion of the dairy industry.

Refrigerated slaughterhouse (mid-19th century)

The American meat suppliers **Daniel Holden** and his brother were the first to consider keeping meat in refriger-

Two illustrations depicting the Cincinnati Slaughter House in 1873.

ated slaughterhouses. As eastern American cities grew rapidly, transporting live cattle by road and then by train could not keep up with demand. Slaughterhouses were therefore built in the farming areas, and smoked or salted (cured) meat was processed. This kind of operation, however, could only be done during winter. Summer curing with ice was inaugurated in 1853 in Chicago, but ice reserves took up too much space in the plant. The Holden brothers' refrigerated installation came at the right moment. Equipped with a Carre absorption machine, the system was first installed in the Fulton slaughterhouse in Texas, whose intake capacity was hundred steers per day.

The cow bag.

Cream separator (1877)

The cream separator was invented in 1877 by the Swede **Gustave de Laval**, who received a gold medal at the 1879 Royal British Exposition for his invention. The machine worked in the following manner: Milk was placed in a chamber heated to 86° F, which is the best temperature for separating the cream. From there the milk went through tubes to a container that was spun by a steam engine. The centrifugal force thus created separated the cream, which, being lighter, settled in the center of the container. The milk, being heavier, was pushed to the outer part. Before Laval's separator came into use, cream was separated with small strainer dishes. (The centrifugal principle is also used in numerous industrial sectors and in biological laboratories.) The New American Butter Separator, manufactured by the Wahlin Firm of New York in 1895, was derived from Laval's invention and allowed for continuous production of butter.

Barbed wire (1873)

In 1873, **Joseph Glidden,** of De Kalb, Illinois, built the first machine capable of producing barbed wire in large quantities. This cheap fencing spread very quickly throughout vast regions of western America. Soon after, disputes broke out between farmers installing fences and big stock breeders who were accustomed to herding their cattle cross-country as they pleased.

Automatic drinking trough (1912)

The first automatic, individual drinking trough appeared in 1912 in a catalogue of the **Rassman Company** (Beaver, Wisconsin). Up to that time,

drinking troughs had to be filled by hand several times a day. The Rassman drinking trough maintains a constant water level and functions according to the communicating vessel principle. When an animal drinks from the trough, a reservoir compensates for the water consumed.

Unconfined stalling (1913)

Unconfined stalling is attributed to three American zoo technicians. In 1913, **S. B. Buckley,** of the University of Maryland, conducted the first experiments in raising milk cows at liberty in large sheds. Then in 1934, J. R. Dice of North Dakota State College, conducted similar experiments. In 1944, S. A. Witzell, of the University of Wisconsin, conclusively proved that cows produced more milk when they moved about freely. Milk cows had always been tied up in stalls for long periods. They were sheltered from bad weather, but they lacked exercise. Today, they commonly move about freely in a paddock with a shed where they can take shelter.

Computerized herd management (1979)

In February 1979, the American firm **Agri Electronics** developed AGRI 80, a computerized system that allows strict management of cattle herds. Which type of feed must be bought and in what quantity? When should animals be bred or sold? The computer answers these questions and many others and, after analyzing the cattle, gives all the information necessary to obtain the greatest productivity.

Waste converter (1979)

In 1979, the **Waste Conversion Company** of Seattle had the idea of trans-

forming chicken waste back into food for the chickens themselves. Approximately 20% of the food consumed by chickens is not digested and comes out in their excrement. It was to recuperate this substantial unused fraction that the system was invented.

Inflatable animal cushion (1982)

In 1982, the Australian firm **Kellybuilt** invented an inflatable bag for supporting cows and horses in the standing position. The animal is pushed onto the deflated bag, which is then inflated. This bag has many advantages. It is easy to transport and is very comfortable for animals. It can also be covered with a small tent in case of bad weather. Horses with injured legs or cows too weak to stand can therefore recover in the best of conditions. A smaller cushion functioning on the same principle was invented by the Englishmen T. B. Snell and P. S. Kerton in September 1981.

Fish breeding

Artificial spawning grounds (antiquity)

The first fish breeders were probably the Chinese. Early on they placed artificial spawning grounds in stream beds. These were wicker screens onto which the fish deposited their eggs and milk. After the eggs were fertilized the breeders collected them in buckets and they were then sold by vendors to peasants, who then sold by vendors to peasants, who seeded their ponds and pools with them.

Fish pools (1st century A.D.)

The Roman patricians were lovers of rare fish and often built fish pools, sometimes right in their feast halls.

Seneca (1st century A.D.) tells of guests gathering around a pool to delight in the color changes in dying red mullet. Canals were built to connect the sea or streams with the pools in the Roman villas. Fish that swam up the canals during spawning season were made captive by sluice gates and reproduced in the pools. The food for the fish so raised was expensive, even causing the ruin of some rich families. Moray eels, kept by Caesar (among others), who offered them to his friends, actually caused the loss of human life. Certain rich Romans did not hesitate to feed their slaves to the eels to satisfy their own taste for the meat of these fish.

Artificial fertilization of fish eggs (1420)

The invention of artificial fertilization of fish eggs is attributed to **Dom Pinchon**, a monk of the Reome abbey, near Montbard (in the Cote d'Or region of France). He was concerned, as were all his contemporaries, about providing Christians with food for the numerous "abstinence" days. In 1420, he had the idea of fertilizing trout eggs. He did this by squeezing the eggs out of several female fish and the milt out of male fish, which he placed in water that he later agitated. He then put the fertilized eggs in a wooden case closed at either end by a wicker grill and covered with sand at the bottom. The case was put in gently running water and all that was left to do was to wait for the hatching. Dom Pinchon's work remained unknown at the time and no one continued it. In 1763, the German naturalist Jacobi "Reinvented" Dom Pinchon's box, baptizing it the "hatching box."

Spawning cases (1701)

In 1701, the Swedish naturalist C. F. **Lund** established the fact that fish looked for warm and shallow water during the spawning season, and that perch and roe eggs developed better when they remained caught in the branches of juniper tress along banks. He created a spawning case. It was wooden and spacious but not too deep, and made to be immersed in slow-moving warm water near a bank. After he had garnished the case with juniper branches, Lund introduced female and male fish who had well-developed eggs and milt. After two or three days, the eggs were deposited and fertilized. The sides of the case would then be let down to spread out the juniper branches and allow the eggs to develop properly.

Tuna net (Middle Ages)

This stationary fishing device is composed of nets and posts that drive the fish to a "death chamber." It is used in the South of France for tuna fishing. However, the tuna net originated in the Adriatic, where it was invented in the Middle Ages. After the tuna swam into the basins constructed in the lagoon by the fishermen, the basins were closed off by reed fences supported by stakes, which formed a veritable labyrinth. The fish could get through these fences on the way into the basin by applying just a slight bit of pressure with their heads, but could not get back out again. This system acted, therefore, as a veritable trap.

Oyster farming (2nd century B.C.)

The first oyster farmer recorded by history was a Roman named Sergius Orata. He built an oyster bed on his property, in Lake Lucrin, near Naples, at the beginning of the 2ndd century B.C. Orata built fish ponds that communicated with the sea but protected the oyster brood from waves. The young oysters were provided with posts to which they could cling and grow in proper conditions of temperature and light. Sergius Orata's know-how was such that he made a fortune selling his oysters. His contemporaries said of him, "He could grow oysters on a roof." Lake Lucrin disappeared in 1583 after an earthquake and a volcanic eruption, common occurrences in that area.

6. Foods

Sugar (antiquity)

The extraction of cane sugar goes back to earliest antiquity, and the plant was probably first cultivated in India: The Greeks and Romans referred to sugar as "Indian salt" and "honey of India." The Christians probably brought it to the West during the Crusades. It has been proved that during the 12th century there were mills in Sicily that produced "honey canes." Since the earliest times the Chinese knew not only how to extract cane sugar but also how to refine it, an operation that was developed much later in the West.

Yogurt (antiquity)

Originally from Asia Minor (Persia and Turkey), yogurt first appeared, according to legend, during biblical times—revealed to Abraham by an angel and supposedly responsible for the patriarch's long life. It appeared in France towards 1542. Suleiman the Magnificent gave some to King Francis I, who was suffering from intestinal problems. However, the sultan's doctor (and messenger) left with the secret of how to make it. Not until World War I did the recipe, rediscovered in Bulgaria, reapppear in France. Yogurt was not produced industrially until the 1950s.

Potatoes (16th century)

In 1554, the potato was introduced to Spain by the Spanish conquistador Pizarro on his return from the New World. In 1537, a detachment of Spanish infantry had discovered potatoes stocked in the Incan city of Sorocata. This tuber had been cultivated for a long time by the Indians in the Andean Cordillera in Chile and Peru, but it was unknown to the Mayans and Aztecs of Mexico. At the beginning of the 17th century, the potato spread from Spain to the British Isles, to Belgium by the end of the 17th century, to Austria and Germany by the beginning of the 18th century.

Popcorn (before 1630)

British colonists in Massachusetts Bay Colony first encountered popcorn at their first anniversary dinner, February 22, 1630. Massossoit's brother, Quadequina, gave them a deerskin sasck filled with popped kernels of corn.

Crescent rolls, or croissant (1683)

The crescent roll was invented in 1683, in Vienna, by the Pole **Kulyeziski**. The city had been under siege by an immense Turkish army led by Kara Mustafa. The famished Viennese were finally saved by Charles de Lorraine and the king of Poland, John III, Sobieski. Kulyeziski, having taken a decisive part in the final victory, was given the stocks of coffee abandoned by the routed Turkish army and was authorized to open a cafe in Vienna. This he did, and to accompany his coffee, he had a baker make small milk bread rolls in the shape of crescents to commemorate the victory over the Turks. They were immediately successful.

Sandwich (1762)

John Montagu, fourth count of Sandwich, is said to have invented the sandwich in 1762. A devoted gambler, one day the count refused to leave his gambling table for lunch. His cook prepared a small snack for him consisting of a slice of meat between two slices of buttered bread. The sandwich rapidly became very popular throughout the British Isles but did not spread to continental Europe until the following century.

Chocolate bars (1819)

The Swiss **Francois-Louis Cailler** was 23 years old when he made the first chocolate in bar form in Vevey, Switzerland, in 1819. Small-scale production of chocolate had started in France and Italy after the Spanish brought back the recipe for chocolate from their conquest of the New World. It was at that time a drink prepared from cacao beans that were pulverized and roasted.

Cocoa powder (1828)

Cocoa powder was made for the first time by the Dutchman **Conrad Van Houten** in Amsterdam in 1828. He succeeded in extracting the fat from crushed cocoa seeds, thus obtaining a soluble powder used to make chocolate.

An alkaloid can also be drawn from cocoa: theobromine, (or "food of the gods"), which is a powerful diuretic.

Chewing gum (1869)

Since time immemorial, man has chewed leaves, herbs, or other plant material. The Mayas, for example, like many other people of Central America, chewed the chicle/solidified sap) of the sapodilla for centuries. British colonists in America learned from the Indians of New England to chew the resin of the spruce tree.

In 1848, the American **J. Curtis** marketed spruce resin for the first time, spruce resin in the state of Maine. Toward 1860, was abandoned for chicle because advantages as a chewing material (chicle holds flavorings like mint or anise better). It seems that the first person to market chicle was the American T. Adams.

On December 18, 1869, **William F. Semple**, of Ohio, patented "chewing gum," a well-proportioned mix of rubber and other ingredients. There followed a great increase in the use of chewing gum and, starting in 1900, it was manufactured industrially.

Margarine (1869)

Margarine was invented in 1869 by the Frenchman **Hippolyte Mege-Mouries**, following a contest launched by Napoleon III to come up with a replacement for butter. An artificial butter that would be economical and conservable and would not go rancid represented an uncontestable advantage for the nutrition of the army and navy.

Mege's method consisted in processing animal fat (essentially tallow) from which he obtained a paste whose color and consistency were close to butter and which did not have a disagreeable odor. He christened his product margarine because of its pearly color (pearl is margaron in Greek). Later, thanks to improvements in Mege's procedure, margarine was made from vegetable fat.

Artificial flavoring (1874)

In 1874, two Germans, **Dr. Wilhelm Haarman** and **Professor Ferdinand Tiemann**, synthesized vanillan, the principal component in vanilla husk. Two years later, Karl Reimer conceived a chemical compound reproducing the entire flavor of vanilla.

Saccharin (1879)

Saccharin was discovered by **Constantin Fahlberg**, an American who was working under the direction of Professor Ira Remsen at Johns Hopkins University in Baltimore. He stated the results of his work in an article entitled "On the subject of liquid toluesulphochloride," published February 27, 1879. Recent studies seem to prove that saccharin can be hazardous to health but only if consumed in extremely high quantities.

Ice cream sundae (1890)

A simple retail merchant in Wisconsin by the name of **Smithson** invented the ice cream sundae in 1890. This American specialty was born out of a scarcity of ice cream. Not receiving deliveries on Sundays, Smithson was often short of merchandise on that day. It is said that this is what prompted the idea of reducing his portions of ice cream and adding chocolate sauce or fruit syrup. The mixture was so successful that his customers asked him for the delectable Sunday ice cream even on the other days of the week. When a puritan-minded citizen criticized the sacrilegious use of the word Sunday, the Lord's day, to designate something so profane, Smithson modified the spelling.

Corn flakes (1898)

Corn flakes were popularized in 1898 by the American **William Kellogg**. Before that, the American Henry D. Perky of Denver, Colorado, was the first (1893) to have the idea of making ready-to-go cereals for breakfast. Having met a man who nursed his stomach pains by eating boiled wheat soaked in milk every morning, Perky came up with a product made from wheat, which he called shredded wheat. In 1895, a doctor from a sanatorium in Battle Creek, Michigan, outdid Perky by making "flakes" of wheat, which he named "granose" flakes.

Ice cream bars (1922)

The American, **C. K. Nelson**, of Onawa, Iowa, was the first to have the idea of coating a bar of ice cream with chocolate. He named it Eskimo Pie and patented his invention on January 24, 1922.

Frozen prepared meals (1945)

In 1945, an American was the first to offer precooked frozen meals to airline passengers. But frozen meals like those served to airline passengers

Frozen meals were invented in 1945.

soon gained a wider audience among all those hypnotized by their television sets. A number of American firms studied the idea before coming out with the famous TV dinner which developed spectacularly from 1954 on. In 1960, there were 215 million trays of frozen meals prepared in the United States.

Powdered butter (1952)

Invented in 1952 by research scientists at the Institute of Margarine, in Moscow. By simple kneading in water, this product provides a butter that is quite satisfactory. It can be conserved for two years; it contains a bit less fat than "ordinary" butter (81% compared to 89%) but is slightly higher in casein. It can be used as is in prepared dishes provided that they are cooked at no less than 226°F (100° C).

Cheese substitute (1982)

The research laboratories of the **Milk Marketing Board** in Great Britain developed, in 1982, the technology for producing a cheese substitute made from low-cost fat, casein, water, and spices. The overall cost of these ingredients is less than half that of the milk necessary for making the same quantity of real cheese.

Edible snail shell (1982)

In March 1982, **Roger Petoit**, director of the French company Torino, patented a process for making snail shells out of dough. They are available in salted or sugared versions according to whether they are to be stuffed with real snails or a totally different ingredient. The shell stands cooking as well as freezing. With a lower price than real snail shells and

the head-turning slogan "Eat the snail . . . and the shell, too," Torino hopes to inundate France and the world with this invention.

Low-calorie delicatessen dishes (1983)

It was for the French company Morey and Son that **Dr. Robot** invented, in 1983, a low calorie liver pate and a low calorie gelatine. Delicatessen dishes containing less fat than traditional pates had already been attempted. The liver pate invented by Dr. Robot has 201 calories per 100 grams, as compared to 300 for a classic product, while his gelatine only contains 198 calories as compared to 400 for a traditional gelatine. And that is not all. He is perfecting a low fat pate, for diabetics and anyone who must avoid consuming sugar.

7. Beverages

Coffee (15th century)

According to legend the stimulating qualities of coffee were discovered by a Moslem goat herder in Yemen. Having grazed on the red fruit of the small coffee tree, his animals could not sleep the following nights. Coffee was first introduced to Europe in the 16th century. It came from the Arabian coffee shrub, a strain originally found in the high valleys of Ethiopia. Coffee was first drunk in Italy around 1640. Great Britain followed in 1652. Eight years later, a sailing ship from Egypt moored in the port of Marseilles, France, with the first cargo of coffee. By 1720, 380 varieties of coffee were on sale in Paris.

Champagne (1688)

In 1688, **Dom Pierre Perignon** (1638—1715), a monk at the abbey of Hautvillers, France, invented champagne. He was not only a brilliant winetaster but also a chemist and physician.

The wine of Champagne was already famous in the time of Clovis (466—511 A.D.) and it is said that Saint Remy offered him a barrel of it, promising him victory so long as there remained a drop of wine in the barrel.

Dom Perignon developed the sparkling wine that we know today. Thanks to his extremely fine palate, he dared to blend several wines, going against all the rules of winemaking. Thus, champagne comes from a number of different wines carefully mixed together. Furthermore, Dom Pe'rignon understood and harnessed the principle of carbonization. Champagne does not ferment completely during the weeks following pressing. During the winter the ferment becomes less active and then, in the spring, it picks up again. Dom Perignon had the idea of sealing the wine in tightly corked bottles. This air-tight enclosure conserves the carbon dioxide produced by the spring fermentation and makes the wine "sparkling." There were still many problems to be resolved: nearly half the bottles exploded from the pressure.

Champagne corks (ca. 1690)

When **Dom Pierre Perignon** invented champagne in 1688, wine bottles were stoppered with oiled hemp. This rudimentary stopper, however, couldn't withstand the gaseous pressure (of up to 12 pounds) exerted by the carbon dioxide inside the bottle. He therefore developed the use of corks that were held in the bottles by wire.

Carbonated mineral water (1741)

In 1741, carbonated mineral water was invented by the Englishman **William Browning** of Whitehaven, Cumberland. He had the idea of adding carbonic acid (which produces bubbles) to ordinary spring water and then bottling it. When the bottle was opened, gas bubbles appeared.

Powdered milk (1805)

The production of powdered milk was first undertaken in 1805, by the Frenchman **Parmentier** (1737—1813). However, it was the German Grunwald who carried out the first industrial tests in 1855, and it was not until the 20th century that the powdered milk industry

Edible snail shells—dough.

Powdered milk.

started to develop, first in the United States, then in other industrialized countries.

Concentrated milk (1827)

Even though the Frenchman **Nicolas Appert** (1749—1841) had the idea of concentrating milk as early as 1827, the process was not applied industrially until thirty years earlier. It was only in 1858 that an American, Gail Borden, set up the first concentrated milk factory in the United States. Up until 1844 concentrated milk was always sugary. However, Mayenberg's work permitted the production of non-sugary concentrated milk from that time on.

Chilled carafes (1866)

The idea of chilling carafes is attributed to the Frenchman **Edmond Carre**, who perfected a refrigerating machine (see Refrigeration: Water vapor refrigerating machine) for this purpose in 1866.

Instant coffee (1867)

The first attempts at the production and marketing of instant coffee took place in the United States as early as 1867, but without great success. The Swiss company Nestle acquired the lion's share of the instant coffee market by creating its famous Nescafe in 1937. In 1938, Nestle took out a patent for its instant tea. In 1966, Nestle refined its instant coffee and took out a freeze-drying patent on it.

Coca-Cola (1886)

Coca-Cola was invented in 1886 in Atlanta, Georgia, by a pharmacist, **John Pemberton**, then fifty years old. Pemberton was fascinated by soda fountains, which offered a choice of drinks to the customer. He decided to develop a syrup that would be original and thirst-quenching. Working relentlessly in the back room of his shop, he produced a mixture containing, notably, cola nut extract, sugar, a little caffeine, decocainized coca leaves, and vegetable extracts. (The syrup's exact composition is still kept secret.) By chance, a few months later, an assistant served a customer Coca-Cola mixed with soda water: This was the little touch that made the drink a success.

To market his invention, Pemberton formed a partnership with D. Doe and Frank Robinson. Robinson was endowed with exquisite handwriting, rich in tapered strokes. The way he wrote "Coca-Cola" was so elegant and original that it was decided to use his handwriting for the graphic designating the product. Today, the trademark, written in red letters, still uses Robinson's script.

In May, 1985, executives of Coca-Cola decided that it was time to reformulate their beverage for broader appeal to contemporary tastebuds. Amidst much fanfare 'New' Coke was introduced and the old formula was retired. Years of research and taste tests yielded a sweeter, less fizzier Coke.

Coca-Cola drinkers across America were outraged. The executives had tampered with an American institution and many Coke drinkers were boycotting the new taste and, even worse, switching to the rival Pepsi.

By July the executives got the message and retreated, somewhat. The old formula was brought back. Now there would be two Cokes, "New" and "Classic". For the time being, peace again reigns in the kingdom of Coke.

Coca-Cola bottle (1913)

With the immense success of Coca-Cola, there came many imitations. In July 1913, the company's managers realized that the only way to put an end to this was to give Coca-Cola a bottle that would be absolutely original. They entrusted the task to the famous glass-maker of Terre Haute, Indiana, **C. S. Root**, who charged one of his assistants, Edward, with the task of gathering all possible information on the beverage. Edward discovered an illustration of a kola nut, which he copied. The design could not be used as it was, since a bottle of this design could not be stood upright. The base of the nut was therefore cut off. The technical director of the factory, Samuelson, reproduced the truncated nut in glass and decorated it with vertical lines. Samuelson named the bottle after a dress that was in style at that time: the "sheath dress." The prototype was registered at the American patent office, and a trademark on the bottle made the fortune of C. S. Root.

Straws

The drinking straw was invented on January 2, 1888, by the American **Marvin Chester Stone**, of Washington, D.C. His hand rolled straws were made of paraffined paper. Faced with a growing demand, he invented, in 1905, a machine to roll his straws.

Frozen orange juice concentrate (1947)

After preliminary studies in 1944 and 1945 by J. L. Heid, an American citrus cooperative in Lake Wales, Florida, produced frozen orange juice concentrate in 1947. The juice is concentrated at 66° F by means of a heat pump. The concentrate obtained is packaged, then frozen. This process was immediately successful and was soon applied, though to a lesser extent, to other fruit juices: pineapple, grape, strawberry, etc.

Instant ice cold drinks (1976)

The American company **Chill Car Industries** perfected in 1976 a process by which canned drinks can be refrigerated instantly. When the can is opened, a small capsule inserted in the container releases a fluid that until then was held under pressure. The sudden release of the fluid causes the temperature of the beverage to drop considerably.

Instant hot beverages (1977)

In 1977, the German company **Pozel** invented a beverage vending machine with coffee, milk, tea, soup, etc., all in self-heating containers. Unscrewing the glass cap of the beverage sets off an exothermic chemical reaction. The heat produced by this reaction is transmitted to the beverage by means of a coil. This process permits the consumer to have a hot drink at any time and in any place.

8. Preserving

Preserving jars (1795)

In 1795, the Frenchman **Nicolas Appert** (1749—1841) discovered a method (later called appertization) for preserving food that revolutionized the

Top: Canned food around 1855.
Bottom: An early sardine can.

food processing industry. Appertization, still much used more and more today, consists in sterilizing food in air-tight containers. The first containers were not cans but jars covered by five layers of cork. Appert won a contest organized by the French government; his prize was 12,000 francs. Thanks to this money, he was able to perfect his process and apply it industrially.

The Ministry of the Navy tested Appert's products in 1804. Samples were sent to Brest and kept there three months before being tested. The results were convincing: The maritime chief commissioner reported that the string beans and peas, prepared with and without meat, had the freshness and pleasant taste of fresh vegetables. Following this, Appert cultivated a garden farm in Massy where he grew vegetables. After harvesting them they were immediately sent to a nearby factory he had built to be preserved in jars.

Tin cans (1810)

In 1810, **Pierre Durand** patented a metalled vessel (notably tin) for preserving food. The patent was bought for 1,000 pounds by the Englishmen Bryan Donkin and John Hall, who combined Durand's preserving process with Appert's. Tin cans were first made in 1812 in a preserving factory built at Blue Anchor Road (Bermonsey). The first samples were tast-

ed by eminent navy officers, Lord Wellesley in particular, the future duke of Wellington.

Tin cans opened by a key

The American **J. Osterhoudt**, of New York, invented on October 2, 1866, a tin can with a key fixed onto the top. The key is simply loosened and then turned to open the can.

Industrial freezing (1924)

In 1924, the American **Clarence Birdseye** launched industrial freezing. He had a few predecessors in the United States. Around 1905, on the east coast, strawberries, ice cream, and other products were being frozen by the coldpack process, which consisted of freezing packaged sweet food in brine at 2° F. In 1913, Mary Pennington tried to freeze kernals of corn in Minnesota.

Birdseye visited Labrador during the years 1912 to 1915 and observed that the Eskimos preserved frozen fish and game. In 1924, he created the Freezing Company which, using the methods of the time, froze up to 500 tons of fruit and vegetables per year. He later realized that the freezing process needed to be accelerated and in 1929 invented a double-band freezing device. The product was chilled simultaneously on two sides through contact with two concave metallic bands containing cold brine. Finally, in 1935, he invented his multiple plate freezer, still in use today.

Dehydration freezing (1949)

Conceived by an American named **Howard** in 1945 and patented in 1949, the dehydration process completes the techniques of freezing. It is known by the technical name dehydrofreezing and consists in partially dehydrating the food product with a conventional heating system and then immediately freezing it. Approximately 50% of the water is eliminated and a lighter product that keeps

well and rehydrates easily is obtained.

The advantage of dehydrofreezing is that it reduces the mass to be transported. It is used for various fruits and vegetables (sliced carrots, potatoes, etc.), which can be used in soups and sauces or other preparations.

Freeze-dried food (1946)

In 1946 and 1947, the American **E. W. Flosdorff** demonstrated that the process of freeze-drying, which was already known and used, could also be applied under proper conditions to products such as coffee, orange juice, or meat. Also called cryodessication or cryosublimation (dessication or sublimation at low temperature), freeze-drying achieves dehydration through refrigeration: The water content solidifies faster than the other elements in the product and is eliminated in the form of ice.

Freeze-drying itself was invented by the Frenchmen **Arsene d'Arsonval** and **F. Bordas** in Paris, in 1906, and rediscovered independently by an American, Shackwell (1851—1940), in St. Louis, Missouri, in 1909. The process was first applied medically: BCG vaccine and penicillin were freeze-dried. A Swiss laboratory had already developed freeze-dried coffee in 1934. It was not until 1955 that freeze-drying entered the food industry, where it was applied to Texas shrimp and Maryland crabs.

Special plastic packaging (1946)

Since controlled atmosphere chambers (cold rooms) are costly, the possibility of replacing them with easier-to-build enclosures was studied. In 1946, **Platenius** had the idea of using plastic envelopes to keep fruit in a confined atmosphere favorable to their preservation. The first perforated plastic packages were used in the United States in 1955.

An industrial freezing plant.

the arts

VISUAL COMMUNICATION

Egyptian writing around 1600 B.C.

Egyptian writing was deciphered by Champollion in 1822.

1. Writing

The origins of writing

The invention of writing seemed so extraordinary that more than one civilization claimed to have obtained it from its gods (the Egyptians, from the God Thot, for example). However, although the ritualistic and symbolic drawings found in later paleolithic caves (some of which are more than 15,000 years old) bear the seeds of the basic principles of writing, historically the three most ancient writing systems (all appearing around 3000 B.C.) of which traces remain are: cuneiform and hieroglyphs (in the Middle East) and Chinese ideograms (in the Far East).

Sumero-Akkadian writing (4th milennium B.C.)

Originally approximate representations of the objects denoted (semi-pictographic), cuneiform, invented by the Sumerians in the 4th and 3rd millenia B.C., is the oldest writing system documented. The term *cuneiform* refers to the extremely angular aspect of the signs, generally printed on clay tablets with a beveled reed. Borrowed by the Akkadians around the middle of the 3rd millenium B.C., Sumerian writing, already difficult, became even more complicated. Helped along by the Babylonian civilization and its conquests, however, it spread all over the Near East. It included roughly 20,000 ideograms.

Chinese writing (ca. 2000 B.C.)

Tradition has it that Chinese writing was invented by Chinese emperors during the 3rd millenium B.C. The most ancient documents discovered in Honan, date from the second half of the 2nd millenium B.C. (Yin dynasty). Koreans, Japanese, and Annamese later adopted aspects of Chinese writing.

Chinese is the only ancient writing system still in use today. Used by one-fifth of the world's population, it has hardly changed in 4000 years. Though ancient Chinese has broken up into a large number of dialects, its script remains completely understandable by all, despite its age, and its characters have the same meaning no matter where they're used. Chinese writing is made up of 50,000 signs.

Alphabet (1500—1000 B.C.)

The discovery of Ras Shamra and Byblos, two ancient Middle Eastern

The first form of written expression.

Summerian writing was very complicated.

Egyptian hieroglyphics.

cities, confirms that the alphabet was invented by the Phoenicians, whose archaic right-to-left writing style began circulating in the 10th century B.C. However, the alphabet is generally agreed to have appeared first in the second half of the 2nd millenium. The consonant-based Phoenician alphabet has 22 signs, and it is the origin of most of our modern alphabets.

Greek alphabet (ca. 1000 B.C.)

The Greeks borrowed their alphabet from the Phoenicians and adapted it to their language (late 2nd, early 1st millenium). However, the fundamental innovation of the Greek alphabet, which served as a link between the Semitic and Latin alphabets, was the introduction and rigorous notation of vowels. From the 4th century on, Athens adopted the Ionian alphabet as its official alphabet and, after 500 B.C., writing went from left to right. The first two letters of the Greek alphabet, *alpha* and *beta*, are the root words of the word *alphabet*.

Egyptian writing (ca. 300 B.C.)

The appearance of hieroglyphics, the most ancient and characteristic form of Egyptian writing, coincided with the unification of Egypt, around 3000 B.C. This form was used until the 3rd century A.D. Hieroglyphs originated as sacred engraved signs containing sayings of the gods. They continued the practice of ideographic and phonetic writing.

Hieratic writing, simplified in relation to hieroglyphics, was invented at the same time and used primarily by the priesthood. It lasted until the 3rd century A.D.

Demotic writing came from hieratic writing and represented an advanced stage of spoken language. Developed at the beginning of the 1st millenium B.C., it was used until the 5th century A.D. Seven of its letters became part of the Coptic alphabet, used in Egypt until its gradual decline after the Arab conquest (641).

Latin alphabet (ca. 100 B.C.)

The Latin alphabet comes from the Greek via the Etruscans. Latin, an Italic dialect, was essentially the language of Rome, who as conqueror imposed it on the whole Italian peninsula and then on all of the Western world of that period. One century before Christ, the Latin alphabet had 23 letters, to which were added the Y and Z of the Ionian alphabet. Latin writing underwent many modifications afterwards, but it was with this 1st-century alphabet that the first phase in the history of Western writing came to a close.

The simplicity of the Latin alphabet and its adaptability to widely varying languages allowed it to spread rapidly around the world, ultimately to the point of adoption of Latin characters in Turkey, Latinized notation of certain African dialects, and even attempts to Latinize spoken Chinese.

Arabic writing (ca. 500 A.D.)

After Latin, Arabic is one of the great international writing systems, adopted in the Moslem world not only by the Arabs themselves, but also by other languages, such as Persian, which is spoken in Iran. Of Semitic origin and close to Hebrew, Arabic is associated with Phoenician writing. The alphabet has 28 letters, all consonants, with vowels and diphthongs indicated by symbols above or below the letters. The first inscription in Arabic dates from 512—513.

Cyrillic alphabet (862 A.D.)

The origin of the Cyrillic alphabet is the *glagolite*, invented around 862 by the philosopher Constantine (and called Cyrillic from Cyril, Constantine's name as a monk) in order to convert the Moraves. This glagolitic alphabet (from *glagol*, "the word," in ancient Slavic) was modified around the beginning of the 10th century by the introduction of 24 Greek letters, giving birth to today's Cyrillic alphabet. The modern Russian alphabet derives from a simplified version of this Cyrillic alphabet. It consists of 33 signs.

The religious schism in the early 11th century roughly divided the Slavic world into two gigantic groups: the Russians, Ukrainians, Bulgarians, and Serbs adopted the Cyrillic alphabet along with Greek Orthodoxy; the Poles, Czechs, Slovaks, Slovenes and Croats adopted the Latin alphabet along with Roman Catholicism.

Chinese ideograms.

Stenography (1602)

Stenography (from stenos, "narrow," and graphein, "to write") is a phonetically based system of abbreviated and simplified writing consisting of signs that allow one to notate speech as fast as it is spoken. Although the Greeks and the Romans had already used similar systems, it was the Englishman **J. Willis** who wrote the first treatise on shorthand (1602). Later on other methods were developed, such as the systems of Taylor (1786), Pitman (1837), Duploye (1834), and Gregg (1888).

Cryptography (secret writing)

Cryptography, or secret writing, uses a cipher to protect the content of a message (frequently diplomatic or military) through two basic techniques: transposition and substitution.

The first coding apparatus, according to Plutarch, was apparently the Lacedaemonians' *scytale*, but if one

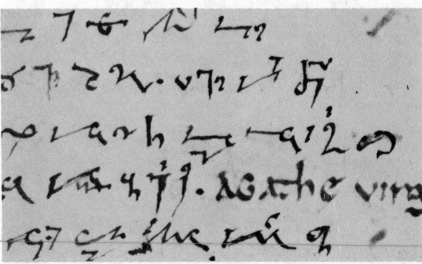

Tiro, Cicero's secretary, invented this shorthand system.

believes Suetonius, it was the Romans who invented the first cipher based on substitution, the so-called Julius Caesar method. Not until the end of the 14th century, however, does one find the invention of the first manual of ciphers for diplomatic use: the *Liber Zifforum* by Gabriel de Lavinde, the Pope's secretary.

The "Grid" and the "Indecipherable Square" (1560)

Around 1560, B. de Vigenere popularized **Abbe' Tritheme's** invention: the indecipherable square. This system, like Julius Caesar's, is the basis of numerous more recent systems, such as that of the English admiral Sir Francis Beaufort, developed in 1857.

As for the "Grid," dear to spy story readers and invented by the mathematician **Geronimo Cardano** (1501—1576), it has almost completely fallen into disuse.

However, thanks to electronics, cryptography has grown considerably since World War II.

Braille alphabet (1829)

In 1829, the Frenchman **Louis Braille** (1809—1852), professor at the Institute for the Blind, invented a writing system for his students' use. He had lost his own sight at three years of age by poking out his eyes with his father's tools. Afterwards, determined to have an education, he developed, with Valentin Hauy, a system in relief, adapted from the Morse Code used by the Marines. He needed more than 10 years of research before coming up with a real alphabet.

2. Writing Materials

Papyrus (before 1800 B.C.)

The **Egyptians** are usually credited with the invention of papyrus. The stem of a reed cultivated in the Nile Valley constituted the raw material. Scholars who accompanied Bonaparte on his Egyptian campaign discovered, in 1798, sheets of papyrus dating back to 1800 B.C.

Ink (ca. 2500 B.C.)

The **Chinese** invented ink around 2500 B.C. It was made from glue vapor and aromatic substances. Black- and red-ink writings drawn on papyrus documents with a calamus and even a quill have been discovered in Egyptian hypogea (underground sepulchers) dating from the same period.

Louis Braille.

Parchment (2nd century B.C.)

Legend has it that the inhabitants of Pergamum, in Asia Minor, invented parchment during the 2nd century B.C. The oldest example of written parchment is a fragment dating probably from the end of the first century. Its use became widespread in the 4th century A.D. Parchment is made from the skin of sheep, goats, or lambs (vellum).

Paper (2nd century A.D.)

The invention of paper, made from plant fibers reduced to a paste, was certainly the work of the Chinese. But it was not widely used until the 2nd century A.D. In the 9th century, the Arabs promoted its use in Europe. The oldest known documents are Buddhist texts from the 2nd century. The missal from Silos (near Burgos, Spain) is the oldest manuscript written on European paper. It dates from the 11th century. By the end of the 14th century, many European countries made their paper from wood from the forests. It was in 16th-century Europe that the first paper capable

A paper-making machine around 1900.

of being printed on both sides was developed.

Pencil (16th century)

In 1564, the discovery of graphite in the Cumberland area of England (near the Scottish border) led to the invention of graphite pencils, first introduced in France during the reign of Louis XIII. The interruption of economic relations between France and England in 1792 led the French engineer Jacques-Nicolas Conte to develop a pencil made of graphite and clay wrapped in cedar. Soon the demand extended beyond France's borders and became world-wide.

Eraser (ca. 1750)

Made from a rubber base that erases pen or pencil marks, the eraser is believed to have been invented in the mid-18th century by a Portuguese physicist named **Magalhaens**, or Magellan (1722—1790), who perfected numerous instruments for use in physics and astronomy. An eraser was mentioned for the first time in 1770 by the British chemist J. Priestley. Though the term *rubber* was commonly used until 1850, the word *eraser* was found in 1778 in an encyclopedia. Today, plastic and synthetic rubber are commonly used to manufacture erasers.

Fountain pen (modern, 1884)

Nobody knows exactly how far back the revolutionary idea of adding an ink reservoir to a quill pen goes. In her *Memoirs*, Catherine the Great of Russia noted that she used an "endless quill" in 1748. Was she referring to one of the first pens? Between 1880 and 1900, fountain-pen inventions proliferated: More than 400 patents were registered. In 1884, **L. E. Waterman** founded in New York the first fountain-pen industry, after having solved the delicate problem of maintaining a constant, even flow of ink.

Ink Cartridge (1935)

We owe the invention, in 1935, of the ink cartridge to **M. Perraud**, president of Jif-Waterman. This small plastic cylindrical reservoir, which is screwed onto the body of the ink pen, was an immediate success and replaced, on a worldwide basis, traditional piston-based pens.

Ball-point pen (1938)

The ball-point pen was invented in 1938 by **H. Biro**, a Hungarian journalist. During a visit to the print shop of the magazine for which he wrote, Biro was impressed by the advantages of an ink that dries quickly. He proceeded to make a prototype of a pen based on the same principle. To escape the Nazi threat, he settled in Argentina in 1940 and there developed his invention. He patented it on June 10, 1943, and his pens were sold in Buenos Aires starting in 1945. H. Martin, an Englishman, found in this invention the answer to a problem he had been trying to solve: how to supply pilots who flew at high altitudes with an instrument that would write. He began producing ball-point pens for the RAF in 1944. Soon afterwards, the Biro ball-point pen was universally adopted. Its distribution began in 1948.

3. Painting and Drawing

Origins (30,000 B.C.)

Should one suppose that the first base used to paint on was—as it still is in the case of certain tribes—the human body? Archeologically, the most ancient vestiges of this art are figures and signs painted on cave walls in paleolithic times—from about 30,000 to 12,000 B.C. Handprints (sometimes outlined with the most distant ancestor of the modern airbrush, a hollow bone from rich red ochre powder was blown) and markings made by fingers that had been dipped in dyes then developed into designs and drawings of animals that doubtless had some magical meaning.

Painted cloth (ca. 2000 B.C.)

Cloth (linen, cotton, and hessian), today far from the most likely base for painting on, is a very old form of art material: A fragment of painted linen found in Egypt dated from the 12th dynasty (between 2000 and 1788 B.C.).

Particularly for larger paintings, cloth offered a cheaper and much lighter base than wood. However, its usage, together with oil painting didn't become more widespread until late in the 15th century and, particularly, in the 16th century. In the Middle Ages, it was only used for painted banners for processions.

Oil painting (1420)

The most important technical event in the history of painting was the invention of oil painting at about the beginning of the 15th century, which today is still surrounded by an aura of mystery. Quick-drying oils were known to the Greeks, and texts prior to the 15th century make reference to their use (which was sporadic because of difficulties with their stickiness). Vasari, writing in 1550, reported that painters at the end of the Middle Ages were looking for a new medium that would give colors greater brilliance and depth, dry less quickly, facilitate the nuances and range of colors available, and offer better resistance to humidity when dry.

It seems that it was the **Van Eyck** brothers who, perhaps thanks to Jan Van Eyck's knowledge of alchemy, successfully used oils for the first time in about 1420.

4. Printing

Origins

Many things were already being printed in China under the T'ang dynasty (618—907): magic books, scholastic manuals, etc. A copy of *The Diamond Sutra* printed in 868 was discovered in the caves of Dunharang. It bears the name Wang Jie, who is, if not the first printer, at least the first known printer. In 1041, the Chinese Picheng developed movable characters of heat-hardened clay. As a solution to their fragility, one of Picheng's successors developed movable letters in lead.

Under the Song dynasty (960—1279), Chinese printing prospered. In the year 1000, a huge Buddhist canon of the most prestigious dynastic stories was published.

Gutenberg press (ca. 1440)

The Mongol invasions and the relations that Saint Louis and the papacy

Writing instruments as illustrated in an 18th century French encyclopedia.

productivity by ten, reaching 3000 sheets per day.

The first printing press run by a steam machine was developed by the German F. Koening in 1811. Later, Koening, in association with A. Bauer, constructed the first cylinder press, which was followed by many others. In London, notably, a single machine printed overnight the 4000 copies of the London edition of *The Times*.

Lithography (1796)

In 1796, the Polish typographer **Aloys Senefelder** invented lithography, a method of printing by transfer. He realized that a drawing done with a soft-lead pencil on limestone (*lithography* comes from the Greek word for stone, *lithos*) is water-resistant. However, the unmarked stone absorbs water. If a coat of ink is spread over the surface of the stone, it doesn't stick to the wet spots, only to the dry or greasy areas. The stone has only to be placed in a press and it reproduces the initial drawings. Senefelder rapidly improved his method. Instead of water, he used a solution of gum arabic and nitric acid, which is completely impervious to printer's ink.

Heliogravure (1875)

The heliogravure printing process was invented by the Austrian **Karl Klietsch** in 1875. The form of what is to be printed is engraved by chemical processes (capture of an acid through a sensitized gelatin, on which the form is indicated by a photographic process) on a metal plate. The recesses in the metal plate are filled with ink. The surface is wiped and the ink remaining in the recesses is left on paper.

An industrial process, heliogravure expanded rapidly and was used especially for highprint-run products: magazines, catalogs, wrappings, etc.

Monotype (1887)

In 1887, one year after the marketing of the linotype, the American **Tolbert Lanston** devised the monotype. This composing machine casts and composes isolated characters that are assembled in the order assigned to them. Lanston's machine was capable of composing 9000 characters an hour; correction of a text was easier than on the linotype because each character could be changed. On the other hand, linotype offered a better alignment.

---

had with the Khans facilitated the introduction of printing in the West. Starting in 1400, wooden plates engraved in relief were used to reproduce on paper a combination of text and illustration; this technique was called xylography.

Around 1440—1450, the German printer **Johann Gensfleich** called Gutenberg (born in Mainz between 1394 and 1398, died in 1468), developed with his associates the technique of movable characters. In addition, he perfected the material necessary for the quality and conservation of characters: an alloy of lead, antimony and tin.

In 1455, in Mainz, Gutenberg printed the *Biblia sacra latina*, called the 42-line (per page) Bible. It was the first Latin edition of the Bible in movable characters. His business partner, Johann Fust, took him to court for repayment of a loan earlier advanced, got complete possession of the movable type characters, and took all the

Bible's profits. Gutenberg was able to start up a new printing business by 1460.

In 1470, in Paris, the Sorbonne presses printed the first book in French. The manuscript copy was mislaid in the early 16th century.

The printing press invented by Gutenberg remained in use for four centuries. Starting in the 18th century, a series of transformations allowed the technique to move forward.

The Frenchman Francois Didot (founder of a printing dynasty that shone till the end of the 19th century) replaced wood with iron for the *stone* (horizontal table that receives the text settings). He developed the "single-passage press," then doubled the printing speed. In 1737, the "Didot points" gave printing characters a base unit of measure and hence standardized their dimensions and bulk.

The Englishman C. Stanhope got rid of wood completely in presses and used metal instead. He multiplied

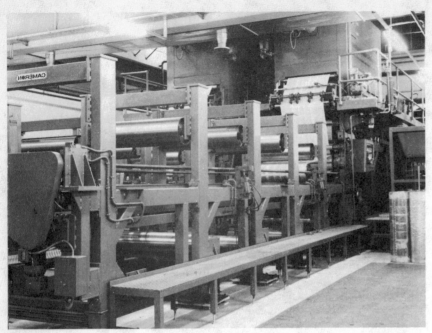

The Cameron Printer prints a complete book without human intervention.

Offset

Invented by the American lithographer **Rubel** in 1904, offset came out of the lithographic process he perfected. The word *offset* covers the same transfer technique (direct apposition) as lithography. Offset is not done on stone but on a sheet of zinc. This transformation allowed print-runs to reach industrial dimensions.

Cameron press (1968)

The Cameron press is a revolutionary machine for printing books. In fact, this machine, developed by the research director of the Cameron Machine Company of New Jersey, **Charles Aaron**, can print and bind a book without human intervention. While production-line printing was first invented by Stoud-Bridgeman, a Canadian company, Charles Aaron was the first to develop the automatic collation system that is a distinct feature of the Cameron press. The first machine became operational in October 1968, at the Kingsport Press in Tennessee. Currently, the various Cameron machines in use print more than 300 million books a year.

5. Publishing

Poster

Origins

The first known poster dates from 3000 B.C. It is an Egyptian papyrus, discovered in Thebes, offering a re-

ward for finding an escaped slave. In Greece, news meant for the public was painted on wooden tablets and exposed in the public squares. The Romans created the *album* (the "white"), a wall decorated with cornices and pediments and divided into whitewashed rectangles on which the *scriptor* (writer) wrote in charcoal or ochre. Sales, rentals, shows, and festivities were most commonly featured. Even personal commentary was found in Pompeii—graffiti by a woman noting that she started a tapestry on December 26; laudatory remarks on the qualities of a gladiator or a slave, etc. These inscriptions themselves were noted in other comments—for example: "I am amazed, O wall, that you haven't crumbled under the weight of all the rubbish that you have been covered with."

Printed poster

The poster as we know it today was born when printing it became possible. The first known poster was printed by the Englishman **W. Caxton** in 1477. It announced the thermal cures of Salisbury and already bore the injunction "Do not tear apart." In 1482, the Frenchman Jean du Pre realized the first illustrated poster on the occasion of the great pardon of Notre-Dame.

Modern poster (early 19th century)

The first truly modern posters appeared at the beginning of the 19th century thanks to the technique of

lithography. They proliferated to such an extent that, in 1839, they were no longer allowed to be hung on walls that were privately owned in England.

The illustrated poster, that including a brief text and a simple but striking image, not just a typographical message, was invented by artists and coincided with the birth of modern art. The Frenchman **Jules Cheret** (1835—1932) is considered the inventor of the modern illustrated poster, initially confined to the entertainment field. His color poster for Sarah Bernhardt (*La Biche au Bois*) in 1867 marked the birth of a new graphic art form. The Frenchman Toulouse-Lautrec (1864—1901) was the first to use large colored surfaces that immediately caught the eye. In 1891, he produced his first poster (*La Gouline*, the Glutton) for the Moulin Rouge. Leonetto Cappiello, a French painter of Italian origin (1875—1942), was the first to really tackle the utilitarian poster (an equestrian poster for Klaus chocolates, in 1903).

Comic strips

Origins (1500 B.C.)

If prehistoric rock paintings are the first attempts at telling stories through pictures, *The Book of the Dead* from the ancient Egypt (1500 B.C.) constitutes the first example of a figurative story whose successive scenes are ordered in superimposed strips with the text situated inside the picture. After the Egyptians, the narrative picture became mute, i.e., without text. This was true for the Parthenon frieze (5th century B.C.), Trajan's column (113 A.D.), and the Bayeux tapestries (11th century).

Classic comic strip (1896)

In 1896, *The Yellow Kid* by **R. F. Outcauld** appeared in the *New York World*. It was the first classic comic strip. It consisted of a series of color drawings with balloons that contained the texts. Not until *Les Pieds Nickeles* by L. Forton, in 1908, was the balloon technique used occasionally in France, and not before *Zig et Puce* by A. Saint-Ogan, in 1925, does one find a strip that meets all the criteria of the technique.

Comic book (1933)

The comic book as we know it today originated in the American daily press, in particular, *Funnies on Parade*. Printed on the same printing presses and the same paper as the daily, this book appeared in 1933 when it

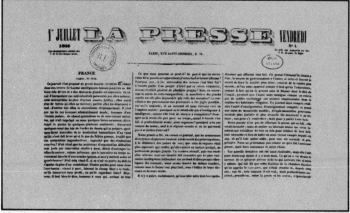

La Presse, published in France in 1836 by Emile de Giardin, was the first popular newspaper.

The first published newspaper in Flemish in 1605.

was discovered that the classic newspaper page, folded twice, could be used as an 8-by-11-inch brochure. This first trial was not sold but offered by certain brand-name products such as Canada Dry. The first comic book to be sold and published periodically (from May 1934 on) was *Famous Funnies.*

Library (1700 B.C.)

The first libraries appeared in Chaldea in 1700 B.C. The first books were baked clay tablets. In 540 B.C. Pisitratus endowed Athens with the first public library. As for the Alexandria library, founded in 284 B.C., it was damaged by the fire that Julius Caesar inflicted on the entire city and finally destroyed in 391 A.D. when the Enperor Theodosius I ordered all pagan structures razed.

Book publishing

Veritable publishing houses already existed in ancient Greek and Roman times. Athens and Rome boasted printed works of which several hundred copies were published. These came straight from the houses of the "ancient publisher." It was, of course, only with the invention of the industrial print shop that publishing really took off, and the retail bookstore came soon after. The latter was born in London with the bookstore of Wynkin de Worde, successor to Caxton, the publisher of the first book in London in 1477. But it was not until the late 16th century that bookstores began to specialize in selling books from one field or another, and it was only then that publishers charged them with the task of distributing their products.

Cookbook *(62 A.D.)*

The cookbook dates back to the famous treatise *De Re Coquinaria* published by the Roman gastronome Apicius in 62 A.D. It described the feasts of the Emperor Claudius I (10 B.C. —54 A.D.) and his successive wives, Messalina and Agrippina.

Pocket book *(1939)*

In 1939, the New York publisher Simon & Schuster invented the pocket book, a paperback book that could be carried in a pants or coat pocket.

Newspaper (1605)

The first gazette with a regular periodicity appeared in Antwerp in 1605. It was called the *Wettliycke Tigdinghe* and came out twice a month. Directed by **Abraham Verhoeven**, it was written in French and German.

The first two weeklies were published in German. The *Aviso Relation oder Zeitung*, published by Julius Adolph von Sohne at Wolfenuttel in Saxony, Germany, was published for the first time on January 15, 1609. Issue number 37 of its competitor, the *Relation: Aller Furnemmen und Gedenckwurdigen Historien*, published in Strasburg by Johann Carolus, is famous—it gives an account of the invention of the telescope by Galileo. This paper came out in 1622. The first English-language paper was the *Corrant of Italy, Germany, etc.*, printed in Amsterdam by George Veseler between 1620 and 1621. In France, Theophraste Renaudot (1586—1653), under the patronage of Cardinal Richelieu, first published his Gazette on May 16, 1631. It had four pages and a print run of 1200 copies.

The Daily *(1650)*

The German **Thimothee Ritzsch** from Leipzig invented the daily. His paper, the *Einkommenden Zeitungen*, came out every day between July and September 1650.

Classified ads *(1657)*

The first classified ads were published by the Englishman **Thomas Newcome** in his *Public Adviser*, which was published from May to September 1657. Principal columns included boats, property to sell or rent, doctors, lost persons or objects. The cost of the ad depended on its content, not its length. "Marine looking for work" was thus less expensive than "House for sale."

Mass-circulation papers *(1833)*

The low-cost nespaper was invented by the American **Benjamin Day**, who launched "yellow journalism" in 1833 with *The Sun*, which sold for two cents. In England, the *Daily Telegraph*, at one penny, was launched in 1855.

6. Office Equipment

Duplicating machine (1773)

The duplicating machine by which several copies of a document can be reproduced, was invented in 1773 by the famous inventor **James Watt** (1736—1819) to expedite the voluminous business correspondence of the factory he directed in Birmingham.

Carbon copy *(1806)*

Carbon paper, used to obtain several copies from one document, was invented by the Englishman **R. Wedgewood**, who patented it on October 7, 1806. The process he described employed a thin paper saturated with ink and dried between sheets of blotting paper.

Stencil (1877)

The first duplication process that used a stencil was developed in 1877 by the American **Thomas Edison** (1847—1931). In 1881, the Hungarian D. Gestetner invented and marketed a duplicating machine that used a wax-coated stencil, the "cyclostyle." In 1888, he marketed the first stencil for a typewriter.

Typewriter (1808)

An ancestor of the typewriter made its appearance in 1660 with the "two-quill manual system" of the Irishman William Petty. On January 7, 1714, Queen Anne of England granted a patent to Henry Mill for a "machine capable of replacing handwriting by the printing of letters similar to those used in print shops." However, the machine was never built.

In 1808, the Italian **Pellegrini Turri** made, for love's sake, the first typewriter. It was invented for use by his friend the Countess Caroline Frantoni, who was blind. The archives of Reggio, Italy, have conserved 16 of their letters written between 1808 and 1810, but no details about the construction of this machine have survived. Throughout the century numerous European inventors created machines to help the blind, but all of them remained at the prototype stage.

First modern typewriter (1833)

The first modern typewriter, likewise conceived for the blind, was invented in 1833 by a man from Marseilles, **Xavier Progin**. It featured a circular basket of bars bearing characters that converged and struck at the same point. A few models were marketed.

Artwork executed with a typewriter.

First marketed typewriter (1870)

In 1870, a Danish pastor, **Malling Hansen**, had the first typewriter built that was intended for the market. Built by Jurgens, a mechanical company in Copenhagen, it was marketed in October 1870 under the name Skrivekugle.

"Literary piano" (1874)

Encouraged by his friend Carlos Glidden (a controller of the Milwaukee, Wisconsin port), **Christopher Latham Sholes** plunged into research on a typewriter that he called the "literary piano." Encouraged by his associate,

James Densmore, he built more than 30 models before coming up with one that finally succeeded. They sold it to the American firm Remington Small Arms of Illion, New York, which marketed the Remington Model 1 in 1874. Until that time, the firm founded by Philo Remington (1816—1889) had specialized in the manufacture of rifles.

Portable typewriter (1889)

In 1889, **George C. Blickensderfer**, from Erie, Pennsylvania, built the first portable typewriter. The "blick" could be stored in a carrying case.

Electric typewriter (1901)

Dr. Thaddeus Cahill, an Iowa-born inventor, developed an electric typewriter in 1901. But his company folded after having built only 40 models at a unit cost of $3925! The American R. G. Thomson worked 11 years to develop the "Electromatic," successfully launched by IBM in 1933.

Electronic typewriter (1965)

In 1965, the American company **IBM** launched the 72BM, the first typewriter with a memory, which used a

The Brother EP20 and an 1895 typewriter.

magnetic tape. In 1978, Olivetti (in Europe) and Casio (in Japan) marketed the first electronic typewriters with rapid-access memories. They featured "type wheels" rather than "balls." In 1980, Olivetti created the Praxis 35, the first portable electronic typewriter. In 1982, the Japanese company Brother invented a miniaturized electronic typewriter. Truly portable (2.3 kilograms), it is completely silent due to its system of thermal printing by points. The Brother EP20 has a small screen that allows one to correct before printing. It also serves as a calculator.

Photocopy (1903)

Photocopy refers to the process of rapid reproduction of a document by the instantaneous development of a photo negative. It was invented by the American **G. C. Beidler** in 1903. Beidler, a clerk in a litigation office, noticed the constant need for copies of documents, so he developed a machine for replacing laborious manual or typed copies and patented it in 1906. The first photocopy machine was marketed by the American company Rectigraph in 1907, but it was not until the 1960s that the photocopy became commonplace.

Xerography (1938)

On October 22, 1938, the American **Chester Carlson** (1906—1968) produced the first xerographic image (from the Greek xeros, "dry," and graphein, "to write"). Carlson first studied the possibilities of photoconductivity as revealed by the Hungarian Paul Selenyi. He began his experiments in the kitchen of his apartment in Jackson Heights, Queens, New York. That's where he discovered the basic principle of xerography. He baptized this principle *electrophotography*, perfected it, and patented it. It consists of dry photocopying on untreated paper. Between 1939 and 1944, 20 different companies refused to develop his patents. In 1944, Memorial Battelle Institute, a non-profit foundation in Columbus, Ohio, signed an agreement with Carlson and started developing xerography. In 1947, Battelle signed an agreement with a small photo company, Haloid, which later became Xerox. In 1959, the first xerographic photocopy machine was marketed, the Xerox 914. It produced 6 copies a minute.

Thermo-Fax (1944)

The inventor of the first "bathless" photocopy process (1944), **Carl Miller**,

was decisively inspired by watching snow melt under dead fallen leaves. The snow melted more quickly there than in the open air, because the dark color of the leaves absorbed more radiant energy than the white of the snow. This phenomenon (heat absorption by a dark surface) led the inventor, after several years of patient research in the 3M laboratory, in Saint Paul, Minnesota, to develop the Thermo-Fax technique. Infrared rays

are used to reproduce the original directly, without a negative and in the presence of light, on treated thermosensitive paper.

Paper clip (1900)

The paper clip, a metallic clip that allows one to attach sheets of paper together, was invented in 1900 by a Norwegian, **Johann Waaler**, who patented his invention in Germany.

MUSIC

7. Instruments

Origins

It is hardly possible to retrace the origin of most musical instruments, since they are often the end products of a long evolution and of traditions whose roots go deep into the history of mankind. However, research undertaken to perfect them and multiply their sound possibilities puts instrument makers on a level with the greatest inventors.

Stringed instruments

Harp (2nd millennium B.C.)

The harp is one of the oldest musical instruments. It stems from the primitive musical arc. After enjoying great popularity in the Middle Ages, it fell into disuse until the 18th century, at which time a German instrument maker, Hoohbrucker, picked up on an idea put forward by a Tyrolean craftsman around 1660 and devised pedals to control a series of hooks that allowed the strings to be shortened. In 1786, the Frenchman S. Erard modified this system and, in 1801, put together the first double-movement harps. This principle is generally associated today with the Erard harps.

Harpsichord (14th century)

The legendary precursor of the harpsichord is the monochord of the Greek

inventor Archimedes (278—212 B.C.) from Syracuse. The monochord was created to study the relationships between the lengths of sounding strings that furnish the notes of the overtone series. The modern instrument originated in the 14th century. (the word *harpsichord* first appeared in 1631.) Until the mid-18th century, the instrument was primarily assigned the role of basso continuo, later that of soloist. It belongs to the family of plucked strings, including the spinet and the virginal.

Violin (1529)

The violin evolved from medieval and Renaissance stringed instruments such as the rebec and the lira da braccio. It appeared in France in 1529, but it was in 17th- and 18th-century Italy that the art of instrument makers, such as N. Amati and especially N. Stradivarius (1644—1737), brought the violin to its apogee. The most famous violin is the Greffuhle, made in Cremona in 1709 by Stradivarius.

Bow (Middle Ages)

The bow as we know it today is the culmination of an evolution begun in the Middle Ages. The Italians Arcangello Corelli (1653—1713) and Giuseppe Tartini (1692—1770) modified it. The current model was established around 1775 by the Frenchman F. Tourte, who defined its material, proportions, and weight. Since then, it

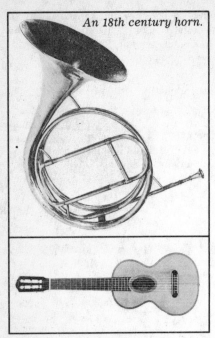

An 18th century horn.

A 19th century guitar.

chanate, he introduced between the strings of the piano various materials (such as plumb lines, corks, wood, rubber, metal) that cushion the sound and bring about variations in timbre. The effect is reminiscent of the sound of Javanese orchestras.

Wind instruments

Reed instruments

The sound of certain wind instruments is produced when reed or metal strips are made to vibrate. The clarinet, saxophone, krummhorn, oboe, English horn, bassoon, contrabassoon, and bagpipe are all reed instruments. An instrument can have one or several reeds.

Flute (neolithic)

The recorder and the (transverse) flute, like the oboe, clarinet, piccolo, English horn, bassoon, contrabassoon, and saxophone belong to the woodwind family.

The flute is one of the oldest instruments. Its origins go back to prehistoric times. Several sheep tibias punched full of holes (usable as recorders or flutes) have been preserved; they date perhaps from the neolithic era. Examples of their rep-

Cossineau's harp.   *A clarinet.*

has been subject to only minor modifications. Tourte was called the Stradivarius of the bow.

Piano (1698)

The ancestor of the piano (an instrument whose strings are struck) is the *echiquier*, a psaltery fitted with a keyboard that triggers small hammers, but the inventor of the pianoforte is B. Cristofori (Florence, 1698). To this first keyboard, capable of producing varying dynamics, were added numerous modifications and improvements: G. Silbermann perfected the system of hammers, J.A. Stein invented pedals (1789), etc. But it was the Frenchman Erard who, with the invention in 1822 of the double escapement that allows a note to be repeated, can be considered as the actual creator of the modern piano.

Guitar (modern, 1850)

The guitar's origins are disputed. In any case, everyone agrees that it was born along the Mediterranean coast. After a long evolution, the modern guitar was developed by the Spanish instrument maker A. de Torres (around 1850). He was particularly concerned about embellishing the sonority, improving the upper sounding board, and standardizing the length of the strings.

Prepared piano (1938)

The prepared piano was invented by the American composer **John Cage** (born in 1912). In 1938, for his Bac-

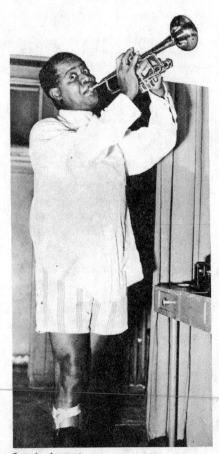

Louis Armstrong.

resentation later on, by 3rd-millenium Sumerians and Egyptians, have also been found.

From the Middle Ages to the 18th century, the recorder was favored. Only in the 17th century did the flute begin to replace it. Lully was the first to write for flute (1681). However, flute-making was not perfected until the 19th century, when the German Theobald Boehm modified the holes and conceived a rational system for the keys. The Boehm system is today still the best woodwind mechanism and has contributed to making the flute a featured orchestral instrument.

Horn (antiquity)

Like the trumpet, the tuba, and the trombone, the horn belongs to the brass family. It was for a long time used in war and hunting. During ancient and medieval periods, horns were carved from animal horns, elephant tusks (whence the terms *olifant, ivorn horn*), and seashells. It is said that Alexander the Great had a metallic horn whose sound carried for 12 miles (18 kilometers).

The horn was introduced into the orchestra in the 17th century. In 1760, the German horn player Hampl placed a wad of cotton in the horn's bell to modify the pitch and to obrain more veiled sonorities. The horn hence became chromatic. In 1815, the German Stoelzel invented the (chromatic) valve horn, the only type used today.

Trumpet (2nd miliennium B.C.)

The trumpet is a very old wind instrument. It has been known since antiquity. A bronze trumpet dating from the 2nd millenium B.C. was discovered in Egypt.

The trumpet was straight for a long time, then curved into an S-shape in the 15th century. The art of rolling the tube into a loop became known only in the 16th century. The value trumpet, which increased considerably the instrument's chromatic capacities, appeared in approximately 1815. The valve trumpet's invention is attributed to the German Stolzel.

Oboe (2000 B.C.)

The instruments from the oboe family were known in Egypt before 2000 B.C. Like numerous instruments of this type, it comes from the double-reed *aulos* whose invention the Greeks attributed to Minerva and even to Apollo. In any case, despite its early beginnings, not until the refinements by Frederic Triebert in the 19th century does one find the oboe in its definitive form.

Organ (200 B.C.)

The oldest organs discovered by archeologists date back to 200 B.C. They were fitted with a sort of large panpipe with two pumps that sent air into the tubes (hence the name "hydraulic"). They were invented by the Greek **Ctesibios** of Alexandria, or by **Archimedes**, in Syracuse.

Clarinet (ca. 1670)

In Nuremberg around 1670, the German **J. C. Denner** invented the clarinet. It is thus one of the more recent instruments in the symphony orchestra, but its origins are quite ancient. Its ancestors are the *arghoul* of ancient Egypt, the Greek *aulos*, and, more recently the pipes of the 17th century. After a series of modifications, the Boehm key system (*see* Flute) was applied to it, and the clarinet attained its technical perfection.

Bassoon (1671)

The bassoon was probably first used in an orchestra in 1671. It is a double-reed instrument whose form comes from two bombardons (a wind instrument of Celtic origin) and resembles that of a bundle of sticks (whence *fagot*, the term still used by the Germans).

Accordion (1829)

The accordion can claim a large number of ancestors: the Chinese *sheng* invented by a legendary queen, Nyu

This saxophone goes back to 1867. The instrument was patented by A. Sax in 1846.

Wa, in the middle of the 3rd millenium B.C.; closer to us, the *eoline*, the *handaoline*, and the *harmonica* of Buschmann. But we owe the invention of the accordion to the Austrian **C. Demian**, who patented it on May 6, 1829.

Saxophone (1846)

In his attempts to improve the bass clarinet, the Belgian **Adolphe Sax** (1814—1894) invented a new instrument, the saxophone. Patented in 1846, the saxophone had its first successes in military fanfares. Its rich sound aroused the enthusiasm of musicians such as Berlioz and Rossini, and the importance that jazz has accorded it since about 1920 is well known.

Harmonica

Two different types of instruments are meant by this term. The first is actually a nonwind instrument, the glass harmonica, invented by **Benjamin Franklin** around 1761 after hearing, in 1746, C. W. Gluck's *Concerto pour 24 verres accordes a l'eau de source* (Concerto for 24 Glasses Tuned to Spring Water). Mozart composed a small quintet for flute, oboe, viola, cello, and glass harmonica. Its place in the orchestra is currently occupied by the celesta, invented by Mustel in 1868.

The second harmonica, also called the mouth organ, is a wind instrument with free reeds and was invented in 1857 by the German organ builder **Matthias Horner**.

Percussion instruments

Man gave free rein to his inventive genius in the field of percussion. Instruments that are struck, scraped, shaken, or banged together are so numerous that they cannot all be cited. Their origins often go back to prehistoric times. Such is the case of the scraper and several kinds of rattles, which in addition to their rhythmic function, undoubtedly served a religious and magical role as well.

Percussion instruments—drum, tam-tam, maracas, tambourine, linga, cymbalum, etc.—are to be found all over the world and particularly in folk music. The richness and color of their sound has often opened the door to the symphonic orchestra. Many composers have written for percussion: The glockenspiel was first used by Handel in 1783; Bela Bartok wrote for the celesta; Saint-Saens introduced the xylophone into the orchesstra in the *Danse Macabre*; and Ravel, the slapstick, in his *Concerto en sol*.

Mechanical and electronic instruments

Music box (1796)

The music box is a mechanical musical instrument and was probably invented in Geneva in 1796 by the watchmaker **A. Favre**. It uses a pin-studded cylinder process similar to that used for mechanical chimes, which were produced in the 14th century.

Barrel organ (1800)

This instrument, probably invented around 1800 by **Barberi**, was meant to accompany wandering musicians. Though similar to the Limonaire organ, invented in 1880 by the Limonaire brothers, the two instruments should not be confused.

Player piano (1887)

The player piano was born in 1887 in the United States. It had a huge success, particularly in the Western saloons.

Electronic organ (1930)

Its electromechanical ancestor was developed in 1895 by the American Thaddeus Cahill. However, the pioneers of the electronic organ are the Frenchmen **Coupleux** and **Givelet**,

Percussion instruments give rythym to an orchestra.

Electronic music (contemporary, 1951)

In 1913, the American engineers Meissner and Armstrong imagined using a triode lamp as an oscillator capable of producing sounds. The German **J. Mager** applied this idea to the construction of instruments that could not be realized with traditional means. He invented the spherophon, among other things. Numerous other instruments came into being later on: the theremin by the Russian engineer and physicist **L. Theremin** in 1920; the trautonium by the German **F. Trautwein** in 1930. But it was under the impetus of Herbert Eimert that electronic music as we know it today appeared (Cologne, Germany, 1951).

Electroacoustic music (1953)

In 1953, the union of concrete and electronic means of production resulted in a new form of music baptized electroacoustic. The German composer **Karlheinz Stockhausen** (born in 1928) is one of the first who worked in this area. In his *Gesang der Junglinge*, timbres associated with the songs of young children are constructed synthetically from simple electric generator oscillations.

who, around 1930, invented an oscillating circuit organ with sonorities resembling those of the classical organ. Unfortunately the number of oscillating circuits (about 80) made it rather unstable. A better performing model was presented in 1943 by Constant Martin.

Electromagnetic organ (1935)

The inventor of the electromagnetic organ was the American **Laurens Hammond** from Chicago. A former clockmaker, ruined by the Great Depression, Hammond decided to convert the unused cogwheels he had in stock to make an organ to be used for playing light music (around 1935).

Synthesizer (1949)

In 1949, a Frenchman, **J. A. Dereux**, took out a patent for a device (a synthesizer) allowing amplification of the acoustic properties of the harmonium. It was a first step that, after the invention of the electrostatic organ with open reeds (presented at the Paris Fair in 1950), and its perfection with phonic wheels (1952), led to the invention of the recording organ with recorded sounds (1958).

The French composer Jean-Michael Jarre gave the first synthesizer concert in China.

8. Musical Composition and Performance

Composition

Composition, as used here, is closely linked to notation and form that is, to the faculties of reading, writing, and constructing and assembling sounds. The Italo-French musician Jean Baptiste Lully (1632—1687) initiated a veritable musical revolution at the court of Louis XIV when he demanded that a corps of violinists play exactly what was written. He was thus the first composer in the modern sense of the term. Indeed, for centuries notation had been very imprecise and interpretation left to the inspiration of the performers.

Notation (5th century B.C.)

The earliest example of notated music dates from the 5th century B.C. and was discovered in Greece.

The history of musical notation is long and complex. First Greek, then Roman alphabetical signs were replaced by *neumes* (8th century A.D.), a kind of musical stenography of mediocre precision. Then a system of aural and visual analogs was used to improve this system by placing the neumes higher or lower according to the sound's pitch, and by transcribing the general melodic line into a sort of garland. Later on, this neumatic system underwent many modifications (the addition of letters from the alphabet, a system still in use in England and Germany; distribution of the notation around a line that fixed the fa tone). Only in the 15th century did half-notes and quarter-notes come into general use and give birth, in the late 16th century, to our current system.

Staff (early 11th century)

The introduction of the staff in music writing is credited to the Italian monk **Guido d'Arezzo** (ca. 990—ca. 1050). He advocated the use of four lines on which different keys and colors served as points of reference. The earliest staves appeared in the 10th century as syllables placed between horizontal lines in order to represent the pitch of the sounds.

Automatic composition (1956)

In 1956, in the United States, **L. Hiller** and **L. Isaacson** composed the Lliac Suite for string quartet with the help of a computer.

Automatic music composition, based on the kinship between music and mathematics, requires a computer. The idea of a "composing machine" based on such a kinship had been conceived already in the 18th century and has been credited to Mozart, but not until the second half of the 20th century and the invention of computers did the idea lead to authentic technical realizations.

Concert (paid, 1672)

The first paid concert, given before an audience of 42, was organized in 1672 by the Englishman **J. Banister**. But since the word concert implies that an audience was brought together specifically to listen to a well-defined musical program, its origins go back to the mists of time.

Orchestral conducting (19th century)

This specialty is an invention of the 19th century. Until then the composer himself was most likely to conduct the orchestra, as did Mendelssohn and Berlioz in the 19th century. Towards the middle of the century, the conducting was increasingly entrusted to specialists such as the Frenchmen Charles Lamoureux (1834—1899) and Edouard Colonne (1838—1910).

Conductor's baton (1815)

The German **L. Spohr** was the first to conduct an orchestra with a baton (the Frankfurt Opera, between 1815 and 1817). Indeed, it was only in the 19th century that it was felt necessary to make the conductor's gestures more visible and precise. In the 17th and 18th centuries, a cane was used to tap out the beats. Lully conducted his *Te Deum* with such a cane in the *Eglise des Feuillants* on January 8, 1687. He tapped his foot with such force that he hurt himself. Gangrene set in and killed him a few weeks later.

Tuning fork (1711)

The tuning fork was invented in 1711 by the Englishman **J. Shore**. It is a small forked instrument that reproduces an "A440," the frequency used as a reference for tuning instruments.

Metronome (1816)

The invention of the metronome, a device that allows the exact determination of tempo, has been credited to an Austrian physician and friend of Beethoven, **J. N. Maelzel**, who patented it in 1816.

Negro spirituals (18th century)

The first Christian songs sung by blacks in the United States were heard during the 18th century at prayer meetings, the only kind of meetings the slaves were authorized to hold. These songs were called spirituals. In the 19th century, these songs, until then spontaneous, were more or less codified; the spiritual was later defined as a musical genre at Fisk University in Nashville, Tennessee. Martin Luther King, Jr., in one of his books (*The Tempest of Conscience*, 1967) stresses that most of the time Negro spirituals contained coded messages that allowed blacks to communicate with each other without their masters knowing about it.

Jazz (1885)

Jazz has its origins in the sacred and secular songs of black America. The birthplace of jazz is New Orleans, Louisiana. It appeared around 1885, when the first jazz orchestras were created. The word *jazz*, initially spelled "jass," turned up around 1910. Its etymology has been explained in various ways. It may have come from the French verb *jaser* or from the name of a singer, Jazzbo Brown; or it may have been an onomatopoeia used to encourage the musicians. "Jazz it," which is thought to have come from a scornful expression used by white racists to indicate the "awful" music of blacks, seems the most likely explanation.

Blues (1909)

The blues, originally a kind of monologue sung by sad and resigned black slaves, spread throughout the United States at the time of the Civil War (1861-1865). This gave Northerners the chance to get to know it. Its official recognition as a musical genre dates from 1909. During an election campaign for the mayoralty of Memphis, one of the candidates, E. H. Crump, hired a black musician, W. C. Handy, who composed a piece in blues form for Crump. When Crump was elected, W. C. Handy became famous and his piece, baptized "Memphis Blues," became history's first official blues.

9. Sound Reproduction

Phonograph (1877)

The inventor of the phonograph is incontestably the American **Thomas**

Indians listening to a gramophon.

which time the American company Audio Fidelity and the British companies Pye and Decca, thanks to numerous technical advances, issued the first commercial stereo records.

Long playing record *(1947)*

The long playing record was invented in 1947, in the United States, by **Peter Goldenmark**, who was working for CBS. It was to replace the 78 rpm record. An LP is a record whose rotation speed is 33⅓ rpm. It contains 100 grooves per centimeter (as against 36 for the 78 rpm) and a 30-centimeter side lasts from 25 to 30 minutes (as against 4½ minutes). The patent was taken out under the initials LP (long playing). The first LP recordings were the Mendelssohn *Violin Concerto*, Tchaikovsky's *4th Symphony*, and the musical comedy *South Pacific*.

Compact disc *(1979)*

The compact disc is different from traditional records in that its engraving is no longer analog but digital. This means that the traditional groove no longer exists. Now there is a series of pits that contain binary bits of information (a string of 1's and 0's). For exact reproduction, this transcription requires the PCM technique (*see* above). Oddly enough, audio compact discs appeared on the scene after video discs. Most existing players derive from videodisc players.

Philips' compact disc, presented to the 62nd Congress of the AES in March 1979 and the following month to the press, used a 14-bit code. At the same time, Sony had also made technological innovations in this area. In August 1979, the two companies united efforts in the hope of producing a single model that might have some chance of setting a worldwide standard. It was decided to keep Philips' compact disc (derived from its optic laser videodisc player), but the resolution was changed to 16 bits, and protection against reading errors was increased through more efficient coding. This work was brought to term by two research teams, one led by Joop Sinjou (Philips), the other by Toshi Tada Doi (Sony). The concrete result was a 12-centimeter disc containing an hour's worth of music on one side (one side only for the moment). It has in fact been adopted as standard worldwide.

Edison (1847—1931). On August 12, 1877, he realized a device that worked with a cylinder and patented it on February 17, 1878. The sound quality left a lot to be desired, and the invention was forgotten for a time.

Gramophone *(1887)*

The phonograph was not marketed until the German researcher **Emile Berliner** emigrated to the United States and, in 1887, developed the technique of recording on wax-covered zinc disks. The player was baptized a "gramophone." After returning to Germany, Berliner founded the Deutsche Gramophone Gesellschaft (1898).

Stereo records *(1933)*

The first stereo records were produced in 1933 by the British firm **EMI** (Electric and Musical Industries). The research, directed by the physicist Alan Dower Blumlein, culminated in the 78 rpm stereo recording. When an artist records, he sings into two microphones. Each mike is independent.

By recording the electric signals from the two mikes onto one record, it is possible to reproduce a stereophonic effect in the loudspeakers. The record groove thus carries two different bits of information. The left-hand signal is engrave on the inside of the groove, the right-hand signal on the outside.

The work of Blumlein and EMI remained experimental until 1958, at

An early juke box.

Tape recorder (1898)

The principle of the tape recorder was worked out theoretically in 1888 by the Englishman Oberlin Smith.

Ten years later, the 20-year-old Dane **Valdemar Poulsen** concretized the invention. His device recorded and reproduced sounds by residual magnetization of a steel wire. Nothing much came of the presentation of his telegraphon at the 1900 World Fair in Paris. It was not until 1935 that AEG, a German company, made (applying the same principle) a plastic tape device, the magnetophon, that ran at 7.6 meters per second.

Magnetic recording tape (1928)

The first magnetic recording tape was made in 1928 by the German **Fritz Pfleumer**. Magnetic tape is a thin strip of plastic, or sometimes paper, one side of which is covered with a ferromagnetic substance in powder form (usually an iron oxide). Recording on this tape involves conserving signals in the form of a variable permanent induction (magnetization) in the ferromagnetic environment.

Already in September 1888, the Englishman Oberlin Smith, in an article entitled "Several Possible Developments of the Phonograph," suggested using strips of fabric containing iron filings. And in 1898, the Danish physicist V. Poulsen carried out the first recording on tape, with his *telegraphone*. Not until the beginning of the 20th century, however, was Poulsen's work taken up again, notably by Pfleumer, who received German patent no. 500900 in 1928 for his idea of covering a strip of paper with a magnetizable iron powder. Despite the failure of his experiments—the paper didn't have enough resistance—Pfleumer offered his invention to a German electrotechnical company, AEG, which concluded that only the chemical industry could use it. The idea was subsequently entrusted to BASF. The BASF chemists suggested replacing the paper strip with a strip of cellulose-acetylene. This was first tried out in 1932. Three years later, BASF presented the first usable magnetic tape at the Berlin Fair.

Premagnetization of tape recorders (1921)

This process was discovered in 1921 by two U.S. Navy officers, **W. L. Carson** and **G. W. Carpenter**, who patented it in 1927 without quite grasping its full significance. Not until 1940 did H. J. von Braunmuhl and Dr. W. Weber, who worked in radio broadcasting in Hitler's Germany, rediscover the process and figure out all the advantages that could be drawn from it for tape recorders of that era. Patented by them in 1946, thus after the war, pre-

Gaumont's Chronophone was invented in 1910.

magnetization was credited to them retroactively, dating from 1940.

The process consists in adding a good-value, high-frequency current (100 kHz, for example) to the musical electrical signal to be recorded. This current's action, called premagnetization, results both in a decrease in distortion and an increase in the recording's dynamic level.

First public recording using a tape recorder (1936)

This took place at the BASF factory in Ludwigshafen, Germany, on November 19, 1936, when the London Philharmonic Orchestra was in town.

First audio cassette (1963)

The first audio cassette was presented by **Philips** in 1963 at the Berlin Radio TV exhibit. During the same period, Grundig came out with a cassette that had less success, perhaps because Philips had forfeited any claims on its patent in order to encourage universal adoption of its system.

Dolby noise reduction (1967)

The first noise reducer, made to improve the signal/noise ratio of taped material, was the work of the American **Ray Dolby** in 1967. Intended for professional usage, this noise reducer was named "Dolby A" when its inventor, several years later, brought out a simplified version intended for general public use, named "Dolby B." Since then, more efficient versions have come out of the Dolby Laboratories, in particular the "Dolby C."

Microcassette (1976)

This is an audio cassette of a smaller format than the Philips minicassette, patented in 1976 by the Japanese firm Olympus. Used in electronic "aide-memoires" (pocket tape recorders), the tape runs at a reduced speed (1.2 or 2.4 centimeters per second). Even though it cannot be used as a hi-fi system, the microcassette is being talked about again for numerical recordings (PCM) on account of its being equipped with a tape whose magnetic field is metallic, incorporated under vacuum by evaporation.

Walkman (1979)

The Sony Walkman is a portable stereo cassette player equipped with headphones. It was developed in April 1979 by the Japanese **Shibuara Plant**, a Sony engineer in Tokyo. Its technical name is the TPSL2.

Photographic recording of sounds (1901)

The photographic recording of sound on sensitive tape is attributed to a German, Ruhmer, on the one hand, and to an Englishman, Duddel, on the other. They developed it simultaneously in 1901. They used the "singing arc" phenomenon, known since 1892 and studied by Thomson and Simon.

Loudspeakers

A loudspeaker is a transducer, that is, an apparatus that converts one form of energy into another. In the case of the loudspeaker, the musical message's electrical energy comes from an amplifier and is first converted into mechanical energy, which is then transformed into sound energy. The type of loudspeaker used almost exclusively today is the dynamic loudspeaker.

Dynamic loudspeaker (1924)

The first patents for dynamic loudspeakers date from the 19th century: Ernst Wermer, from Siemens, Germany, registered a description on December 14, 1877. That of Britain's Sir Oliver Lodge dates from April 27, 1898. At the time, music had no elec-

trical source that would allow such a device to work. In 1924, **Chester W. Rice** and **Edward W. Kellog**, both from General Electric research laboratories, took out a patent on a dynamic loudspeaker and, at the same time, built an amplifier that could supply a power of one watt to their device, called the "Radiola Model 104." The following year, it was marketed together with a built-in amplifier at $250. The coil, connected to the membrane, is thrust into a radial magnetic field. A variable current coming from the amplifier goes through the coil and sets it in motion. The motion is analogous to the cur-

rent that incites it. The movement of the coil causes the membrane to move back and forth, and that generates the sound waves.

Pulse code modulation (1926)

Judging from the description given in the patent, pulse code modulation seems to have been invented by **Paul M. Rainey** (November 30, 1926). Sample, quantification, coding . . . it's all there. PCM was reinvented in 1939 by A. H. Reeves and again during World War II by Bell Laboratories for application to secret telephone conversations.

PHOTOGRAPHY

10. Still Photography

Camera obscura

A sunbeam reflected off any object and passing through a small hole in a dark box projects onto the opposite wall of the box the inverted image of the object. This is the principle on which the camera obscura is based. In 1593, the Neapolitan Giambattista della Porta (1538—1615) used this well-known phenomenon to reproduce engravings. The box had a small hole in it, named the "Stenope" (a word that means "narrow view" in Greek). In front of the hole, on the outside, he placed the engraving to be reproduced. The image appeared inside the box on white drawing paper he had placed there and he had only to trace its outline. The Stenope was later replaced by a magnifying glass, which made the image brighter. Later, the camera obscura was improved and equipped with an inclined mirror that reflected the image onto a horizontal piece of glass on which the drawer placed a translucent piece of paper.

Glass lens

The French philosopher and mathematician Rene Descartes formulat-

ed, in *La Dioptrique* (1637), the laws of refraction. In so doing he established the basic principles of modern optics. Glass lenses (and glasses as we know them) had been made for a long time already, but only empirically: No one could explain how they worked. Convergent convex lenses group light rays and enlarge the image. These are magnifying glasses. In order to disperse light rays, divergent concave lenses are used. An object seen through such a lens seems smaller.

A magnifying glass does not give a perfect image. The flaws, called aberrations, are corrected by combining different lenses. Such combinations form, in particular, the lenses of cameras and movie cameras.

Photographic lenses

Field lens (1830)

Dullong, an Englishman, worked on chromatic aberration (1757). On the basis of his work, the French optician **Chevalier** developed the achromatic meniscus (1830); this was the lens used in the first cameras. Chevalier's lens didn't reconstitute light very well and every photo required a long exposure time, which made portraits difficult to do. That is why it is called the field lens.

Portrait lens (1840)

The portrait lens was created by the Austrian **J. Petzval** in 1840. It diffused 16 times more light than the field lens and was perfectly suited to photographing people; it reduced considerably the exposure time. The fact that it took in an angle of only 20° was of little importance. Since it captured an inverted image, the person whose picture was to be taken had to compensate—for example, hang his military decorations on the right so that they might correctly appear at left in the portrait.

Principal types of lenses

• *Rectilinear lenses* consist of two symmetrical groups of lenses; the back element corrects the distortion of the front element. The rapid rectilinear lens, invented in 1866 by the German **Dallmeyer**, was the classic example of this. Astigmatic flaws, that is, those in which vertical lines move in relation to horizontal lines, could not yet be corrected.

• *Anastigmatic lenses* corrected aberration by a new type of glass that greatly deflected highly refractive rays. **Doctor Schott** from Jena, Germany, was the first to use these in 1888.

• *Three-element lenses*, composed of two convergent achromatic elements and separated by a divergent element, were developed in 1893 by **H. D. Taylor**. They corrected astigmatism quite nicely. The well-known Tessar de Zeiss was from this family of lenses.

• *The telephoto lens* is composed of two elements that are spaced apart, one convergent in the front, the other divergent in the rear, providing a kind

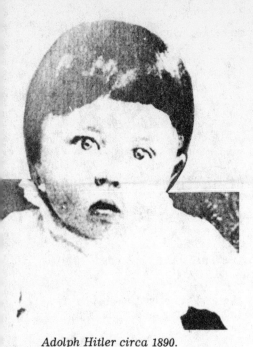

Adolph Hitler circa 1890.

of long-distance view. The first model was patented in 1891 by the German **Thomas Rudolf Dallmeyer**, a London resident.

• *Wide-angle lenses*, unlike the telephoto lens, have their divergent element in the front. The panoramic liquid lens conceived by **Thomas Sutton** in 1859 was the first. (A water-filled sphere replaced the glass lens.)

• *Transformable lenses* are those employing a double set of lenses that can be moved in relation to each other, so their focal distance is modifiable. Chevalier's combined glass device was one example of this. This idea culminated in the zoom lens, allowing a wide-ranging variable focal distance.

• *Supplementary lenses* first appeared in 1895. When placed in front of the fixed lens, these additional lenses allowed shots to be taken closer up.

• *Antireflection coating* neutralized the loss of light due to the reflection of rays on the lens. This system was thought up by **H. D. Taylor** and was patented on December 31, 1904. It consisted of dulling the surface of the lenses. It had been discovered that used lenses gave better light than new ones because their dull surface absorbed light rays rather than reflecting them. In 1935, Karl Zeiss made an antireflection coating by using magnesium fluoride. Since 1945, all lens surfaces are treated, which means that more of them can be used in the camera lenses without darkening the image captured.

Diaphragm (1840)

In 1840, the diaphragm was just a simple disk with a hole whose diameter was big or small. It was placed in front of the lens to control the entry of light beams into the camera.

Waterhouse diaphragm (1858)

In 1858, a plate with a slit in the mounting between the two optic elements was introduced into rectilinear and two-element lenses. This was the waterhouse diaphragm.

Iris diaphragm (1880s)

Today, modern lenses are equipped with iris diaphragms, popularized since 1886. The iris diaphragm is composed of a set of small plates mounted in the shape of a crown. These move apart from the center and back together again and thus increase or decrease the size of the aperture through which the light enters, just as the iris does in the eye. By reducing the diameter of the light beam, the diaphragm keeps out imprecise marginal rays. It also diminishes the quantity of light, and only lets the parallel central rays through, which gives a clear image. Loss of light is compensated for by an increase in the exposure time.

Automatic diaphragm control (1939)

Developing a diaphragm that opened automatically according to the light received was the work of the Frenchman **Martin**, who patented his invention in 1939.

Photosensitive mediums

Negative (1835)

The first negatives were obtained in 1835 by the English physicist **Henry Fox Talbot** (1800—1877).

The negative, or negative print, represents black areas of the photographed objects as white and vice versa. This allows one to print the desired number of positive prints—that is, photos. In 1835, Talbot obtained after-image drawings. For example, he exposed a piece of lace on a sheet of paper made light-sensitive with silver chloride, which blackens on contact with light. When he exposed the ensemble to light, the paper blackened except where the lace had covered it. Talbot then fixed the image with a concentrated sea-salt solution. By then putting this "negative" in contact with a blank sheet of light-sensitive paper and exposing it again, he obtained a positive print. At the time, Talbot had to spend more than an hour to get a negative.

Talbot's first cameras, built in 1840, were simple wooden boxes on the bottom of which he placed his light-sensitive paper. The camera used an ordinary lens. These devices captured an image; Talbot's wife baptized them "mousetraps."

In 1839, the Frenchman Hippolyte Bayard also began trying to obtain direct positives in the darkroom by using light-sensitive silver paper. Also going through the stage of photographic afterimages, like Talbot, he succeeded in obtaining unique direct positive images in only 130 days, and

A giant photographic camera designed by the American engineer Lawrence in 1918. It weighed over 1200 pounds.

they were of such quality that they became the object of the first photographic exhibition (June 24, 1839).

Daguerreotypes (1839)

The daguerreotype is named after its inventor, the Frenchman **Jacques Daguerre** (1787—1851), who in 1839 perfected this primitive photographic process by which an object's image was fixed on a metallic plate. This achievement crowned the work begun by the inventor of photography, the Frenchman **Nicephore Niepce** (1765—1833), who died before his associate Daguerre produced his first photographs.

The metallic plate was copper and silver-coated and treated with iodine vapor (as well as burnished so that it would have a mirrorlike effect). A light-sensitive surface was obtained in this way. This light-sensitive plate was then placed in the camera at the focal point of the lens, which projected onto it the image of the object to be photographed. After a long exposure time, which excluded portraits, the plate was removed. The image developed when the plate was exposed to the vapors of heated mercury. The photographer fixed the image in a bath of ordinary salt.

Calotype (1840)

In 1840, **Henry Fox Talbot** discovered that after a 20-second exposure to a strong light, an image invisible to the eye was created in the paper. He managed to make it appear by bathing his paper in a wash composed of gallo-nitrate and silver nitrate. This was the calotype, which allowed an unlimited number of proofs to be printed on paper with relatively simple materials.

Albumen-on-glass process (1847)

The albumen-on-glass process,invented in 1847 by the Frenchman **Abel Niepce de Saint-Victor**, the nephew of Nicephore Niepce, consists of a light-sensitive plate covered with an emulsion of egg white, potassium iodide, and silver acetonitrate. Its exposure time is from 5 to 15 minutes.

Collodion process (1849)

In 1849, the Frenchman **Le Gray** light-sensitized his negative materials with a collodion-base preparation, that is, one made of powdered cotton dissolved in ether. Nonetheless, it was the Englishman **Scott Archer** who first promoted the use of this process in 1851. Though the process enjoyed great success, Archer, who had not taken out a patent, died poverty-stricken.

Gelatino bromide process (1871)

It was the English doctor **Richard L. Madox** who first suggested, in 1871, using a gelatin emulsion composed of cadmium bromide and silver nitrate. This light-sensitive mixture was perfected in 1878 by his compatriot Harper Bennet, who lowered the exposure time to 1/25th of a second.

Dry plate (1878)

In 1878, a young American amateur photographer of 24, **George Eastman** (1854—1932), made for his personal use plates covered with a dry gelatin. The damp plates used up to that time had to be coated with an emulsion and then immediately be exposed and developed in this damp state. Dry plates could be exposed and developed at the whim of the photographer. In 1879, Eastman developed an emulsion-coating machine that allowed him to make dry plates in quantity. On January 1, 1881, in association with a manufacturer of coach whips, Henry A. Strong, George Eastman founded the Eastman Dry Plate Company in Rochester, New York.

Celluloid roll film (1889)

The first celluloid roll film to be used commercially was developed by **George Eastman** and marketed in 1889. Already in 1884, Eastman and William H. Walker had invented a container for rolls of negative paper. In 1885, they developed the film Eastman America. Unlike negative paper, it was a thin film that used paper only as a temporary support for the emulsion. The paper was eliminated after development and a thin negative film remained, which was then mounted on glass for the production of prints.

In 1888, Eastman launched the trade name Kodak. He wanted a concise word that could be spelled and pronounced without difficulty in any language. His slogan was already "You press the button, we'll do the rest."

The existence of celluloid roll film made it possible for Thomas Edison, in 1891, to realize the movie camera and projector.

Kodak film-disk (1982)

On February 3, 1982, Kodak introduced a new type of photographic medium: the film-disc, a revolution when compared to traditional film and even cassettes. An ultraflat plastic cartridge contains a film-disk that turns in front of an exposure window situated behind the camera lens.

The material of the Kodacolor HR

A negative print by Talbot (1840).

film-disk, thicker than that of traditional films, is a thick and rigid "Estar."

New photographic emulsion technology (1983)

The American, French, and English research laboratories of the **Eastman Kodak Company** announced in January 1983 that they had invented a new photographic emulsion technology. Modifying the form and sensitivity of silver haloid crystals allowed photographic efficiency to be increased considerably. Thanks to the flattened "T Grain," 1000 ASA color film with a better definition can be obtained.

Color photography (1849)

Direct color photos were first attempted by the Frenchman **Edmond Becquerel** (1820—1891), who succeeded, in 1849, in reproducing photographically all the colors of the light spectrum. But the fixing of direct color photos proved to be unrealizable.

In 1855, the famous physicist James Clerk Maxwell (1831—1879) put forward a theory on photographic reproduction of colors. He laid down the basic principles of the additive three-color process. It involved sending light through colored filters that screen rays of all the other colors (e.g., a red filter lets only red rays through). By adding three colored negatives (red, blue, and green), one manages to reproduce all the shades of color.

In November 1868, the Frenchman Louis Ducos du Hauron obtained patent no. 83061 for his work Les couleurs en Photographie, solution du probleme (Colors in Photography, Solution to the Problem), in which he describes the basic processes of all the modern techniques, such as additive synthesis and subtractive synthesis.

But, in both cases, he had to use the intermediary of the three monochromate plates.

Autochrome plates (1904)

Following the work of Ducos du Hauron, **Louis Lumiere** (1864—1948), the inventor of filmmaking, published in 1904 the result of research that had led him to discover a way to obtain a color photo on one plate, by simply operating in the same way as for black and white. He thus dropped the technique of superimposing three monochromes obtained through selection, and grouped all the properties on a single plate. Autochromes were launched in 1907.

Color films (1912)

In 1912, the German chemist **Hans Fischer** discovered that colors could be formed by oxidation or by "copulation" with other chemical substances. All modern color film processes are based on this general principle.

The Gasparcolor process, invented by Bella Gaspar, is based on the destruction of colorants built into the film. The patent was bought up by Kodak. It was improved by using masks to absorb the parasites and culminated in Eastmancolor in 1950.

Kodachrome, the first color film for amateurs, appeared in 1935. It had been developed by two American musicians, Leopold Mannes and Leopold Godowsky. Since they hadn't very much money, they obtained aid from Kodak, which helped them develop their device. The "copulators," which must be combined with chemical substances in order to bring out all the colors, are found in the development bath and not in the film itself.

16mm Agfacolor, launched in Germany in 1936, and Gevacolor are direct applications of Fischer's system.

Infrared photography (1942)

Kodak was the first to develop a means of photographing the invisible infra- red rays: the Kodak Infrared film, in black and white, in 1942, and the Ektachrome Aero, also in 1942. The latter was the first color film sensitized by infrared rays and was used during the Korean War to detect camouflage. Infrared photography allows atmospheric haze to be better penetrated: Spy planes use it at very great heights. Marine currents can be localized, thanks to it, volcanoes can be watched over, thermal charts can be established and, by covering the flash with a black filter that stops all but infrared rays from penetrating, a subject can even be photographed without his knowledge, in the pure James Bond tradition. In November 1983, the IRAS satellite photographed our galaxy using infra-red rays and transmitted back images that, among other things, allowed a tenth planet up until then invisible to be pinpointed.

Modern cameras

First instant development camera (1948)

In November 1948, the American **Edwin Herbert Land** marketed in Boston the first instant development camera, the Polaroid Model 95. Land had founded, at the age of 28, the Polaroid Corporation, which specialized in making sunglasses and polarizing filters. Installed in Boston in 1937, the Polaroid Corporation prospered and its president reinvested the profits in research.

During photography's early years, and particularly at the time of wet plates, exposure and development were done instantaneously. However, at the time, heavy equipment was necessary and it was very difficult to transport. Around 1880, a small-format professional camera, the Dubroni, met with success. Land's much later invention was truly revolutionary. The Polaroid 95 produced sepia-colored photos. It had a light sensitivity of 100 ASA and development took one minute.

Instant photography (1972)

In May 1972, when **Edwin Herbert Land** described the Polaroid SX-70 to the Society of Photographic Scientists and Engineers, he could have felt that he had fully realized his dream of an "instant" camera. The SX-70 was a pocket camera and lighter than its predecessors. It had a reflex view finder and, thanks to the progress in electronics, was entirely automatic. Finally, a color print was delivered that didn't leave traces on the photographer's hand. The print developed all by itself, in sunlight or shade. The image appeared immediately.

Faced with the Polaroid Corporation's success, Kodak, which in 1947

The American George Eastman became interested in photography while employed as a bank clerk. The cartoon at the right illustrates his passion for photography.

CHRONOLOGY OF COLOR PHOTOGRAPHY

1849	Frenchman Edmond Becquerel succeeds in reproducing all the colors of the solar spectrum photographically.
1855	Scottish physicist James Clark Maxwell lays the basis of the trichromatic additive.
1868	Frenchman Louis Ducos du Hauron describes the process of additive and subtractive synthesis.
1907	Frenchman Louis Lumiere invents autochrome plates.
1912	German chemist Hans Fischer discovers that colorants can be formed by "copulation" with other chemicals.
1935	Kodachrome, the first color film for amateurs, is produced by two American musicians, then taken over by Kodak.
1936	Agfacolor and Gevacolor appear, both deriving from Fisher's system.
1950	Eastmancolor, an improved version of Gasparcolor, invented by Bella Gaspar.

had refused to take an interest in Land's invention, developed an instant development camera that functions according to a similar process.

Automatic focusing

The automation of camera functions has been total since techniques for focusing from a distance have become an integral part of the camera. Different principles exist, the most widely used being that which uses a CCD (charge coupling element). Another system, produced by Polaroid, uses a sonar beam that, when set off, sends out a series of ultrasonic beams. A third system, exclusive to Minolta (1983), uses an infrared ray to automatically determine the distance between the camera and object and takes care of the focusing.

Filmless camera or Mavica (1981)

The Japanese company **Sony** (Tokyo), which created the Mavica-Magnetic Video Camera, presented it to the press on August 24, 1981. This camera uses a magnetic disk and constitutes a revolutionary invention in the realm of photography: An electro-magnetic system has replaced the chemical mixtures; it needs no film and thus neither development nor printing is necessary. It is the size of a traditional reflex camera equipped with a 35 mm. lens. The Mavica can store 50 images in color. The recorded images can be seen immediately on a screen through a viewer. One can then obtain paper copies. This camera takes up to 10 shots per second.

Laser reproduction of slides (1981)

In 1981, the American firm **Laser Color** developed an electronic slide reproducer that functions (without op-

Edwin Land, inventor of the Polaroid, developing an instant picture, his portrait, in 1948.

The 1948 Polaroid 95 and today's SLR680.

tics) according to a laser-read process. The operation is carried out in three phases. The document is screened by three monochromatic laser beams; a reader placed behind the document records the residual light and transforms it into electromagnetic pulses, which are then stocked on diskettes; the computer (programmed according to the characteristics of the film) sends the pulses back to the three lasers that reexpose the new material.

11. Motion Pictures

Magic lantern (1654)

In 1654, the German Jesuit scholar **P. Kircher** had the idea for a magic lantern. It was a small projector for transparent views. The 18th and 19th centuries made considerable use of these lanterns, which enchanted young and old, at home or in public. Their production was industrialized in 1845. The views were hand painted on glass plates. In 1895, these images were printed by decal transfers. Our photo slide projectors are nothing more than perfected magic lanterns.

Fantasmagoria (1798)

In 1798, the Frenchman Robertson, whose real name was **Gaspard Robert**, opened in Paris the first public projections: the Fantasmagorias. He became master in the art of making fantastic luminous images appear in the dark. In a semidarkened room he indulged in a pseudo-incantatory production, then he made the darkness total. At that point, hideous monsters and ghosts swept over the terrified audience, making a horrible racket. In fact, he projected his painted subjects on glass plates behind a screen of translucent percale, which camouflaged the operators and machines. The secret was kept for eight years, during. which time the show had an amazing success.

Thaumatrope (1825)

The English doctor **J. A. Paris** invented in 1825 a remarkable toy that he baptized the Thaumatrope. It was a simple, small piece of cardboard that one rotated between one's fingers with two strings. On each side was a different design, but the two were complementary. For example, a horse on one side and a horseman on the other; when one rotated the piece of cardboard rapidly, the horseman mounted his horse. This little toy used the principle of "persistence of vision" on the retina, without, however, creating movement.

Phenakistiscope (1833)

The phenakistiscope was imagined and constructed by the Belgian physicist **Joseph Plateau** in 1833. For the first time, an image could be made to move. This device is the ancestor of motion pictures. It was a cardboard disk bearing a crown of drawings that constituted a sequence of exposures of a subject: a dancer, for example. The circumference of the disk was crenelated with openings that separated the exposures. The disk was rotated on an axis facing a mirror. This invention allowed Plateau to demonstrate his theory concerning the persistence of vision: "If several objects, differing slightly from one another in form and position, are presented successively to the eye during very short but sufficiently frequent intervals, the impressions they produce on the retina will be linked together without being confused, and one will imagine seeing a single object that gradually changes in form and position." The principle of motion pictures was born.

Plateau eventually lost his sight because he subjected his retina to too many experiments.

Stroboscope (1833)

The stroboscope, akin to the phenakistiscope (see above), was also made in 1833, but it was the Austrian **Simon Von Stampfer** who developed it, independently of the research of Plateau. The principle of the stroboscope is used today to give the appearance of immobility or slowness to objects animated by a rapid periodic movement. Hence, in industry, moving machines can be studied with precision. Nightclubs often use this instrument to "animate" a dance floor.

Praxinoscope (1876)

The praxinoscope, based on the same principle as the phenakistiscope (see above) but more elaborate, is an optical device invented in 1876 by the Frenchman **Emile Raynaud** (1844–1918), who took out a patent on it a year later. Inspired though it was, this toy was limited to reproducing a kind of perpetual movement that was always identical, since it revolved around a maximum of 12 images.

Chronophotography (1874)

In 1874, using a photographic "revolver" of his own invention, the French astronomer **Jules Janssen** obtained a series of negatives of Venus passing in front of the sun. His camera, adapted to a telescope, was baptized the "revolver" by Janssen because of its mechanism, which rotated intermittently. The astronomical "revolver" is the ancestor of all our movie cameras.

After years of work the English-American photographer Eadweard Muybridge succeeded in demonstrating, in 1878, that a galloping horse does indeed lift all four legs completely off the ground. (The horse set off a series of cameras as it galloped along the track.)

In 1882, the French physiologist Jules Marey (1850–1905) developed chronophotography by constructing a repetitive-action "photographic gun" that allowed him to take successive shots of a bird in flight. At the seashore in Naples, Marey chronophotographed the flight of sea gulls. Peasants who saw him spoke about a crazy man armed with a gun who aimed at birds without ever shooting and seemed delighted to come back from hunting empty-handed.

Fixed plate chronophotograph (1882)

Since the images given by his chronophotographic "gun" (see above) were too small and since the inertia of his plate was difficult to control, Marey realized, in the same year (1882), the fixed plate chronophotograph. This same machine allowed one to analyze a man's walk, the fall of a body, the high jump, boxing, the flight of birds, etc. But though the speed of the shutter's rotation could be regulated, the images often overlapped. Marey fixed this by inventing chronophotographic flexible film.

Tachyscope (1889)

In 1889, the German **Ottomar Anschutz**, from Lissa, invented the electric tachyscope, which was the percursor of modern stroboscopic photography.

Kinetograph (1889)

In 1889, the American **Thomas Edison** (1847–1931) and his team invented the first sound movie camera, the kinetograph. This first movie device had sound because Edison, inventor of the phonograph, had taken that as his point of departure. This device, constructed in total secrecy, gave birth to the kinetograph, which was patented in 1891. It used perforated 35 mm. film, approximately the same as that used today. W. K. L. Dickson, a collaborator of Edison, shot the first films for him in the world's first studio. It was more or less a shed lit by large windows and constructed on a pivot. The whole thing turned, following the

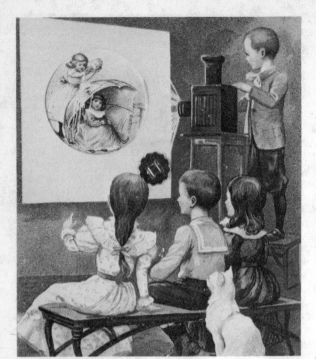

Above and left. Viewers intrigued and entertained by the magic lantern. Below. A Fantasmagora puppet.

path of the sun; thus the scenes, played inside, received a constant flow of light. The first films shot in this studio were shown on the kinetoscope (see below).

Kinetoscope (1891)

Invented by the American **Thomas Edison**, who patented it at the same time as the kinetograph (see above), in 1891. The kinetoscope was marketed in 1892. It was a sort of viewer to show the films shot with the kinetograph. A single spectator could see a scene of animated images on a continuous loop. But this system admitted too little light for it to be suitable for projection purposes. This is why the Lumiere brothers outdistanced Edison when they projected a large luminous image on a screen. However, Edison had adopted his formula for commercial reasons: He believed private showings to be more profitable potentially and thought he could sell more kinetoscopes than projec tors.

Luminous pantomimes (1892)

On October 28, 1892, three years before the Lumiere brothers' historic show, the Frenchman **Emile Reynaud** gave the world's first animated show: the luminous pantomimes. Seated in front of the screen in total darkness, the spectators saw the decor light up in such a way that they took it to be a magic lantern image. But suddenly luminous characters appeared on stage and started gesticulating: Pierrot, Arlequin, Colombine, they were all there, acting like brilliant and unreal little marionettes. Sound effects accompanied the scenes.

The decor was projected the wrong way around onto the screen by a magic lantern. As for the characters, Reynaud lightly painted all of their successive positions on a long perforated translucent strip. It was the world's first animated cartoon. Reynaud was also the originator of the perforated strip that today's cinema uses. However, the artist refused to use photographic processes because he thought his drawings were more beautiful.

Emile Reynaud, a fertile inventor, had to compete with the cinematograph (see below). He was bound to the Musee Grevin by a draconian contract and wore himself out at work. In despair, he destroyed his machines and died in misery at the Ivry hospice in 1918.

Cinematograph (1893)

In 1892, **Bouly**, a French engineer, invented and patented a sort of movie camera. It was an instant development camera that automatically took negatives that captured sequences of movement. The cinematograph (1893) was the second device that he realized. Like the first, it allowed shots to be taken. Moreover, it could be used to project the negatives that it had recorded. This was the beginning of the cinema. Unfortunately, Bouly's shots have been lost. The Lumiere brothers immortalized Bouly's device by using its name for their invention.

Movie projector (1894)

Credit must be given to the American **Le Roy** for having invented the first projector that brings together the essential elements of our current models. On February 5, 1894, at 16 Beekman Street, New York, he publicly showed two films from Edison's kinetoscope.

Lumiere brothers' Cinematographe (1895)

On February 13, 1895, the Lumiere brothers, Louis (1864—1948) and August (1862—1954), took out a patent for "a device for obtaining and viewing chronophotographic prints." They baptized it the Cinematographe. It was a black box with a mechanism that guaranteed the intermittent pull-through of 35 mm. perforated film and the rotation of a shutter. The device, equipped with a shooting lens, recorded images during the instants when the film had stopped (15 times per second), with its shutter open.

During the dead time, with the shutter closed, the film changed position. A series of negatives was thus obtained. Then the lens was taken off and the camera was loaded with the negative sequence, which was put in contact with blank film. It was faced toward the light, which exposed the positive film.

The loaded camera was then equipped with a projection lens and placed in front of an electric lantern. The beam of light went through the film and the lens. In the dark, the camera became a projector. On the screen, a succession of immobile images unrolled, separated by moments of black. However, the images lingered on the retina and the spectator saw an animated image. This illusion was, and still is, motion pictures.

The first public showing occurred on December 28, 1895, in Paris. This new form of entertainment attracted crowds, and its success permitted the growth and development of the motion picture industry. One might add to the Lumiere brothers' glory in mentioning that their films, which were the first moving picture films, had real artistic value: The very first, the entry of a train into the La Ciotat station, was taken at such an angle that the spectators were terrified.

Movie camera

The Frenchman **Charles Pathe** (1867—1957) broke down the Lumiere brothers' Cinematographe into two distinct elements (the camera and the projector) and made them independent. Around 1904, the camera's speed was made variable in order to allow speeded-up and slow-motion films. Sixteen images per second were used for ordinary shooting, but 24 per second were necessary for sound films (see below).

Sound movies (1902)

In 1902, the Frenchmen **Baron** and **Gaumont** were the first to develop a sound reproduction system that was synchronized with an animated projection. But the film in itself remained silent.

Contribution of the phonograph

Since sound recording had been realized in 1877 by Edison's phonograph (*see* Phonograph, p. 93), the problem consisted in rigorously synchronizing the sound with the projection of the film. When Edison tried to couple his first movie camera (*see* Kinetograph, p. 101) with a built-in phonograph in 1889, the synchronization was not successful. The Pathe process, designed by the engineers Joly and Mendel, only tried to make a phonograph work in the movie theater. The phonograph's sound was transmitted by telephone to the projectionist, who, sitting in his booth, only had to turn the handle according to the rhythm he heard.

Emil Reynaud presents his Optical Theater and the first animated cartoon.

Automatic electric synchronization (1902)

In 1902, the Frenchmen **Baron** and **Gaumont** built an automatic electric synchronization system using the phonograph and the projector, but the phonographs were not powerful enough. Sound amplification was a difficulty resolved in 1910 by the Frenchmen Decaux and Laudet for the Gaumont company. They modulated a flow of compressed air. In 1912, this process was used in Gaumont's large movie theater in Paris. At the time, projectors had record players situated directly behind them and the sound and the image were me-

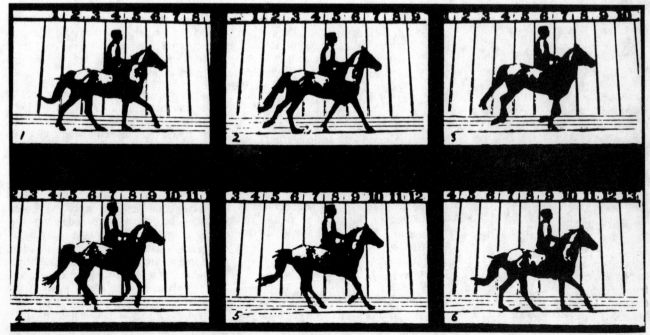

In 1882, Muybridge takes pictures of each movement of a galloping horse.

chanically synchronized. In the front a sort of block was adapted to play back the sound for photographic sound track films, the only system which was to survive. (See Photographic reproduction of sound, p. 000.)

The first "talkie" was *The Jazz Singer*, an American film produced by Warner Brothers in the United States in 1927. Several of its sequences had sound added by means of 33 rpm records.

Optical Sound Track (1929)

In 1929, the first truly talking film came out: *Hallelujah*, an American film produced by Metro-Goldwyn-Mayer and directed by King Vidor. On the film itself, next to the images, an optical soundtrack was engraved through photoelectric processes.

Slow motion (1904)

Two students of Jules Marey, the Frenchmen **Lucien Bull** and **Henri Nogues**, invented slow motion filming in 1904. They understood that by shooting more frames per second than one normally would shoot, one obtains a longer film. Then, projecting this longer film at normal speed,

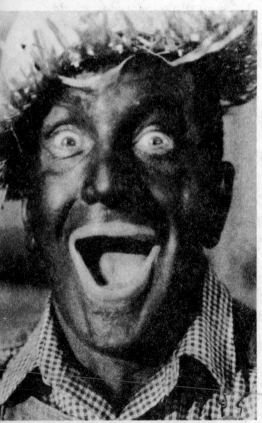

Al Jolson in the first sound film, The Jazz Singer.

King Kong's image has become familiar to generations of film-goers.

events are seen to take place less rapidly than they do in reality. That's slow motion. In 1904, Bull, under-director of the Institut de Nancy, succeeded in taking 3000 images per second. Later on, Nogues and Bull thought up a new process: The film is immobilized and scanned by the subject's intermittent image, which is lighted by a succession of high frequency electric sparks. Hence, at 25,000 images per second, they obtained a series of photos of a rifle bullet going through a glass bulb. Such films are rarely destined to be shown. Rather, they are studied image by image, particularly for industrial research. Progress in electronics now allows one to take several million images per second.

Wide-screen (1927)

In 1927, the French physicist **Henri Jacques Chretien** (1879—1956) devised a hypergonar whose patent, bought by the French company Pathe-Nathan in 1931, was bought up again in 1952 by the American company 20th Century-Fox, which called the process CinemaScope. The first wide-screen film was *The Robe* (1953).

The process is simple. An anamorphic lens squeezes the image—much in the same way as do certain distorting mirrors—to half its normal size. At the time of the projection, the lens works in reverse and "unsqueezes" the image so that it is seen in its proper proportions. The process has been perfected, thanks in particular to the use of a 70 mm. anamorphosized film (since 1963).

Movies in living color

The stencil process (1900)

The first color films were hand painted, image by image. Then, starting in 1900 and until World War I, Melies and **Pathe** used a mechanical exposure technique: the stencil. The colors have remained fresh on the rare films that have come down to us.

Additive processes

Additive processes consist of combining three primary colors (blue, green, and red).

• *Lenticular system*. The most perfected additive process stems from an invention patented in 1908 by the Frenchman **Berthon**. First developed by the French company Keller-Dorian, it was bought up in 1925 by Kodak, which derived Kodacolor from it in 1928. The Agfa company launched a comparable product in 1932, Agfacolor. But it was necessary to use complex equipment and several filters, and a lot of light was lost.

• *Successive projection of three monochromes*. Numerous inventors became attached to a process consisting in projecting three monochrome images in succession and playing on the persistence-of-vision phenomenon. Unfortunately, the image changes position after each color, so a precise geometric super-imposition of the three images cannot be realized.

• *Simultaneous projection of three monochromes*. The process of simultaneous projection of three monochromes had a certain success. Pagnol's film *La Belle Meuniere* was shot in Rouxcolor in 1948, the very year that the Roux brothers invented this process. It did not offer very satisfying results.

Subtractive processes

The subtractive processes are based on the principle of varying light absorption by colored layers. The film is thus made up of images in three primary colors.

• *Technicolor*. This idea goes back to an invention by the Frenchman Leon Didier in 1903, but the production of matrices had already been conceived by the American Warnake in 1881, and it was the Americans who realized the first color film.

Technicolor can be compared to a three-color process exposure. A simple gelatin film is exposed successively to three matrices that contain coloring elements. The real problem consists in superimposing the three primary images with precision. The American Herbert Kalmus set up the first Technicolor laboratory in 1915, in a wagon at the Massachusetts Institute of Technology, in Cambridge, Massachusetts.

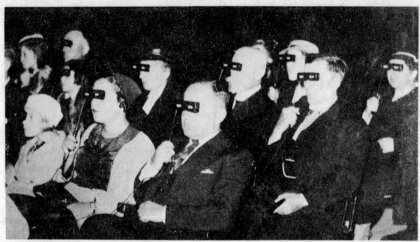

Spectators at a 3-D movie.

Trick photography and special effects

Since the very beginning of the cinema, its creators have resorted to trick photography and special effects to evoke the dream and the imaginary. The processes used have been so numerous that it is impossible to mention them all. Several names and several films, however, highlight the history of the cinema, among them notably the Frenchman Georges Melies (1861—1938), a veritable magician of the cinema who had a considerable number of productions, including *Le Diable au Couvent* (1899), *Les Voyages de Gulliver* (1902), *Le Palais des Mille et Une Nuits* (1905), *La Fee Libellule* (1908).

Animated cartoon (1893)

The first drawn images that one saw move were those of the phenakistiscope (*see* Phenakistiscope, p. 101) in 1833. Around 1850, the Austrian Von Uehatius and the Frenchman Duboscq drew, in the same manner, a series of successive poses arranged in frames set in a crown shape in a glass disk. But one always saw the same gesture repeated. **Emile Reynaud**, in 1893, was the real pre-cursor of the animated cartoon (*see* Luminous pantomimes, p. 102).

The first film shot in Technicolor used a two-color process. It was *The Gulf Between,* with Grace Darmond and Niles Welch, in 1917. The first film shot in three-color and in Technicolor was Walt Disney's animated cartoon *Flowers and Trees,* produced in 1932. The first full-length film shot in three-color and in Technicolor was *Becky Sharp,* realized by Rouben Mamoulian in 1935.

3D (1894)

Starting as early as 1894, attempts were made in the direction of the 3D cinema. Efforts at developing a lenticular system (*see* p. 104) were very quickly abandoned, except in the Soviet Union, where experiments were tried out in the forties.

The anaglyphic principle was considered to be more promising: the superimposition of two images, one green, the other red, viewed through bicolored glasses, giving the illusion of a third dimension, depth. It was in the thirties that the first commercial 3D films were produced, in the form of "shorts."

But it was the Polaroid system, based on the principle of light filtering, that sparked off the great epoch of the 3D cinema. The first full-length film was released in Germany in 1937. The Gunzberg brothers were to import into the United States their application of the process developed by Dr. E. H. Land, which permitted producer Arch Oboler to shoot *Bwana* in 1952. The film was shot with two cameras; the two reels had to be projected by two machines that were strictly synchronized with each other. And, of course, the spectators had to wear Polaroid glasses. A good hundred films were produced in this way.

In 1966, Arch Oboler shot *The Bub-*ble with the use of a special lens, the polarisator. This device did away with the inconveniences of double projection. Today, even if there is a return to 3D films, the future seems to belong to the holographic process.

The heroes of Star Wars; C3PO and R2D2.

Film cartoon *(1908)*

J. Stuart Blackton and **Emile Courtel** (called Emile Cohl) invented the animated cartoon. In 1908, Cohl showed a 36-meter-long film entitled *La Fantasmagorie*.

In 1914, the American Earl Hurd invented a system that avoided redrawing the decor each time. The fixed parts were drawn on a piece of cardboard and the mobile parts were drawn on transparent sheets of celluloid called "cels."

In 1915, an American caricaturist created a comic character nicknamed Mentoultant. American productions continued to be extremely rich and they attracted most of the creative artists working in this sector. Among the most famous are Pat O'Sullivan and his *Felix the Cat*, the Max brothers and Lou Fleisher, who created the vamp *Betty Boop*, and, of course, *Popeye the Sailor* by Max Fleisher and *Mickey Mouse* by Walt Disney (1927).

Animation by computer *(1960s)*

It was at the beginning of the sixties that animation techniques by computer were first experimented with, mainly for space research. The cinema got hold of it some twenty years later, notably for the shooting of the film *Tron* (1982) by Steven Lisberger. To start with, an artist draws an object from three different angles— from above, below, and from the side. Then the three drawings are programmed in a computer, which records the dimensions of the object and puts it on scene in space, in perspective, by means of a control screen. A different piece of information is given to each of the luminous points (of which there are more than two million) determining the color and texture of the object. The actors work on a stage that is completely empty and bare, which will later be furnished on film image-by-image by the computer.

Oscars (film awards) *(1928)*

In 1928, the Americans **A. Zukor**, **L. B. Mayer**, and **C. Laemmle** created the Academy of Motion Picture Arts and Sciences. Since then, this academy has bestowed awards every year on the best directors, artists, productions, and technicians. When the trophy's sculptor presented his work, L. B. Mayer's secretary exclaimed, "Oh, it's amazing how much he resembles my Uncle Oscar!" And the famous award was so baptized.

media and communications

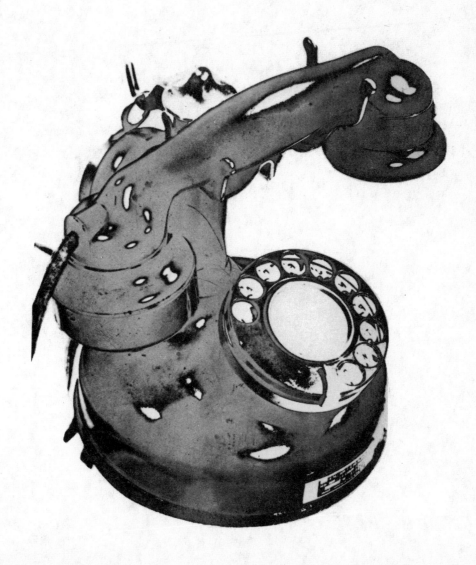

RADIO

Radio waves

Radio broadcasting, formerly called wireless, operates by means of electromagnetic waves (called radio-electric waves where radio is concerned) which travel though space and carry messages. However, it was first necessary to discover these waves; this was the work of James Maxwell; next, it was necessary to know how to generate them, Heinrich Hertz's discovery; and finally, one had to know how to receive them, Edouard Branly's discovery.

Maxwell's equations (1865)

The English physicist **James Clerk Maxwell** (1831—1879) demonstrated that light is the result of electromagnetic vibrations of a certain wave length. His theory was put forth in 1865 and led to the study of the propagation, reflection, and diffraction of light. It showed how electromagnetic waves other than those of light could be propagated.

The Edison effect (1883)

In 1883, **Thomas Edison** (1847—1931) invented the first incandescent light bulb (see Incandescent bulb, p. 170).

This bulb reveals the Edison effect: A metal heated until red-hot emits an electron cloud. Radio tubes would make use of this effect. Until the appearance of radio tubes, the wireless system operated on an all-or-nothing basis. A uniform signal was, or was not, transmitted, and this was why the Morse code was the only available means of transmission.

The radio tube would permit the transmission of sounds.

Stereophonics (1881)

The first stereo transmission took place at the initiative of the aviation pioneer **Clement Ader**, during the first World Electricity Fair in Paris, in 1881. For the occasion, Ader had invented a stereophonic telephone system. Each evening it allowed an enthusiastic public to hear the programs presented at the Paris Opera, 3 kilometers away.

Stereophonic and high fidelity recording had their origins in the work of Western Electric and Bell Laboratories in the 1920s during research on sound motion pictures. In 1931, Bell Labs developed the gold-sputtering technique for making master discs, and in 1933 the company performed one of the first demonstrations of high fidelity and stereophonic sound—recordings of concerts and rehearsals of the Philadelphia Orchestra conducted by Leopold Stokowski.

In 1931, the English physicist A.D. Blumlein began to develop the modern stereophonic system.

Wave oscillator, emitter, detector (1888)

In 1888, the German **Heinrich Rudolf Hertz** (1857—1894), the 31-year-old professor of physics at the Karlsruhe Polytechnical School, detected and made use of electromagnetic waves. After learning of Maxwell's theories, Heinrich Hertz made a rudimentary emitter. He was interested in the following phenomenon: A Ruhmkorff coil, which generates a high voltage current, causes sparks at the terminals of a ball spark gap, and these sparks induce oscillations, which give birth to waves. In order to demonstrate these waves, Hertz built his "resonator," which was a simple conductor ring with a gap in it, placed

By 1945, balloon antennas made possible the monitoring or jamming of radio broadcasts.

rather close to the Ruhmkorff coil. The waves induced current which was observed in the appearance of sparks at the gap point. Hertz thus verified all of Maxwell's theories. For the first time, waves, subsequently called hertz waves, were made and then detected at a distance: 4 to 5 meters maximum.

Radioconductor or coheror
(1888)

In 1888, the Frenchman **Edouard Branly** (1844—1940), doctor of science, doctor of medicine, and professor of physics at the Catholic Institute in Physics, discovered radioconduction.

After two years of research and fine-tuning, he presented his radioconductor, also called the filing tube, to the Academy of Sciences on November 24, 1890.

A property of this apparatus allowed it to become a conductor of an electric current when acted upon by electromagnetic waves. It was made up of iron filings contained in a glass tube between two electrodes, and from a distance it set off electric bells, telegraphic relays, etc. Using this very sensitive apparatus, Branly demonstrated that wave detection was possible from a distance of several dozen meters and through solid objects. The wireless telegraph was born. Branly witnessed the developments growing of his research: He was lucky enough to see the birth of and progress in radio broadcasting and the beginnings of television.

The Englishman Oliver Lodge renamed the apparatus "coheror," the name by which it is known today.

Syntony (tuning) of circuits
(1894)

In 1894, the Englishman **Oliver Lodge** (1851—1940), professor at the University of Biringham, introduced a new idea, that of tuning. Lodge wanted to verify and reperform Hertz's experiments. He was in possession of Branly's coheror, which was more sensitive than Hertz's resonator. Today it seems obvious that the receiver must be adjusted to the wave length of the transmitter that one is trying to receive. By applying the work of Lord Kelvin, Lodge first carried out this tuning. (It was called syntony for some time.)

As a result, the sensitivity of the reception increased and the quality of the transmission received was improved, even before tuning became indispensable for selecting the desired transmission because of the increased number of transmitters.

It would not be fair to omit mention of the Croat Nicolas Tesla (1856—1943), who carried out most of his work in the United States. A pioneer of alternating current, then of high voltage current transformers, he concentrated his efforts on the transmission of energy via waves. Like Lodge, he was a promoter of syntony.

The antenna (1895)

In 1895, the Russian **Alexander Popov** (1859—1906) invented the antenna.

In order to improve both transmission and reception, radio waves must be diffused into space and then received under optimal conditions. The antenna (or oversized conducting wire) improves reception.

In 1894—1895, Popov, assistant professor at the Cronstadt Naval Academy, used the methods of Branly and Lodge to detect approaching storms. He noted that sensitivity was greater when one held up a large vertical wire to receive the waves created by lightning. A lightning rod was the first antenna.

The correspondent of numerous scientists and researchers, Popov built a

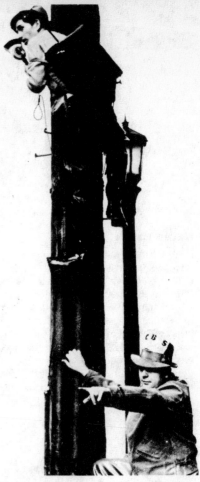

A reporter during the heyday of radio news.

receiver that he applied to telegraphy. The first radio-electric Morse connection over a rather great distance (250 meters) was carried out by Popov: *Heinrich Hertz* were the first words transmitted, on March 24, 1896.

While Alexander Popov is truly the inventor of the antenna it should be noted that on January 12, 1891 (four years earler), Edouard Branly had noted that by equipping his apparatuses with long metallic stems, reception was improved.

FIRST RADIO BROADCASTS IN THE UNITED STATES	
Jan. 13, 1910	Demonstration by Lee De Forest: retransmission from the Metropolitan Opera House in New York and his home of Caruso singing *Cavalleria Rusticana*.
Oct. 1915	The Western Electric Company sent a transmission from Arlington, Virginia.
Nov. 1920	In Pittsburgh, a small station belonging to the Westinghouse Company broadcast the presidential election results.
July 2, 1921	In Pittsburgh, the RCA station broadcast the boxing match of the century: Carpentier vs. Dempsey.
1922	AT&T became the first company to sponsor a radio broadcast and operated radio station WBAY, later called WEAF.
1923	The first radio network was formed when Boston station WNAC and New York City's WEAF were connected by wire to broadcast the same program.

Crystal set (1910)

In 1910, the work of two American researchers, **Dunwoody and Pickard**, on crystals led to the invention of the crystal set, which was the first radio. Galena is a lead sulfur crystal that, combined with some simple elements, permitted thousands of amateurs to build their own wireless sets and to receive the first radio broadcasting transmissions.

Diode (1904)

Marconi's collaborator **Sir John A. Fleming** (1849—1945), created the diode in 1904.

He placed a plate in front of a heated wire (the filament) in a vacuum tube. The diode was the first radio tube, but it did not much advance the wireless sets of the time.

Radio tube (triode/Audion) (1907)

Lee De Forest (1873—1961), an American, invented the first triode in 1907; he names it the Audion.

He introduced a grill between the filament and the diode plate. The Audion thus comprises a heated filament, which becomes luminous and emits electrons; a metallic plate, which attracts them (the anode); and finally, between the filament and the plate, a metallic wire, in spiral, zigzag, or mesh form, which is the auxiliary electrode. With the Audion, the radio was equipped with an extremely sensitive apparatus. Due to the auxiliary electrode, it became possible to gauge transmission power with precision and thus to transmit voice vibrations, music, and other sound in all their subtle nuances.

Lee De Forest's invention is the basis not only for radio but also for television, radar, and the first computers.

Variable frequency receiver (superheterodyne) (1917)

In August 1917, and in October 1918, the Frenchman Lucien Levy (1891—1965) patented his superheterodyne.

At the time, it was very difficult to achieve proper receiver adjustment, since receivers were equipped with a great many buttons and tuning in to a different frequency involved complicated manipulations. Levy's invention constituted a great step forward. The superheterodyne, or variable frequency receiver, allows the search for stations via a single button. It considerably simplifies receiver adjustments and also facilitates their manufacture.

Presently, 99% of radio and television receivers, radar connections via satellite, etc. operate employing the principle of the superheterodyne.

Car radio (1922)

The invention of the car radio can be attributed to the American **George Frost**, who, in 1922 at the age of 18, installed a radio in a model-T Ford. The first industrially produced car radio was the Phildo Transitone, manufactured by the Philadelphia Storage and Battery Company, in 1927. By 1993, 100,000 cars had been equipped with car radios in the United States.

Frequency modulation (FM) (1933)

The American **Edwin Armstrong** began studying the principle of frequency modulation in 1925. This consists in modifying the frequency (or wave length) of a transmission to adapt it to the rhythm of sound variations.

He studied the principal transmission and reception circuits in 1933 (his first patent was taken out on January 24, 1933) and he demonstrated that noise can be decreased by increasing the frequency band (contrary to what takes place in amplitude modulation).

In 1938 General Electric installed the first transmitter of this type in Schenectady, New York, and this was followed by the "Yankee Network." FM was also adopted in 1939 for police-patrol radios in the United States.

The transistor (1947)

A new era in radio technology began in 1947 when three AT&T Bell Laboratories scientists—**John Bardeen, Walter Brattain and William Shockley**—demonstrated the results of the work that would win them the Nobel Prize in physics in 1956. They had invented the transistor, which replaced the vacuum tube, revolutionized the field of electronics and founded whole new industries.

Transistors detect, amplify, rectify and switch currents. They are tiny (millions can be mounted on a fingernail-size microelectronic chip), relatively cheap, and use very little power. These properties have made possible modern computers, space flight, hearing aids, electric guitars and many other common electronic devices.

The first "transistor" radios appeared in 1955.

TELEVISION

Reproduction of fixed images by means of pendulum (1843)

On November 27, 1843, the Englishman **Alexander Bain** (1810—1877) applied for a patent for his "Method for transmitting copies over a distance, by means of electricity." This system of reproducing a fixed image from a distance was a precursor of television.

A design, drawn in a nonconducting ink on a special conducting piece of paper, was traced by a stylus that descended slowly and was activated sideways by a pendulum. When the stylus pressed over the ink, the current was temporarily interrupted.

A wire carried the electrical current from the first pendulum to a second, identical pendulum. This second pendulum was also equipped with a stylus, which, governed by the same current, reproduced the zigzag movement of the first. The stylus was applied to paper impregnated with potassium ferrocyanure; the paper darkened upon passage of the electrical current and thus reproduced the initial design.

Theoretically the Bain system was workable. Unfortunately, although the two pendulums were identical, they were not synchronized. Thus, transmission of the design was unpredictable, and Bain's trials did not meet with success.

Synchronization of transmission and reception

In 1848, the Englishman Bakewell attempted to use signals in order to synchronize the two pendulums. The image to be analyzed was placed on a rotating cylinder which moved sideways along a master screw. At each turn of the cylinder, a synchronization stop put the cylinder receiver in phase with the transmitter. These experiments were not successful either.

Pantelegraph (1862)

In 1862, an Italian physicist, **Abbe Giovanni Caselli** (1815—1891), was really the first to succeed in reproducing fixed images that had been transmitted over a distance. He added a simple but very efficient synchronization system to Alexander Bain's pendulum. His pantelegraph was used by the French postal authorities between 1865 and 1870 to transmit fixed images from Paris to Marseilles.

Selenium

In 1817, the Swede Berzelius discovered selenium. Then, in 1873, two English telegraph engineers, May and Smith, working in Valentia, Ireland, demonstrated the resistance variations of selenium under the influence of light. Selenium is the basis for the manufacture of photoelectric cells.

Since May and Smith's discovery, researchers have had a means of transforming luminous signals, and therefore images, into electrical signals.

Nipkow disc, electric telescope (1884)

The German **Paul Nipkow** (1860—1940) was a student when he patented his electric telescope in 1884—he had discovered how to divide images into lines.

A perforated disc, laid out in spiral fashion, turns in front of the image to be analyzed and successively reveals all the points on a line basis. Electrical switching was no longer necessary: Here, optical analysis was the governing technique and one cell was sufficient. This device, called the Nipkow disc, was the foundation of mechanical television and was used until 1935.

First experiments with mechanical television (1923)

The Scottish engineer **John Baird** (1886—1946) was one of the pioneers

in television. In 1923 he applied for a patent based on the use of the Nipkow disc and performed his first mechanical television experiment (with eight lines).

Mechanical television uses a Nipkow disc with a photoelectric cell (or one of its derivatives) to record images; and another to reconstitute the images(receive). The second Nipkow disc is in synchrony with the first and connected to a neon light.

In 1927, the Bell System sent live TV images of Herbert Hoover, then Secretary of Commerce, by telephone lines from Washington, D.C. to New York City—the first public demonstration in the United States of long-distance television transmission. A narrow beam of light rapidly scanned Hoover's face through tiny apertures in a rotating disc. The light reflected from his features was picked up and converted into electrical signals by specially designed photoelectric cells. These signals, corresponding in intensity to the reflections of the fright and shaded facial areas, were then transmitted to New York, where the signals were displayed in rows of tiny dots across the face of a neon glow lamp used as a picture tube.

In 1928, Baird transmitted images of London to New York via short waves (45 centimeters), 34 years before the Telstar satellite transmitted images across the Atlantic. In 1930, he marketed the Televisor, the first general public television; it had very few buyers.

Video camera (1923)

The video camera converts the images captured by a lens (which is iden-

tical to that of a photographic camera) into electrical signals. The first video camera (or television) was the result of the work of the naturalized Russian-American **Vladimir Kosma Zworykin** (1889—1982). He arrived in the United States in 1919. Zworykin conceived of an electronic analysis procedure that led to the creation of the iconoscope. (For this he well deserved the title "father of television" that was subsequently accorded him.) After having applied for a patent for the procedure in December 1923, Zworykin performed an initial demonstration of his invention in the RCA electronic research laboratories. This led to RCA's making the first experimental transmission of electronic television in 1933 from a transmitter installed at the top of the Empire State Building in New York, with a picture resolution of 240 lines.

Working at the same time as Zworykin, the American engineer Philo Farnsworth (1906—1971) invented an analysis tube, the "Dissector," which used the principle of the Nipkow perforated disc electronically. Less sensitive than the Zworykin iconoscope, which used a photosensitive pattern, the Dissector was perfected by another American, Allen du Mont, but it was abandoned at the end of the 1930s because it was not as good as the igonoscope.

COLOR TELEVISION

Origins (1928)

In 1928, the Scot **John Baird** built a color television. Here again, mechanical television was involved. The first

The "vidiwall" of Hachette-Opera, Paris, 1984.

FIRST IMPORTANT TELEVISION BROADCASTS

1927	USA	First long-distance TV transmission, between New York City and Washington, D.C.
1928	Britain	John Baird transmits images of London to New York, via short waves
1930	Britain	Beginning of regular BBC television transmissions
April 14, 1931	France	First demonstration of mechanical television at the Ecole Supérieure d'Electricité of Malakoff (with studios in Montrouge, in the suburbs of Paris) by René Barthélemy
1931	France	Transmission from Toulouse to Le Havre, by Henri de France
Nov. 1935	France	First regular transmissions by the PTT (Post and Telecommunications Authority), transmitted from atop the Eiffel Tower
1936	Britain	First BBC 405-line programs
Sept. 1937	France	Beginning of daily 455-line transmissions
1940—1944	France	Cessation of French transmission
1946	USA	First TV transmission by coaxial cable, New York to Washington
Nov. 20, 1948	France	Picture resolution set at 819 lines (system of H. de France)
1949	Belgium	RTB establishes four channels
1949	Italy	RAI produces the first transmission in Turin (625 lines)
1950	USA	The Bell System design and builds the first mobile pickup unit
April 25, 1950	France	Beginning of regular transmissions using 819 lines; transmitter atop the Eiffel Tower. Installation of a transmitter on the belfry in Lille. Installation of the first Paris-Lille hertz beam
1951	USA	The first coast-to-coast TV transmission, a Bell System telecast of the Japanese Peace Conference in San Francisco
June 25, 1953	Europe	Birth of Eurovision: transmission of the coronation of Queen Elizabeth II of England (images received in Great Britain, France, Germany, Belgium, the Netherlands, Denmark)
1956	USA	The first video recorder is unveiled
1962	USA	First US-Europe link-up via the Telstar satellite
April 1964	France	Inauguration of the second channel (625 lines)
July 21, 1969	USA	First moonwalk by American astronauts (Armstrong and Aldrin) in Worldvision: 600 million viewers
Dec. 31, 1972	France	Inauguration of the third channel (625 lines and color)

demonstration took place at 133 Long Acre, in London, in Baird's studios. One saw a policeman's helmet, a man sticking his tongue out, a bouquet of roses, red and blue scarves, and a lit cigarette.

NTSC procedure (1954)

In 1954, Bell Labs scientists on the National Television System Committee helped develop commercial standards for practical nationwide color telecasting.

SECAM procedure (1956)

Henri de France developed the SECAM (sequential color with memory) procedure in 1956. It was adopted in France, and the first SECAM color transmission between Paris and London took place in 1960.

Ampex engineers in 1954, working on one of the first videorecorders.

In the early 60's, pay television invaded bars and taverns.

PAL procedure (1962)

In 1962, a German, Professor Walter Bruch, developed a PAL (phased alternate line) system. The first transmissions of color television using the PAL procedure began in Germany in June 1967.

The three procedures for color television (NTSC, PAL and SECAM) are all based on the same principle, using cathode tubes made up of a multitude of juxtaposed elements that give red, yellow, or blue light via fluorescence. The tube comprises three electronic "cannons," each of which produces only a single color, due to a system of perforated masks.

Trinitron color tube (1968)

The Japanese company Sony developed Trinitron color tubes in 1968. The principle had previously been studied by Thompson.

Only one electronic tube and the center of a wide lens are used. There is also a grid for the selection of colors, which permits a greater number of color beams to reach the screen in a more organized manner.

ADVANCES IN COLOR		
1928	Britain	John Baird experiments with a Nipkow disc comprising three sets of holes equipped with colored filters.
1929	USA	Bell Labs develops a system for color TV.
1937	France	The engineer Valensi defines the principle of "double compatibility," which all current procedures follow—color television sets must be capable of receiving unmodified black and white broadcasts, and black and white sets must be capable of receiving color broadcasts in black and white.
1949	USA	David Sarnoff develops the first color tube, the Shadow Mask, in the RCA laboratories.
1951	USA	CBS develops a noncompatible sequential system.
September 1951	France	First experiment with color television at the Surgical Congress in Paris
1954	USA	Adoption of the NTSC system
1954	USA	The first network color telecast, showing the Tournament of Roses parade
1956	France	Henri de France presents the SECAM procedure.
1960	France	First SECAM transmission between Paris and London
1962	Germany	Walter Bruch develops the PAL procedure.
October 1, 1967 (2:15 P.M.)	France	First color broadcast
1967	Germany	First PAL broadcast

VIDEO

Video is to image what audio is to sound. An electric video signal contains all the information necessary to modulate an electron beam in a cathode tube and make it sweep from one side of the tube (or screen) to the other in order to make an animated black and white image appear. This is also true of the tri-chrome cathode tube (having three tubes and therefore three electron beams), which allows an animated color image to be obtained. Similarly, an electrical audio signal is converted by an intermediary mechanism (such as a loudspeaker) and heard as a sound signal.

In general, the video signal carrying the image is accompanied by an audio signal, so that the pictures one sees are not silent pictures. Sound registration on engraved discs commercially preceded the use of magnetic tape, but the opposite was true for recording images—images were first recorded on and read off a magnetic tape by means of the video set before being fixed on videodiscs.

Video recorders

First video recorders

The first attempts to record images on magnetic tape for professional purposes were undertaken at the beginning of the 1950s. At that time, several American companies used as inspiration the data provided for the first videotape recordings. These appeared in the United States in 1948, and work was subsequently begun to develop longitudinal recording systems. Such systems required unwinding speeds that were high in relation to the fixed magnetic heads—on the order of 10 meters per second—and this required a prohibitive quantity of tape and an excessive coil weight. Nonetheless, this was the choice of Mincom, a subsidiary of Scotch 3M (which in 1951 was the first to demonstrate recorded video images), of RCA (which did the same thing in 1954), and of Bing Crosby Enterprises. In each case, the demonstration was inconclusive as regards the quality of the reproduced images. This was due to several causes, in particular:

• Difficulty in obtaining constant tape speed, with fluctuations incompatible with the precision required for the reproduction or recording of a television signal;

• The necessity for a video signal to cover at least 18 bytes in order to provide an acceptable image, while the procedures used, without signal modulation, permitted only 10 bytes.

The Ampex solution (1955)

At the end of 1951, and at the instigation of **Alexander M. Poniatoff** (1892—1980), founder and president of the California Ampex Corporation, a team started work on the solution of the videotape problem. Heading up the team were Charles P. Ginsberg and Charles E. Anderson. They were joined in August 1952 by a 19 year-old student whose name would later be passed along to posterity for his noise-reduction systems: Ray M. Dolby. From the outset, the solution of mobile tape reader heads was explored: The initial heads moved about 64 meters/second, while the magnetic support had a much slower speed (76 centimeters/second). This resolved the problem of tape consumption. Nevertheless, it should be noted that the first trials, in October 1952, reproduced an image that was identifiable only with difficulty. In order to improve the image, the team devised

A magnifying television screen.

the four-head drum system in September 1954; the heads were separated by 90 degrees and rotated at high speeds, while the tape unwound perpendicular to the drum, pressing against it by means of an identical groove. The tape then took on the form of this groove, thanks to an air-induced depression. (In fact, the system of rotating heads had already been tested, some years previously by Scotch 3M, but in a different form. Several months later, Anderson developed a procedure for FM recording (at low unwinding speed) and this appears to have provided good results, although it had no confirmed theoretical basis. This allowed the team to give a very convincing demonstration of their work to the Ampex directors on March 2, 1955. The video-tape recorder, with added improvements, was presented in April 1956, under the name of Ampex VR 1000, at the Chicago Congress of the National Association of Radiotele-broadcasting, with great success.

First retransmission of a recorded television transmission (1956)

This took place on November 30, 1956, on an Ampex VR 1000. That day the CBS studio at Television City in Hollywood recorded the "Douglas Edwards and the News" program broadcast from New York, in order to retransmit it 3 hours later.

Video magnetic tape (1956)

The first videotape marketed was developed in 1956 by two researchers from the American firm 3M, **Mel Sater** and **Joe Mazzitello**. Working day and night with the means at their disposal, the two men succeeded in presenting their invention the very day that Ampex Corporation placed the first video recorder on the market. This first "Scotch 179" reel was 2 inches wide, nearly 800 meters long, and weighed 10 kilograms!

First color video recorder (1958)

The first color video recorder must also be credited to Ampex. It was presented in 1958, two years after the first video recorder, under the name VR 1000 B. It was followed in 1963 by a transistor version, the VR 110.

Meanwhile the Japanese had been working on video recorder technology:

Portable video cameras and recorders are now affordable for family use.

• In 1958, Toshiba announced the first single-head video recorder.

• In 1959, JVC developed the first two-head video recorder, the KVI.

• In 1962, Shiba Electric (now Hitachi), in cooperation with Asahi Broadcasting, presented a professional transistorized video recorder.

• In 1964, Sony marketed the first video recorder for the general public.

• In 1964, Philips placed the VR 650 video recorder on the market in Europe.

• In 1965, Shiba Electric marketed a small portable video recorder.

Videocassette recorders (1970)

In addition to professional video recorders produced for television stations and industry, video recorders for the general public began to be developed, first with an open reel. However, since 1970, the emphasis has been on models in which the tape is contained in cassettes, as in audio models. In October 1970, Philips unveiled its VCR videocassette recorder, which was put on the market in April 1972, while JVC and Matsushita were jointly developing in Japan the standard U, a cassette with a 3/4-inch tape. Little by little, this came to be considered the top-of-the-line product, and it is now rated as meeting professional standards.

The Philips VCR was followed by the Long Play model, produced by Philips in August 1977, and by a derivative model, the SVR, produced by the German company Grundig. The subject of research since the beginning of the 1970s, the longitudinal video re-

corder (LVR) models were introduced by BASF and by Toshiba. The first of these comprised a 600-meter tape and 72 tracks, read at the speed of 4 meters per second, one after the other. The LVR reader head moved vertically in order to read each of the tracks, then jumped to the next when the preceding track had stopped unwinding. The second model made use of an endless tape read in a spiral.

Mini video recorders (1982)

These use smaller cassettes and can be used in standard-size video recorders with an adapter. They are intended for use with portable systems, everything having been done to reduce the weight and volume. Models are:

• The JVC minicompact (VHS standard), 2 kilograms (2.4 with accumulator), with a 30-minute minicassette. This machine appeared in spring of 1982 and has been marketed since the autumn of that same year.

• The Grundig minicompact (VCC 2000), 2.1 kilograms with a 120-minute minicassette. Presented at the 1982 Hifivideo in Dusseldorf, this mini video recorder was marketed in 1983.

Video camera

Video cameras for the general public were derived from the professional models invented by Vladimir Zworykin (see p. 112).

Camera with built-in videocassette recorders (camescope) (1980)

In July 1980, **Sony** presented a prototype camera with a built-in video re-

corder. The videocassette on which images were recorded was placed directly in the camera. Smaller than a packet of cigarettes, it was the tiniest videocassette ever produced.

Sony was followed in September 1980 by Hitachi, and in February 1981 by Matsushita; these two firms presented their own models, which were also only prototypes.

The disparity among models and a desire to standardize the manufactured products led to an agreement among the top five producers (see below), but before this agreement was reached, another Japanese company, Funai, exhibited a model of a camera with a built-in videocassette recorder at the CES in Las Vegas in January 1982. What was particularly notable about this cassette was that it was absolutely identical to those used for portable video recorders.

Below, top, Sony's Betamovie, Bottom, Children preparing to enjoy a videodisc program.

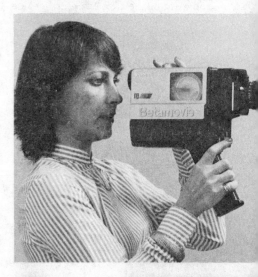

Agreement of the Big Five
(1982)

On January 20, 1982, five major video producers (Hitachi, JVC, Matsushita, Philips, and Sony) signed an agreement to work together on a camera with a built-in videocassette recorder (8 mm. video camescope) using a 9 x 6 x 1.4-centimeter cassette and allowing a film capacity of one hour at the outset.

Betamovie (1982)

This camera, with a built-in videocassette recorder, was presented by Sony at the Japan Electronic Show in Tokyo in October 1982. Since then, all the companies that have decided to use the Betamax standard have alsos opted for this camescope; it weighs 2.5 kilograms and uses standard Betamax cassettes, providing up to 3 hours and 35 minutes of recording.

Videodisc

The precursors

The first attempts to store recorded video signals on a disc date back to 1927. The Scot **John L. Baird** had the idea, and he printed discs following the format already used for the phonograph. These discs were then reproduced at a speed of 78 rpm. The system's tape, reduced to 5 kilohertz (the images one sees using current systems require several megahertz), nonetheless permitted viewing of images with a picture resolution of 30 lines (15 points per line), at a rate of 12.5 images per second.

An initial attempt to produce a commercial version used the Phonovid procedure, proposed in 1965 by the American company Westinghouse. This used a microgroove disc for information storage and provided good quality images. Unfortunately the Phonovid could only store 200 fixed images per disc, and several more years were necessary before animated reproductions became possible.

Compact disc (1972)

Two different version exist, one with a mechanical and one an electrical reader-guiding system.

The first system was created by **RCA** (1972). In this version, the information was inscribed in the form of pits following a spiral groove. The disc surface was metallized and the signal read in the form of capacitive electrical variations between the reader and the disc surface. The RCA videodisc was launced, at great expense, throughout the United States

on March 22, 1980. It proved a commercial failure, because of inherent disadvantages in comparison to laser-read discs put on the market at about the same time.

The second system was presented by **JVC** (1976) under the name of VHD (video high density disc). This is also a compact videodisc, however, without the guide-groove. The VHD is simply encoded with pits that contain both the video information and control signals designed to guide the reader.

With both systems, program viewing for a single disc could be as long as two hours. The disc always had to remain in its protective jacket—disc and jacket were both put into the reader apparatus. The VHD is compatible with the NTSC, PAL, and SECAM color transmission systems.

Optical videodisc (1972)

The **Philips** videodisc makes use of an optical reader made up of a gas laser and a photodiode. Reading thus takes

place without any mechanical contact with the surface of the disc, which carries the information in the form of metallized pits which are used in conjunction with a servomechanism to guide the reader. Reading is done by reflecting the laser beam off the surface of the disc and the disc requires no particular protection.

Presented in 1972, the videodisc was first marketed in the United States in 1980 by MCA, a subsidiary of Philips, and by Philips itself in Great Britain and in Germany beginning in 1982.

The Thomson-CSF videodisc, which appeared in 1975, also makes use of an optical reader, but uses a transmission mechanism: Laser and photodiode are separated by the transparent disc, which has a thickness of 150 microns. This videodisc was essentially reserved for professional and industrial use. This system requires no metallization of the disc surface, but it does require permanent protection of the disc.

POST AND
TELECOMMUNICATIONS

The post (6th century B.C.)

Cyrus the Great (558—528 B.C.), the king of Persia, is the supposed inventor of the post. Cyrus had conquered an enormous empire, and individually dispatched messengers bearing letters and other information were no longer sufficient. The king therefore organized a postal service made up of relay stations located at regular intervals, with personnel responsible for tending to the couriers' horses after a reasonable daily run.

The Romans copied this system and under the reign of Augustus (27 B.C. to 14 A.D.) they created the *Cursus publicus*. Military routes were marked out with mutationes, or relay stations, which could provide fresh horses, and mansiones, or inns, reserved for official travelers.

The monks' post

By the Middle Ages, the Roman postal system had long since disappeared. As Europe became dotted with mon-

asteries, it was deemed necessary to link those belonging to the same religious order. Correspondence was relayed on a roll of parchment, a *rotula*, on which the first abbe wrote his message. At each of the succeeding abbeys, other messages were added. The *rotula* could become quite long. The *Saint Vital Rotula*, which announced the death of one abbe, was 9.5 meters long and 0.25 meters wide.

Mail boxes (1653)

Toward 1650, four post offices were operating in Paris, linking the city with the provinces and abroad. However, the inhabitants of the city had no organized means of communicating among themselves. **Jean-Jacques Renouard de Villayer** remedied this lack by creating the "small post." In 1653, he had "wall" boxes placed at the corners of the main streets of the capital, wherein everyone could drop their letters on condition that they

The Saint Vital Rotula.

were wrapped in a ribbon that said "postage paid." The mail was picked up three times daily.

The envelope (17th century)

In 1820, an inhabitant of Brighton, England, named Brewer announced that he had created the envelope. However, several envelopes dating from 1615 are preserved in Geneva. At that time, the letter was folded and wrapped with a silk thread; a wax seal held together the two extremities. Later, the letter was enclosed in a piece of folded white paper, on which the address of the receiver was written. It was only in 1841 that the Frenchman Maquet began the industrial manufacture of envelopes in Paris.

Directories (1785)

The first city directory was for the city and suburbs of Philadelphia, between 1st Street to the north and Maiden Street to the south, 10th Street to the west and Delaware Street to the east. Published by John Macpherson on October 1, 1785, it included 6250 names and addresses.

Stamps (1834)

The Irishman **James Chalmers** printed the first stamp in Dundee, in 1834, but it was not until 1838 that stamps were used, following the British postal reform carried out by Rowland Hill.

Initially, the receiver and not the sender of a letter paid for the service. This is why Renouard de Villayer, counselor at the Parliament of Paris, instituted the "post paid" notice in 1653. This notice, which was not joined to the letter itself, was

only valid for the Paris "small post," and it was necessary to buy it separately from a postal center. Thus it was not a big success. Nonetheless, the idea of freeing the receiver from having to pay for letters sent to him was born.

The first adhesive stamps, which went into use in Great Britain on May 6, 1840, were the penny black and the twopenny blue.

Perforated stamps *(1854)*

The first apparatus for separating stamps was invented in 1847 by the Englishman **Henry Archer**. It could only make slits. Its inventor perfected it one year later, and the machine could then perforate a series of small holes. The first perforated stamp was the penny red, issued in February 1854.

Post card (1861)

The post card was invented in Philadelphia in 1861 by **John P. Charlton**, who obtained a copyright for it and then sold his rights to a stationer named Harry L. Lipman. The latter published the cards with a picture and the words "Lipman post card, patent pending."

The prestamped post card was conceived of by the Austrian Emmanuel Herrman, of the Neustadt Military Academy in Vienna. The first of these was made available on October 1, 1869. It was off-white in color and had a 2 kreuzer stamp.

Electric telegraph Origins

Origins

In 1827, the German Carl August von Steinheil discovered that a single electric wire with a ground circuit

sufficed to constitute a transmission line. The Italian Alessandro Volta had developed the first electric battery (see Volta battery, p.318), in 1800, and the electric energy it produced allowed messages to be sent. In 1833, the Englishman Michael Faraday (1791—1867) discovered that electric currents could be altered by magnets. If variations were produced according to a code common to the transmitter and the receiver, two persons could exchange messages with each other.

Schilling: Russian telegraph (1832)

In 1832, the Russian diplomat Schilling had the idea (already put forth in 1820 by the French physicist Ampere) of using the deviation produced by the passage of an electric current over a magnetic needle as a telegraphic signal. In 1837, the Czar appointed a committee to express its opinion on the experimental installation of the Schilling telegraph in Saint Petersburg. Unfortunately, Schilling died.

Cooke and Wheatstone: The English telegraph (1837)

In March 1836, William F. Cooke, a young officer demobilized from the Indian army, had a Schilling electric telegraph at his disposal. As soon as he arrived in England, he began to

Samuel Morse on board the Sully in 1832.

The Paris telegraph office in 1851 where messages were routed and received over 200 lines

perfect it. Faced with technical difficultues, he appealed to the famous English physicist **Charles Wheatstone** (1802—1875), professor at King's College, London. In June 1837, Cooke and Wheatstone applied for the first patent for a telegraphic signal—the signal was given by hand. The first trial linked Euston and Camden, 1.2 miles apart. They offered their invention to the British rail authorities.

On January 1, 1845, a telegraphic message was instrumental in the arrest of a murderer (coming from Paddington) as he got off a train in London. When he was hanged, the newspapers enthusiastically hailed the formidable efficiency of the electric telegraph.

Morse: American telegraph (1837)

In 1837, the American painter and dilettante **Samuel Morse** (1791—1872) created a telegraph. Working with Alfred Lewis Vail, whose father owned a small electromechanical factory, he developed a working model.

The signal transmitted materializes through the rhythmic interruption of an electrical current for longer or shorter intervals corresponding to the dots and dashes of the Morse code(a - .-; b - ..; etc.). These dots and dashes can be either heard or read on paper tape.

It took Morse several years of perseverance to obtain from Congress the $30,000 he needed to create an experimental line between Washington and Baltimore. The debate in Congress, in February 1842, was not conducted in a very serious manner, and the decision to finance the telegraph was ratified by a small majority: 89 to 83. Seventy members of Congress neglected to vote. The Washington-Baltimore line was inaugurated in January 1845.

The wireless telegraph

First wireless connection (1895)

The first wireless transmission was accomplished by the Italian physicist **Guglielmo Marconi** (1874—1937), who in 1895 succeeded in transmitting a signal a distance of 2400 meters. Marconi was 21 years old at the time. On October 26, 1898, the Frenchman Eugene Ducretet (1844—1915), assisted by his colleague, Ernest Roger, was able to make a "wireless" connection between the Eiffel Tower and the Pantheon, a distance of 4 kilometers. The following month (also in Paris) he made a 7-kilometer connection between the Sacre-Coeur, on Montmartre, and the Church of Saint Anne, rue de Tolbiac. France was not immediately interested in Ducretet's experiments. In September 1899, Tissot, a lieutenant in the French navy, made a 42-kilometer connection at sea, between the Stiff Lighthouse and Virgin's Island. Following that success, Eugene Ducretet sent him this telegraph: "When you get back, tell the Minister that we can do as well as the English, with some money. Sincerely, Ducretet."

First connection between England and the Continent (1899)

On March 28, 1899, **Marconi** made the first telegraphic transmission between Dover (England) and Wimereux (France), 50 kilometers apart. Aware of what he owed to Edouard Branly for the success of his experiment, he sent the following telegraph: "Mr. Marconi sends Mr. Branly his respectful compliments by means of wireless telegraph across the English Channel; this marvelous result is due in part to the remarkable work carried out by Mr. Branly." This historical telegraph is preserved in the museum of Radio France in Paris.

Marconi had studied in Italy, where he achieved his historic first in 1895. But his compatriots were not much interested in his work. Then his Irish mother sent him to London, where he was presented to the Minister of Post and Telegraph, Sir William Preece, who promised him help. In 1897 the young emigre founded Marconi's Wireless Telegraph and Signal Company. He transmitted the results of a regatta for the Prince of Wales by telegraph, and this was the beginning of his fame. Without knowing it, he had given the first sports report using wireless telegraph.

First transatlantic connection (1901)

Marconi made the first wireless connection across the Atlantic on December 12, 1901, with the collaboration of the Englishman J. A. Fleming (1849—1945). They transmitted a message between Cornwall (southwest England) and Newfoundland, 3400 kilometers away. Marconi received the Nobel Prize in physics in 1909.

Underwater telegraphic cable

Three things are necessary in order to have a functioning underwater telegraphic cable: an insulated electric conductor; a solid cable that can rest against the sea floor; and specialized ships that can install the cable.

Insulation (1847)

In 1847, the German engineer Werner von Siemens developed machines that could coat telegraph cables with gutta-percha.

Solid cable (1850)

After the rupture of the first telegraphic cable under the English Channel, on August 28, 1850, a cable made of four copper wires, each with a diameter of 1.65 millimeters, was used. The cable's wires were twisted into a single strand that was covered with tar-cloth and reinforced with 10 galvanized iron wires, each with a diameter of 7 millimeters.

Cable installation ships (1865)

After 1865, the first installation ships, which were simple steamboats equipped for the purpose, gave way to bigger, specially equipped ships that had cable tanks, winches for unrolling the cable, etc. The English ship *Great Eastern*, the largest of the time, was the prototype of the installation ship.

First cable under the English Channel (1851)

In 1851, the first large-scale telegraphic cable was laid between Dover and Calais by the English ship *Blazer*. The English technician Jacob Brett was the principal architect of this operation.

England, fearing the loss of its "splendid isolation," did not encourage the project. The engineer George

Cable being readied for laying in the Blazer Hold of a cable installation ship.

Stephenson even orchestrated a campaign of detraction. Finally, Louis-Napoleon Bonaparte financed Jacob Brett. In 1850, the first attempt to lay a cable under the Channel met with failure—the cable was cut by fishermen who thought they had discovered an alga whose heart was filled with gold!

First successful transatlantic cable (1867)

On July 27, 1867, the first successful transatlantic telegraphic cable was laid. The American Cyrus Field (1819—1892) was the promoter of the operation, which nearly cost him his entire fortune. Previous attempts in 1857 and 1858 had met with failure. A third attempt seemed at first to have succeeded. By August 5, 1858, 3240 kilometers (1944 miles) of cable had been laid between Ireland and Newfoundland. At 2:15 P.M. that day, the first underwater telegraphic message crossed the Atlantic. It announced to the *Niagara*, one of the two cable-installation ships, that the other installation ship had arrived in Ireland. In America, the news was welcomed with extraordinary waves of enthusiasm. But on September 3, 1858, at about 1:00 o'clock, a handling error by an operator put the cable under 2000 volts of tension and this put it out of order. In 1865, after the American Civil War, Field made another abortive attempt to install a transatlantic cable. It measured 3700 kilometers (2200 miles), but, alas, it broke.

The telephone

As with the telegraph, the candidates for the title of inventor of the telephone are legion, and numerous countries claim their man invented it. Many researchers described a system in theory before 1876. Nonetheless, it is generally agreed upon that the first operational telephone was made in the United States in 1876. Almost simultaneously, two men produced a workable prototype: **Alexander Graham Bell** and **Elisha Gray**. But it is Bell (born in 1847 in Edinburgh, he later became an American citizen; he died in 1922, in Canada) who is historically credited as inventor of the telephone.

On exactly the same day, February 14, 1876, the two rival inventors applied for a patent at the New York patent office. Gray came two hours after his competitor. It was these two hours that swayed the judges during the course of the enormous lawsuit that, several years later, pitted the two rival telephone companies

against each other. Bell's patent was not free from criticism. A crucial issue concerned the description of the variable resistance device used. In fact, Bell had added the description of his device in the margin of his patent application. He maintained it was because he realized at the last moment that he had forgotten to include it in his text. The lawyers for Gray maintained that having heard of Gray's patent, Bell purely and simply copied it. The judges decided in favor of Bell. This fortuitous outcome marked the beginning of the rise of the Bell Telephone Company, today AT&T, which was, for a long time, the most powerful company in the world.

The Bell telephone (1876)

At the request of Western Union Telegraph, Bell worked at improving the telegraph. On June 2, 1875, a decisive event took place. Bell in one room, his assistant, Thomas Watson, in another, were both assiduously at work, as always, on their telegraph. As usual, nothing worked as it should. Suddenly Watson made a wrong step: The poor contact of an adjustment screw, which was too tight, transformed what should have been an intermittent current transmission into a continuous current. Bell, at the other end of the line, distinctly heard the sound produced by the drop of the contactor. It was an accident, but Bell, obsessed by the idea of a wordtransmitting machine, screamed to Watson, "Don't touch anything!"

Bell spent the autumn and the winter making calculations. On February 14, 1876, he sent his silent partner, Gardiner Hubbard, to apply for a patent in his name for an apparatus designed to "transmit voice or sounds" (the harmonic telegraph had still not been abandoned). Still the apparatus did not work; it was not until March 6, 1876, that Bell succeeded in transmitting intelligible words to his collaborator: "Come here, Watson, I want you."

The telephone was born. The world would learn of it when Bell presented it at the American Centenary Exhibition in Philadelphia, in June 1876.

Public telephone (1889)

William Gray, of Hartford, Connecticut, was granted patent number 408,709 on August 13, 1889, for an apparatus that allowed coin-operated telephoning. The first telephone thus equipped was installed in the Hartford Bank. In 1891, William Gray, Amos Whitney and Francis Pratt formed a company that installed pay telephones in department stores.

Bell's first telephone.

Telephone exchange

Manual exchange *(1878)*

In 1878, the first manual telephone exchange was put into operation. It was invented at the same time as the telephone by **Bell** and by **Gray**.

Among the initial subscribers were James Garfield, who became President of the United States, and the famous writer Mark Twain.

The first telephone exchange operators were men. It was several months before the "telephone ladies" appeared on the scene.

Automatic exchange *(1891)*

In 1891, an American undertaker from Kansas City, **Almon B. Strowger**, applied for a patent for the first automatic telephone exchange. Strowger had discovered that his competitor's wife, an operator at the local manual telephone exchange, was the very first to learn of deaths in the city. Doubtlessly, she was directing the calls sent to Strowger's enterprise to her husband. It is thus easy to understand Strowger's interest in an automatic telephone system.

Strowger produced his system using two techniques partially developed by Bell System research scientists: a dial device that did not work (patent dated 1879, "Connoly-Mac-Tighe") and a primitive exchange credited in 1884 to Ezra Gilliland (it could only handle 15 subscribers).

Based on his patent, Strowger founded the Automatic Electric Company. Later, under licenses from Strowger, AT&T improved the system and made the automatic telephone commercially viable. The Strowger system could link up to 99 subscribers.

Telephone dial *(Rotary system)* *(1923)*

On May 18, 1923, the French engineer **Antoine Barnay** (1893—1845)

applied for a patent for a system automatically linking telephone subscribers. This was the first appearance of the rotary dial, once found on almost all telephones. The dial allowed variable electric impulses to be emitted. An exchange registered and decoded the signals, searched for the subscriber, called him and established the link-up.

An improvement over Strowger's system came in 1905 when AT&T scientists invented a method called translation that converted dial pulses from decimal to nondecimal form, thus paving the way for a more efficient method of switching calls known as common control.

Push-button dialing began in 1941 in Baltimore, but was too expensive for general use until the advent of low-cost transistors and associated components brought Touch-Tone service into being in 1963. This service opened the way for new applications of the telephone, including its use for gaining access to computers.

Electronic exchange *(1960s)*

Development of electronic telephone exchanges has been taking place since the 1960s at many companies. These systems, which employ the stored program control principle, are programmed and directed in a manner similar to the way a digital computer is controlled. This makes it relatively simple to add new services and features.

The first electronic switching system was tested at Bell Labs in 1958 and field tested in Illinois in 1960. The first public electronic exchange was put into operation in New Jersey in 1965.

The central telephone switchboard in Paris, 1878.

● *The English laboratories.* In 1962, a completely electronic telephone exchange was put into service in the north of London. However, the technology of the time (utilizing memories and semiconductors but with limited possibilities) forced the British Post Office engineers to abandon the task and to concentrate on semielectronic systems. At the end of the 1970s, the British went completely electronic, with the system "X."

● *The French CNET.* In 1957, the first trials on a French electronic exchange were carried out under the direction of Joseph Libois, the head of the CNET (National Center for the Study of Telecommunications). At the time the CNET held important pat-

Before they went underground, telephone lines were a dominant feature of the landscape.

ents relative to the MIC system (see below).

Joseph Libois, together with Andre Pinet, launched the Plato project (Lannion Prototype for instant switching, with numerical organization).

In January 1970, in Perros-Guirec (in Brittany, France), the world's first temporal electronic switching device for public network subscribers was put into operation. In 1977, at the International Conference in Atlanta, it became evident that "time switching" would soon be placed on the world market. CNET was two years ahead of its competitors.

Automatic switching uses the MIC system (modulation via coded impulse), and thus multiplexing—a quasi-infinite number of communications use the same support (cable or hertz frequency). Each one is coded, in the form of numerical signals at the time of sending. Then, it is decoded upon arrival. The numerical signals can represent both sounds (radio broadcasting and telephone) and images (television).

Lightwave communications
(1880)

Lightwave (or fiber optic) communications systems transmit voice, data and video signals as pulses of light through hair-thin glass fibers, rather than as electric charges through copper wires. Such systems have enormous capacity for carrying information, and their use continues to spread.

The first use of light beams to transmit messages was demonstrated by **Alexander Graham Bell** in 1880, and he considered this his greatest invention. His Photophone was capable of sending "sound" a short distance—thus anticipating current work in lightwave systems—but it had no practical application at the time.

It was not until the 1970s that the full potential of lightwave systems began to be realized. Advances in glass fibers, lasers, light-emitting diodes and photodetectors led to fullscale experiments at Western Electric's Atlanta facility in 1976. A year later, the world's first lightwave system to carry voice, data and video went into service in Chicago.

Mobile communications
(1924)

The Bell System provided mobile communications to New York City police cars in 1924, but the modern era began in 1946 when the first commercial mobile telephone service went into operation in St. Louis.

In 1978, AT&T Bell Laboratories began testing in Chicago a new type of mobile system known as Advanced Mobile Phone Service. This system divides an area into contiguous hexagonal "cells," each of which has its own radio transmitter. As a vehicle travels from one cell to another, electronic switching equipment transfers a call to the new transmitter without interruption and permits re-use of radio frequencies. This cellular radio system began going nationwide in 1981.

Microwave radio (1940s)

Microwave radio, the dominant carrier for long distance telecommunications, had its origins in work done by Bell Laboratories on radar systems during World War II. **Harold Friis** and his associates developed the horn antenna now used everywhere in microwave relay towers. They also put together the basic elements of a microwave system that was one of the more closely guarded wartime secrets. Secrecy was lifted in a public demonstration between New York City and Neshanic, New Jersey, in 1945, and two years later the first experimental circuits for domestic use were installed.

Transatlantic telephone cable
(1956)

The first long-distance underwater telephone cable was installed in 1952 between Key West and Havana, Cuba. Four years later, a 48-circuit cable connected the United Kingdom, Canada and the United States. This was a joint venture by AT&T, the British Post Office and Canadian Overseas Telecommunications. In 1963, a cable between England and the United States transmitted 138 simultaneous conversations. There are currently five transatlantic cables in use, and a lightwave cable is planned for 1988.

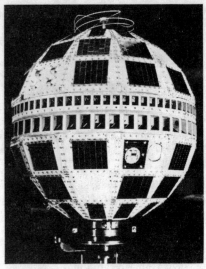

A model of Telstar 1.

Telecommunications satellite
(1960)

By 1945, the English engineer Arthur C. Clarke had published the first theoretical analysis of an artificial satellite system in the technical review *Wireless World*.

On August 12, 1960, **NASA** launched the first American telecommunications satellite: Echo I. This was still only a big balloon, 30 meters in diameter, whose metallized surface reflected the waves without either amplifying or directing them. Echo I did not endure the action of meteorites very long.

Telstar (1962)

On July 10, 1962, **NASA**, under contract to AT&T, launched the first really efficient civil telecommunications satellite: Telstar. It was put into orbit by a Delta rocket. It contained 1064 transmitters and 1464 diodes and was fed by 3600 solar cells.

Telstar was able to transmit either telephone conversations or television images between the American station in Andover, Maine, and the European stations in Goonhilly Down, Great Britain, and Pleumeur-Bodou, France. On July 12, 1962, these two stations competed for the television images transmitted from the other side of the Atlantic. The American images registered on the Pleumeur-Bodou control screens without any problem, while the Goonhill Down screens remained completely blank on that day.

games, toys, and sports

GAMES AND TOYS

Alquerque (1200 B.C.)

The earliest ancestor to checkers is found in the Egyptian temple of Kurna, which was built 1200 years before Christ. The game was introduced in Spain by the Arabs, who called it El'quirkat. It appears in the *Book of Games* by Alphonso X (1283) and was quite popular in medieval France.

Two players are opposed, each controlling 12 men of different color over a board made up of 25 intersecting lines. Each player moves a man to a free, adjacent intersection or jumps over an opposing man behind which lies a free intersection. A piece so jumped over is captured. The rule of "doubling" an opposing piece, to avoid capture, was formerly in custom, but it is no longer common today. Forgotten for a long time, Alquerque is once again a popular "thinking man's" game.

Perilikatuma

This is a variation of Alquerque practiced in Sri Lanka. Although one can't assert a common origin with contemporary checkers, Perilikatuma has an identical set of rules. However, the board, giving each player 23 men, is different from our checkerboard. As in Alquerque, the pieces are placed on the intersections rather than on the squares defined by them.

Ball games (antiquity)

The first ball players were probably the Greeks. The ball, an inflated cow's bladder, was meant to be taken to a given point while dodging the counterattacks of the opposing team. This ball game, the ancestor of modern-day football, was adopted by the Romans. With the Roman conquests, the game penetrated Gaul and Britain, but the word *ballon* (French for ball) did not appear in France until 1549, as a derivation of the Italian word *ballone* (a dialectic deformation of *pallone*, meaning "big ball").

Bridge (1850)

A derivative of whist, bridge first appeared around 1850 in Istanbul. Bridge differs from whist in that one of the four players' hands, the dummy, is exposed on the table. The game, played by two teams of two partners (each partner sitting opposite the other), was popular on the French Riviera in the 1870s and by 1880 was called by its present name.

In the beginning, tricks were assigned values according to the suit: Spades were worth 2 points, clubs 4 points, diamonds 6 points, hearts 8 points, and no-trump 10 points. Fifty years later a variant called "auction" bridge was king. It required teams to bid competitively to name the trump suit; and it also increased the value of spades over clubs, diamonds, and hearts.

A French variation of bridge, *plafond* ("ceiling"), was the first to award points for reaching a declared contract. However, attaining the declared contract in plafond was not sufficiently compensated (one could obtain game points for tricks won under or over the declared contract), and "contract" bridge filled this shortcoming. For a long time only very experienced, good players played contract bridge, but the latter won out shortly after 1945 and plafond disappeared.

Bridge was and continues to be the object of numerous studies of a scientific nature (probability, tactics). An American, Ely Culbertson, revolutionized the game around 1925 by inventing the system of counting points in a hand before beginning the bidding, which allows a player to assess the strength of his hand.

Casino games

Throughout history, governments have had various attitudes toward gambling. The Roman emperors were so addicted to gambling that their subjects followed suit. Nero bet up to 400,000 sesterces on one roll of the dice. The first Christians, under the reign of Justinian, tried to stem the tide without much success.

Las Vegas

Gambling is illegal in many countries, such as the Soviet Union, Canada, India, Brazil, Japan, and everywhere in the United States except Nevada and New Jersey. Las Vegas enjoyed its unique legal status for 43 years. In 1905, it was little more than

The Tumbler's Fair during the reign of Louis XIV of France.

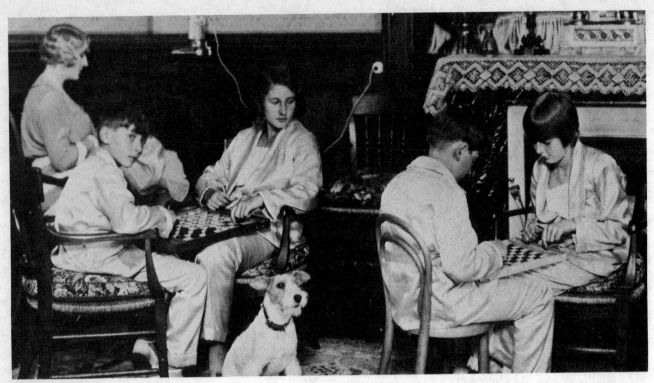

Checkers players in the 1930's.

a rude village of tents. Ruined by the Great Crash of 1929, Las Vegas counted no more than 8,000 inhabitants in 1940. Since then it has become one of the gambling capitals of the world. Today it is a city of 175,000 people and 250 hotels, each with one or two casinos, some of which have as many as 800 slot machines. One can play 24 hours a day, year round, at roulette, blackjack, baccarat, and craps—except for the minute preceding New Year. For that minute everything stops—the electricity is shut off throughout the whole town! The annual income from gambling is about $2.5 billion.

Roulette *(1780)*

Mentioned in Diderot's *Encyclopedia*, roulette evolved into its present form around 1780. This game of chance uses a fixed concave plateau, the interior of which has a cylinder (wheel) 22 inches (56 centimeters) in diameter. The periphery of the cylinder is divided into 37 squares, numbered from 0 to 36. Each number has a color, red or black, except 0. A small ball is thrown into the moving cylinder and gamblers bet on where it will land.

Checkers (ca. 1720)

The ancient Egyptian game of Alquerque progressively evolved into what is known today as checkers. In France, during the regency of Philip

of Orleans (1715—1723), an officer of the royal guard changed the checkerboard and also made the rules much more complex. He increased the number of squares to 100 and brought the number of playing pieces, or "men," on each player's side to 20. Without affecting the popularity of chess, checkers thus became a game of unlimited strategy and tactics. In American checkers, players generally use a checkerboard consisting of 64 squares and play with 12 men each.

Chess (ca. 500 A.D.)

The earliest mention of chess goes back to the Persian writer Karmanak (590—628 A.D.). It appears that the game originated in the north of India around 500 A.D., but the current rules were established in Europe around 1550. The expression *checkmate* is a derivation of the Arabic phrase *al shah mat*, meaning "The king is dead."

Crossword puzzle (1913)

Crossword puzzles, as we know them, were invented by an English-born American journalist, **Arthur Wynne.** Wynne worked on the amusements supplement of the *New York World* and was always looking for new puzzle ideas. He remembered a Victorian-era game, the Magic Square, which his grandfather had taught him. In redesigning the square by

interspersing black squares in the grill and adding a list of 32 definitions, he invented crossword puzzles, the first of which appeared in the weekly supplement to the *New York World*'s December 21, 1913, edition. His definitions were very simple and purely descriptive.

Dice

The Greek philosopher Plato (428—348 B.C.) attributed the invention of dice to the Egyptian god Thot, the god of knowledge and magic formulas. Etymologically, the Greek word *kubos* (which means "die" and, by extension, "cube") leads to the Arabic word *kaab*, which refers to an ossicle. The English word *dice*, from the French *de*, comes from the Latin *dare*, meaning "to give"; by extension, the latter meant "to throw," as to throw something on a table.

The Latin historian Tacitus (55—117 A.D.) wrote that certain Germanic peoples staked their freedom on the game and would docilely submit themselves to serfdom if they lost. St. Ambrose (340—397) noted that some people gambled their fingers or limbs and, if they lost, would allow those parts to be cut off without protest. When they had nothing else to stake, they would even gamble away their lives.

Diplomacy (1958)

Of all the games that require alliances and negotiations between contestants, Diplomacy is the best known. It was created by an American, **Alan B. Calhamer**, in 1958. The number of devotees continues to grow despite the appalling disputes it sometimes causes between friends. Several divorces are even attributed to it! The game board, however, is simple enough: the political map of Europe in 1900. Each of seven players controls many pieces representing army and navy forces. The aim of the game is to be the first person to control 18 of the 34 strategic regions in Europe. No player can do it alone—hence, one must form alliances—but one player can be the sole victor. In other words, alliances must sooner or later be betrayed.

The combat phase, reduced to simplest terms, is preceded by the most tantalizing part of the game: negotiation. For 15 minutes the players discuss and try to persuade their adversaries to form alliances, however temporary. Any player may lie, bluff to catch others out, or double-cross. The moment of truth comes when each player writes down, in secret, the placement of his pieces on the board. When the moves are examined, the true intentions and betrayed promises of each player are exposed. The moves are made, and each player wins or loses ground and the next round begins.

Dolls (antiquity)

The origin of dolls is as ancient as the first little girl. In prehistoric times, children built dolls. The first dolls were made mainly of wood or clay and, more rarely, of wax and ivory. The Greeks and Romans already had dolls with movable limbs. In 19th-century Europe, Saxony produced most of the torsos for dolls made of papier-mache. Nuremberg and London, among other cities, specialized in porcelain dolls. Paris specialized in dolls' clothes, which were sold in places as far away as China.

Talking doll (1880)

Thomas A. Edison (1847—1931), the inventor of the record player, was the first to make a talking doll. Perfected around 1880, the doll said, "mommy, daddy," using a recording of a real human voice.

Barbie doll (1958)

Barbie was created in 1958 by Mattel. It was the first doll to have an adult's physiology and a whole wardrobe of miniature clothes. Ken, Barbie's boyfriend, appeared in 1961. In 1963, Barbie's best friend, Midge, came onto the market; then Skipper (Barbie's little sister), Allan, etc.

In 1976, at the time of the United States' bicentenary, Barbie dolls were put into time capsules, which were then sealed and buried. The capsules will be opened in 2076, the date for the tricentenary.

Cabbage Patch dolls (1983)

Undoubtedly the result of a curious genetic manipulation in the doll kingdom, the Cabbage Patch Kids are a new "race" of dolls that is all the rage in America. Born on November 12, 1983, in Haywood, California, the Cabbage Patch dolls neither walk nor talk and they have a funny-looking round head. They are even a bit cross-eyed, but people love them and snatch them up nevertheless. Two hundred thousand Cabbage Patch dolls see the light of day each week.

Each doll is unique, which perhaps explains this extraordinary craze. Coleco, the firm that produces them, has entrusted the manufacture of these dolls to a computer. They have succeeded in producing a million and a half of them. Each doll is sold with both a birth certificate and a certificate of adoption, both made out in due and proper form. In February of 1984 a medical center was even opened up in Florida for "sick Cabbage Patch dolls"—with an operating room!

Dominoes (before 2450 B.C.)

The National Museum of Baghdad, Iraq, has in its collection objects made of bone, discovered at Ur, in Chaldea, which date from around 2450 B.C. and which archeologists compare to present-day dominoes. Yet it was not until the middle of the 18th century that the game appeared in Europe and reached England, via France, around 1795. The term *domino* derives from the name given similar black clothing worn by priests in winter over their white surplices (which may have come from the Latin saying *Benedicamus domino*—"Let us bless the Lord").

Frisbee (1947)

Apparently **Yale University students** invented the Frisbee in 1947. After having finished eating a certain brand of canned pork, the students enjoyed throwing the empty can, which was flat and round. Fred Morrison patented a plastic thing looking like the can in 1968.

German Klicky Dolls.

Barbie in an array of clothing and accessories.

The frisbee has become a universal sport. Even dogs can play, as demonstrated in San Francisco.

A director from the Wham-O Manufacturing Company, of San Gabriel, California, decided to manufacture a plastic model after having seen the students play. He called it Frisbee, after the name of the canned pork manufacturer, the Frisbee Company, of Bridgeport, Connecticut.

Go (ca. 1000 B.C.)

Tradition has it that the game of Go was invented 3000 years ago by the Chinese emperor **Yao**, to develop the habit of reflection. It is a game of static strategy—once a man (or playing piece) has been played, it cannot be moved again. Go was introduced into Japan by way of Korea around 750 A.D., and it was in Japan that the game developed rapidly.

From its debut, the game was closely tied to the most profound political and moral ideas of Asian peoples. The military saw in it an educational means of initiation into the art of war. And apparently Buddhist priests in the 16th century played the game with fervor and great skill. The 17th century marks the apogee of the game; a Buddhist monk and champion of the game, Hon Inbosansa, founded an official academy of Go in Edo (Tokyo).

Heads or tails

Beginning with the era when Greece reigned over the Mediterranean basin, the game of heads or tails was widely played in public places. As today, the game allowed a player to make a simple choice (i.e., "heads, yes; tails, no"). In addition, the game

Greeks played a hoop game.

was played as one of chance; each player in turn guessed the outcome of a throw of a shell—white on one side, black on the other—which was tossed up into the air with a rotating flick.

Romans were the first to use a coin to play the game, which they called *caput aut navis* ("chief or ship"). In Gaul, the game was referred to by its Latin translation. Depending on the coin in circulation in a particular country, the game was called different names: "flowers or saints" in Italy; "Castile or Leon" in Spain; "heads or tails" in England. If the term "heads" designates the face often found on one of the sides of a coin, we can explain the origins of the term *pile* (French for "tails") by the fact that representations of ships commonly figured on the reverse side of French coins. *Pile* is the archaic term in French that designated a ship's pilot.

Hex (1942)

Hex is the prototype of linking games—games in which each of two opposing players tries to reach the opposite side of the board by forming an unbroken chain with his pieces. Hex was created in 1942 by **Piet Hein**, a Danish research scientist in the Neils Bohr Institute of Physics. It quickly became popular under the name Polygon. In 1948, a student at Princeton University, John F. Nash, "reinvented" the game. It then became known as Nash. In 1952, the game was put out by Parker Brothers Inc. (producer of Monopoly) and called Hex. The game is played on a board made up of 121 hexagonal spaces arranged in the shape of a diamond, 11 x 11. The players choose the opposing areas they must reach and move their pieces, or different-colored marbles, in an attempt to form a continuous chain.

Hula-hoop (1958)

The plastic Hula-hoop was invented in 1958 by **Richard P. Knerr** and **Arthur K. "Spud" Melvin**, owners of the Wham-O Manufacturing Company, San Gabriel, California. In six months, 20 million Hula-hoops at $1.98 were sold in the United States.

Kites (2nd century B.C.)

The invention of kites goes back to the 2nd century B.C. and was the work of the Chinese general **Han Sin**. They were first magical and ritualistic instruments in the Far East. A kite signified duality—the shadow or spirits of the dead, given to the power of the wind; its string, the tenuous link with the world of the living.

This airborne device was used in ancient China to measure distances, as well as to transmit messages by its shape and by its colors. Beseiged troops or prisoners could easily see one in the sky.

Lego bricks (1955)

Lego bricks, as we know them today, came onto the market in 1955. They were invented after World War II by a Dane, **Ole Kirk Christiansen**, a former carpenter who had become a game manufacturer. The word *lego* is derived from the Danish expression *Leg-godt*, which means "play well." The biggest Lego contruction in the world is Legoland, a veritable amusement park built with Lego bricks in Billund, Denmark. Since its opening in 1968, Legoland has received more than 750 million visitors.

Life (1970)

This extraordinarily rich mathematical game was created in 1970 by **John Horton Conway**, who was influenced by the work of the famous mathematician Von Newman on the possibility of "reproduction" by machines.

The game is played on a chess board. The player places his pieces on the squares of his choice. Each piece represents a living cell. A group of pieces is called a "base population." Like a living organism, certain cells will survive, certain cells will die, and others will be born. The game seeks to explore the generational development of a base population according to the original placement of pieces. Certain cell groups disappear after a couple of generations, while others stabilize. John Conway's Game of Life has spawned variations that oppose two players defending different-colored cell groups, or base populations. The last of these was brought out in the magazine *Games and Strategies*.

Liquid crystal games

Progress made in the field of component miniaturization has led to the development of "liquid crystal" games. They are manufactured by several firms in Japan and Hong Kong. Small, wafer-thin devices that can be held in the palm of the hand, liquid crystal games are controlled by microprocessors similar to those found in microcomputers. Liquid crystal games are primarily games of skill and reflex, involving egg-throwing, drops of petroleum, octopuses, etc.

Lottery (16th century)

Lottery games originated in Genoa, Italy, and are attributed to **Benedetto Gentile**. He simply transformed the process of designating city council members into a game. Ninety wooden balls or tokens, numbered from 1 to 90, are put into a bag. Players divide 24 rectangular cards among themselves; each card has 27 squares, 15 of which have numbers on them. Each player places an agreed upon sum of money into a basket. Then, one by one, the balls or tokens are picked out of the sack. As the numbers are picked out and called, players cover the corresponding squares on their cards. Each player takes a part of the pot according to the outcome.

The Magic Wand, invented by Texas Instruments (1982), fascinates children. As they move the wand over pictures a voice identifies the object.

Magic Wand (1982)

Magic Wand is a revolutionary educational game for young children. It was invented by **Texas Instruments** in 1982. By means of an optical pencil, which the child passes over a "code bar," the Magic Wand pronounces words or musical notes. For example, if a drawing of a truck is on a page and the child passes the optical pencil over the code bar word under the truck, the device says the word *truck*. The same applies to letters and numbers, and there are also short songs that the device can read. The possibilities are limitless.

Mah-jong (1850)

Unlike many other games, Mah-jong does not find its roots in antiquity, even though it was created in China.

It appeared in 1850 in the town of Ning-Po. Its anonymous creator wanted to change the traditional Chinese playing cards to dominos, known as "tiles" by the players. The game was made popular the world over during the long period of colonization in the 19th and early 20th centuries. Anglo-Saxon countries were the most open to the charm of Mah-jong. It has been the rage for the last three years in the United States, particularly on the West Coast. It is basically a card game, the aim of which is to form combinations, such as pairs, three of a kind, four of a kind—as in gin rummy. Mah-jong cards don't use Western faces (clubs, hearts, spades, diamonds). The three suits are bamboo, characters, and circles, and the master cards are the four winds,

the four dragons, the four seasons, and the four flowers. The relative unpopularity of Mah-jong in Latin countries is probably due to its lack of clearly defined rules: There are hundreds, if not thousands, of variable and optional rules.

The first part of the game, as it's played in China, consists of negotiations between the players to establish which rules are to be used. Each player seeks to have the rules he knows best used. It's rare for Western games to have this element of negotiation. This aspect of the game, which is a source of pleasure in the Orient, too easily becomes a source of conflict in the West.

Marbles (antiquity)

The game of marbles seems as ancient as humanity. In Greece they played it with small rounded bones, acorns, chestnuts, and even olives, which were aimed and tossed at a small hole. The Romans adopted the game but replaced acorns with walnuts and hazelnuts, to such an extent that the latter became the symbol of childhood. Marbles were first made of wood or metal; they were roughly spherical and relatively large.

Mastermind (1977)

Mastermind was invented in 1977 by an Israeli, **Mordecai Maerowitz**. It is a game of logic. A player arranges four playing pieces in any order he wants and conceals them. His opponent has to deduce their arrangement by successive moves, rearranging his own corresponding playing pieces.

Mickey Mouse (1928)

On November 18, 1928, a small animated character, a smiling mouse with big round black ears, appeared on the screen of the Colony Theater, in New York City. The character was named Mickey Mouse, and it was the beginning of a glorious "career" that has never waned. The first Mickey Mouse stuffed animal came onto the market in 1930, when the character's spiritual father, the artist **Walter Elias Disney** (born in Chicago, 1901; died in Burbank, 1966), better known as Walt Disney, sold the reproduction rights to a toy company.

Microcomputer games (1980s)

The 1980s mark the era of a great revolution in the world of games: the advent of microcomputer games. Nearly all well-known microcomputers, including Apple II, Atari, Tandy model TRS 80, and Texas

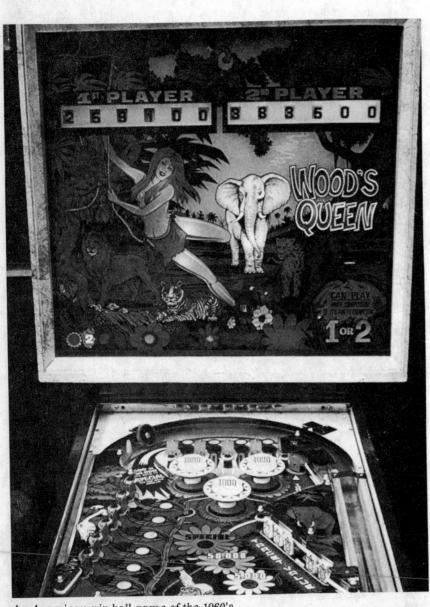

An American pin ball game of the 1960's.

Instruments' TI 99, offer ever-increasing numbers of games of skill. Games are available on standard tape-player cassettes or on discs. To play, a person needs the microcomputer itself—sometimes called the central processing unit (CPU); a TV screen hooked up to the computer; and a loader, also hooked up to the computer, to read the cassette or disc. Games can be played by two or more people, in which case the computer serves as the referee; or one person can play against the game program. The computer amounts to a formidable electronic adversary, since its powerful memory allows it to control hundreds of variables in a time interval impossible for a human being (close to 500,000 elementary operations per second). All the classic games of skill and reflection (chess, checkers, backgammon, Go, Othello) are well suited to this new form.

The most amazing microcomputer games are no doubt the simulation games: economics games, war games, and adventure games. One of the more sophisticated games, Timezone, uses six two-sided discs and allows the player to travel through time and space. Mastering the game requires at least two or three months' practice. Other computer games invite players to be Napoleon! By simulating battle during the era when orders were transmitted at the speed of a galloping horse, the games require players to anticipate the considerable ineffectiveness of commands.

The United States currently produces 99% of all game programs.

Monopoly (1933)

Monopoly was invented in 1933 by an American, **Charles Darrow** (1889–1967), during the Great Depression that followed in the wake of the Crash of 1929. He invented a game of real estate investment at a time when he himself was unemployed. He used the names of streets in Atlantic City, New Jersey, where he went on vacation. Monopoly has been an enormous success and has been translated into numerous languages. Parker Brothers, which manufactures the game, has sold more than 80 million of them.

Murder Party (1920)

Under the influence of games of pretense and games in which players assume roles, Murder Party is once again deservedly popular. It was created in the United States in 1920, the

Paul Newman plays poker; a game invented 3,000 years ago.

heyday of the detective novel. It's played in the following manner: The leader of the game hosts a party made up of about 12 people. He pulls out as many cards from a card deck as there are players, making sure to take one King of Diamonds and one Jack of Spades. Each player draws a card at random. The one who draws the Jack of Spades becomes the assassin; the King of Diamonds becomes the detective. The other players become witnesses. Each player slips his card into his pocket. The leader of the game announces that he is going to turn out the lights in the next minute. As soon as it's dark, the assassin has 10 seconds to choose a victim, grab him by the shoulders, and then move away from the "scene of the crime." As soon as a player is grabbed by the shoulders he must fall to the ground with a cry. After another 10 seconds the leader of the game switches the light back on. The detective then shows his card and the search begins. He proceeds in the manner he sees fit to discover the assassin: group interrogation, face-to-face interviews, questioning of the

witnesses, etc. When he thinks he has found the "perpetrator," he fingers him. The suspect must then show his card. If it's the Jack of Spades, the detective has won; if not, he has abused justice.

Othello (1974)

Goro Hasegawa, a Japanese, invented Othello in 1974. This game of strategy, which pits two players against each other, consists not in eliminating an opponent's men by removing them from the board (as in chess or checkers) but in turning them over to take possession of them.

Othello, which also exists in an electronic form, is the most widely played game in Japan after the game of Go.

Poker (ca. 1920)

Contrary to conventional wisdom, poker was born in Chicago during Prohibition. The origins of poker go back 3000 years to a hybrid game, *As*, played in Persia. Pairs, three of a kind, full house, and four of a kind were all used, as well as one of the chief elements of poker itself, the bluff. It was introduced in Europe during the Crusades and developed under the name of *Primero* in Italy. *Bouillotte* and *Brelan* are French variations, which the French colonists brought with them to Louisiana. These games moved up the Mississippi with the conquest of the West and a new variation of an old game was born—poker.

Puppets (antiquity)

Originally puppets were created to take part in religious festivals in Egypt, China, etc.

Puzzles (1760)

Puzzles, images glued to supports and cut into irregular pieces that a person fits together, originated simultaneously in about 1760 in France and Great Britain. In the beginning, they were education games. In 1762, in France, a man named **Dumas** started selling cut-up geography maps, which students (and others) reassembled. **John Spilbury**, an Englishman, glued a map of England onto a thin layer of mahogany; his puzzle was cut up according to the boundaries of each county, and each county was sold separately. Spilbury died at the age of 29, in relative obscurity, his game never achieving success.

Subsequently, puzzles became historical. In 1787, the Englishman William Darton published a puzzle with the portraits of all the English kings from William the Conqueror to George III. Each rectangular piece consisted of a portrait in the form of a medallion, with captions noting the principal events of each king's reign around it. The rectangular pieces to the puzzle were about the same size, and the object was to assemble the puzzle by memorizing the succession of the kings. In 1789, Wallis simplified the game and published a colored historical board of English history. His puzzle served as a model for subsequent ones, which demanded more patience and observation.

Role games (1970s)

The latest rage among the more popular games, role games began in the United States in the 1970s and are organized by many game associations. The game is very often itself the reason for the formation of a club.

One of the players, the organizer, uses his imagination to create an entire universe: He draws a map of a region, complete with roads, waterways, and houses; on more detailed maps he notes the number of rooms in buildings and situates furniture in them, etc. In his universe, the organizer also hides treasures and people, and to the latter he ascribes behavioral characteristics or personalities. All this preliminary work can take days, weeks, or even more time. When the organizer finishes creating his scenario, he invites friends, each of whom will play the role of one of the invented characters for an evening.

The game begins in the form of a dialogue between the players and the organizer. The latter describes the surroundings each player "sees" and asks the player for a reaction or response. A player responds by interpreting his assigned character and giving the organizer a reaction. The organizer, in turn, indicates the consequence of the player's reaction; he determines the consequence of the player's reaction by referring to the rules prescribed for this universe. The goal is for each player to come out alive from the universe in which he has found himself; or better, to improve the "qualities" of his "personality" by accumulating "experience points." The main characteristic of these games, apart from their 6- to 10-hour duration, is the possibility offered to the player of reusing the personality in another scenario.

Dungeons and Dragons (1973)

The oldest and best known of the "new formula" role games (*see* Murder Party), Dungeons and Dragons was created by an American, **Gary Gigax**, a salesman and a fan of simulation games. Bored with playing war games, Gigax and a friend, Dave Arneson, thought up a simulation game in which players were not confined to moving playing pieces on a game board.

English literature, notably the works of J. R. R. Tolkien (author of *Lord of the Rings*), as well as magic, inspired Gigax to create hundreds of characters with very precise characteristics. After having refined the rules of his game, he tried in vain to get it manufactured by major American game companies. Turned down repeatedly, he became a part-time shoe repairman and used his spare time to write new rules and get the game manufactured on his own, in 1973. TRS Hobbies, of which Gigax is today president, is the biggest money-earner of all game companies.

Killer (1983)

The role-playing game of the year 1983! It resembles the much earlier game Murder Party and has its origins in a short story by the great science-fiction writer Robert Sheckley. Published in *Galaxy* magazine in 1953 and entitled "The Seventh Victim," the story inspired a film by the Italian director Carlo Ponti. In turn, the film gave American college students the idea of playing out assassination themes, each more diabolical than the preceding one.

Each participant is given the name of his "victim" and the "arms" he can use. Among the arms, which for the most part consist of gags and party novelties, are confetti (which is supposed to represent boiling oil) and sneezing powder (which, sealed in an envelope addressed to the victim, is supposed to be deadly poison). The winner is the player who manages to outlive the others. The game is currently developing a following at universities in Europe.

Rubik's cube (1979)

In 1979, a Hungarian chess fanatic and professor at the University of Budapest, **Erno Rubik**, invented a cube puzzle brain-teaser. This game of logic first interested colleagues and mathematicians, but shortly afterwards the head of Ideal Toys, an American game company, found the game irresistible.

The cube is made up of 27 smaller cubes, and each of the six main sides of each cube is a different color. Each large side has nine squares, and the player can change the position of each square independently (except for the square in the center of a side). At the start, each side of Rubik's cube is the same color, but in a matter of seconds the player moves the squares around to mix up the colors on each side. The object of the game is then to reassemble the squares as quickly as possible so that each side is again all the same color.

Scrabble (1948)

In 1948, **James Brunot**, from Newton, Connecticut, patented the game of Scrabble. The source of his patent was a game called criss-cross, in which players had to make up crosswords on a piece of cardboard using wooden letters. Criss-cross was invented in 1931 by Alfred M. Butts.

Senet (2nd milennium B.C.)

Senet was the most popular game in ancient Egypt. Pharaonic culture even attributed to the game the virtue of being played by the spirits of the dead. Papyrus, or ancient scrolls, testify to the game's popularity among peasants and artisans as well as the pharaohs themselves. A representation of the playing board figures on the tomb of Ramses II (1182–1151 B.C.) and the game itself has been found in the sepulcher of Tutankhamen (1347–1339 B.C.). The game consisted in a race across the playing board, with each player having five playing pieces. Details of the game are not clear, but apparently the game resembled backgammon. If this is so, Senet may be considered the oldest known board game of any sort, that is, a game where players vie to be the first to reach a point across a playing board.

Silly Putty (1947)

Toward the end of the 1940s, researchers at General Electric discovered a synthetic substance they thought might be able to replace rubber. Like rubber, the substance was soft, elastic, pliable, and moldable; but shortly afterward, GE realized that the material had no real industrial potential. In 1947, an advertising agent from New Haven, Connecticut, **Peter Hodgson**, bought $147 of the substance from GE and hired a Yale University student to make 100-gram balls and package them in plastic cubes. In 1949, Hodgson sold his cubes by mail, advertising them in a client's toy store catalog. Silly Putty was an instant success and the substance, which resembles chewing gum but does not stick, was appreciated by mothers and nervous executives alike.

Simulation games (1980s)

They make up what may be called the great new game fad of the 1980s. The common element among all these games is the desire to have players enact a realistic scenario—indeed, to reenact an episode of a real event, past or present. The game board is supposed to represent as faithfully as possible the place where the action takes place; similarly, the playing pieces that are moved about represent the characters or an entire group of individuals.

In this context, the simulation game author Steve Jackson scored a great success with Rain on Iran, which went on the market shortly after Americans were taken hostage at the embassy in Teheran. One player moves pieces representing the revolutionary guards; the other player moves pieces representing American soldiers. The gameboard represents the plan of the embassy.

Solitaire (17th century)

Solitaire appeared in the 17th century and was perhaps invented by a man named **Pelisson**, who was imprisoned by Louis XIV, or by a mathematician trying to relieve the boredom of stagecoach travel. The Latin poet Ovid (43 B.C.—28 or 29 A.D.) had already alluded to a game like solitaire: "One has some little balls at one's disposal on a table which consists of hollowed-out cavities: To be able to move a ball, one has to remove one among them."

Sowing games

Sowing games may be said to symbolize all African games and must be considered as the most-played games in the world. *Mancala* is the generic term that designates them. In fact, Mancala is nothing more than a North African variant of a similar game, Awele, played in the Ivory Coast (West Africa). These games all have in common one fundamentally different characteristic when compared to North American and European games: In checkers and chess the players dispute over common territory, each player moving his own pieces; in sowing games, the players divide the territory into two, and the playing pieces themselves are used in common by both players.

From Africa, sowing games spread to the West Indies, to Louisiana during the era of slavery, and to the Indian subcontinent and Indonesia via the traditional trading routes between East Africa and India. The common principle in such games is for a player to take playing pieces in one of his squares and distribute them, one by one, into successive squares (or more precisely, small holes). Since the playing pieces are most often a variety of grain or seed, a player's move has good reason to be called "sowing."

Awele

Awele is one of many games, very common in the world, consisting of twelve playing spaces (or small holes). Steeped in tradition, it is a difficult and fascinating game of strategy.

The game is played on a wooden board or even with holes made in the

During the 18th Century, swings were quite provocative as they revealed women's legs.

ground. The playing board generally has two rows of symmetrical holes. Each of the two players, designated "north" or "south," has as his territory one of the parallel rows of holes— half the board. The number of playing pieces is a multiple of the number of holes, and the object is to capture as many of the pieces as possible. A rule called "feeding" (*wari*), stipulating that a player must place a "seed" in any empty space of his opponent, influences the outcome. A player loses if he does not "feed" his opponent.

Speak and Spell (1978)

Speak and Spell was launched in the United States by **Texas Instruments** in 1978. The device's voice, which uses a word synthesizer, asks a child to correctly write one of 142 words on a small screen. The words are divided according to their difficulty into four categories. Each word is chosen randomly by the robot. It pronounces one by one the letters that the child writes down on the screen.

Swings

The origin of swings is mythical: In order to atone for a crime of murder, the Roman god Bacchus condemned the first grape pickers to observe solemn ceremonial games to honor King Icarius. The games consisted of swinging on a rope strung between two trees.

According to legend, Bacchus had given the gift of wine making to King Icarius of Laconia. The king, wanting to thank his vintners, had them drink their fill of wine. Soon drunk, the vintners thought they had been poisoned and killed the king in anger.

For a long time afterwards, swings retained a symbolic and religious character. Thus, swings were commonly associated with images of rain, fertility and springtime. In the 18th century, swings were considered as an erotic game because they exposed womens' underclothes.

Tarot

No one knows who created tarot or when or where it originated. In 1457, St. Antoine, in his *Treatise on Theology*, makes the first known reference to tarot. In any case, tarot is considered as the precursor of modern card games. No one knows either what the word means or what language it comes from: The Italians transmitted it under the name *tarocco*. Some say the origin of the word is Egyptian (*tar*: way; *ro*: royal); others claim it has its origins in Hebrew

Jungle Hunt video game.

(a derivation of *Torah*, the Law). An anagram of the word gives *rota*, meaning "wheel" in Latin—the wheel of fortune or destiny, as Arcane X of the Marseilles Tarot, engraved around 1500, suggests. The Frenchman Court de Gebelin rediscovered tarot in 1781.

Teddy bears (late 19th, early 20th century)

Americans and Russians both claim paternity of the stuffed teddy bear.

For a long time, wooden toys representing bears and sleds were made in Russia. It is said that **Czar Nicolas II** (1868—1918) gave gifts of toy bears covered with long hair to his hosts. He offered one of these to President Loubet of France upon the signing of the Franco-Russian Treaty of 1892.

Another anecdote relates that President Theodore Roosevelt's son had a live pet bear. When the bear died, the boy was so stricken with grief that a handyman came up with the idea of making a toy bear for him. He made the bear with bits of plush stuffed with rags. The President's son's name was Teddy, which explains the name "teddy bear."

Ideal Toy Corporation marketed the first American teddy bears at about the same time that a German firm, Stieff, of Gingen (Wurtemberg, West Germany) began to sell the same thing in Europe.

Track (1980)

The urge to carry simulation to ever more realistic heights has brought

The hero of the film Wargames picks an electronic lock.

about games that, perhaps, stretch the limits of the definition of games. This is the case with Track, which could as well be called Manhunt. Developed in the United States in 1980, it is firmly in the "Western" tradition. The participants mark off a game zone of several hundred yards by several miles. The aim of the game is to be the only survivor. Everyone is armed; the "weapons" are highly specialized projectiles: plastic balls filled with red ink normally used to brand cattle! The well-protected participants enjoy the atmosphere of the game and "kill each other" to the last. Any contestant showing signs of red ink is eliminated.

Ur

In 1920 the English archeologist Sir Leonard Woolley discovered among the ruins of the ancient city of Ur (a Sumerian city in Mesopotamia) a game considered the oldest in the world. The specimen of the game-board, now on prominent display in a room at the British Museum, was dug out of a tomb in the royal cemetery. Though it is considered to be a distant cousin to backgammon, nothing allows us to establish firmly a direct link.

Video games (1970s)

At the beginning of the 1970s, game arcades (primarily closed to minors and recently banned in many Southeast Asian countries) contained mostly pinball machines and an occasional early video game like Breakout. Gradually more electronic games were introduced, and on June 16, 1978, a Japanese company, Taito Corporation, launched Space Invaders, a milestone in video arcade games. A tremendous success, the game pits a player against rows of advancing Martians in close formation.

The year 1980 marks the beginning of the era of video console games. Each system employs its own electronic console capable of using interchangeable cartridge programs. Videopac by Philips, VCS by Atari, and Intellivision by Mattel are among the most popular console systems. Microcomputers and video consoles came onto the market at about the same time.

Pong (1972)

The first video game, Pong was invented in 1972 by an American engineer, **Noland Buschnel**, who was 28 years old at the time. A relatively unsophisticated video game by today's standards, it was a sort of electronic tennis game. But Buschnel persevered and his game spread rapidly in arcades. He formed his own company, Atari. In 1976, Warner Communications bought Atari for $28 million. The same year, Atari launched game cartridges for home consoles. Video games entered the home market.

Pac-Man

By 1983, Pac-Man was the uncontested king of video games. It was the result of collaboration between two companies, **Namco** (Japan) and **Midway** (United States).

Using a lever control, the player maneuvers a small yellow circle that gobbles up everything in its path. The game consists of eliminating various characters controlled by the game program and called "phantoms"; they move about within a labyrinth. Initially, the phantoms go after the "pac-man" and destroy him if they catch him. But if the pac-man succeeds in gobbling up one of the colored discs in the labyrinth, it is the pac-man who, under the player's control, for a limited time hunts the phantoms.

Hundreds of thousands of Pac-Man games, including illegal pirate copies, have been sold.

Wargames (1780)

Wargames simulate military battles that may or may not have a historic character. The majority of the 300 or 400 war games produced today have the following characteristics: The game board (16 x 24 inches) represents a terrain, real or imaginary, with all the appropriate topography and vegetation. It is divided into hexagonal spaces. Each player controls 15 to 250 cardboard squares, which represent a person, a group, or an entire regiment. Each piece is endowed with numerous characteristics that qualify its particular capacities for movement and strength. By turns, each player moves all or just some of his pieces, as if engaging in a maneuver. Enemy pieces that find themselves on adjacent spaces engage in combat, at the end of which they are either eliminated or forced to retreat. While chess and Go are symbolically war games, Wargame is far more realistic. Wargame, in its modern form, was invented by **Helvig**, Master of the Pages to the Duke of Brunswick, in 1780. In 1798, the military author Georg Vuiturinus set down the most elaborate rules. In 1837, the general-in-chief of the Prussian army, Von Moltke, incorporated these rules in military formation. All nations followed this example after 1870. Eventually the game was popularized in the United States by Charles Roberts in 1953. The mass production of the game Tactics allowed its author to create the Avalon Hill Society, which today controls the marketing of Wargame.

Whist (18th century)

This card game, of English origin and the true ancestor of bridge, was described for the first time in 1743 by Edmond Hoyle in a treatise he devoted to it. Contrary to bridge, however, whist does not involve any system of calls. It gave birth to a number of important variations, among which should be cited Dummy, Humbug, and above all, the American variation Boston. The last variation, bridge, has won out over all the others. The word *whist* refers to the prohibition against speaking among the players. From its birth in the 18th century it was highly popular and remained so until 1867, when its enthusiasts turned to bridge.

SPORTS

Auto racing (1894)

Early competitions

The first automobile road race took place in France on July 22, 1894. It was run over a distance of 78 miles, from Paris to Rouen. The previous record average speed of 13.6 mph was broken in this race by a prototype steam automobile driven by Count Dion. But no official time was kept, and the first prize was split between Panhard and Peugeot.

The first road races where official time was kept were the Paris-Bordeaux-Paris (730 miles), in June

The Jamais Contente *(1849), the first car to go over 100 Km/h.*

of 1895, and the Paris-Marseilles-Paris, in 1896. During this last race, Emile Levassor was seriously injured during a collision with a dog. He was hurled out of his car and died a year later as a result of his injuries.

Gordon-Bennett Cup (1900)

This annual cup was founded by the American journalist and philanthropist J. Gordon-Bennett, whose aim was "to promote motoring progress in Europe." It was awarded from 1900 to 1905. Each country could enter only three cars, and the drivers were chosen by their respective national automobile clubs. The cup was won the first two years by two different French makes, the Panhard and the Levassor. But in 1902 it went to Napier, a British make.

Monte Carlo Rally (1911)

The most famous of all auto rallies was created in 1911 by **Vialon** and **Le Boucher**. They got the idea for the rules from an Italian bicycle race and launched an open trial for "horseless carriages." The principle of the rally was to reach Monte Carlo in seven days, from several different departure points. Though most of the 23 contestants for the first competition left from Paris, others chose Berlin, Vienna, and even Brussels. At a time when motor vehicles were not equipped with four-wheel brakes or snow tires, and antifreeze and car heaters did not exist, the average speed was set at between 6 mph minimum and 15 mph maximum. Vehi-

cle classification was based, among other points, on the number of passengers a car could carry, as well as its comfort and the state it was in when it arrived.

By 1912, the number of cars taking place in the rally had jumped to 88. One of them even started out from St. Petersburg in Russia!

Le Mans 24-hour endurance rally (1923)

The first 24-hour endurance race at Le Mans was organized on May 25, 1923, on the initiative of three men: **Georges Durand**, **Emile Coquille**, and the journalist **Charles Faroux**. Their aim was to create a road testing ground for mass-produced "improved" cars. The time was set at 24 hours to push the machinery to its maximum and especially to "put a hard test to the lighting devices." In front of an audience of nearly 40,000 spectators, and in a driving rainstorm, 33 cars—representing 16 different French makes and 2 foreign makes—took the starting line. Only 3 of them finished the race. The Amilcar, Berliet, Bugatti, Delage, Salmson, and Voisin astonished people with their resistance, but it was the Chenard & Walker driven by Lagache and Leonard that finished in front, covering a distance of 1370 miles at an average of 57 mph.

Long distance rallies (1907)

At the beginning of 1907, the Parisian daily *Le Matin* launched a fantastic challenge: "Are there any volunteers

for linking Peking to Paris?" The race began in Peking on June 10, 1907, and the winning Itala, driven by Prince Borghese, reached Paris two months later on August 11 after surmounting unbelievable difficulties.

The following year, an even crazier race took place: New York to Paris via Alaska and the Russian Empire! The prize was carried away by an American Thomas Flyer. In 1909, the American Transcontinental Race from New York to Seattle was won by a Ford.

Baseball (1840)

According to honored legend, baseball was invented in 1839 by Abner Doubleday in Cooperstown, New York. It is doubtful that this is so. The game actually has its origins in England, where around 1750, young men were playing it as a simplified form of cricket. The principles of the game were established around 1840. In 1845, one of the founders of the first baseball club, the Knickerbocker Club of New York, defined the rules of the game. His name was Alexander Cartwright, and he invented the diamond-shaped playing field. Up until then the game had been played on square or pentagonal fields. The first game played according to Cartwright's rules took place in Hoboken, New Jersey, between the New York Nine and the Knickerbockers, who lost 23-1.

Basketball (1891)

Basketball is one of the few sports that did not come about as a result of a slow evolution. In 1891, **James Naismith**, a professor at the International YMCA College in Springfield, Massachusetts, decided to develop a sport that could be played indoors, at night or in winter. He nailed two baskets to the opposite walls of a gym and set down the rules. The first game took place on January 20, 1892. The teams were made up of seven players each, and the game was played over three 20-minute periods.

Boules—bocci—carpet ball (antiquity)

A form of bocci was played by the Greeks, who called it "spheres," as a form of healthy relaxation for the body. It consisted in throwing balls as far as possible. Building upon this, the Romans added an additional rule, requiring the players to come as close as possible to a fixed point, near or far.

4 women beat the 30,000 Km record in 1937 with a Matford-Yacco.

A Citroen in the Gobi Desert during a Paris-Peking Raid in 1931-32.

nents over the course of his career.

In modern times, the first great champion was the Englishman James Figg (born in 1696). He was so talented with his fists that he had no need of any tactics to defeat his opponents. Figg's style of fighting became so popular in England that he opened the first boxing school in 1719, at Tottenham Court Road, London. He retired at the age of 34 without having lost a fight.

Another Englishman, Jack Broughton (1704—1789), established the rules for fighting barefisted. He set the dimensions for the ring and prohibited blows below the belt.

The eighth Marquis of Queensbury, John Sholto, together with the boxer Arthur Chambers, invented the rules used today: the wearing of gloves, a break after each 3-minute round, a 10-second count for a floored boxer, etc. In the first heavyweight fight conducted according to the Queensbury rules, John L. Sullivan defeated Dominick McCaffery in Cincinnati, August 19, 1885.

Cycling

Bicross (1972)

Cross-country cycling goes back to the very first cycle races. But the bicross, a small cycle having 20-inch wheels, without suspension, and unbelievably manageable, was born in 1972 in the United States. Today, more than 150,000 of them are sold annually in the United States.

Mountain bike (1973)

The mountain bike, or cross-country hiking bicycle, was created around 1973 in California. Aficionados of sports, adventure and space fooled around with their racing bicycles in order to be able to take the slopes of the neighboring mountains. Little by little, entrepreneurs began to develop these so-called mountain bikes. Today almost all builders represented on the American market offer this kind of bicycle.

Full-wheel bicycle (1984)

Designed by an Italian, **Professor Dal Monte**, an international specialist in aerodynamics, this unique and futuristic bicycle (which cost 2 billion lire in investment) was used by the Italian Francesco Moser in his attempt to beat the speed record, which had been held since 1972 by the Belgian Eddy Merckx, at 49.481 km/h. In Mexico, on January 19, 1984, following a long preparation in which researchers and technicians were associated, Francesco Moser rode 50.808 kilometers in one hour.

Bowling (5200 B.C.)

Bowling first became popular 5200 years before Christ—nine pins and a ball of stone were found in the tomb of an Egyptian child. The "bowling alley" was covered by three arches through which the ball had to pass, in a manner similar to croquet. Much later, in 4th-century Germany, bowling was a sort of religious rite. Sinners would place a wooden pole (*Kegel*) at the end of an alley paved with stones in their church cloister. Then they would hurl a ball; if they succeeded in knocking down the pole, they were absolved of their sin. Over the centuries it must have become more and more difficult to obtain absolution, because the number of pickets to be knocked down increased, varying from 3 to 17. It

was probably Martin Luther who fixed the number at 9.

The first sheltered wooden bowling alley was built in London in 1450. The tenth pin appeared in Connecticut in 1845, in order to circumvent a law prohibiting the game of bowling with nine pins.

Boxing (3000 B.C.)

Boxing, the sport of "fist fighting," is quite old. A decorative fresco in Iraq, dated at about 3000 B.C., shows boxers with their fists wrapped in pieces of leather. In ancient Greece—where boxing was called *pancrace*—the gloves were lined with pieces of iron, and often the contest was to the death. It's calculated that the champion Theagene, of Thaso, must have killed 1425 oppo-

Sylvester Stallone as Rocky.

Eric Heiden

Four days later, unaffected by his previous exploit, Moser set off once again and improved his own record, this time making 51.151 kilometers in one hour.

Fencing (1383)

The first fencing association was established in Germany in 1383. It was known as the Marxburger Guild, of Lowenburg. The Egyptians had used masks and blunted swords for training, but it was the Germans who really made fencing a sport. The *coquille* (or box), which protects the hand, was invented by the Cordoban Gonzolo, a captain in the Spanish army, in 1510.

Football

The forerunner of all football games is soccer. But it is rugby, an offshoot of soccer, which served as the departure point for American football.

Rugby was invented in 1823 when a frustrated William Webb Ellis at Rugby School in England suddenly picked up the ball during a soccer game and ran with it to the goal line. The idea of carrying the ball and kicking it was soon accepted and English soccer purists formed an association to separate soccer from rugby.

Princeton and Rutgers Universities pioneered intercollegiate soccer in the United States in 1869. Over a few years the game evolved closer to Ellis' rugby.

In 1880 at Yale University, a halfback named Walter Camp urged the rules committee to reduce the players from fifteen to eleven per side, and to substitute scrimmage for rugby scrum. Giving unhindered possession of the ball to one team until it is snapped back by the center to the quarterback is considered the most important rule in American football.

Golf (15th century)

Golf was invented in Scotland, and it gained such rapid popularity that in March 1457, King James II was obliged to ban the sport. He figured that his subjects devoted too much time to the game, when they could have been doing something more useful, such as practicing archery or horseback riding. However, golf remained the national sport. Queen Mary Stuart (1542—1587) was the first woman to play the game. It was she who created the first great golf course, Saint Andrews, which still exists.

The first golf balls consisted of a leather wrapping stuffed with feathers. Balls made with a rubber core were invented in 1902.

An American football game.

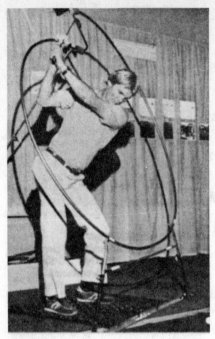

Hansen's device to improve one's golf swing.

A game of golf, including an electric golf cart.

Thanks to air-cushionned shoes, this horse can jump even on very hard pavement.

Handball (1915)

Handball first appeared in Germany in 1915. It's a version of soccer played with the hands, or, more precisely, it is a reverse of soccer. Thus, only the goalkeeper is allowed to use his legs or feet to stop the ball. Handball has been included in the Olympics since 1936.

Hang gliders (1948)

In 1948, an American, **Francis Melvin Rogallo**, devised a supple and flexible wing made of woven metal covered by a silicon-based coating. His invention interested diverse groups, notably NASA, before being abandoned.

There were numerous attempts by man to fly with birdlike imitations of wings. Leonardo da Vinci (1452—1519) was the first to sketch a design for a parachute and a wing, a device that would allow man to realize one of his oldest dreams: to fly. Throughout the centuries inventions multiplied, and attempts often ended with tragic results. One of the pioneers of gliders, Otto Lilienthal, a German, died in a crash during a flight test.

Ten months before Rogallo, in January 1948, an American named Wanner patented a kite whose shape resembled that of hang gliders, but his design was lost in obscurity.

From 1964, more operational hang gliders came onto the scene. Bill Moyes, an Australian who had worked for NASA, made a hang glider whose wing measured 4.5 square meters. His glider was drawn by a boat, and the standard parallel bars of the era were replaced by a trapeze. On July 4, 1969, his associate, Bill Bennett, took off on skis, towed by a boat, then let go of the tow line to fly over the Statue of Liberty. Thanks to the inventor Dave Killbourne, hang gliders can now take off by themselves and no longer need to be towed. Today, free flight has become an international sport.

Horse racing (1400 B.C.)

Beginning about 1400 B.C., the ancient Hittites of Anatolia gave themselves up to frenzied horse racing. Horse racing became an Olympic sport in 648 B.C. As for the English, they were already importing Arabian horses under the Roman occupation, and their first race took place in 210 A.D. at Netherby, in Cumbria. English horse racing in its modern form began in 1600 in Newmarket. Each horse carried a uniform weight of 140 pounds.

Ice sailing (1790)

In 1790, an American, **Oliver Booth**, fitted out the body of a sailboat with long blades. In the winter he sailed this ice boat along the frozen Hudson, near Poughkeepsie, where he founded the first Wind Sailing Club. The largest ice boat ever constructed measured 68 feet in length (21 meters) and carried 1000 square feet of sail. It was built in 1870 for John E. Roosevelt, of New York. In 1938, sailing on the ice of Lake Winnebago, in Wisconsin, John D. Backstaff attained the fastest speed of any vehicle propelled by wind: 153 mph (230 km/h).

Ice skating

Ice skating has existed for centuries in northern Europe, notably Scandinavia, Holland, and Flanders. At first, blades made of bone were fitted onto wooden soles; probably at the beginning of the 17th century these blades became metallic. In the United States, in 1850, the first true skates appeared, with steel blades. Immediately after the European tour by the American Jackson Haines, the American-style skates were adopted. The first world championships took place in 1896. In 1924, ice skating became an Olympic contest.

Skating rinks (1876)

The Englishman **John Gamgee** is, it seems, the first to have installed ice-skating rinks, one in London and one in Chelsea, in 1876. The London rink had a surface area of 400 square feet (40 square meters) and the one at Chelsea, 1000 meters. They used Pistet sulphur oxide machines and glycerine water. The New York Skating Rink, built in 1879, measured 860 meters.

Judo (1882)

Judo was invented in February 1882 by the Japanese **Jigoro Kano**. Born in 1860 in Mikage, in the province of Settsu, Kano devoted himself to jiu-jitsu. But he wasn't very strong, and he sought to compensate for this handicap by developing his body and his spirit. He studied at the University of Tokyo and became interested in all forms of jiu-jitsu. In the end, he perfected a method of attack and defense which brought victory over

an opponent by using suppleness rather than strength. He began teaching the sport in February 1882. Though a moral and even a mystic discipline, judo has been very popular, first in Japan, and then throughout the world.

Land sailing (antiquity)

Apparently, Egyptians and Romans used land-sailing vehicles. In the 6th century A.D., the Chinese Emperor Liang Yuan Ti gave the first detailed description of such a vehicle: "Kaotshan Wu Shu succeeded in constructing a land-sailing vehicle which, in a single day, was capable of transporting 30 men over several hundred kilometers."

The Flemish mathematician Simon Stevin, called Simon of Bruges (1548–1620), constructed land-sailing vehicles with wooden wheels that attained about 45 mph. It is said that the axles caught fire.

In the 19th century, maintenance

Peggy Fleming practicing in 1968.

men for the Kansas Pacific Railroad used "sail-cars" to check the rails. They were able to cover up to 85 miles in four hours. Two Belgians, Frank and Ben Dumont, launched land-sailing as a sport in 1910; their vehicles were faster than the automobiles of the era. Starting in 1925, the design for contemporary land-sailing vehicles—two rear wheels and one directional front wheel—was established. In California deserts it is common to see large land-sailing vehicles go faster than 75 mph.

Marathon (1896)

The marathon, a long-distance race, was invented during the first modern Olympics, in Athens, in 1896. Its distance was permanently fixed in 1924 at 42,195 meters (26 miles, 280 yards). The race was baptized "marathon" in memory of the soldier who ran from Marathon to Athens to announce the victory won by the Athenian general Miltiades over the Persians in 490 B.C. The soldier died of exhaustion on arriving at the foot of the Parthenon.

Motorcycling

Supercross (1972)

In 1972, at the Los Angeles Coliseum, aficionados of cross-country motorcycling were able to attend the astonishing spectacle of an indoor cross-country race.

ATC (1974)

Honda launched its ATC (all-terrain cycles) in the United States in 1974. These were large tricycles with large wheels, designed for fishermen, farmers, and riders. However, racing them very quickly became popular.

The Olympics (modern, 1896)

Modern Olympics were organized by a Frenchman, **Pierre de Fredy**, baron of Coubertin (1863—1937). The inaugurating ceremonies were held in Athens on April 6, 1896. Since then the Olympics have been held every four years, except for interruptions during the two World Wars. In 1924, the first winter games took place in Chamonix. In 1928, against the wishes of Coubertin, the games were opened to women.

Ancient Olympics (1450 B.C.)

Just who invented the first Olympics is unknown. They probably date back to 1450 B.C. Afterwards, they were held every four years. At first, they

The arrival of a famous marathon soldier.

were religious ceremonies. The first recorded Olympic games took place in Olympus, in 776 B.C.. The only contest was a 200-meter race, which was won by Corobius, a cook from Elis. The pentathlon appeared in 708 B.C., during the 16th games. It was made up of five contests: jumping, racing, the discus throw, the javelin throw, and wrestling. In 688 B.C. boxing was added, and in 680 B.C., chariot racing. Other games were held in Greece, notably in Corinth and Delphi. The Olympic games were finally banned by the Christian Emperor Theodosius, in 393 A.D.

Women's Olympics (600 B.C.)

At first the Olympics were restricted to men, and it was even forbidden for women to watch the competitions, on pain of death. In 600 B.C., a sportswoman by the name of Hippodemie organized the first games for women, on the occasion of her marriage to Pelope. These games, called the Herae, were also held every four years, alternating with the Olympics.

Athletic nudity (720 B.C.)

At the time of the 15th Olympic games, in 720 B.C., a runner named Orsippus of Megare lost his loincloth during the race, which he won over all the favorites. His victory was attributed to his nudity; afterwards, all the athletes chose to run in the nude in order to improve their chances for victory.

Roller skates (1759)

In 1759, a Belgian manufacturer of musical instruments, **Joseph Merlin**, invented roller skates. Invited to a ball at the Carlisle House in London, he had the idea of making a gliding entrance while playing the violin at the same time. Unfortunately, he hadn't thought about the problem of brakes and he crashed into the mirror at the end of the grand entrance, smashing the mirror and his violin to pieces and wounding himself seriously. His skates each had two large wheels.

Roller skates achieved noble status with *The Prophet,* an opera of five acts given at the Paris Opera in 1849. In the third act, a quadrille of ice skaters (in reality, roller skaters) on a frozen lake drew rave reviews.

In the sporting world, the ice skater J. Garcin invented roller skating at the beginning of the 19th century, at first simply for summer training. In 1863, an American, James L. Plimpton, of New York, patented the first roller skates with four wheels.

Rowing races (antiquity)

The origin of rowing races goes back to ancient times. Virgil described a rowing race in the *Aeneid.*

Rowing has been an Olympic sport since the Paris games of 1900, with these principal competitions: skiffing; double-sculling; two rowers, with or without a helmsman; four rowers, with or without a helmsman; and eight rowers, with a helmsman. The most famous race is the one that has taken place every year since 1829 between Oxford and Cambridge universities. The teams consist of eight rowers and a helmsman.

A 1982 national opinion survey projects 39 million Americans are roller skaters.

Skaters shown in a Washington, D.C. rink circa 1880.

SAILING

Yachting (17th century)

In the 17th century, the most prosperous country in Europe was Holland. Her navy of 3500 ships was twice the size of the French and English navies combined. Canals being the chief means of communication for the Dutch, the sailboat replaced the coach. Exiled in Holland, the heir to the British throne, the future Charles II (1630—1685), enjoyed this form of travel and after the Restoration (in 1664) established yachting in England. Within six years there were no fewer than 17 yachts, since his court soon followed in his footsteps. Following the Dutch example, Charles II organized yachting races, the first one pitting the royal yacht *Jamie* against the yacht of his brother and future successor. The king won. Yachting, the sport of kings, spread throughout Europe.

Louis XIV built mini-navies to sail on the waters of Versailles, and had a Venetian gondola hand-carried over the Alps. This new sport enticed the upper middle class as well; and in County Cork, Ireland, in 1720, the Water Club was created, whose descendant, the Royal Cork Yacht Club, is the oldest yacht club in the world.

In 1755, the brother of George III, the Duke of Cumberland, founded the Cumberland Fleet, which organized the first yacht regatta (the word is of Venetian origin). The Fleet became, in 1830, the Royal Thames Yacht Club. In 1815, just after Waterloo, 42 people, among them 24 nobles, founded the Yacht Club in London, which became in 1833 the highly selective Royal Yacht Squadron.

The first French yacht club was the Regatta Society of Havre, created in 1840; while in the United States, the New York Yacht Club was founded in 1844. The sport of kings became democratized, but the largest yacht

ever (125.6 meters) still belongs to the British Crown, although a yacht of 130 meters is under construction for an Arab prince.

Catamaran

The catamaran appeared several thousand years ago off the coast of Coromandel, India; it was essentially a raft made of three tree trunks of different length. Its name comes from the Tamil words *katu* (to attach, fasten) and *maran* (wooden trunk). The term is now applied only to boats having two similar hulls. These derive more immediately from the great canoes, each hewed from a single trunk and temporarily joined to another, used by Polynesians from Tahiti to Hawaii to cross the ocean. When the Europeans discovered Polynesia, catamarans were the official means of travel.

In 1663, the Englishman William Petty, having learned of the Pacific style of sailing, built two catamarans and raced them successfully against single-hulled yachts. Unfortunately, his following effort, *Experiment*, sank with 50 crew members during a tempest off the Gulf of Gascogne in 1665. Petty made another try 20 years later, but the lack of maneuverability of this *Double Bottom* doomed the effort.

The first American catamaran was the *Double Trouble*, built by John C. Stevens in 1820: Unfortunately, its name turned out to be prescient.

The first person to come up with a seaworthy and manageable sporting catamaran was Nathanael G. Herreshof, working from 1876 to 1881. At the time, these catamarans were the fastest in the world. The hulls were equipped with a kind of independent suspension. The cost and complexity of these vessels, combined with a prohibition against racing them against other ships, impeded the development of boats with two hulls. The first modern, high seas catamaran was launched from Hawaii in 1947 by Brown, Kumalae and Choy. It was called *Manu Kai*, measured 12 meters in length, and was the first sailing vessel to exceed 20 knots. The most popular catamaran model in the world is the *Hobie Cat*, first built in 1967.

Trimaran (1786)

The first boat with three hulls, derived from the Polynesian *latakois*, was built by the Scot **Patrick Miller** in 1786. It was essentially a steam-propelled trimaran with a paddle wheel between each hull. In

1868, the first inflatable boat to cross the Atlantic, the American sailing ship *Non-Pareil*, had three hulls. This crossing was to prove the efficacy of "tri-hulls" as rescue boats. Moreover, around 1890, the three-hulled *La Mouette* served as a rescue boat in Madagascar.

Around 1943, a Russian immigrant to the United States, Victor Tchechett, decided to improve ships with three hulls; he coined the word *trimaran*. It wasn't until the end of the 1960s, however, with the designs of Derek Kelsall and especially those of the American Dick Newick, that trimarans were really refined.

Prau (1933)

Around 1933, **Nathanael Herreshof** thought of applying the Pacific prau boat technique to sailing. These amphidromes (the technical term), whose origins are in Malaysia, are laterally asymetrical. Instead of tacking in a classical way, they reverse direction, the stem becoming the stern and vice versa. While Herreshof built models, an Englishman, Rod Macalpine-Downie, actually built a prau boat, the *Crossbow*, in 1970; it attained 31.09 knots in 1975. The main hull measured 18.3 meters, while the windward hull measured only 4.6 meters. The crew ran the length of the structure between the two hulls to maintain the sailboat's balance.

The only test of a Pacific prau on the open seas was made by Frenchman Guy Delage, who, in November 1982, capsized at the starting line of the Route de Rhum regatta.

In 1968, Dick Newick, an American, invented the Atlantic prau: It also changed directions to turn, but instead of a balance as counterweight on the windward side, it has a leeward float. In fact, Newick's Atlantic prau could be considered a trimaran, minus the windward arm and float; or a catamaran whose windward hull rigging had been moved. This particular solution allows for a much lighter sailing boat and, theoretically, higher speeds. Unfortunately, a sudden change of wind direction on the float-side can capsize an Atlantic prau. If, on the open seas, praus have to be more than 15 meters long to guarantee sufficient stability, the boats are nonetheless interesting since, when compared to any other multihull boat of equal length, they are almost half the price and almost 50 percent lighter.

The first prau for cruising, designed by a Belgian, Daniel Charles, and launched in 1981.

Polyester boats (1948)

Though fiberglass has been commercialized since 1938, the first polyester resin-reinforced fiberglass boats appeared only after World War II. At first, the resin had to be catalyzed under pressure and required heating: A hull measuring 8 meters required vapor-heated molds weighing 11.5 tons. In 1947, resin catalyzing at ambient air temperatures became available; and in 1948, at the New York Boat Show, the first pram (a small rowboat), the "Beetle," was presented as the first civilian example of the technique. In Europe, the first polyester-reinforced fiberglass dingy was the "420," which appeared in 1958.

Wood epoxy boats (1951)

Though wood is a high-performance material, its mechanical resistance changes according to the wood's degree of humidity. Polyester resin does not adhere well to wood and is too porous to be an effective waterproofing agent. Another resin, epoxy, more costly and difficult to work with, does not have these disadvantages. The first epoxy-coated wood boat was made in 1951, in India, by the Italian **Renato Levi**. In 1968, three Americans, Meade, Jan and Joe Gougeon, perfected the wood epoxy-saturated technique. By using this technique as both a glue and a way of waterproofing, it is possible to use high-performance but moisture-sensitive woods like poplar or red cedar. Numerous high-performance multihull boats are built with wood epoxy.

Centerboard and keel (15th century)

Sails on the large, ancient square-rigged boats could not be sufficiently brought in to the centerline to allow effective tacking. When sails that could be drawn in to the centerline came onto the scene (the jib and mainsail), it became possible to sail upwind. Meanwhile, it became necessary to modify the hull with surfaces that would prevent the boat from drifting sideways—keels.

At the beginning of the 15th century, the Dutch equipped their boats with lateral centerboards, two streamlined surfaces which were mounted when tacking. A few hundred years earlier, the Incas equipped their rafts with *guaras*, small centerboards that were driven down vertically between two balsa logs.

Familiar with the Inca system, a British officer, Lieutenant Shank, in 1774 in Boston (then an English colonial city), built a small boat equipped with a vertical centerboard running the whole length of the keel. In 1781, Shank designed a yacht with five vertical centerboards; built for the commodore of the Cumberland fleet, the *Cumberland* won several regattas. But the design was impractical; the five centerboards effectively cut a ship's living space in half—each centerboard required a shaft through the middle of the boat. In 1809 in France, an English captain, Molyneux Schudlam (who was languishing in Napoleon's prison in Verdun), invented a ballasted pivoting centerboard, but this went unnoticed. On the other hand, three American brothers, Joshua, Henry and Jacocks Swain, patented a nonballasted pivoting centerboard in 1811 which was an instant success. An Englishman, Uffa Fox, popularized the planing centerboard sailboat, capable of attaining amazing speeds. In 1928, Fox's *Avenger* won 52 out of 57 races. Since then, the hulls of such sailboats have changed very little; a sailboat like the "505," conceived in 1953, is still among the most technologically advanced.

Keel boats (1840)

The complexity of centerboards and the space taken up by their housing led builders to place fixed centerboards. These fixed ballasted keels first appeared in Bermuda around 1840. In Marseilles, France, boats with nonballasted fixed keels (the *Hovari*) were built towards the end of the 1870s. The first keel-ballasted boat, with a bulb keel, was the 15.47-meter boat *Experiment*, launched in 1880 by an Englishman, E. Bentall. But the first really effective keel of this type was made by Nathanael Herreshof for his boat *Dilemma* in 1891. His sailboat led to a veritable design revolution and prefigured modern monohulls.

Marconi rigging (1895)

Around 1820 the "leg of mutton" sail appeared in the Bermudas. Instead of being trapezoidal like the sails of the time, the sail was triangular. The English imported the design to Europe around 1880, and it became known as "Bermuda rigging." At the same time, it was determined that lengthening the sail influenced the boat's speed. In 1895, a small American yacht owned by **William P. Stephens**, the *Ethelwynn*, had a single-sail Bermuda rigging mast, but the mast was shrouded to give greater rigidity. This streamlined, shrouded rigging was called "Marconi rigging" by analogy with the design of radio antennas. The first great yacht equipped with Marconi rigging was the 34.2-meter-long *Nyria*, designed by Charles E. Nicholson in 1921.

America's Cup (1851)

The oldest sports trophy still in existence is the America's Cup.

Organizers of the 1851 London World Exposition wanted a certain well-known type of New York pilot boat to visit English waters for the event. Rising to the occasion, the commodore and founder of the New York Yacht Club, John Cox Stevens, had a schooner built based on designs by a specialist in pilot boats, George Steers. The boat, the *America*, was underwritten by Stevens and five other investors who wanted to bet with the English, beat them, and win some money—in short, make the affair an adventure that combined sport and profit. Unhappily for the American sponsors, no one in England dared stake the enormous bets proposed by Stevens.

So that the schooner would not have made the voyage for nothing, the English finally came up with an alternative: Instead of betting money on a race, they suggested vying for a silver trophy cup costing 100 guineas and weighing 3827 grams. During a regatta around the Isle of Wight, the *America* managed a shortcut and won by 21 minutes over the second-place *Aurora*. The press heralded the *America* as unbeatable. The New York syndicate sold the boat in Britain and returned to the United States no richer than before; for a time there was talk of melting down the cup to make medals. Finally, though, it was bequeathed to the New York Yacht Club to become a permanent trophy. The cup never once changed hands in twenty-five challenges; but in 1983, the Americans finally lost the cup to the Australians.

The America's Cup races take place just off the coast and rarely cover more than 30 miles. Nevertheless, some of the biggest sailboats of all time have raced for the trophy. In 1903, the *Reliance*, a gigantic boat with 1500 square meters of sails, measuring 62 meters long and manned by 70 crewmen, set the record for the most sails on a single mast.

Solo navigating and solo racing

In 1601, a Dutch surgeon, **Henry De Voogt**, requested a permit to sail solo from Flushing to London; the permit was issued on April 19, 1601, but no one knows whether the surgeon made the voyage.

No doubt the first solo navigators were Eskimos: We know that as early as 1682 residents of islands off northern Scotland encountered them on several occasions; the Eskimos were in kayaks and had traveled some 1700 miles (2500 kilometers).

The first yachtsman to complete a significant solo voyage was Empson Edward Middleton: In 1869, he circumnavigated England aboard a 7-meter yawl. The first man to cross the Atlantic solo was Alfred Johnson, who went from Gloucester, Massachusetts, to Liverpool, aboard a 6.1-meter dory (a flat-bottomed fishing boat) in 1876. His boat is on display in the basement of the Manchester Museum. In 1882, an American, Bernard Gilboy, set out from San Francisco on a "pleasure trip" to Australia aboard a 5.5-meter sailboat. Gilboy was rescued 162 days later, half-starved to death, 160 miles from his destination. He had navigated 7000 miles without putting into port: a record which held for 84 years. The first solo circumnavigation of the globe was made by an American, Captain Joshua Slocum, aboard the 11-meter sailboat *Spray*. It took him from 1895 to 1898 to accomplish the feat.

First race (1891)

The first solo transatlantic race left Boston on June 17, 1891. The two participants were **William A. Andrews**, aboard the *Mermaid*, and **Si Lawlor**, aboard the *Sea Serpent*; both sailing boats measured 4.6 meters. The *Mermaid* was not ballasted and capsized seven times; Lawlor won, but it should be noted that Andrews abandoned the race.

Solo transatlantic, east to west (1960)

In 1958, a World War II hero, **Colonel Blondie Hasler**, proposed organizing a solo transatlantic race against the prevailing winds from Plymouth, Great Britain, to Newport, Rhode Island. A former airplane pilot and map publisher, **Francis Chichester**, learned of the proposition on the eve of being operated on for cancer. A year later, and apparently cured, Chichester found the same challenge stapled onto the same bulletin board! The race began in 1960 and there were four participants. Called the Trans-Atlantic, it is open to all classes of boats (allowing multihulls to prove themselves) and has become the queen of solo races.

Races around the world (1968)

Since the Trans-Atlantic there have been solo races around the world; one of them, in 1968, was nonstop and was won by **Robin-Knox Johnson**, an Englishman. Frightened at the thought of contact with civilization

Walter Cronkite sails along with the crew of Courageous.

again, one of the participants, Bernard Moitessier, a Frenchman, continued on alone to complete a total of one and a half circumnavigations of the globe. On October 31, 1982, John Sanders, an Australian, completed the first double, nonstop circumnavigation of the globe after 420 days at sea.

The smallest boat to ever cross an ocean was the 1.8-meter *April Fool*; on board was Hugo Vihlen, who went from Casablanca, Morocco, to Miami, in 84 days, in 1968.

Skiing

Water skiing (1927)

In 1921, an anonymous skier conceived the idea of having himself pulled by a boat on Lake D'Annecy. Historically, water skiing was created in 1927, in Cannes (on the French Cote d'Azur) by a Norwegian, Peterson.

The first European and world championships were held in France in 1947 and 1949 respectively. There are three different styles or disciplines in water skiing: the slalom, jumping, and figure skiing.

Snow skiing (competitive, 1877)

Skiing is Nordic in origin. It was used as a means of transportation for centuries.

In 1877, the first contests (langlauf and ski jumping) were held in Norway. Shortly after, in the same country, in 1880, Sondve Nordheim had the idea of bending up the front end of the skis. In 1888, a Scandinavian team crossed Greenland and published a story about their expedition. This story gave the Austrian Zderski the idea of shortening the skis and equipping them with metal bindings for hunters. Skiing won over Europe: In 1896, the Frenchman Henri Duhamel founded the Alpine Ski Club.

The first downhill race was held on the Alberg, Austria, on January 6, 1911. It was initiated by a veteran of the Indian Army, Lord Roberts of Kandahar (1832—1914). In 1922, another Englishman, Arnold Lunn, created the slalom, in Mussen, Switzerland. But it wasn't until 1936 that the Alpine (that is, downhill and the slalom) figured in the Olympics.

Soccer (modern, 1863)

Soccer, as it exists today, was born in 1863, when the **Football Association**, founded in London, distinguished it from "football rugby," setting up rules for it and naming it "football

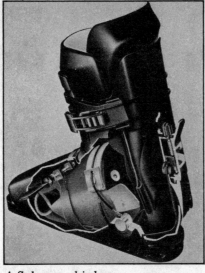

A Salomon ski shoe.

The Futura ski shoe.

association." Except for the goalkeepers, players could no longer use their hands, except when "throwing in."

The Chinese played a comparable ball game around 350 B.C.: tsu-chin. The Greeks and Romans used to kick balls around. The French of the Middle Ages played *a la soule* and the Italians *calcio*.

Two English towns, Derby and Chester, claim to have organized the first soccer game. In any case, the game was part of the Shrove Tuesday festival around 1220. The game remained a Shrove Tuesday tradition up until April 13, 1314, when King

Edward II banned soccer "because too much noise is created in the town by all the shoving and pushing around of the big balls, which may give rise to any number of the many evils that God forbids." However, the edict was not applied.

Swimming (competitive, 1837)

Swimming dates from the earliest times in all civilizations, but the first competitions date from the 19th century: The first meet was held in London in 1837. The Australians held their own meet in 1846 (the Sidney Championships), and shortly after they organized the first world championships.

In 1875, the Englishman Matthew Webb was the first to swim across the English Channel (in 21 hours, 45 minutes). When the first modern

Mark Spitz in 1967.

Jimmy Connors

Yannick Noah

Bjorn Borg

Olympic games were held in 1896, there were three swimming contests: the 100 meter, the 500 meter, and the 1200 meter. Only the 100 meter has remained an Olympic contest. The crawl did not yet exist. It was invented in 1900 and revolutionized swimming. At this time water polo appeared as well.

Tennis (1793)

Origins

Tennis was invented in 1793 by **Walter C. Wingfield**, a major in the English army stationed in India. The Greeks had already played *a la phoeninde*, in which one player tried to upset another by hitting a ball at him after having pretended to hit it to someone else. The Romans played *hapaste*. In France, "royal tennis," a game played indoors, appeared around 1050.

Major Wingfield introduced rules borrowed from the Indian game *Poona* and used grass as a playing surface. The first match was played in the *Pays de Galles*, September 29, 1793. Wingfield had baptized the game *spharistike* after the Greek name for games with balls (*spherique*). The word *tennis* comes from the French: The server would shout "*Tenez!*" (Hold! Attention!) in order to warn his opponent that he was about to serve; the English made this *tennis*.

A friend of Major Wingfield brought his racket and balls with him to Bermuda, while on vacation. There he met a young American woman, Mary Ewing Outerbridge, who was quite enthusiastic about the new sport. In March 1874, she introduced the game to the United States, and the first matches were played at the Cricket Club in Staten Island.

Tie break

In order to limit an overlong game, the tie break, first introduced at Forest Hills by the professionals, was made official at Wimbledon in 1971. With a score of 6 even, it allows a set to be ended by a final game won by the first player to win 7 points (or more), with 2 leading points.

The first big matches (1877)

In 1877, the first Wimbledon match (English International) was won by S. W. Gore in men's singles, and in 1881, the Forest Hills Match (American International, played since 1978 at Flushing Meadows) was won by R. D. Sears. The first Australian International was played in 1905, and the first French International (played at Roland Garros since 1928) was won in 1925 by Rene Lacoste. The Davis Cup was inaugurated in 1900.

Half-court (1970s)

In Australia, during the 1970s, an individual wanted to build a tennis court in his back yard, which was too small for one. So he dreamed up the possibility of a reduced court that would still permit full exercise of the sport.

The half-court is played on a surface of 12.62 meters by 6.4 meters— that is, one-third the size of a tennis court, with rackets of about 50 centimeters, having a very short handle and stringing that is less taut than for usual tennis rackets. The balls used are of the same size, but softer. The half-court method was promoted in Australia by players such as Tony Roche and Allan Stone.

After Australia and the United States, the half-court won over Europe and France. The first Paris Internationals took place in November 1983, at the French Stadium.

Platform tennis (1928)

Platform tennis was born in 1928 on the East Coast of the United States, according to the idea of **F. Blanchard** and **J. Cogwell**, who, by raising the surface, were looking for an easier way to sweep the snow away during the winter. Further, traditional tennis courts demand substantial open space. The principal quality of the platform is that it can be installed everywhere, or nearly. The court is four times smaller than the traditional court. The game is played using a grid racket, whose format is somewhere between a ping-pong paddle

Rene Lacoste

and the usual tennis racket. The ball is solid. The originality of the game is based on the enclosure grill, which provides a play-back surface.

Track and field events (antiquity)

Since ancient Greek times, track and field events have occupied an important place in stadium games. Modern track and field events began their development in Britain, especially after the middle of the 19th century. They were included in the first Olympic games of modern times, held at Athens in 1896.

Trampoline (1936)

It was an American, **George Nissen**, who in 1936 built the first prototype trampoline, the Model T. The name of this sport comes from the Spanish word *trampolin*, which means "springboard."

ULM (ultra-light motorized) flyers (1970)

Around 1970, the hang glider became extremely popular. However, for people living in the plains, it is often necessary to travel great distances in order to practice this sport. For this reason many devotees of individual free flight dreamed of motorizing their Delta wings.

In 1975, the first reliable prototypes appeared almost simultaneously. The American **Homer Kolb** perfected his *Flyer,* which he had been working on since 1970. He is regarded as the father of this new sport. The Frenchman Jean-Marc Geiser made the first trial turns with his *Moto-delta*: a union of a Delta wing and a wagon he equipped with a flatwing, 12-horsepower engine. An

Olympic Runners Carl Lewis and Calvin Smith.

Australian, Gary Kimberly, also flew aboard his *Scout*. The Australian government established regulations for flyers, and Kimberly was then able to commercialize his *Scout*. Various other models emerged soon after.

The first gathering of ULM flyers took place in Brook Fields, Michigan, July 1-3, 1977. John Moody climbed to 3937 feet (1200 meters) in his *Easy Riser* (equipped with a 12-horsepower engine), and he set the record for time aloft: 32 minutes. On May 9, 1978, David Cook was the first to cross the English Channel, going from England to France in 1 hour 15 minutes.

Volleyball (1895)

Volleyball was invented in 1895 by an American, **William G. Morgan**, a gymnastics professor at the Holyoke, Massachusetts, YMCA. The game was at first called mimonette. It became an Olympic sport in 1964 and is now played all over the world.

Wind surfing (1958)

Wind surfing was invented by an Englishman, **Peter Chilvers**, in 1958. But this mechanic for old Rolls-Royces was happy just to sail about by himself, and he never exploited his invention.

In 1964, an American, Newman Darby, had the same idea and mounted a sail on a surfboard. Four years later, two Californians, Jim Drake and Hoyle Schweitzer, who had never heard of their forerunners, also put up a sail on a surfboard—in order to be able to practice their favorite sport in calm seas. But they understood that just a board and a sail were not enough, and they invented the leeway, the articulating joint, and the wishbone, thus giving the wind surfer its definitive form. They took out a patent on March 27, 1968. But they didn't protect their invention in France. Moreover, their wind surfer was made of wood and was heavy and expensive. The Germans and the French innovated and brought the wind surfer to a higher state of development. Today, after important technical advances, and intense legal battles, the premier manufacturer is French—Bic Marine.

In 1982, the Frenchman Christian Marty crossed the Atlantic in 36 days on a wind surfer. He was followed by a boat. In the same year, another Frenchman, Pascal Malka, reached 27.8 knots on a wind surfer so small that, at rest, it sank under a man's weight.

There is a prize of $10,000 for the first wind surfer who succeeds in turning a 360° loop with his board while jumping a wave.

Inflatable wind-surfing board
(1981)

In 1981, the **Hutchinson-Marine Company** perfected Hutchinsurf, an inflatable wind-surfing board. Easily transported when deflated and folded in its sack, the board is convenient for wind surfers who do not have cars.

Aluminum wind-surfing board
(1982)

In 1982, a French expatriate, **Chabiland**, developed an aluminum wind-surfing board for the Hawaii Surf Company. Very sturdy (its builder guarantees it for three years), it can be used by beginners, in regattas, or for free-style.

everyday life

THE KITCHEN

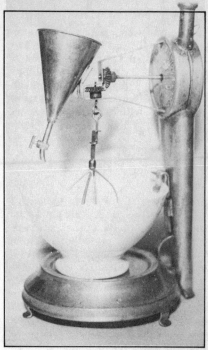

This mixer was powered by water under pressure.

1. Cooking

Beater (1910)

Fred Osius, Chester A. Beach, and L. H. Hamilton marketed the first household electric motor that could be used with alternating or direct current in Racine, Wisconsin, in 1910. They first applied it to the home sewing machine and to professional beaters used for making milkshakes.

In 1923, Air-0 Mix Inc. of Wilmington, Delaware, marketed Whip-All, a portable beater. Finally, mention should be made of Sunbeam of Chicago's Mixmaster, the first stationary (like present day kitchen models) beater to achieve success: 60,000 units were sold in 1930, the year it was marketed.

Coffeepot (1802)

The Frenchman **Descroisilles**, a Rouen pharmacist, invented the coffeepot in 1802. It was then called a *cafeolette* and consisted of two superposed containers separated by a filter. Shortly afterwards, the chemist Antoine Cadet created the porcelain coffeepot.

But the true innovation in the art of coffee-making belonged to the Italian Gaggia, who in 1946 invented the espresso machine. Since then, espresso has become synonymous with good coffee all over the world.

Coffee filters (1908)

A German woman, **Mrs. Melitta Bentz,** wished to improve the quality of her family's coffee. She pierced holes in the bottom of a pewter container, then cut out a disc of absorbent paper which she placed at the bottom of the container. She put the container on the coffeepot, put coffee in the container, and poured in boiling water. This coffee filter produced excellent results. Thus, the Melitta filter was born in 1908.

Coffee grinder (1687)

The invention of the coffee grinder in 1687 gave this new drink more widespread consumption. The electric coffee grinder was invented in 1937 by the Kitchen-Aid division of the Hobart Manufacturing Co. of Troy, Ohio. The first model was the 10A, and it sold for $12.75.

A machine that washed, dryed and sterilized dishes in 1925.

Corkscrew (18th century)

It was the widespread use of water-tight corks toward the end of the 18th century that made the corkscrew indispensable. We do not know who first invented it, but in 1795 an Englishman named **Samuel Hershaw** perfected a model with a moving screw and fixed collar. Lucien Monod, a French inventor, created the world's fastest corkscrew in 1976.

Dishwasher

The earliest dishwashing model, inspired by the same principle as for the washing machine (i.e., a crank-driven paddle or propeller machine), was developed between 1850 and 1865 in the United States. The electric-motor dishwasher appeared in 1912. During the 1920s, small-scale production began in the United States. In 1932, the discovery of a suitable detergent, Calgon, helped promote the development of this household aid.

The automatic dishwasher, also created in the United States, in 1940, was not exported to Europe until around 1960. Its compactness and simplicity compared with earlier models assured it rapid public acceptance.

Electric kettle (1893)

An electric kettle had already been presented in Chicago in 1893. But the heating unit, an enamel-coated wire, was unreliable. It was not until 1923 that a resident of Birmingham, England, **A. L. Large**, invented a kettle whose heating unit was completely buried inside it. Boiling time was reduced by one-third.

Finally, in 1931, another Birmingham manufacturer, **W. H. Bulpitt**, developed a safety switch. When his *Swan* kettle overheated, a fuse blew and power was cut off.

Food processor (1947)

This kitchen appliance, destined to equip millions of home kitchens throughout the world, has its roots in a 1947 design by the Englishman **Kenneth Wood**. Marketed as the Robot Kenwood Chef, it was composed of a powerful and sturdy engine block to which a large number of accessories could be fitted: mixer, citrus squeezer, mincer, slicer and shredder, centrifuge, pasta and ravioli maker, food mill, can opener, etc. Its multiple uses allowed it to singlehandedly replace a large number of small appliances.

In 1963, the Robot-coupe was patented by the French inventor **Pierre Verdun**. This device is simple. It consists of a cylindrical tank inside which a knife revolves close to the bottom and the walls.

Within a few years, the Robot-coupe made a name for itself in the trade. But Pierre Verdun did not stop there. On March 1, 1971, he marketed the Magimix for the "housewife market." This automatic appliance cuts, mixes and kneads. But it did not catch on with the French. Americans, on the other hand, who knew it under the name of Cuisinart, readily took to it. In 1977, 385,000 Cuisinarts were sold in the United States out of a total production of 530,000.

Finally, in 1980, the Robot-coupe was marketed in the United States under its own trademark, with the slogan: *You pronounce it Robot-coop*.

Fork

Distant precursors of the modern fork have been unearthed at the diggings made at the site of *Catal Hoyuk* in Turkey (6th to 3rd millennium B.C.). It then seems to vanish, and the first indication of its "reinvention" is its mention in certain inventories drawn up in the 14th century. It was probably brought to the Occident by the Italians. The 1307 inventory of Edward I, King of England, mentions seven forks, including one in gold. In 1380, the inventory of Charles V of France mentions twelve forks, some of them "decorated with precious stones." Finally, the inventory drawn up at the death of the Duchess of Touraine in 1389 lists two gilded silver forks.

However, it should be noted that these "two-tine" forks were quite rare among the large quantity of silverware. A symbol of luxury and refinement, the fork only came into general use in France at the close of the 18th century, at the same time as individual plates and glasses. It was still absent from humble tables on the eve of the French Revolution.

Gadget (1886)

If we pronounce the word "gadget" correctly, it should be "Gah-jay"

George IV, King of England, loved to eat. Here he is eating with a two-tined fork.

(spelled "Gaget" in French). Surprisingly, though this word seems so American, it's origin is actually French. **Monsieur Gaget** was one of the partners in the French construction firm of Gaget, Gauhier & Cie., which built the Statue of Liberty (designed by the French sculptor Bartholdi). For the inauguration ceremonies in 1886, he got the idea of selling miniature souvenir Statues of Liberty to the Americans. Given the American pronunciation of his name, Gaget's souvenir statues became "gadgets," and the century-old career of this term was inaugurated along with the world-famous statue of "Liberty Lighting the World."

Matches (1681)

The first matches using sulfur and phosphorus were invented in 1681 by the Englishman **Robert Boyle** (1627—1691). They were simple wooden sticks that had been dipped in melted sulfur. They could only ignite after contact with a burning object.

Phosphorus matches were created by Charles Sauria, a pupil at a high school in the French Jura, but only began to be produced industrially in Germany after 1832, at the initiative of Frederick Kammerer. Their chief drawback was that they caught fire with disconcerting ease.

A Swede, Carl Frantz Lundstrom (1815—1888), finally solved this problem. In 1844 he replaced yellow phosphorus by red or amorphous phosphorus. Today's matches descend from this process.

Microwave oven (1947)

An American, **Percy Le Baron Spencer**, while working on the emission of short-wave electromagnetic energy, discovered the microwave oven by accident, in 1946. He was doing research for the Raytheon Company when he noticed that the microwaves had melted a piece of candy in his pocket. As we now know, microwaves directly affect the food they cook by creating within the foodstuff a molecular vibration that produces heat.

Raytheon developed Spencer's observation and marketed the first microwave oven in 1947.

Non-burning hotplate (1984)

This hotplate was invented in 1984 by **Anton Seelig**, an engineer with the German firm AEG. It can heat the contents of a saucepan without becoming hot itself. Because the food is heated by electromagnetic induction, children can place their hands on the

Storyteller, Hans Christian Andersen's "Little Match Girl" offering her wares to passersby.

hotplate surfaces without any danger of being burned.

Plastic bottle (1969)

The Societe generale des eaux minerales de Vittel (Vittel Mineral Water Company) invented the plastic bottle. Starting from the assumption that plastic bottles would be easier for all parties to use—from wholesaler to consumer—four years of research was required to develop the *maxiround* in 1969. They had to discover a plastic that would be perfectly water-resistant and satisfactorily designed. In 1976, the *maxisquare* further eased storage difficulties.

Plate

In Antiquity, guests helped themselves from a communal dish or cauldron.

During the Middle Ages, the tren-

cher and the porringer (used by two people at a time) appeared. In the 15th century, ceramic bowls began to be produced.

In the 18th century, trenchers and porringers, used mainly by the underprivileged classes, were abandoned in favor of individual plates.

Pressure cooker

The inventor of the pressure-cooking method is the Frenchman **Denis Papin** (1647—1714), who in 1680 developed a device intended to soften bones and cook meat in a short while. The problem was that food cooked at 302°F (150°C) lost its flavor, its color, and its vitamins.

In 1927, the French engineer **Hautier** patented the first controlled low-pressure cooker.

Finally, the SEB (*Societe d'emboutissage de Boulogne*, France) Super Pressure Cooker resulted in 1953 from the research of the Lescure brothers (Frederic, Jean and Henri). It was the fruit of observations made on some forty pressure cookers from France and abroad.

Refrigerator (1913)

The Domelre, manufactured in Chicago in 1913, was the first functional household refrigerator. Afterwards, the American **Nathaniel Wales** designed a device that was widely marketed under the name of Kelvinator starting in 1918. The Frigidaire trademark appeared one year later in 1919.

Until 1916, the United States had a choice of a dozen makes of costly and unreliable appliances. Most were composed of two quite distinct parts: an insulated casing, cooled by brine coils, and a chilling unit often placed in the basement. These chilling units made use of sulfur oxide and oxygen. In 1931 the R 12 machine was developed, but it did not enjoy immediate commercial success.

The Swedes C. Munters and B. von Platen succeeded in constructing a silent and functional refrigerator. They filed their first patent in 1920, and developed a condensor device in 1929. Mass production began in 1931 with Electrolux, in Stockholm, and Servel, in the United States.

In 1926, the American company General Electric manufactured a hermetically sealed household unit and, in 1939, it introduced the first dual-temperature refrigerator. This allowed frozen food to be preserved in one of the unit's compartments.

Tefal no-stick frying pan (1954)

The Tefal no-stick frying pan was invented in December 1954. A French engineer, **Louis Hartmann**, had been carrying out research for a number of years on PTFE (polytetrafluorethylene), a compound remarkable for its sliding and non-sticking properties. Hartmann discovered a process whereby one could fix a thin layer of PTFE on an aluminum surface. This was no mean feat, as to this time no metal would stick to PTFE.

A colleague, Marc Gregoire, then had the idea of using Hartmann's process for fixing PTFE on an aluminum frying pan. Gregoire founded, in 1956, the Tefal Company, in order to market this invention which has transformed the life of cooks.

Thermos (1906)

In 1906, the Scottish physicist and chemist **Sir James Dewar** (1842—1923) invented this heat-insulating device in order to preserve liquified gases.

The Dewar container is a double-walled silver-plated glass insulating container in which a vacuum has been created. The result is an almost total elimination of radiation (reflected by the silver-plated double wall), convection, and external heat conduction. The standard version of the Dewar container is the thermos flask.

Toaster (1909)

The first toasters were marketed by the General Electric Company of Schenectady, New York, in 1909. They quite simply consisted of bare wires wound around mica strips.

In 1927, a Stillwater, Minnesota mechanic invented a toaster that could heat the toast on both sides and pop it up when done. In his first toaster, a clockwork mechanism cut off the power and released a spring. Later, in 1930, the system was improved by adding a thermostat which shut off the power when the bread reached a certain temperature.

Possibly this chef would have an easier time if he used a non-stick frying pan.

Tupperware (1945)

The small multicolored plastic boxes that are known throughout the world were the brainchild of the American **Earl W. Tupper**. A former chemist with the American firm Du Pont de Nemours, Tupper had a double brainstorm in 1945. He invented boxes of all sizes and shapes with a plastic lid which, although flexible, provided airtight sealing. Then, he set up an absolutely revolutionary distribution system.

Women held tupperware parties in their homes to which they invited their neighbors, friends and relatives. They played the role of demonstrators/saleswomen of tupperware boxes, and for this they were paid . . . in tupperware boxes.

STYLE, BEAUTY, AND HEALTH

2. Clothing And Accessories

Bikini (1946)

On July 5, 1946, the French fashion designer **Louis Reard** presented in his collection a daring two-piece bathing suit, the *bikini*. The bikini was so named because its creator considered it to be as explosive as the American atom bomb which had exploded four days earlier in the Bikini atoll in the Pacific.

To symbolize this marriage of fashion and contemporary events, the original bikini—worn by the model Micheline Bernardi—was made of a cotton fabric printed with press clippings.

Bloomers (1851)

About 1851, the American **Amelia Bloomer** devised a most original sports outfit for the girls attending her school. It consisted of baggy trousers that were tight at the ankles and a short belted tunic. But the age only had eyes for crinoline, and Amelia Bloomer's invention met with no success.

Button

Amulets carved out of seashells and pierced with two holes have been discovered in the Indus valley. They may have been buttons, whose invention would then date as far back as the third millennium B.C.

A thousand years later, jet buttons were being made in Scotland. We next encounter the button during the Middle Ages, used initially as an aesthetic embellishment and later as a utilitarian object (14th century). During this period, the material it was made of began to be diversified.

Crinoline (1846)

Crinoline, or more precisely the crinoline petticoats worn by women to swell the volume of their dresses, first appeared in 1846. These petticoats were made of a fabric woven of horsehair, this fabric having been invented by the Frenchman **Oudinot** in 1830 in order to stiffen soldiers' tie collars.

But the fabric was expensive, and as the size of petticoats grew ever larger, crinoline soon became a cage stiffened with whalebone and flexible steel hoops. The best-known innovations were the steel underskirt, invented by the Frenchman Tavernier in 1856, and the Thomson American cage of the same period.

Dry cleaning (1855)

The first dry-cleaning laundry was established in Paris, in 1855, by the Frenchman **J. B. Jolly**. He had accidentally stumbled upon the principle of dry cleaning. After spilling a bottle of turpentine on a dress, he noticed that, far from being stained, the dress had been cleaned.

A German from Leipzig, Ludwig Anthelin, made another breakthrough in 1897. He discovered the use of carbon tetrachloride, which was much less flammable. Unfortunately, this product affected the respiratory tract.

It was therefore replaced in turn, in 1918, by trichlorethylene.

Fan

The fan has been known since earliest antiquity. The Egyptian pharoahs were fanned by slaves using giant lotus leaves. The *ripis* of Greek women and *flabellum* of Roman women were trimmed with peacock feathers.

During the Middle Ages, fans were made of parchment, on ivory or precious-metal mounts. The folding fan appeared toward the end of the 16th century.

Knowing use of a fan can be quite seductive.

The French writer Rabelais described the fly or codpiece, but the Flemish artist Brueghel painted it.

Fashion parade (1908)

The first true fashion parade was held in 1908 in London, probably to present Lady Duff-Gordon's collection at Hanover Square. Similar showings were held the same year by the houses of Paquin and Poiret in Paris.

Since 1905 the great fashion houses had already been showing their models at set times for their clients, but without any special ceremony.

Fly

The early *codpiece* appears for the first time in literature during the 15th century. Shakespeare used it in several of his plays. Its etymology derives from a bag worn in front of the breeches.

The 16th century witnessed an exaggerated development of the codpiece, which even served as a pocket. According to certain chroniclers, some lords would slip an apple inside it as a form of boasting. It was only toward the middle of the 19th century that masculine trousers opened by a vertical frontal slit, our modern fly.

Haute couture (1857)

However great a blow it may be to French pride, **Charles-Frederick Worth**, universally acknowledged as the founder of haute couture, was not French. This young Englishman was born in 1825. At the age of 20, he arrived in Paris, the world capital of elegance, and made his debut as a salesman for the Gamelin firm, the well-known dry goods store.

Twelve years later in 1857, with the financial backing of the Swede Otto Bobergh, Charles-Frederick opened a house at 7, rue de la Paix whose name would henceforth be famous: Worth. It was the first haute couture house.

Jeans (1873)

Jeans were created in 1873 by **Oscar Levi Strauss** for pioneers in the West. These hard-wearing trousers were originally cut out of a blue cloth which served for tenting. This cloth had been imported from Nimes, France's traditional production center, hence the term *denim*.

The earliest mention of the word *jean* dates from 1567 and appears to be a corrupt form of the word *genoese*, from Genoa. A twill cotton fabric was also manufactured in Genoa, where sailors wore pants made of this material. Jean is a heavy cotton cloth having an ecru woof and an indigo warp.

Lady's suit (1885)

The lady's suit, a feminine outfit directly inspired by the man's suit, appears to have been invented around 1885 by the English fashion designer **Redfern**. The first model was worn by the Princess of Wales, Alexandra: a plain wool lady's suit consisting of a skirt, a blouse and a strictly tailored jacket. Redfern's stroke of daring: The jacket designed for the Princess of Wales had a suit collar and lapels, features until then reserved for the masculine suit.

Mannequin

Store dummies

The advent of the store dummy coincided with the birth of the department store. The forerunner of the store dummy was the patented bust which the Frenchman **Alexis-Marie Lavigne** created in the 1860s. It was still quite rudimentary.

One of Lavigne's employees, the Belgian Fred Stockman, put to good use the knowledge he had acquired in his architecture courses by founding, in 1869, the Stockman Brothers Company to produce busts and dummies. It offered its customers a full range of dummies. The bodies were complete with joints and the wax heads wore perfectly realistic natural hair. Stockman even built a dummy bicyclist able to pedal for three hours.

Live models (1846)

The idea of presenting clothes modeled by a woman belonged to Charles-Frederick Worth (see *Haute couture*). At the time (1846) a salesman with the Gamelin firm, the future high-fashion designer asked one of the store's pretty salesgirls, Marie Verne, to model the shawls he wished to show to advantage.

Beginning in 1851—after marrying Marie Verne—Worth designed outfits for her whose success was so great that in 1857 he decided to open his own store. As time went by, Mrs. Worth could no longer present all the models by herself, so the fashion designer hired young women to specialize in this work.

Until 1905, mannequins presented designs over tight-fitting black sheaths, with long sleeves and high collars. At first they were nicknamed *doubles* because of their deliberate likeness to a certain high society customer, then *Miss Mannequin.* Finally, the girls who presented the designs were called *mannequins.*

Miniskirt (1965)

The miniskirt was created by the dress designer **Mary Quant** in her store, Bazaar, on King's Road, London in the spring of 1965.

Almost simultaneously, the French fashion designer Courreges was creating a line that was very constructed, quite short and futuristic. Wearing miniskirts, opaque tights and small helmets, fashionable women resembled astronauts. But this was a haute couture collection, whereas Mary Quant's skirts were enthusiastically taken up by a generation of young women.

Needle

The needle dates from prehistoric times. It was then made of fishbone or a piece of bone with an eye pierced in the middle or at the tip of this rudimentary implement. Needles have been uncovered in the oldest archeological strata at Troy. Homeric Greece was acquainted with ivory needles. In ancient Egypt, copper and bronze needles were used, and in ancient Rome iron needles served for sewing.

In 1370, the town of Nuremberg was already a production center for burnished iron needles, whose invention has been ascribed to the Arabs. This type of needle only gained general circulation much later on: about the middle of the 16th century in France and during the 18th century in England.

New look

On February 12, 1947, with the backing of the textile manufacturer Boussac, **Christian Dior** launched a totally new collection that would be chris-

tened the *new look.* Reacting against wartime austerity, the fashion designer presented models wearing long wide skirts, with slim waists and full busts. He had adopted and accentuated an idea launched by Balmain and Rochas two years earlier. The *new look* was so novel for its time that it drew the wrath of certain women's groups.

Pin (1817)

In 1817, the American **Seth Hunt** informed the Patent Office of his invention of an automatic machine for manufacturing one-piece pins, with body, head, and point. His machine began operation in 1824, when Samuel Wright filed a patent in England. England had been the home of the industrial pin since John Tilsby had founded the first large pin works in Gloucester in 1625.

Hunt's machine was improved in 1838 by the Englishmen Henry Shuttle Worth and Daniel Foote Taylor, from Birmingham. Their pin was more functional and less dangerous.

Hatpin (1832)

In 1832, the American **John Ireland Howe** patented a machine for manufacturing one-piece hatpins. This machine's performance was amazing. Ten pin workers could produce some 48,000 hatpins per day, using 121 lbs (55 kg) of metal. (A skilled worker could manually produce only twenty per day.)

Safety pin

The invention of the modern safety pin is attributed to the American **Walter Hunt,** who developed it in 1849. It appears that fibulas and brooches, used in ancient Crete to attach draped clothes, were made according to a related principle.

Raincoat (1748)

The first attempts to waterproof clothing and footware were made by the French mathematician, astronomer, and engineer **Francois Fresneau** (1703—1770). In 1748, he delivered a paper to the Academy of Sciences in which he related that the boots and overcoat he had coated with latex sap had been waterproofed. Fresneau stated, however, that the rubber could not be used untreated and it was vital to be able to dissolve this resin on the spot and render it liquid in order to shape new objects.

Finally, it was the Scotsman Mackintosh who in 1823 was the first to succeed in producing a waterproof

The New Look was introduced by Christian Dior in 1947.

cloth that could be used to manufacture clothing. His cotton fabric, imbued with a mixture of rubber and turpentine, maintained its full flexibility. The *mackintosh* would thereafter designate the waterproofed overcoat worn by men during the 19th century.

Rayon (1895)

The name "rayon" was applied in 1936 to a fiber already known by the name of *viscose*. This material, produced in 1895 by two Frenchmen, **Charles Crosset** and **Ernest Bevan**, was obtained by dissolving cellulose extracted from cotton waste in caustic soda, then in carbon disulphide.

Synthetic fibers

The earliest studies made on the production of synthetic fibers (obtained by a synthesis, notably, of coal and petroleum) date from 1922. This research led, after 1930, to:

1931: The PA 6-6 polyamide (nylon) was produced by Carothers and marketed by Du Pont de Nemours in 1938 (United States).

1939: Plumkett synthesized polytetrafluorethylene (Teflon), which began to be produced in 1954.

1941: Klatte and Corbiere developed vinyl polychloride (Rhovyl), marketed in 1943.

1943: Rein developed polyacrylonitrile (Orlon), which began to be produced immediately.

1944: The PA II polyamide (Rilsan) was invented by Zeltner-Genas and produced in 1950.

Sewing tape measure (1847)

On September 3, 1847, after four years of research, the French fashion designer and professor of fashion design, **Lavigne**, filed a patent for a machine to print ribbon intended for making supple sewing tape measures for seamstresses.

Shoe (Neolithic times)

The Roman Museum contains a pair of Egyptian shoes made of papyrus which go back to 4000 B.C. But the invention of the shoe goes back even further, most certainly to Neolithic times.

The invention of shoes specifically designed for the left foot or the right foot only goes back to the middle of the 19th century. Up until that time, people had to have their shoes made to measure if they wanted them to be the right size.

Snap-fastener (1855)

The snap-fastener was invented in 1885 by a Grenoble (France) manufacturer, **Paul-Albert Regnault.** This metal snap-fastener was first used as a fastening device in the local glove industry.

Stockings

In ancient Rome, women wore bandages wrapped around their feet and legs. These served as stockings. Men then adopted this use and it became widespread in Europe up until the middle of the medieval period, when bits of fabric replaced bandages.

During the 16th century, probably in Spain, the idea arose of making stocking breeches independent clothing items and of knitting them like gloves. At the end of the 16th century, the Englishman W. Lee's invention of the stocking loom revolutionized their production. A little later, the Frenchman Fournier began producing silk stockings in Lyons. Cotton stockings had appeared by the middle of the 17th century.

The hosiery market was radically transformed after the development of nylon in 1938, by a research team at the American company Du Pont de Nemours (headed by Dr. W. Carothers). The first nylon stockings were sold in Europe in 1945.

Suspenders (1840)

In 1840, the Rouen (France) artisans **Ratier** and **Guibal** invented modern suspenders, whose resistance and flexibility were due to the placement of rubber straps at the tips of strips of cloth.

Marlene Dietrich and her famous garters in "The Blue Angel."

Tie

The tie seems to be of military origin. Roman soldiers already wore the *focale*, a kind of tie, around their necks. But it was not until the 17th century that use of this scarf, first worn by Croatian soldiers, became general.

The wearing of ties is generally attributed to Croatian soldiers, but two accounts exist. The fashion may have been brought into France around 1600 when the French, who were fighting alongside the Swedes during the Thirty Years War, found it handy to use these tied scarves. The second version traces this fashion to 1668, when a regiment of Croatian mercenaries arrived in France. Officers' ties were made of silk or muslin, and the ends, which trailed down to the chest, were decorated with tassels.

Top hat

The French and English both claim paternity of the top hat. A painting by the French painter Charles Vernet (1758—1836) titled *Un Incroyable de 1796* depicts a dandy of his time wearing a stovepipe hat. According to other accounts, this hat first appeared on the head of its inventor, a Strand haberdasher by the name of John Etherington, on January 15, 1797. An issue of the *Times* from that period bears witness that when John Etherington left his shop, his extraordinary hatpiece drew a crowd and the curiosity of onlookers quickly degenerated into a shoving match. Etherington was summoned to appear in court before the Lord Mayor for disturbance of the peace.

Velcro (1948)

Velcro is a Swiss invention whose discovery dates from 1948. Returning from a day's hunting, the engineer **Georges de Mestral** often noticed that thistle blossoms clung to his clothing. Under the microscope he discovered that each of these blossoms contained minute hooks allowing it to catch onto fabrics. It then occurred to him to fix similar hooks on fabric strips which would cling together and serve as fasteners.

Eight years were needed to develop the basic product: two nylon strips, one of which contained thousands of small hooks, and the other even smaller loops. When the two strips were pressed together, they formed a quick and practical fastener. The invention was named *Velcro* (from the French *velours*, velvet, and *crochet*, hook). It was patented worldwide in 1957.

Zipper (1890)

Around 1890, the American **Whitecomb Judson** devised a quick zipper system based on interlocking small teeth. The idea was ingenious but its practical application not simple. Judson filed his patent in 1893 and entered into partnership with a lawyer, Walter, to found a company.

In 1905, machines for manufacturing zippers were in operation, but production was far from perfect. It was not until 1912 that Judson's invention provided full satisfaction to its users: this due to the improvements made by the Swede Sundback.

The American zipper was marketed by the Goodrich Company, which used it on boots.

3. Beauty And Cosmetics

Cold cream (1911)

In 1911, a Hamburg pharmacist named **Beiersdorf** invented a cream for skin care. It was marketed by the firm he had founded in 1882. Snow-white, the product was called Nivea cold cream.

This modern cosmetic would supplant the heavy beauty creams then in use and be adopted all over the world on all levels of society. By combining for the first time a nourishing and moisturizing action, it opened a new era and allowed the Beiersdorf Company to expand its market and develop its product line.

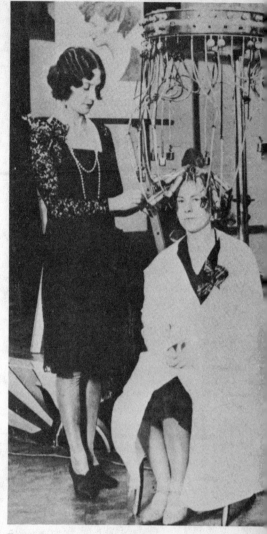

Getting a permanent in 1929 was a rather complex affair.

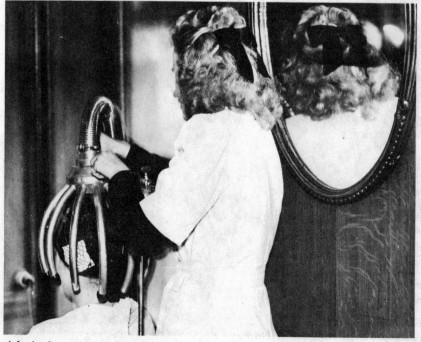

A hair dryer around 1945.

Hair styling

Hair set (1870)

The Frenchman **Lentheric** invented, in about 1870, the modern hair set. Waviness was no longer achieved by heating the hair with a curling iron but by hot-air drying. Before drying, Lentheric would set the hair in waves by using special strand curlers.

In April 1882, the Frenchman Marcel opened a shop on 2, rue de l'Echelle in Paris. He handled the curling iron with such dexterity that he soon was considered the king of the wave.

The Parisian hairdresser Rambaud realized during the 1920s the benefits to be obtained by combining the techniques of the permanent and the hair set. After cutting and permanenting the hair, Rambaud curled it, fixing it with small hair pins. Drying occurred under a hair dryer blowing hot air. Results were amazing. Rather than the permanent ringlets produced by the hot permanent used alone, soft and loosely waving hair was now obtained.

Hot permanent wave

On October 8, 1906, **Nestle**, a German hairdresser living in London, demonstrated a new method of curling hair. Each lock was steeped in an ammonia solution and curled around a safety pin. Then the ammoniated lock was heated intensely with an iron. Nestle later emigrated to the United States,

where his invention enjoyed a huge success under the name of *permanent waving*.

In 1914, a disciple of Nestle, Eugene Sutter, developed the first *Eugene* apparatus in which the hair-dryer contained 20 heaters.

In 1919, the Frenchman Gaston Boudou, the founder of the Gallia firm, improved Eugene Sutter's device by making it safer to use. He also invented the automatic roller.

In 1925, the Frenchman J.M. Leclabart filed a patent for his invention of a wire device, the *Perma standard*. This device was manufactured by the newly created Perma Company, in partnership with Alsthom-Thomson.

In 1927, Franz Stholer developed the *Wella* device.

In 1928, Schwarzkopf invented *curling tablets*: reducing and neutralizing agents.

In 1934, Leclabart brought further progress to the hot permanent wave technique. He invented a wireless device, the *Regina*. The irons, actual heat accumulators, were heated on the device, then placed on the curlers.

Cold permanent wave

The invention of the cold permanent was preceded by several attempts to develop a warm permanent.

On December 10, 1934, an English professor, **Speakman**, decided to use sodium sulphite at 104°F (40°C) as a reducer.

On June 16, 1941, the Americans MacDonough and R. Evans filed a patent that contained an important innovation with respect to Speakman's process. The mineral reducer was replaced by an organic reducer: thioglycolic acid and certain of its salts.

On October 2, 1945, the L'Oreal laboratories (L'Oreal was founded by Eugene Schueller, a chemical engineer) filed a French patent. The principle was the same but the permanent wave was set by a traditional peroxide. This invention gave rise to a new product, L'Oreal, which, due to the action of thioglycolic acid, brought about a "lasting curl," eliminating the complicated electrical apparatus used in hot permanents.

In 1967, the *Miniwave* appeared. This formula produced a soft permanent wave through the combination of certain chemical components and an original hair-set device.

Heating hair-curlers (1960)

They were invented by the Dane **Aren Bybjerg Pederson** in 1960. His first model was baptized *Carmen*.

Hair dye (1909)

The first conclusive tests were made in 1909 by the French chemist Eugene Schueller. He founded the *French Harmless Hair Dye Company*, which became, in 1910, the L'Oreal Company. The dye's base was paraphenylene-diamine, nicknamed by the trade "para." It was called "The Oreales."

In 1927, the invention of *Imedia*, a dye made from organic coloring matter, revolutionized the hairdresser's art by the variety and naturalness of the shades available. It represented a considerable advance for covering white hair.

1953 witnessed the birth of *Imedia Creme*, which "in a single operation" both bleached and dyed without the need for two separate products.

Also dating from 1953 was a new type of hair dye called *Rege Color*. It imparted shiny reflections using a cosmetic base in order to highlight the hair. This product would revolutionize the hair-dye market.

Electric combing iron (1959)

Two Frenchmen, **Rene Lelievre** and **Roger Lemoine**, were responsible for the invention of the *Babyliss*, a heating-combing iron which enjoyed sensational success. Introduced in 1959, this electric-powered device was basically intended for women who wished to achieve a smooth and bouffant hairstyle while sparing themselves the ordeal (not always crowned with suc-

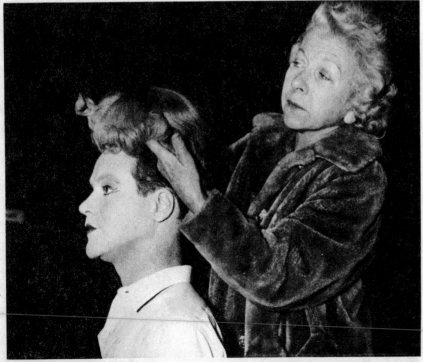

A wig being added to Jack Lemmon for his transformation in the film "Some Like it Hot."

For women who care, applying lipstick properly is an important aspect of their makeup procedure.

cess) of using unheated rollers. It also saved women a trip to the hairdresser.

Hair sprays (1960)

New concepts in hairstyling appeared in 1960. This was the period of frizzy hair. L'Oreal developed an original formula based on polymers synthesized in its laboratories. The *Elnett* hair spray produced an invisible netting to hold the hair in place and could also be removed "with a simple brushing."

Hair dryer (1920)

The two earliest models appeared in Racine, Wisconsin in 1920. They were the *Race*, made by the Racine Universal Motor Co., and the *Cyclone*, made by Hamilton Beach. They were manual models. In the winter of 1951, Sears, Roebuck & Co. marketed the *Ann Barton*, the first helmeted model for home use.

Shampoo (1877)

The term "shampoo" originated in England in 1877. It was derived from *champo*, a word in Hindi (the official federal language of India) which means "to massage" or "to knead." Originally brewed by hairdressers, shampoos were made by boiling soft soap in soda water. But it was not until after the Second World War that shampoos came into general popular use.

Lipstick

Early in the 17th century, women used to color their lips with ointments that were lightly scented and colored by dark grape juice and alcamet juice. Later cerates were obtained

from wax and oil. (The oil of roses ointment used for preventing chapped lips still operates on this principle.)

In the 20th century, chemists came to the aid of cosmetologists. Lipsticks were produced that were easy to mould but hard and harmless for the mouth's mucous membrane.

Perfume

Eau de Cologne

The **Farina** family is credited with the invention of eau de Cologne, a lemon spirit blended with bitter orange and bergamot oils. Jean-Marie and Jean-Baptiste Farina, Italians living in Cologne in 1709, created an aqua admirabilis which enjoyed a tremendous success among the French soldiers stationed in that city during the Seven Years War.

Another tradition holds that a monk, Paolo Feminis, passed on the secret to his nephew Jean-Antoine Farina, who moved from Cologne to Paris in 1806.

Perfume bottles of some of the famous creators of fashionable scents.

In 1863, the first modern perfume factory was built at Levallois-Perret (in the Paris suburbs). Two cousins, Armand Roger and Charles Gallet, founded it after taking it over from Leon Collas, who in turn took it over from Jean-Marie Farina. The wheel had come full circle.

L'Air du Temps

L'Air du Temps, the leading perfume of Robert Ricci, was marketed in 1947. A flowery blend, *L'Air du Temps* evokes youth and romance, as symbolized by its dove-stoppered bottle designed by the master crystal engraver Lalique.

Azzaro Femme

Loris Azzaro, born in Tunis in 1933, marketed *Azzaro Femme* in 1970, a subtle blend of rose of Bulgaria, tuberose, ambergris, musk, sandalwood, etc. Its bottle, designed by Pierre Din-

In 1854 Guerlain's Imperiale Cologne was honored by appointment to the royal palace of Napoleon.

and in the Art Deco style (a glass bubble suspended in a black bakelite circle), is displayed at New York's Museum of Modern Art.

Chanel N° 5

Chanel N° 5 is a perfume that resulted from the collaboration between Ernest Beaux and Coco Chanel in 1921, and which Chanel baptized with her favorite number.

When the Americans arrived in Paris, a double row of GIs formed on rue Cambon to bring home N° 5 to women in the States. And, Marilyn Monroe once replied to a journalist who asked her what she wore at night: "Chanel N° 5"!

Miss Dior

In 1947, **Christian Dior** (1905—1957) marketed *Miss Dior*, the first cre-ation of the Christian Dior Perfume Company. A blend of rose and jasmin, this perfume sought to dethrone the boudoir perfumes to which women had grown accustomed. Its bottle, a Baccarat cut-crystal jar, also contrib-uted to Miss Dior's success, and the fragrance was soon distributed throughout the world.

Shalimar

Pierre and **Jacques Guerlain** created Shalimar (*temple of love* in Sanskrit) in 1925 after a rajah passing through Paris told them the story of the Shali-mar gardens in Lahore.

It was in these gardens full of sweet-smelling blossoming trees from all over the world that the Shah Jahan, the Emperor of India in the 17th century, and the lovely Mumtaz Mahal shared an exceptional love story. Their love was so remarkable that after the death of his beloved wife, the Shah Ja-han built the magnificent Taj Mahal mausoleum as her memorial.

Solid stick perfumes

This invention was registered in 1944 by the French company Bruno Court.

Razor

Obviously no one person can claim to have invented the razor. Man has al-ways shaved, with seashells, shark's teeth, and later, bronze blades.

The razor, properly speaking, dates from the 12th century. The steel razor was created at Sheffield, England in the 18th century.

Safety razor (1901)

In 1901, the American **King Camp Gillette** patented the safety or mechanical razor, whose distinguishing feature was its double-edged replaceable blades. He marketed his product through the company he founded in Boston, in 1903. The Gillette Safety Razor Company has undergone slow but steady expansion since then.

Electric shaver (1917)

The first electric shaver was developed by the American **Schick** in 1917. The first model was marketed in 1928.

It was followed by a Swiss shaver, the Harab, marketed around 1939.

Disposable razor (1975)

The disposable razor was invented in 1975 by the French firm, Bic, run by Baron Bich. He had the simple idea that a half-blade sufficed, and with the savings so produced a handle that, along with the blade, could be thrown away.

Rouge (1912)

In 1912, the French firm Bourjois created pastel cheek make-up, presented in tiny round compacts. This was the first rouge.

In 1890, Bourjois had launched cheek make-up in an elegant carton box. This was the ancestor of the 1912 pastel ground-powder make-up, whose exclusive technology was due to the action of its binding material and to the variable concentration of its metallic pigments.

Soap

Soap appears to have been invented by the Sumerians. A clay tablet dating from 2500 B.C. has been discovered among the ruins of Sumer. It describes the manufacture of soap. Soap reappeared later in Rome, among the Gauls, and in Islamic countries.

A revolution in soap manufacture occurred in Marseilles, thanks to the application of the discoveries of Lavoisier and Priestley. In 1791, Nicolas Leblanc obtained caustic soda by treating sea salt with sulphuric acid. He founded, in the Paris suburb of Saint-Denis, the first soda works, the *Franciade*. His masterpiece was a soap bust of Louis XVIII made of Leblanc soda, which he presented to Louis' brother, the Count of Artois, the future Charles X, with the inscription: *He removes all stains.*

In 1823, the French chemist Michel-Eugene Chevreul showed that

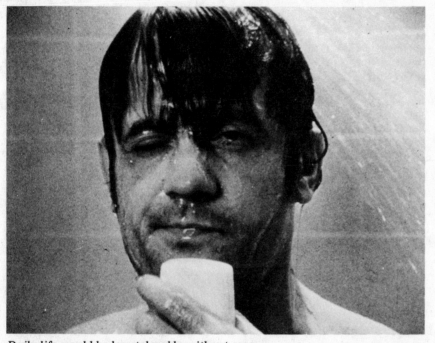

Daily life would be less tolerable without soap.

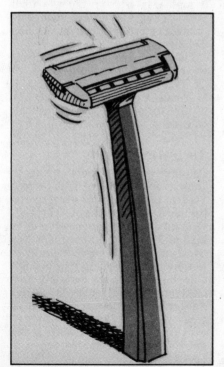

The disposable razor, 1975.

Director Sydney Pollack making a rare screen appearance as a theatrical agent who suddenly has to handle a transformed client (Dustin Hoffman) in "Tootsie".

soap resulted from a specific chemical combination and discovered a new raw material for soap-making: olein.

In 1873, further progress was achieved when the Belgian engineer Solvay developed ammonia soda.

4. Dieting

The science of dietetics came into being about 1900. But in certain cases, where diet alone has proved to be ineffective, surgery has had to take over. About 1975, a method was developed for reducing the length of the large intestine to reduce the assimilation of food. Developed by Drs. **Shibata** in Japan, **Payne** in Great Britain and **Lewis** and **Hull** in the United States, this surgical intervention can allow for a weight loss of up to nearly 200 lbs.

Operations for reducing the size of the stomach also exist, pioneered by Drs. Mason and Ito. The "Magic Bubble" (1984), developed by Drs. Lloyd and Garren in the United States, actually replaces the intake of food itself . . . with air!

Fruit-eater's diet (1910)

Developed by the Swiss **Dr. Bircher-Brenner,** the plan consists of eating nothing but fruit and is recommended for not more than one week. Dr. Bircher-Brenner was also the inventor of musli, a fruit and cereal mixture popular among European health food advocates.

Weight Watchers (1961)

Jean Nidetch was a homemaker in Queens, New York who once weighed 235 lbs. before she finally found a diet that allowed her to lose 80 lbs. She felt the need to share this experience with her friends and began talking to them about it.

This was the birth of the first Weight Watchers Club. The method consists of meetings where each person learns how to change his eating habits, not only to lose weight, but to stay slim.

Dr. Atkins' diet revolution (1963)

The American **Dr. Atkins'** plan consists of cutting out all carbohydrates (sugar and starch) from the diet, an effective method for losing weight very quickly.

Scarsdale diet (1978)

Developed by the heart specialist of the Scarsdale Clinic in New York, **Dr. Herman Tarnower**, it consists of increasing the intake of proteins in the diet. This diet is effective for losing 8 or 9 pounds.

Mayo diet (1978)

Developed at the Mayo Clinic in Minnesota. It provides for a weight loss of 15 to 18 pounds over a period of two weeks.

HOME AND LEISURE

Adhesive bandage (1882)

A century ago, **Paul Beiersdorf,** a pharmacist in Hamburg, Germany, patented a method for the manufacture of plaster-coated bandages. Known as *Hansaplast*, they were the ancestors of today's adhesive bandages.

Aerosol spray (1926)

The Norwegian **Erik Rotheim** invented aerosol in 1926. He discovered that one could spray a product by putting a gas or a liquid into the container to create internal pressure.

On August 22, 1939, Julian S. Kahn of New York invented the disposable spray can. But this idea did not receive its first commercial application until 1941. Two Americans, L. D. Goodhue and W. N. Sullivan, produced a spray can insecticide.

Air conditioning (1911)

The American **Willis Carrier** invented air conditioning in 1911. In 1902 he had studied the regulation of air humidity in a Brooklyn printing house.

By 1904, this led him to devise an air-conditioning system containing an air-cleaner facilitated by water spraying. This system is still in use today. Continuing his research, in 1911 Carrier devised an air-humidity graph enabling one to make a rational estimate of air-conditioning installations.

Artificial ventilation

Air conditioning can naturally be traced back to artificial ventilation, and usually mention is given to the Italian **Agricola,** who in 1155 described ventilation systems in mines.

It is likewise known that Leonardo da Vinci addressed this problem. But mention should also be made of the water-evaporation cooling system used in ancient Andalusian homes, whereby air entered after passing through the vegetation and water fountain in the patio.

Earliest appliances

In 1919, the first air-conditioned movie house was opened in Chicago simultaneously with an air-conditioned department store in Brooklyn, Abraham and Straus.

Individual room air conditioning was envisaged in the United States after 1926 by **H. H. Schutz** and **J. Q. Sherman.** A patent request was filed in 1931 for an air conditioner placed on window ledges.

Air mattress (1478)

The first written account of what we know as an "air mattress" dates from 1478. At that date, however, it sported the more elegant name of "wind bed." William Dujardin, upholsterer to the king of France, mentioned it by that name in a bill to Lord de La Motte Desguy in 1478. But this type of inflatable mattress must certainly have been in existence since the Middle Ages, when it served as a hunting accessory. Made of waterproof waxed canvas and equipped with an air valve like modern mattresses, it allowed the hunter to remain in his hiding place without getting dirty. He

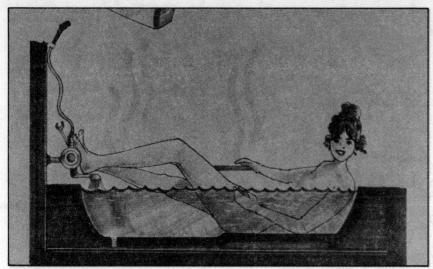

The Foottouch spares one from having to exert oneself in order to adjust a faucet during a relaxing bath.

could lie on the ground comfortably and still be protected from moisture.

Automatic water shut-off

The engineer **Robert Dalferth** filed a patent in 1982 for a simple method allowing one to save a lot of money on one's water bill.

This Perlator (marketed by the German company Gorodal) consists of a small tube and a valve in synthetic material, which, plugged onto a simple faucet, releases water only when pressed by a finger. As soon as one lifts one's hand, the valve shuts off the water.

Baby carriage

The American **Charles Burton** built the first baby carriage in New York in 1848. But pedestrians protested because those who pushed the carriages—doubtless through lack of experience—often bumped into them. Burton left for London, where he continued building carriages, notably for Queen Victoria, Queen Isabel of Spain, and the Pasha of Egypt.

Folding baby carriage

In 1906, a Parisian, **E. Baumann**, struck by the difficulties his fellow countrymen encountered in housing baby carriages inside their cramped accommodations, had the idea of creating an easy-to-store folding model. Christened "The Dream," this folding cart was described as follows in the Catalogue for Arms and Cycle Manufacture in Saint-Etienne: "Handsome bent and varnished wood, cane trimming, two 20 in. (50 cm) rubber wheels and four small 5.9 in. (15 cm) iron wheels. . . . Movable unbleached linen-fringed canopy. Length: 25.6 in.

(65 cm), height: 47.25 in. (120 cm), weight: 20.4 lbs (9.3 kg).''

Landau baby carriage

This type of pram appeared in Landau, Germany right after World War I. This four-wheeled carriage originally had two facing seats inside that were placed parallel to the axles. Accessories added to the landau's comfort: a sunshade to replace the hood during very hot weather, a mosquito net, an adjustable umbrella stand, a spare wheel, etc.

Bathtub

The Greeks and Romans used bathtubs made of marble or silver. During the Middle Ages, there were simple wooden tubs.

In 1679, Louis XIV ordered two marble bathtubs for the bathroom apartments. Hard to heat, they were lined with cloth and lace, which made up the bath equipment. The forerunner of the individual bathtub with heated water was the steam bath for hydrotherapeutic purposes conceived by the French surgeon Ambroise Pare (circa 1509—1590) during the second half of the 16th century.

In the 18th century, metal bathtubs (already used the previous century) increased in number. In France in 1760, the Parisian coppersmith Honel proposed to lease bathtubs with a cylinder for heating water. All these bathtubs were made of brass, and consequently were too expensive for "bourgeois" use. At the end of the 18th century, a special varnish enabled one to dress sheet metal and thus produce bathtubs at a reasonable price. These soon replaced the traditional wooden or marble tubs.

The first hotel where every room was equipped with a bathroom was apparently the Mount Vernon Hotel, in Cape May, New Jersey, in 1853.

Foottouch bathtub faucet

In 1982, a Chinese inventor from Taiwan, **Liou Shu-lien**, presented a device, the *Foottouch*, sparing those who enjoy basking in their tub in maximum comfort the effort of having to leave the water to adjust the faucet. The *Foottouch* uses a system of wheels in place of faucets, and one can maneuver the wheels without any difficulty with one's foot.

Carpet sweeper

The carpet sweeper was invented by the American **Melville R. Bissel**, who patented it on September 19, 1876.

The owner of a china shop in Grand Rapids, Michigan, Bissel suffered from an allergy to dust caused by the straw used in wrapping his goods. To remedy the problem, he devised a sweeper fitted with a cylindrical brush that swept the dust into a bin. He began to manufacture it with the help of female workers who worked at home to produce the various parts that Bissel would assemble in his household.

In view of the apparatus' local success, they founded the Bissel Carpet Sweeper Co. to market their product. Bissel has passed into posterity, since the carpet sweeper is often referred to as a *bissel*.

Central heating

Hot-air thermal-bath central heating

Used in China since ancient times in peasant dwellings where the entire family would sleep on a slab heated

The novelist George Sand (1804-1876) smoked cigars and cigarettes while her male companions were seasick during a voyage.

by a hearth placed below it, this heating system was also much utilized by the Romans in their thermal baths as well as their private homes.

The hypocaust found in Roman baths was a sort of underground furnace which radiated out in vertical square-sectioned pipes. Branch pipes fitted into the main pipes helped to spread the heat throughout all the bathing areas. To keep the fire going, slaves regularly tossed turpentine-coated metal globes onto the hypocaust floor. Since the hearth's floor sloped, the flaming globes emerged at the other end after spreading equal heat all along their path.

Hot-air heating installation

Seneca (lst century A.D.) mentions the existence in certain Roman homes of "tubes embedded in the walls for directing and spreading, equally throughout the house, a soft and regular heat." A sketch recovered in the laconicum, i.e. the public-bath steamroom, enables us to understand its mode of operation. Numerous pipes are pictured surrounding a closed hearth. The hearth's smoke heated the rooms as it traveled through the pipe.

This system resurfaced in 19th century Europe under the name of "cellar heating installation." Slits placed in the heating aperture allowed one to open or close, as desired, the entry of hot air into each room.

Steam heating installation

Less impressive, but able to carry heat further and more surely, the steam heating installation represented an advance over the hot-air heating installation. The first known model is the one installed by the famous Scottish inventor **James Watt** (1736—1819) in his factory on the outskirts of Birmingham.

Hot-water heating installation *(1777)*

Already known by the Romans who occasionally used this central-heating system, the "modern" hot-water heating installation was applied for the first time in France by the architect Bonnemain in 1777 in a castle near Pecq (Paris area).

The Bonnemain heating system consisted of a tank containing the hearth and the boiler. From the boiler branched out a vertical pipe which rose to the highest point to be heated in the building, bent at a right angle, then passed through the different floors. The cooled water returned to the lowest point in the boiler. An open tube, located on top of the circuit, brought the hot water into contact with the air. To regulate the heat, Bonnemain devised a lead rod which, when plunged into the boiler's water, would by its expansion or contraction modify the air intake.

In 1851, the first large private home to enjoy central heating and modern plumbing belonged to the Baron Mayer Amschel de Rothschild, who lived in England. In 1899, the radiator, made up of assembled parts, appeared in the United States.

Electric radiator *(1892)*

The first electric radiator was patented by the Englishmen **R. E. Bell Crompton** and **J. H. Dowsing** in 1892. They attached a wire on a cast-iron plate and protected the entire structure with an enamel coating.

In 1906, the American Albert Marsh, of Lake Country, Illinois, invented a nickel-and-chrome alloy which could be heated white-hot without melting. This resistant alloy proved to be an ideal component in constructing electrical heating units.

Finally, in 1912, the Englishman C. R. Belling, of Enfield, Middlesex, developed a fireproof clay around which could be wound a nickel-chrome alloy wire. The same year he built the first "Standard" radiators.

Cigarette

The cigarette was invented by beggars in Sevilla, Spain at the end of the 16th century. They had the idea of rolling the tobacco salvaged from cigar butts inside a small cylinder of paper. Cigarette use spread to Portugal, Italy, England, and France, where it made its official debut in 1843.

The *cigarettotype*, the first machine to manufacture cigarettes, was invented by the Frenchman Le Maire in 1844. The first industrial cigarette factory opened in Havana in 1853.

Clothes dryer

The first clothes dryer was invented by the Frenchman **Pochon** in 1799. Called *ventilator for drying clothes*, this manual apparatus, pierced with holes, was driven by a crank.

Ecological electronic rodent weapon (1980)

A German, **Axel Gringmuth**, and a Canadian, **Donald Swaby**, invented in 1980 the ultimate weapon to combat rodents. They studied the effect of the Earth's magnetic fields on the rodents' psyche. The result of a decade's research and a $300,000 investment are plain to see: Their machine emits waves which disturb the rodents' mental processes and cause them to flee.

Electronic fish

The German **Wolgang Berge**, manager of the water utility system in the town of Goppingen, near Stuttgart (West Germany), since January 1982, has used fish to monitor the quality of the water consumed by his fellow townsmen.

The species chosen has the particularity of being equipped with a small "radar." A bio-electric organ located in its tail transmits 400 to 800 vibrations per minute in the water, depending on its environment. If the latter contains harmful substances, the vibrations drop to 200. It is only neces-

sary, therefore, to install electrodes inside an aquarium to count the electronic impulses and, consequently, to monitor the water's cleanness.

Electronic mosquito swatter

The *Skeeter-skat* was developed in the United States in 1973. The inventor, whose starting point was the principle that gravid female mosquitoes (the only ones who bite to feed on blood) flee males, had the idea of designing a small cigarette-package-sized apparatus which one could wear and would emit untrasonic sounds of the same frequency as those made by male mosquitoes. The device remains inconspicuous since the human ear cannot distinguish these ultrasonic sounds.

Elevator (1743)

Origins

The earliest known elevator was built under Louis XV at Versailles in 1743. It was installed on the building's exterior, in a small courtyard, and enabled the king to mount from his apartment (on the first floor) to that of his mistress, Madame de Chateauroux (on the second floor). A system of counterweights allowed one to maneuver the device without special effort.

Mechanical elevators (1829)

The first public elevator was built in

the Regent's Park Coliseum in London in 1829. It could hold ten passengers. Tourists were invited to go up to visit a copy of the dome of Saint Paul's cathedral and admire a "panorama" depicting London. For six pence, they could take the elevator up.

On March 23, 1857, the first elevator installed in a department store began operation. The American Elisha Graves Otis built it for E. V. Haughtwout & Co., a five-floor store on Broadway. It cost $300.

Hydraulic elevator (1867)

The Frenchman **Leon Edoux** (1827—1910) installed two 68.9-feet-high (21 meters) elevating devices allowing visitors to mount to a panoramic platform on the top of the engine showroom at the Paris Exposition of 1867. He baptized it the elevator.

Each lift cage was supported by a hydraulic piston that plunged into an underground cylinder. The advent of hydraulic piston elevators, whose use began to spread in the United States after 1879, multiplied traveling speed twenty times over that of Otis' 1857 device, which traveled 8 in. (20 cm) per second. Their development was slowed by the problem of digging very deep cylinders. Yet, Edoux would create an elevator in 1889 for the Eiffel Tower that traveled 525 feet (160 meters)!

A famous cigarette maker, Carmen, interpreted here by Julia Migenes-Johnson in a film by Francesco Rosi.

Brigitte Bardot and Roger Vadim pose in a hydaulic elevator on their wedding day.

Electric elevator *(1887)*

The first electric elevator was built by the German firm **Siemens and Halske** for the Mannheim Industrial Fair in 1887. It rose to 72 feet (22 meters) in 11 seconds. In one month it carried 8,000 passengers to the top of an observation tower overlooking the fair.

The first electric elevator rising to over 656 feet (200 meters) was built in New York in 1908.

Escape chute

The Frenchman **Gerard Zephinie** imagined an original method for evacuating 25 to 35 people per minute in the event of fire. The escape chute was patented in 1972 and is manufactured by the Otis Company. It consists of a concentric fiber sheath able to reach several stories. Fire victims slide into it; their fall is braked by their body's friction against the elastic walls.

Fire extinguisher (1866)

The first fire extinguisher using a chemical base, and not water, was invented in 1866 by the Frenchman **Francois Carlier**. His extinguisher contained bicarbonate of soda and water. A bottle filled with sulfuric acid was attached inside near the cap. To utilize the extinguisher, one had to break the bottle with a needle, thus freeing the sulfuric acid. A chemical reaction then produced carbonic acid, which forced the water out and helped put out the fire.

In 1905, the Russian Alexander Laurent hit upon the idea of mixing a solution of aluminum sulphate and bicarbonate of soda with a stabilizing agent. The bubbles so formed contained carbonic acid. They floated on oil, petroleum, or paint, and prevented contact with the air, thus with oxygen.

Flushing system

As early as 1595, the English poet **John Harrington** had invented a practical water-flushing system for cleaning toilets, but his invention did not end the reign of chamber pots.

It was not until 1775 that the British inventor Alexander Cunnings patented a flushing system. In 1778, another Englishman, thirty-year-old Joseph Brahama, invented the ball-valve-and-U-bend method still used today. But it was not until the end of the 19th century, with the advent of running water and modern plumbing, that the flushing system found its way into apartment housing.

Garbage disposal unit

The simple trap door system appeared in 1919 and consisted of cement-asbestos tubes. Improvements were made during the twenties and led, in 1929, to the development of the sink with waste-disposal unit. The electric garbage disposal unit appeared in the United States in 1938.

Jukebox (1906)

The first public coin-operated gramophone was installed by the American **Louis Glas**, at San Francisco's Royal Palace on November 23, 1889. It was the cylinder gramophone invented by Thomas Edison twelve years earlier. It had four speakers, each fitted with a slot intended for the 20 cents required to hear a tune.

But the jukebox proper, that is, a public phonograph playing a selection of records, was invented by the Gabel Company in 1906. Its model, the *Automatic Entertainer*, was 5 feet (1.5 meters) high; it was dominated by a large ear trumpet. The owner of the cafe had to rewind it. But this device ceased to be manufactured in 1908 because of its mediocre musical quality.

The electric gramophone, which appeared on the American market in 1926, was of good acoustical quality. The Swedish immigrant Justus P. Seeburg, the king of the electric piano, realized that this invention opened up an enormous market and in 1928 he brought out his first jukebox, the *Audiophone*, which offered a choice among eight records.

But the inventor responsible for a practical and efficient record-changing system is B. C. Kenyon. He sold his method to the AMI Company, which used it on all its machines for thirty

years. In 1941, Ed Andrews designed a vertical record-changing device. This invention was purchased by Seeburg, who manufactured, in 1948, the MIOOA, which offered a choice of 100 records and a futuristic decoration. This was the post-World War II jukebox. The MIOOB was marketed in 1950; it was the first apparatus to play 45's.

Lighting lamps

Origins

Lighting lamps existed as early as the 3rd millennium B.C. Small dishes from this epoch have been found which held oil and a wick. Its forms varied depending on the epoch and the locality, but the principle remained the same.

Glass lamps appeared in the 18th century. In 1784, braided wicks replaced ordinary cotton wicks.

Oil lamp (1804)

The Swiss **Aime Argand** (1755—1803) was the inventor of the oil lamp. The first Argand-style lamp was manufactured in England in 1804. It was a draft oil lamp containing a glass chimney and a braided wick shaped like a hollow cylinder. The oil supply was located on a level above the wick. Argand's system eliminated smoke while also increasing light intensity.

Safety lamp (1814)

Sir Humphry Davy (1778—1829) invented the safety lamp for miners in 1814. A distinguished chemist, he had done much research on gases. Firedamp, which explodes on contact with a flame, was extremely dangerous for miners. This is why he invented a lamp with a wire gauze covering that prevented the flame from having contact with the exterior.

But only a flameless lamp was perfectly safe. In 1811, Davy discovered the phenomenon of the electric arc (see *Arc lamp*). It was based on this work that the French physicist Changy developed an electric lamp for miners in 1858.

Arc lamp

The first effective arc lamps were invented in 1847 by the Englishman W. E. Staite and improved the following year by his fellow countryman, W. Petrie.

As early as the end of the 18th century, it was known that when one applied two thick-sectioned conductors to a battery's poles, a powerful spark flashed between them: the arc. In 1811, the English chemist Sir Humphry Davy had demonstrated the luminous prop-

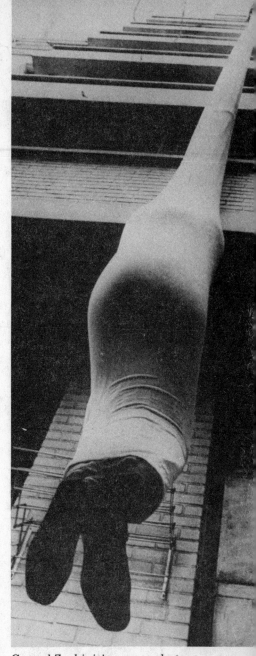

Gerard Zephinie's escape chute.

erties of the electric arc. The first arc-lamp model was *Davy's egg*, which remained at the experimental stage. As for Staite's lamps, they wore out quickly. Different inventors tried to solve this problem.

Jablochkoff candle (1876)

In 1872, the Russian **Paul Nicolaievitch Jablochkoff** was given the task of lighting, as far as possible, the Moscow-Kursk railroad link. The czar was to travel this route and the lights would prevent potential terrorists from taking advantage of the dark. Jablochkoff was led to study the arc lamp, which he improved decisively by creating the *Jablochkoff candle* in 1876.

It consisted of two small parallel carbon tips which were embedded in insulating material so that only the ends were exposed. These tips were supplied with alternating current in order to burn at the same time, while the insulating material progressively melted (like a real candle). The distance between the tips remaining the same, the arc continued to light. This invention opened the way to the electric lighting of public places, then streets and houses.

Earliest public lighting

The first square with electric lighting was the Place de la Concorde in Paris, where arc lamps were installed on an experimental basis in 1841 by Deleuil and Archereau.

The first street lit by electricity was rue Imperiale in Lyons (France), where Lacassagne and Thiers installed arc lamps in 1857.

On December 13, 1878, twenty *Jablochkoff candles* were electrically powered on the Victoria embankment in London. They had been installed by the General Electric Company of Paris. They were supplied by a Gramme dynamo: Power was transmitted by a cable buried beneath the road. But it was not until the incandescent lamp was developed that electric lighting came into general use, for arc lamps produced harsh and tiring light, consumed enormous amounts of electricity and were extremely expensive.

Incandescent lamp (1878—1879)

The incandescent lamp was invented simultaneously during the 1878—1879 period by the famous American inventor **Thomas Alva Edison** (1847—1931), at Menlo Park, New Jersey, and by **Sir Joseph Swan** at Newcastle-upon-Tyne in England. After suing each other for patent violations, the two inventors decided to join forces and founded a company together in 1883.

The principle of the incandescent lamp rested on a physical law set forth by the Englishman James Prescott Joule (1818—1889), called the *Joule effect*. If one passed a sufficiently powerful current through a sufficiently resistant conductor to bring it to white heat, the thermic energy would be converted into luminous energy. The practical problem was to find a filament able to withstand high temperatures without burning, even inside a bulb in which a proper vacuum had been created. Experiments had been made as early as 1860, but the filaments burned in contact with the air. In 1865, H. Sprengel invented a pump enabling one to create a satisfactory vacuum inside the bulbs.

On December 18, 1878, J. Swan presented the carbon-filament lamp, but was unable to give a public demonstration because the filament had already burned. On November 1, 1879, Edison patented his lamp, but his carbonized cotton filament also burned too quickly. The two inventors produced simultaneously, in October-November 1880, marketable lamps with cotton filaments that were steeped in carbon. Subsequently, the filaments were improved. In 1898, the German Auer von Welsbach invented the osmium filament.

Tungsten filament (1910)

The American **William David Coolidge** of the General Electric Co. (Schenectady, New York) succeeded in 1910 in drawing out tungsten into thin filaments for producing incandescent-lamp filaments. These filaments, which solved the problem of resisting high temperatures, were patented on December 30, 1913.

Neon tubes (1909)

The neon tube was invented by the French physicist **Georges Claude** (1870—1960) in 1909. It directly applied the principle of electric discharge in gases. The first neon tubes were used to illuminate the Grand Palais in Paris on December 3, 1910. Unfortunately, the neon tube diffused a reddish light.

A publicist, Jacques Fonseque, instantly recognized the profit to be derived from it. He persuaded Claude to sell him its right of use. In 1912, the first neon sign was put up on Boulevard Montmartre in Paris. It read: *Le Palace Coiffeur* (The Palace Hair-

The disruption caused by the installation of gas lamps throughout Paris dismayed pedestrians.

dresser). Subsequently, it was discovered how to modify the coloring, either by changing the gas used or by placing powder inside the tubes.

Fluorescent tubes

Fluorescent tubes were developed after research by numerous scientists, including Crookes (1872), Cooper-Hewitt (1900), and Risler (1923). The first fluorescent lamps were installed in cities in 1933.

Fluorescent tubes made use of the properties of certain materials to emit light when subjected to ultraviolet radiation. This radiation is produced by an electric discharge in a gas at low pressure.

The "Litec" non-filament and non-electrode bulb (1965)

In 1965, the American **Donald Hollister** developed this non-electrode fluorescent lamp. It resembles an ordinary incandescent lamp, but the filament is replaced by a minuscule electromagnet that receives the current via an electronic device built into the bulb's base.

Despite its undisputed advantages (a 10-year lifespan, consumption 3 or 4 times less than a standard light bulb), Hollister's bright idea did not enjoy immediate success. It was only with the oil crisis (1973) and the campaigns launched in favor of energy savings that an American governmental agency, the ERDA (Energy Research and Development Administration), showed interest in the Litec bulb and signed a contract with Donald Hollister's company (Lightning Technology Corporation).

The SL lamp (1980)

The SL lamp was invented by the Philips laboratory in Eindhoven (Netherlands) in 1980. Its principle is the same as for the fluorescent tube. Yet this is an extremely modern lamp. It only uses 25% of the energy required for an incandescent lamp to produce the same light, and it lasts 5 times longer. In addition, it can be fitted onto all incandescent-lamp installations. In other words, it can be used in place of any screw-in bulb.

Kleenex (1924)

The first disposable paper hankies were produced in 1924 by the Kimberley-Clark Co. of Neenah (Wisconsin) under the name of *Celluwipes*. The product was later renamed Kleenex-'Kerchiefs and then shortened to Kleenex.

Lightning rod (1752)

The American **Benjamin Franklin** (1706—1790) invented the lightning rod in 1752. A printer by trade, later the founder of a newspaper, Franklin was always fascinated by electrical phenomena. It was his discovery of the power of peaks, then of the electrical nature of lightning, that led him to the invention of the lightning rod.

Linoleum

Linoleum is a floor covering that was invented by the Englishman **Fred Walton** in London in 1864. For a long time, oilcloth had been used to cover floors. The chemist Galloway tried to obtain a similar product by mixing cork and rubber shreds and called it *kamptulicon*.

Walton then developed *linoxine* from an oxidized linseed-oil base to replace the costly rubber in the kamptulicon. By mixing this product with a cork powder, he obtained a flexible surface which he reinforced with hessian. Linoleum was born.

Mirror

It is not known who first had the idea of offering this indispensable item. In any event, the Greek philosopher Aristotle (384—322 B.C.) mentions in his writings the existence of glass mirrors lined with burnished metal.

In the 16th century, Venetian mirrors, obtained by glass-blowing and comparable to our mirrors of today, were famous. Their invention has been ascribed to the del Gallo brothers, who are said to have mentioned them in a petition of 1503.

Pinball machine

The pinball machine's forerunner was *English billiards*, or *bagatelle*, which Dickens described in *The Pickwick Papers* in 1836. It consisted of a tilted board, studded with brass nails meant to deflect the glass billiard balls struck with a wooden cue.

About 1929, the American John Sloan, who published a sheet for fairground entertainers and slot machine operators in the Chicago area, spotted a building caretaker playing a modified version of another simple billiard game of the period. The player shot the billiard balls across the board with a spring. Sloan proposed this idea to In and Outdoor Games Inc., which marketed it under the name of *Whoopee*.

In 1931, a young Chicago businessman, Richard T. Moloney, convinced his two partners to begin manufacturing a new product: the bagatelle board, with its eleven holes, would be covered with a plate of glass and seven glass balls shot by a spring with a striking device. It was a success and within a few months he had sold 50,000 models of his *Ballyhoo*. The game later received technical improvements: It became electric in 1933, the year that *Rockelite*, created by the American Bally Company, presented its first illuminated scores.

Samuel Gensberg, a brilliant Polish-born businessman who arrived in the United States at the age of 18, invented (about 1938) the first electric billiard game: the *Beamlight*. He would never sell to his competitors the secret of the bulbs that constituted the charm of his Beamlight. They were quite simply the colored gelatine capsules used for horse medicine!

The word *flipper* appeared in 1947 with Chicago Coin's *Bermuda* and Gottlieb's *Humpty Dumpty*, which enabled players to flip the balls back and to control them.

Pipe

Pipes were already used in prehistoric times. Archeologists have unearthed various models in clay or iron. In Mexico, it was long customary to hand out ready-stuffed pipes at the end of meals to the guests present.

In Europe, the English were the earliest pipe fans. At first made of terracotta, pipes began to be made of wood starting in 1790.

Portable workshop (1968)

The *Workmate*, a workshop for do-it-yourself buffs, was developed by the American **Ronald P. Hickman** and his partner, **Michael J. Roos**, in 1968. This workshop, which is extremely stable and quite safe, is equipped with all the accessories and tools found on a traditional workbench. However, since it is comparatively light and can be folded up, it is easy to transport from room to room inside the house, and even outdoors. Marketed by Black and Decker since 1977, the Workmate has already been sold by the millions throughout the world.

Robot (1982)

Hero 1 is the first household robot put on the market specifically with the general public in mind. It was invented in 1982 by the American firm Heathkit. This company, which specializes in technical education, invented this robot for pedagogical purposes. With Hero 1, young people can enjoy themselves while also learning about robot technology. They can also program it to do many household chores; it can walk, grasp objects and speak.

One of the first sewing machines.

Sauna (1st century B.C.)

The principle of the Sauna was conceived by the Finns in response to a need for relaxation and body hygiene.

The most recent ethnographic research has shown that its origins go back about 2000 years. At that time, it was nothing more than a simple hole in the ground containing a pile of stones that could be heated. Water was thrown over these stones and transformed into steam. When the sauna became a building adjacent to the main house, it also served as a place for the drying of meat and grain.

Self-cleaning gutters (1980)

The American **Raymond Zukauskas**, in 1980, solved the problem of gutters that clog up with dead leaves. The gutter is fitted along its entire length with a gently sloping aluminum roofing over which rainwater slides the leaves toward the ground. The water in turn falls, through a natural effect of tension, into the gutter along a narrow passage placed for this purpose between the gutter and the ledge of the roofing.

Sewing machine (1830)

In 1830, the French engineer **Barthelemy Thimonnier** invented the first sewing machine to operate in a regular and practical manner. It included all the features found in today's machines.

The following year, Thimonnier founded a firm in Paris, but he ran up against violent opposition from workers. A riot led to the destruction of 80 machines operating in his workshops. Thimonnier had to flee to his native town of Amplepuis and resume his trade as a simple tailor.

Concurrently with the Frenchman's research, an American, Walter Hunt, in 1834 designed a double-thread shuttle machine whose crochet hook Elias Howe replaced a dozen years later with a perforated needle.

In 1845, Thimonnier and his new partner, Magnin, filed a second patent request for a machine that could carry out 200 stitches per minute. Then, on August 5, 1848, they filed another request for a metal machine that could perform chain-stitching, called the *stitch-embroiderer*. Shortly before that, Thimonnier filed an English patent that was almost immediately sold to a Manchester firm.

The Englishmen Morey and Johnson, in 1849, purchased an American patent; this American machine had a crochet needle, like Thimonnier's. In 1851, the London Exhibition took place. By an incredible stroke of bad luck, Thimonnier's stitch-embroiderer arrived in London two days after the jury's examination had taken place. Thimonnier died a ruined man in Amplepuis in 1857.

Household sewing machine (1851)

In 1851, the American **Isaac Singer** of Boston, Massachusetts was the first to construct and market a sewing machine for household use. His needle was the same as the needle on Howe's machine.

Solar heating

Individual portable solar water heater (1977)

With the help of the individual portable solar water heater, which the GBM Co. (Angers, France) invented in 1977, you can obtain the equivalent of three coffee-cups of hot water. All you need is a little sunlight. It is ideal for making tea or coffee on the beach.

It has the form of a small folded suitcase that measures 10.24 in (26 cm) x 18.9 in (48 cm), and only weighs 1.98 lbs (900 gr).

"Suntrac" solar water heater (1978)

The Suntrac solar water heater was invented in 1978 by the Australian **Dave Little**. It is the first solar-powered water heater that can rotate with the sun. Due to this ingenious device, a solar water heater can heat 536.8 quarts (508 liters) from 70° to 118° F (21° to 48° C) in 12 hours, which is sufficient for drawing a bath. Maintenance is quite simple, so the device represents a very cheap form of energy. It remains, however, more suitable for sunny regions, since each hour of cloudy weather cuts output by 10 percent.

The Suntrac solar heater follows the sun.

The electric vacuum in 1914.

Vacuuming in 1869.

Solid paint (1976)

"Spred," the solid paint, was invented in 1976 by **Alfred Norman Dunlop**, a Canadian paint manufacturer. It is sold in a cardboard package which also serves as the applicator. When the surface of the paint block is pressed against the object to be painted, friction destroys the gelled structure of the paint and it becomes liquified. The same results can be obtained as with traditional liquid paint.

Tricycle for the handicapped

It was invented in January 1981 by **Dave Survis**, president of the American Tricycle Association. The Surry Tricycloped is the ideal means of transport for the handicapped, at least in California where the climate allows it. It does 30 mph (48 km/h), and has a cruising range of 80 miles (128 km).

Vacuum cleaner

The term "aspirator" was applied in the 19th century to a whole range of industrial appliances that extracted air or gas by means of an aspirator. The first patents filed were those of I. G. MacGaffey in 1869, and H. C. Booth in 1901.

The first electric vacuum cleaner was made by the American **J. M. Splangler** in 1907.

Vacuum cleaning robot (1983)

The Japanese company Automax invented, in 1983, a vacuum cleaning robot that does the housework for you. It is equipped with a position sensor developed with the help of Honda. You put it in a room, it "identifies" the room, then cleans it from top to bottom.

Washing machine (19th century)

Replacing the steam boiler, which itself replaced the washtub of the Middle Ages, the washing machine appeared during the 19th century.

Composed of a wooden bin that was filled with soapy hot water, the first mechanical washing machines used heavy blades to stir the washing. This principle of tossing clothes inside a rotating cylinder still governs the operation of modern machines. One of them operated in 1830 in an English laundry. Around 1840, in France, an industrial model with double sheathing, four compartments, and a draining plug was designed; it was driven by means of a crank.

Electric washing machine

The first electric washing machine was invented and developed by the American **Alva J. Fisher**, in 1901.

It was not until the start of World War II that electric vertical-tube machines with built-in turbo-washers or with a vertical axle fitted with blades began to be mass-produced in the United States. Horizontal-drum machines appeared in 1960.

Wrist TV (1982)

It was invented in 1982 by the Japanese firm Suwa Seikosha, a subsidiary of Seiko. More than a gadget, it is a spinoff of research on flat screens. This TV has a liquid-crystal screen measuring approximately 1 in. (25 mm) per side. It is in black and white, but has other qualities: It serves as a watch, a calendar, an alarm clock, and an FM radio.

BUSINESS

Banknotes (1658)

Originally, bankers at medieval fairs delivered registered receipts to their depositors. Then, about 1587 in Venice, it became possible to transfer these receipts through the practice of endorsement. It was this endorsement that helped spread the use of paper money.

The first bank to issue banknotes was the Riksbank of Stockholm in 1658. England followed its example and, in 1803, the Bank of France was licensed to issue them.

Bill of exchange (4th century B.C.)

In the 4th century B.C., the Greeks invented the bill of exchange. Isocrates' (436—338 B.C.) *Discourse on Banking* bears witness: He refers to this means of payment allowing one to travel without taking along large sums of money. Shopkeepers or travelers addressed themselves to their local banker and handed over to him a sum of money. The banker delivered to them a letter in return. Upon presentation of this letter, the banker at the place of arrival advanced them the money required.

The bill of exchange was the forerunner of the bank check.

Cash register

The cash register was invented on November 4, 1879, by the American **James J. Ritty**. He owned a saloon in Dayton, Ohio, and the constant quarrels with his customers exasperated him. During a boat trip to Europe, he noticed a machine able to register the number of times a propeller turned. This machine triggered the idea for his cash register, which served both as a printing-adding machine and a till.

Coin money (7th century)

Coin, that is, the affixing of an official stamp on ingots, was invented in the middle of the 7th century B.C. in Asia Minor, during the reign of the Lydian King Ardys. Coin money then consisted of small ingots made of a gold-and-silver alloy called *electrum*, found in its natural state in the river flowing

World trade has evolved from a barter arrangement to one which uses currency for valuation.

A rather ornate cash register.

for business in 1858, while in New York the *Harper's Building* was built by the architect John B. Corlies in 1854.

Electronic merchandising (1984)

Imagine a department store contained in a few dozen square feet that allows you to buy any product made by any manufacturer without a sales person. All this is now possible, thanks to Video Comp-U-Store, invented by **Stuart L. Bell** and **Walter A. Forbes**.

The entire product display—photos, text and models—is stored on a video disk. The client types in the reference numbers of what he wants, and the article appears on the screen. After making his choice, the client places his order. If the article is in stock, he takes it with him; if not, it will be delivered to him free.

This system is now being installed in three department store chains in the U.S.: Dillard's, Filene's and Woodward and Lothrop. The only item that is available for the moment, however, is one full line of household linen.

McDonalds

Since 1940, two American brothers, **Maurice** and **Richard McDonald,** had owned a hamburger stand next door to their movie house near Pasadena, California. In 1948, they converted it into a self-service cafeteria, while stressing the quality of the hamburgers they served.

By 1952, the McDonalds' hamburgers were known throughout southern California, where they had set up branches. A restaurant equipment contractor, Ray Kroc, offered to sell franchises for them throughout the country. In 1962, when 200 branches had been opened, Kroc bought out the McDonalds' share in the business for $2.7 million.

Memory card (1974)

The memory card is the invention of a self-taught French engineer named **Roland Moreno.** He patented it in 1974 for his own company, Innovatron. After approaching many different kinds of companies, he succeeded in getting certain banks interested in his idea, as well as the Telecommunications Division (DGI) of the French Post Office, Telephone and Telegraph Ministry.

The memory card allows for the storage of protected data in a greatly reduced form on microcards or microchips. The first computer manufacturer to build a machine using this idea was C.I.I.-Honeywell Bull. In its

through the Lydian capital, the Sardes.

The earliest coins made of this alloy resembled flattened croquettes. On one side they bore a square or triangular hallmark and on the other (the obverse) an Assyrian-style lion's head whose nose was embellished with a sort of radiant sphere.

Public relations (1906)

A serious labor dispute in the Appalachian coal basin in 1906 gave **Ivy L. Lee** the idea of making public relations a technique and an occupation. Lee succeeded, through an extremely well-planned information campaign, in turning around public opinion, which had been strongly opposed to the mine owners. Impressed by the remarkable results of this operation, many businessmen and financiers soon took advantage of Lee's services in order to create and sell a favorable public image.

Credit cards

The first organization permitting credit card payment was set up in 1950 by the American **Ralph Scheider:** The 200 founding Diner's Club members could dine on credit in 27 New York restaurants.

In 1958, Bank of America issued the first bank credit card, the BankAmericard.

Department stores (1824)

The first department store was the *Belle Jardiniere* founded in Paris by **Pierre Parissot** in 1824. Parissot's publicity and sales methods, stressing the fact that items were sold "at a set price and for cash," revolutionized business practices which were based on two major principles: discussion between seller and buyer of the item's price, and sale on credit.

The *Crystal Palace Bazar,* the first London department store, opened

Americans exchanging their travelers checks for gold in August, 1914.

machine, the simple passive memory is replaced by a microprocessor which allows the card itself to control access to this protected data. In addition, the microprocessor can perform such operations as the automatic digitization and de-digitization of protected data, thus avoiding the pirating of data on telematic networks.

In the autumn of 1983, a half dozen pilot installations using the memory card were operating in France. Along with systems developed in other countries, such as the RAPIDS system developed by Philips for the U.S. Defense Department, this represents 130,000 cards. Some 4,800,000 cards have been ordered from Bull and from Schlumberger by the French Post Office, Telephone and Telegraph Ministry. In addition, 5000 public telephones are operating on memory cards.

The principal importance of the memory card lies in the security it represents for data processing information. Faced with the twin dangers of fraudulent consultation and fraudulent reproduction of information, the memory card acts as a kind of "data strongbox."

Scotch tape (1925)

An American, **Dick Drew**, invented adhesive tape in 1925. Then a young assistant at the 3M laboratory at St. Paul, Minnesota, Drew asked automobile manufacturers to test the first samples of waterproof abrasive paper. At that time, body builders had to paint cars in two tones. Paint was applied by spray gun, the touchy part being to separate the colors clearly and distinctly. Glued together newspapers were used, but it often happened that when the bands were removed the fresh paint also came off. Drew studied the problem and, with the encouragement of management,

sought a solution: adhesive masking tape. Five years later cellulose adhesive tape appeared, as we know it today.

But why was it called *Scotch*? Because, in order to facilitate its laying-on and taking-off operations, 3M delivered to the body builders a tape, only the edges of which were self-adhesive. Suspecting that this was done to save adhesive tape, the workers began calling the bands *Scotch tape*. The nickname had a bright future.

Secret signature (1982)

The English firm Transaction Security Ltd. marketed in May 1982 a system for identifying signatures in the 50 branches of the Banco Nacional Financial de Mexico. The customer signed 5 times in a row on a small device recording not only the form of the specimen signature, but a tolerance threshold measuring speed and acceleration in carrying out the signature. The device also measured pressure, which is impossible to forge.

A small computer coded all this data and compared them later to those corresponding to the signature of the client who, say, wishes to withdraw money. Of course, the reference code is never 100% identical to the specimen signature, but a tolerance threshold can be set and adjusted depending on the type of transaction or the size of withdrawal.

Stock exchange (1450)

The stock exchange, that is, a place where financial transactions occur, was created about 1450. Until then, merchants and bankers got together at fairs.

In the middle of the 15th century, a family of Bruges bankers, the Van de Bursen, opened its house to these

transactions. Over the entrance portal was a frontispiece depicting three engraved purses (*bourses* in French). This is the likely origin of the French term *bourse*. Antwerp in turn opened an exchange in 1487. It soon became the largest in Europe. In France, the first bourse appeared in Lyons, opening its doors in 1595. In London, Thomas Gresham founded Royal Exchange, which became known as the Stock Exchange in 1773.

Supermarkets (1879)

The first stores of this type were born in the United States in 1879. They were introduced into Britain in 1909 and into France in 1927. In those days, they had a plain and rather drab appearance, not at all like the flashy style that we know today.

The first "hyper-market" chain, with everything under one roof, was *Carrefour*, which opened in France in June of 1963.

Supermarket carts (1937)

In Oklahoma City, the owner of the Humpty Dumpty Store invented the first supermarket cart (then called *cartwheels*) on June 4, 1937. He had remarked that his customers had trouble lugging all their purchases through the different store departments. He converted folding chairs into carts: the feet were mounted on wheels, a basket replaced the seat, and the back served to push the vehicle.

Toothpick production (1872)

The Americans **Silas Noble** and **James P. Cooley** obtained patent no. 123790 on February 20, 1872 for a machine enabling one, "with very little loss and in a single operation, to slice a block of wood into toothpicks ready for use."

Travel agency (1841)

On July 5, 1841, the Englishman **Thomas Cook** (1808—1892) boarded a train. He was not traveling alone. He had organized and was accompanying the first package tour in the history of tourism. It was no small group: 500 members of a temperance society were leaving Leicester to attend a large meeting at Laughborough.

Traveler's checks (1874)

In 1874, the English founder of the famous Cook Travel Agency, **Thomas Cook** (1808—1892), had the idea of offering his customers *traveler's checks*. Confidence in and the success of this new means of payment were immediate: Three large Parisian hotels accepted them as soon as they were created.

One hundred years later, in 1974, the Cook firm, set up all over the world, was issuing some two and a half million traveler's checks.

Watermark

Clear watermark (1282)

The watermark, that translucent impression found on a sheet of paper, appeared for the first time in 1282 at Fabriano, Italy. The Italian word for watermark is *filigrana*, and it originally referred to the filigree work done by jewellers.

It was the need to mark paper indelibly, without taking space from its writing surface, that pushed a papermaker to imagine that a metal wire fixed on the form or screen would leave its impression in the sheet being formed. For centuries, papermakers used this means to mark their products.

Dark watermark (18th century)

Toward the end of the 18th century, driven by the need to protect a currency being increasingly issued in paper, the appearance of the watermark changed. The new possibility was presented of making a dark design appear in the paper. The *assignats* issued during the period of the French Revolution are a good example of the dark watermark, in association with a clear watermark.

Embossed watermark (1848)

By combining the ridges of the clear watermark with the hollows of the dark watermark, an Englishman named **William Henry Smith** invented the embossed watermark in 1848. The art of the watermark was transformed, allowing for more complicated compositions, fine detail work and a richness of colors and half-tones almost unlimited. The first banknote adopting this new technique was the 100-franc bill, series 1862, embossed with the head of the Roman god Mercury.

CRIME

Composite picture

The first person to be arrested by French police, thanks to a composite picture, was Guy Trebert. The criminal was identified on May 20, 1959, more than a year after discovery of the body of his victim, Arlette Donier, in the forest at Saint-Germain-en-Laye (near Paris).

Criminal identification (1879)

The Frenchman **Alphonse Bertillon** invented in 1879 a system for identifying criminals based on anthropometrical data. Certain bones were measured and the form of nose and ear noted. Bertillon applied his own methods when appointed to head the Criminal Records Office at Paris Police Headquarters.

Electric chair (1888)

The electric chair was invented by **Harold P. Brown** and **Dr. A. E. Kennely** in 1888 in the United States. As early as 1773, Benjamin Franklin had electrocuted animals, specifically an 11-pound (5-kilogram) turkey, a lamb, and some chickens, by using six *Leyde jars*, which are condensers.

One century later, Brown and Kennely, researchers at the Edison Company in Menlo Park, New Jersey, believed that death by electrocution was more humane than the guillotine or hanging. They tried their "chair" out on animals.

The first human electrocution took place in Auburn Prison (New York),

These pictures showing a French murderer illustrate how easily one can create a disguise by simply altering various elements of one's persona. These photos had been created for false IDs.

on August 6, 1890. The murderer William Kemmler was seated on a chair, and electrocuted.

Fingerprint identification (1892)

Sir Francis Galton, the English physiologist and cousin of Charles Darwin, is responsible for developing, in 1892, the fingerprint identification system. He had discovered that every individual had different fingerprints. Scotland Yard, at the prompting of its chief, Edward Henry, immediately adopted this identification system, which improved police methods.

Gas chamber (124)

Major **D. A. Turner**, of the United States army medical corps, had the idea of shutting men sentenced to death in a room where, once inside, toxic gas was injected. The first victim was a Chinese by the name of Gee Jon, who had participated in a mutiny and committed murders. He was executed on February 8, 1924; death occurred after inhaling hydrocyanic gas for six minutes.

Guillotine

Contrary to what is usually believed, the guillotine was invented by a member of the Paris Academy of Surgery, **Dr. Antoine Louis,** and built according to his specifications by a German instrument maker by the name of Tobias Schmidt in 1788. The invention was successfully tested on corpses from the Bicetre Hospital, then on live sheep.

Dr. Joseph Ignace Guillotin (1738—1814), elected deputy to the States General in 1789, was acquainted with this invention and believed it would cut short the suffering of those sentenced to death. This is why he strongly urged his colleagues to adopt the new machine. Moreover, out of a desire for equality, he proposed that both nobles and commoners be subject to the same form of execution, which was not the case up to then.

The first execution was held on April 25, 1792, at the Place de Greve (today the Place de l'Hotel de Ville) in Paris. The head of a highwayman, Pelletier, was severed cleanly. This execution was followed during the French Revolution by 2,498 beheadings. Until the day he died, Guillotin protested the use of his name to designate the head reaper.

the bizarre
and the future

THE BIZARRE

Water walkers (1910)

About 1900, several amateur inventors were simultaneously at work trying to make one of man's oldest dreams come true: to walk on water. A series of gadgets to let people move about upright on water were built. But the problem of placing one foot in front of the other with oval-shaped floats seemed insoluble, not to mention downright hazardous.

Then, in 1910, a German inventor came up with the idea of attaching the two floats to each other before putting them in water.

Individual armor

At the start of World War I, the Allied Armies experimented with a "suit" of armor as a means of protection against enemy fire.

These extremely heavy devices became completely vulnerable to the sophisticated artillery developed at the time. However, these odd-looking contraptions may be thought of as precursors to the assault tanks deployed by the British in 1916.

Firing simulator

During World War I the British decided to train their machine gunners by means of a mounted "firing simulator." Positioned in his turret, the pilot was sent at full speed down a narrow-gauge railroad track. When the target—a picture of a German plane mounted on a board—came into view, he was to fire as a machine gunner would in an actual flight.

Given the relatively slow speeds of World War I fighter planes, the simulator was actually an ingenious if not overly elaborate system.

Multiple-line fishing rod (1933)

In 1933, **J. H. Hirst** took top honors in a fishing contest held on the banks of the Ouse River. Six hundred of the best fishermen in the United Kingdom participated in the contest.

Hirst had thought to increase his chances of winning by using a fishing rod of his own invention: a web-like gadget strung on a frame which held about 15 lines. With this formidable rod, he caught 80 fish in one hour. (It should be noted, however, that at that time rivers had not yet been depleted because of pollution.)

Front wheel drive bicycle (1938)

In 1938, the Frenchman **Souhart** invented a bicycle featuring a special handlebar that activated a front wheel traction mechanism. This equipment enabled the cyclist to operate his bike 20% more efficiently on hills and when starting up. It was also useful for developing his arm muscles.

This giant leap forward in bicycle engineering did not prevent observers

Left: Individual armor on a standing man. Top: Firing simulator. Bottom: With his multi-hook rod, J.H. Hirst won the first prize of the Olney (Great Britain) fishing contest on September 25, 1933. He caught 80 fish in only a few minutes.

The Flying Pancake.

from breaking into laughter at the sight of Souhart manipulating his handlebars as he sped down the road. The spectacle was said to be very comic.

Flying pancake (1942)

Working in 1942 for the Chance Vought Company, the American engineer **Zimmerman** built this prototype which was dubbed the C-173 "Flying Pancake".

With its round fuselage and enormous propellers, this airplane was supposed to have two advantages: It would not spin out, and it could fly at widely varying speeds. But building the aircraft was not a simple matter, and when two more powerful prototypes were built for the United States Marines a host of problems were encountered, including the difficult synchronization of the two engines, the awkward positioning of the propellers and the increase in vibrations.

Finally, a single plane was constructed, the XF-5 U-I. It had two 1600-horsepower engines and an anticipated flight speed of between 40 and 500 mph (60 and 750 kmh). In reality, it never got off the ground and was scrapped in 1948. The Navy preferred to channel its development funds into the new jet-engine aircraft.

Phoney cows

During World War II, the German forces tried to deceive Allied pilots by building fake airports where they parked wooden airplanes. They also built decoy cows; these they placed on operational airstrips to make the enemy think they were grazing in empty pastures. The Germans called these mock-ups *attrapen*.

This was an old trick but the Germans used it. They even built dummy land-based weapons, such as tanks and armored cars, made of canvas and false armor-plating. These phoney weapons were calculated to deceive the opposing forces as to the actual size of the enemy they faced.

Go-anywhere car (1951)

Invented by the Frenchman **M. Hannoyer**, this unique "convertible"was introduced at the Paris Automobile Fair of 1951.

The *Reyonnah*, an anagram of its inventor's name, contracts for easier parking. And just about anyone can operate the automobile's contracting mechanism. At the time of its presentation, the vehicle was advertised as follows: "Like any high-performance vehicle, this car holds the road extremely well. Priced like a motorcycle."

Boat suit or unsuitable boat? (1954)

To enable anglers to venture far from the riverbank, or merely to cross a stream without getting wet, the American company Minnoqua introduced a clever watertight suit in 1954 called the *Floater Bubble*.

Equipped with an adjustable strap seat and large rubber ring, the entire outfit weighed no more than 5½ pounds. It could be folded up and carried in a basket.

Unfortunately, the suit did not protect the unwary fisherman against the vagaries of winds and currents, and the angler risked being carried far away from his catch in his lightweight "bubble."

Inflatable airplane (1958)

In 1958, the Goodyear Company built the first truly successful inflatable airplane. About ten of these curious aircraft were made.

Composed of rubberized nylon, these planes could be inflated by an ordinary vacuum cleaner. By tightening the shrouds before flight, one could obtain the necessary structural rigidity to fly the aircraft.

The first model, the *Inflatoplane*, ordered by the United States Navy, was capable of sea landings. In 1971, Goodyear suggested using them as either remote-control target drones or as rescue planes.

Built-in fallout shelter (1959)

The countdown ended on May 24, 1959, when the Obie Construction Company of Pittsburgh, Pa., marketed the first house with its own built-in fallout shelter.

The underground shelter is equipped with four sleepers, a food storage unit and refrigerator, a powerful radio transmitter/receiver, a first aid kit, a geiger counter and an extinguisher. The living space measures three by seven square meters.

In addition, there is a separate generator, heater, air filtration system and a reserve supply of oxygen. The shelter is constructed of concrete and is protected from radiation by an enclosing lead shield.

Shoes for walking on the ceiling (1962)

The American engineer **John F. Heard** created these "zero gravity shoes," which were perfected in 1962 by the Martin Aerospace Company.

In 1938 a farmer struck upon a way to feed four babies at once—directly from the cow.

The soles of these shoes were fitted with tiny suction cups that attached themselves to overhead surfaces. This purely mechanical system allows one to tramp about on the ceiling. For metallic surfaces, one simply changes to shoes with soles that have been fitted with electromagnets.

This equipment was designed to enable astronauts to walk about in a weightless environment.

Amphibious car (1964)

In April 1964, passersby strolling along the banks of the Thames might have seen a sleek amphibious car navigating the river. A British engineer had the idea of transforming a Triumph Herald Sportscar, with a 1250 cc engine, into a vehicle capable of travelling in water.

This particular amphibious car is powered by a retractable propeller. On a dare, a Frenchman crossed the English Channel in this model, of which about a hundred were made.

Lebouder autoplane (1972)

This autoplane was conceived by the French electronics engineer **Lebouder**. In 1972, he built it by himself in his backyard in the suburbs. Originally, the front part of the craft was a Vespa 400, but Lebouder retained nothing of the original Vespa except for the steering wheel and motor. He then built a completely original car adapted to his airplane.

When used as an automobile, the autoplane is steered from the driver's side, and, when used as an airplane, it is flown from the passenger's side by means of a control stick extending from the ceiling. The autoplane's maximum flying time is three hours and it can attain an airspeed of 120 mph (180 kph).

Observation plane (1974)

The British company Edgley Aircraft Ltd invented a revolutionary observation plane in 1974. The shape of its cabin gives three passengers a panoramic view.

One attraction of this airplane is that it is much less expensive than a helicopter. It has an engine, completely original in design, which powers the plane efficiently at low airspeeds.

Four-eyed dogs (1975)

A French optician, **Denise Lemiere**, registered a patent for dog glasses on March 10, 1975.

These glasses are adapted to the visual deficiencies of the canine wearer and are available in five sizes. The

Boat suit.

Top Left: The amphibious car. Top Right: Lebouder Autoplane. Bottom: The Edgley Observation Plane.

glasses are attached by a strap passing under the jaw and buckled behind the head of the pet.

Lemiere hit upon the idea after making sunglasses for her own dog. Then, she diversified the styles: corrective lenses for myopic dogs; protective lenses against wind and dust for dogs who lean out of car windows; high-protection glasses to be worn day and night by dogs recovering from cataract operations.

Motorized Bar Stool (1976)

It was an American named Mike Taylor who invented the motorized bar stool in 1976. It allows you to go out for a drink without getting unnecessarily tired from changing positions. Already original in itself, this bar stool also contains a little "something extra": rear disk-brakes.

Support for buttering dry toast and crackers (1976)

Two Frenchmen, **Jean-Marc Radigois** and **Bernard Pouligny**, get credit for a completely original idea for buttering dry, brittle toast or crackers.

About the size of a small slice of toast, this gadget (actually registered as a patent on October 4, 1976) is made up of a cardboard base and plastic sheet. Breadcrumbs have been sealed between the two layers, and, acting as a cushion by conforming to the shape of the toast or cracker as it is buttered, it prevents them from crumbling.

Bicycle plane (1977)

Paul McCready, an American, solved the difficult problem of airplane fuel conservation in August 1977, by using

the legpower of his partner, Bryan Allen, for lifing off in the first bicycle plane.

On June 12, 1979, he entrusted a second craft to the same pilot-cyclist. After two-and-a-half hours of intensive pedalling, the craft succeeded in crossing the English Channel.

By inventing a solar-powered aircraft, McCready has made a significant advance that will assure Bryan Allen of much less strenuous flights.

Better vision for horses (1979)

On September 24, 1979, **Bernard Haemmerle** registered a patent for mask-mounted glasses that freely adjust to the optical perception and visual field of horses, and at the same time protect their eyes against dust, mud, sand, etc. Corrective lenses or just tinted lenses are available. Different sizes of blinkers may be fitted onto the mask.

According to the inventor, this optical assistance results in better performance on the part of the horse as it feels calmer and steadier in its surroundings with the optical correction.

Solar-heated shoes (1979)

When he filed his patent for solar-heated shoes on November 29, 1979, **Jose Peres-Conde** became the guru of foot comfort to millions suffering from arthritis and rheumatism.

Peres-Conde succeeded in perfecting shoes fitted with a thermodynamic heating system. Equipped with solar batteries and electronically wired, these shoes keep the feet cool in summer and warm in winter.

Anti-Shark Armor (1980)

From now on, frogmen and scuba-divers can venture without—or nearly without, risk in shark-infested waters.

In 1980, an American marine biologist (and shark specialist) named **Jeremiah S. Sullivan** developed a diving suit made of epoxy resin and steel fabric. This fabric can either be woven into a traditional suit or worn over it, like a coat of chain-mail. Placed on the market in 1984, this diving suit has already, so they claim, broken the teeth (in vain) of several experimental sharks.

Cardboard Furniture (1980)

In 1980, French interior designer **Vincent Geoffroy-Dechaume** launched "Cartage," an entire line of light, foldable and resistant cardboard furniture. This is one sure way of solving the space problem in small apartments and on exhibition stands—everywhere, in fact, where furniture is necessary, but there is not enough space to leave it there permanently.

Folded up, these pieces of furniture resemble an artist's sketching portfolio, and they can be stored away just as easily. Cardboard furniture can be taken care of just like normal furniture and resists water, stains and bumps.

Dog Hygiene Belt (1980)

In 1980, a German named **Walter Snippe** designed a hygienic belt for dogs which is destined to improve the cleanliness of our public streets.

This belt passes around the dog's abdomen and is fitted with a ring, which is positioned just underneath the dog's tail. A second ring is inserted into the first one, and a bag is attached to it (in much the same way that a vacuum cleaner bag is attached). The dog dirt drops into the bag and, at your leisure, you simply unfasten the ring, take out the bag and throw it into the trash.

Earth balloon (The "Ballule") (1980)

This craft, the only one of its kind in the world, was invented by the Frenchman **Gilles Ebersolt** (patent filed October 31, 1980).

It is constructed of two concentric vinyl polychloride bubble spheres measuring 2 and 4 meters in diameter. Its operator positions himself inside the smaller sphere and has the option either of pushing or running to make the bubble roll. It also floats.

Hard-boiled egg sheller (1980)

In 1980, the French company Sodetap introduced the *Coquimatic*, a machine that shells hard-boiled eggs.

The eggshells are stripped off by injecting pressurized water between shell and membrane. By virtue of its efficiency—up to a maximum of 1,200 eggs an hour—the Coquimatic appears especially suitable for the institutional kitchens of hospitals, school cafeterias and other establishments.

Multi-surface scooter (1980)

Claude Jacquin, a French architect, invented a multi-surface scooter in 1980.

This clever vehicle may be taken up the slope in a ski-tow with its rider already on board. The adventure begins with the downhill plunge.

This scooter was made possible by the invention of low-pressure Dunemer wheels.

Seat-Belt for Pet Animals (1980)

Invented in 1980 by Nova Founders, an Austrian firm, this belt consists of a strap with one end attached to the car seat and the other end attached to your pet. The pet is attached to the strap by means of a collar, which goes around his neck, and another strap that goes around his chest, just behind his front legs.

Cardboard furniture, invented in 1980 by Vincent Dechaume. The furniture can be folded very easily.

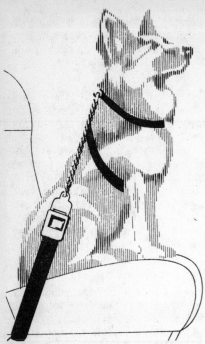

A safety belt for dogs, invented in 1980 by the Austrian firm, Nova Founders.

This seat-belt prevents the pet from jumping onto the lap or the shoulders of the driver, at the same time allowing him to move around freely in his seat. In case of a sudden stop, the belt holds perfectly and prevents your pet from being thrown through the windshield.

Armchair psychiatry (1981)

The robot *Purpose I* was presented to the San Diego public in 1981 as a substitute psychiatrist.

It appears as a comfortably lined armchair shaped like an egg. The patient relaxes inside, and a computer concealed inside the chair is able to have a psychotherapeutic dialog with him or her.

Freudian psychology, hypnosis, meditation, music therapy, gestalt and auto-relaxation are just some of the possible programs available.

Purpose I is superior to the traditional psychiatrist in two ways: It is always there when needed, and it has a perfectly soothing disposition.

Bicycle built for two side-by-side (1981)

In April 1981, the American **Robert Barrett**, of Angola, New York, invented a tandem with its seats arranged side-by-side.

The technical design is actually more complex than one might imagine, but now a couple can pedal the same bike hand-in-hand. It is easily folded and fits into the trunk of a car.

Anti-theft pepper mill (1982)

Introduced in 1982, at the Geneva Inventions Fair, and manufactured by the Swiss company ECD Electronics, this peppermill should lead to a decrease in thefts reported by restaurateurs and hotel managers.

It is equipped with an alarm system that sounds when its light and temperature setting changes abruptly—as for example when the pepper mill is slipped discreetly into a pocket or handbag. Here is something to dissuade the light-fingered!

Artificial pelican beak (1982)

A team of California doctors, headed by **Robert Rooks** and **Rick Woerpel**, invented an artificial beak for the pelican in December 1982.

The physicians had responded when an unknown criminal amputated the beaks of about 30 pelicans. Doctors Rooks and Woerpel provided these birds with fiberglass beaks that allowed them to return to their natural wild state.

Attire for men in warm climates (1982)

Philip Garner, the California inventor and artist, dazzled Hollywood in November 1982 with his new fashion for the businessman living or working in the torrid zones.

The suit jacket is boldly cut in half at the midsection. Now the smart businessman can maintain his cool and always feel confident of being elegantly dressed!

Burglar trap (1982)

In 1982, a prolific Swiss inventor, Emile Munz, devised a surprise trap for thieves. It can be placed either in a safe or on a door knob. This gadget handcuffs the wrists of would-be burglars.

Fork to cool hot morsels (1982)

In 1982, a 14-year-old Belgian inventor, **Eric Van Paris**, devised an ingenious fork to please all children whose parents make them eat hot food.

This apparatus is composed of a rubberized squeeze bulb connected by a tube to a normal fork. By forcing a current of air through the tube to the fork, the steaming morsel is cooled to a palatable temperature.

Hand-powered tricycle (1982)

In 1982, the Taiwanese firm Mang Chiu Industry Company Ltd. invented a tricycle designed to be operated from a standing position.

A child can make the tricycle go by manipulating two belts connnected to the rear wheels. The youngster's feet are placed so that one rests on the rear step connecting the rear wheels, while the other activates a pedal governing the front wheel.

The front foot controls the direction of the tricycle and stops it as well. Named after its inventor, the "Lala Tricycle" is ideal for balance, coordination, and the development of back, stomach and waist muscles.

High-heel shoes without soles (1982)

Working in Hollywood, the inventor and artist **Philip Garner** comes up with the most surprising and offbeat objects. In November 1982, he introduced high-heel shoes without soles.

Evidently designed to do away with the shoe repair business, these "shoes" enable style-conscious women to transform any of their ordinary flat-heeled shoes into formal, high-heeled shoes.

Magic Ash-Tray (1982)

This ash-tray, invented in 1982 by a Frenchman named **Gilbert Moreno**, has the distinctive feature of opening at the approach of a hand, thanks to an electronic device. The device that keeps the receptacle closed in made up of several superimposed metal blades, and it is opened by a nearby electronic detector.

Omniboat (1982)

Yoo Byung Eun, a Korean, researched and perfected an original and economical means of transportation in 1982.

The fruit of his work was a motorboat, but without the usual motor. The Omniboat operates equally well on land or sea, and comes in two models: a motorcycle- or automobile-powered version.

More economical and less cumbersome than a traditional motorboat (measuring 8 x 2.80 meters for the auto-powered version and 4.74 x 1.50 meters for the motorcycle-powered model), the omniboat neatly solves the problem of parking in coastal ports. Its captain simply docks the boat and leaves his road vehicle on board.

Portable lie detector (1982)

In 1982, the Canadian company Thought Technology produced a small machine, the GSR 2 (for Galvanic Skin

Many things that we take for granted in our **Everyday Life** are part of the process of invention. Clocks have been with us for centuries, but the clock-radio, top, right, is a comparatively recent way to wake us. And how much less sparkling would our lives be without Dom Perignon's accidental invention of Champagne in a barrel, middle, right, in 1688. Bottom, right; Distilleries, too, add a taste of excitement to our lives, while a strong cup of coffee, bottom, left, gets our days started after the clock-radio has gently awakened us. Middle, right; Even the carding of wool, which goes back millenia, is still carried out by hand today in many parts of the world.

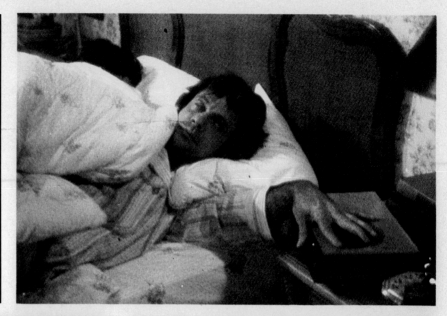

Energy is one of the most critical concerns of our time. This page: Below; since 1855, the Francis turbine has been adapted to innumerable tasks. Much more recently, nuclear power has been in the vanguard; here, bottom, the vapor generator of a nuclear power station. Oil, however, remains the principal fuel of our day, and this well in California, left, pumps industry's lifeblood all twenty-four hours of it. Facing page: Top, left; The refrigeration tower of a nuclear power station. Top, right; Solar power, as typified by this power plant, is also a source of energy for the future. Center, right; The search for oil is never ending and processing (bottom) takes place in huge refineries. Center, left; Gas turbines are also significant sources of energy (photo credit: Alsthom).

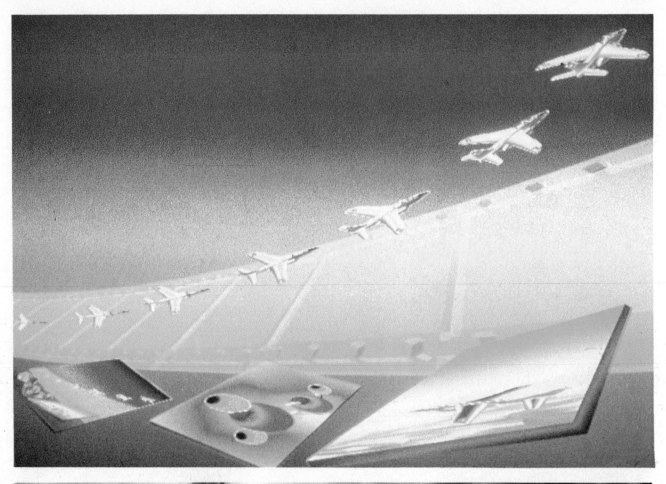

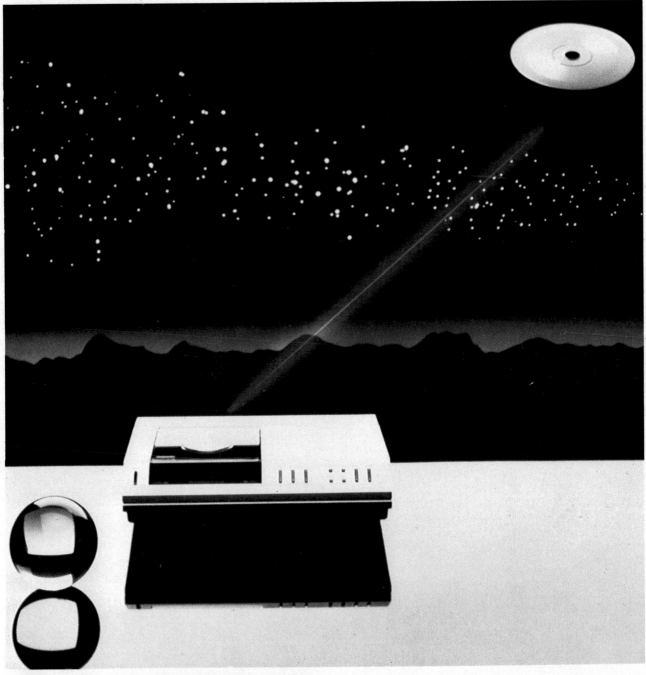

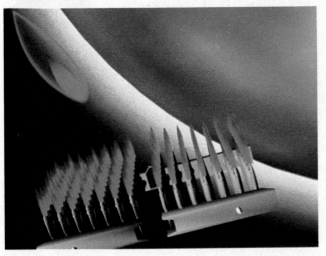

If one were to take a look at the area of **The Arts** that is changing most rapidly, it would probably be seen to be film. Inventions such as the laserdisc, above, in 1979, have opened exciting new possibilities in our ability to view films at home. Imagery in those films, too, has been forever altered through the use of computers. Here, at left, a computer-animated sequence advertises the use of gas in the kitchen. The computer is also utilized in creating 3-D animated imges, such as the simulation of a plane flight, facing page, top. Special effects reached new heights in the film "Quest for Fire" facing page, bottom, in which our ancestors as well as saber-toothed tigers and mastodons were portrayed.

Transportation *comes in many shapes and styles. This page: Below; One of the newest innovations is the anti-theft steering wheel. Bottom; The fastest commercial mode is the jointly-developed Concorde, the supersonic British and French aircraft that has, sadly, been a financial failure. Middle, left; New advances are being made with dirigibles, such as this one that is guided from the ground by verbal orders. Top, left; And what better way to test the tires we ride on than by developing a multi-tire car specifically for the purpose. Facing page: Top, left; The Sun Ray is a very light plane that can land on the ground or the water. Top, right; A classic dirigible, the Skyship. Middle; Subways move large numbers of people about in major urban areas such as Paris. Bottom, left; The modern automobile is now in its second century, as shown by this photo of Delamare and Malandin's 1884 vehicle. Bottom, right; Finally, although it may never have mass applicability, a motorbike that is able to cross bodies of water certainly has its uses.*

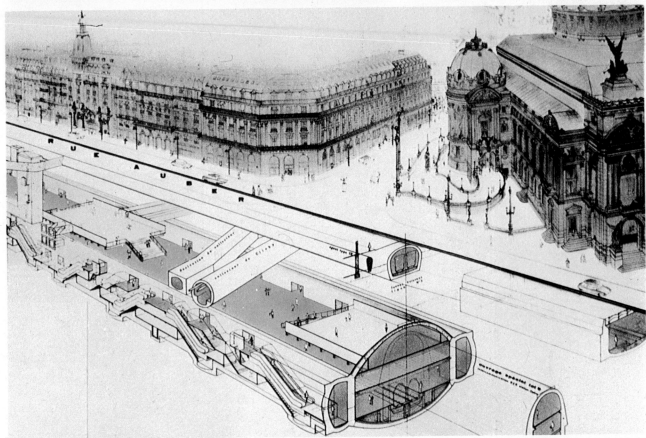

Games, Toys & Sports *form a vital part of our lives, whether as a physical or a mental exercise. This page: Top, left; Power assists abound, such as the Snow Seal, Giovanni Allisio's 1984 invention that allows a skier to go up the trail. Below; And rollerskating has become easier than ever, thanks to these motorized skates. Bottom, right; Even chess has gotten an electronic boost. Bottom, left; Trivial Pursuit dominated the games market in the mid-80s. Facing page: Top; Sic transit caddies, now that the Cadd'x has been invented. It electronically follows the golfer around the course. Bottom; And the newest campus craze? Hacky Sack, a beanbag game played with the feet.*

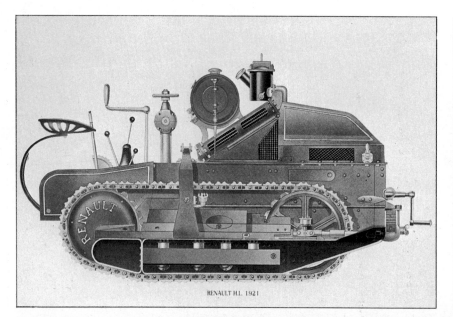

RENAULT H.L. 1921

The well-being of the world's population is dependent upon success with **Agriculture**. Since the dawn of man there has been constant application of thought, labor and technology towards increasing the quality and quantity of agricultural yield. This page, left; a 1921 version of the farm tractor and, below, a stone mill in Corsica. The facing page, top; the largest non-grafted chrysanthemum plant which bears about 680 flowers. Below, left; Alfa-feed, the electronic feeding trough. Each cow wears a medal around its neck. A computer reads the medal and is told how much the cow is permitted to eat and records the amount of feed the cow injests. Right; a turn of the century advertisement for Massey Harvesters.

The fruits of **Science and Technology** can be both beneficial and destructive; mankind, ultimately, has the power to decide how scientific research will be applied. This page, above; a model of a trans-atmospheric vehicle cruising at more than 300,000 feet above Earth. Left; A Gazelle helicopter firing a HOT anti-tank missle. Facing page, top; The atom bomb explodes. Bottom; America's Apollo flight, March 9, 1969.

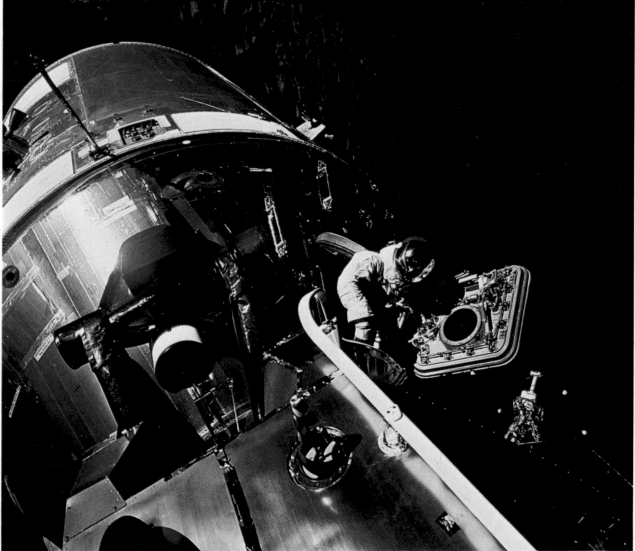

Top, left; A robot works underwater to repair a platform. Below; 56,000 transistors fill a thimble. Bottom, left; three photos show four generations of electronic components compared to a cigarette pack, tubes and transistorized cards. Bottom, right; Bernard Szajner's Albert Einstein android. Facing page, top; an airmotor in Brussels. Bottom; The Wake Imaging system allows one to visualize the shape and structure of the wake shed from airplane components.

Eye glasses for dogs? Many dogs are short-sighted. To correct this problem, French ophthalmologist Denise Lemiere has invented these glasses.

This toy dog's bark is definitely worse than its bite.

Response), that makes use of simple finger contact to measure the variations in skin resistance: The stronger the emotional state of the test subject, the sharper the signal registered.

Properly used during questioning, the GSR 2 can detect lies by revealing the range of emotions of the subject under examination.

Thermometer ring (1982)

In 1982, an American inventor, **Robert Kall**, perfected a signet ring whose transparent plastic stone contains a liquid crystal face that tells its wearer his or her temperature.

Toy dog with bark, not bite (1982)

At the 1982 International Inventions Trade Fair in Geneva, the Korean inventor **Yeu Nam Jin** introduced a stuffed dog which may serve as both a toy and warning device.

This fluffy animal is "trained" to bark like a real dog when approached.

Active Toy for Dogs and Cats (1983)

In 1983, an American inventor named **Harold A. Adler** patented an animated toy for dogs and cats. A ball with a small motor inside is activated by a radio transmitter placed inside the pet's collar. Each time he approaches close enough to the ball, it rolls away, stops and then rolls away again, creating the illusion of a living object.

Aquaspace (1983)

Jacques Rougerie, an oceanic architect, invented the *Aquaspace* in 1983. This catamaran has pontoons made of a transparent cryolite material. It is 20 meters long, 8½ meters wide and carries 28 passengers.

As a marine dwelling it is very comfortable and, so quiet, it doesn't even disturb the fish.

Electronic Home Prison (1983)

The electronic prison, which allows a prisoner to remain comfortably seated on his own sofa while serving his sentence, seems to have been invented simultaneously by two Americans: **Judge Allison** and **Judge Jack Love**.

This "prison" takes the form of an electronic bracelet, attached to the ankle of the prisoner. If he moves more than 1600 feet from his home, an alarm sounds and the prisoner is "brought back to prison" in short order.

Lamps That Blow On and Off (1983)

Turning a lamp out by blowing on it has got to be the height of absurdity —something like an electric candle. But even more far-fetched than that is turning it *on* in the same way!

In 1983, two French inventors, **Alain Domingo** and **Francois Scali**, designed two different lamp models, the "Pianola" and the "Manolo," which work exactly like this. To turn it on and off, you simply blow on it! One sure advantage: no more on-and-off switch means no more shocked fingers.

Listening Thermometer (1983)

Allen-Martin Electronics, Ltd., an English firm, developed an ingenious thermostatic system in 1983. It is sound-sensitive and can regulate the heat, more-or-less, by hearing a voice.

Thanks to a microphone incorporated in the mechanism, this thermostat "hears" the occupants arrive—in an apartment, for example—and then releases the warm or cool air necessary to create the chosen temperature.

Music-Board (1983)

Jogging, break-dance, aerobics, gymnastics, slimming-down, sweating-out, muscle-building, you name it. All these exercises can be done on the music-board, invented by a Frenchman named **Robert Viau** in 1984.

The panel can be supple or rigid, according to your preference, and it allows you to dance and create the music, both at the same time, thanks to an electronic device activated by the feet. Ideal for staying in shape!

Roulis 2000 (1983)

Invented in 1983 by a Swiss named **Paul Henriod**, the "Roulis 2000" is a vehicle positioner for auto builders and service station mechanics.

It allows a man to make repairs on the underside of an automobile while remaining in an upright position. This is more practical and less tiring than working on his back. Just a word of warning: Don't leave anything breakable in your car while it is being repaired.

Solar Bicycle (1983)

Alan Freeman, a retired British engineer, has invented a "bicycle without pedals." A solar panel installed on the handlebars gathers electrical energy which is stored in a solar battery, thus replacing muscle and effort. Experiments on the road have shown that the solar bicycle has an operation range of 30 miles and a maximum speed of 15 mph.

"Flasher" (1984)

It was a French woman, **Carmela Brunet**, who in 1984 invented the "flasher," a small box that goes "bip-bip" when you pass near the man or woman of your dreams—that is, if he (or she) is also tuned in on the same wave-length.

The flasher is about the size of a package of cigarettes and is provided with four coded frequencies. No more hesitating shyness; the flasher takes care of the introductions. But keep in mind that the flasher provides for some *very* special situations, so be careful not to tune in on the wrong frequency!

This lamp can be lit or turned off by blowing on it as you would a candle.

The Roulis 2000, invented by Paul Henriod in order to facilitate the repair of cars.

Pocket-Sized Anti-Atomic Air-Raid Shelter (1984)

Swiss pacifists have chosen a humorous approach. In 1984, they launched a totally new product: the pocket-sized anti-atomic air-raid shelter. After only two months on the market, the product has had a considerable success, and more than 25,000 shelters have already been sold.

The shelter consists of a simple paper bag, and the instruction for its use are quite simple: 1) Unfold the shelter; 2) Put it over your head; 3) Wait for the end!

According to the advertisements, this air-raid shelter is as effective as the Civil Defense program.

Reaction Painting (1984)

Reaction painting was invented by **Prince Jurgen von Anhalt,** Pretender to the throne of Prussia.

Standing on an airport runway, behind the engine of a jet plane, he launches his paint directly into the blast of the engine. The colors fly and spread out "artistically" on a wide canvas held in place by an airport baggage elevator.

Photon Guitar (1985)

The Photon Guitar invented by **George O. Stadnik** and **Merril A. Dana,** of Worcester (MA.) does not play music. It is an instrument which enables individuals to visually express themselves in an instantaneous gratifying way.

The player manipulates a bundle of fiberoptics with one hand, while controlling the color of light passing through the fibers with the other. He can project images on a wall, or on a screen. The Photon can be played to music or not.

Training bell (1985)

The training bell designed by American **Joseph A. Cozzi** is a approach that should magnify your pet's intelligence and personality.

The bell is nearly out of reach for him, and he has to ring it when he is hungry, to signal it is lunch time. You can also train him to express that he wants to play, or to take a walk, with a different number of rings.

Solar boat (1985)

Kenichi Horie, a Japanese, has built a solar boat with which he intends to sail from Hawaii to Japan.

Powered by 1,100 solar cells that produce 1 watt apiece, his boat can make 3 knots per hour. The Sikrinek (the name of the boat, which means Rising Sun in Eskimo language) should take two months for the crossing.

Burial in Space (1985)

Cremation has been a part of human funerary customs since at least the paleolithic period. In the old times, the ashes of the Viking chiefs were dispersed at sea.

Celestis Group, a consortium in both funeral services and aerospace, is now offering much better: for little more than the cost of a conventional American funeral, they will place cremated remains in Space.

The ashes will be sealed in a capsule plated with precious metal and inscribed with the name of the deceased and the symbol of choice. The capsules are placed in an orbital spacecraft that will be sent up to 1900 miles, far from the usual orbital paths. In this environment, the capsules will be preserved forever.

Aquatic treads

The Frenchman **Pierre Jousse** is credited with a new method of water propulsion. His water-treading boat literally rolls on water.

Bathers have nothing to fear since there is no keel or propeller under the hull. This boat comes ashore by changing the rotation speed of its treads, somewhat like a tank.

Here is a boat, finally, that can navigate in shallow water and land on beaches.

"Eclectic" car

The American **George Barris** built his multi-purpose car for the Manufacturers Hanover Trust Company.

Named "Anycar III," it is an automobile of surprises: In addition to its being made up of parts from various models, a small electric cart that doubles as a radiator is hidden under its hood.

Paper-making worms

A Zen priest from Kyoto, Japan, has developed a new technique for making paper. What **Masahisa Ishiko** has done is to divert silkworms from spinning their cocoons; they make silk paper fiber instead.

The process consists of placing the silkworms on a string stretched over a wooden frame. The silkworms then spread the silk over the entire surface of the frame. Total elapsed production time for one sheet of silk paper: three days.

Shoe-light

Blackouts have been with us ever since the advent of modern warfare. The blackout is a wartime measure that involves covering the windows and turning down the lights so that potential enemy bombers cannot use telltale lights to locate a city. To light one's way and to avoid bumping into objects in the dark, an ingenious handyman thought to mount a tiny lamp on one of his shoes.

The solar bicycle and its inventor, Alan Freeman.

Aquatic trends.

THE FUTURE

American Orbital Space Station

Announced by President Reagan on January 26, 1984, the American Orbital Space Station will become a reality in 1992. The four space-shuttles in service in 1985 will be able to make two voyages each month, carrying modules up to 50 feet long and 15 feet in diameter, and weighing up to 30 tons. Teams of 6 to 8 astronauts will work at the space station, with a change-over every two months.

The Orbital Space Station has a triple role: 1) Observation of the Earth and the Solar System, mainly for civil ends, but also military ends; 2) Industrial Manufacturing of pharmaceutical products and the creation of gallium arsenate crystals; 3) Port Installations for the mooring of space-shuttles and the launching of larger ships in the direction of the Moon and the rest of the Solar System.

The 1992 version of the Orbital Space Station will be a "nucleus," and "wings" will be added to this nucleus later on.

Boeing Aerospace Co. is now working on different architectural designs for a permanently manned Space Station in low Earth orbit. A small Space Station could be placed into orbit as early as 1989.

At the same time, the Soviets are currently working on the plans for their own Orbital Space Station, whose elements will probably be carried into space by a modified version of their Cosmos satellites. But the entire operation remains almost totally secret.

Extra-Terrestials

With Voltaire's "Micromega"-the ancestor, in fact, of "E.T."!-we saw the arrival of the first being from another world, a being endowed with a philosophy and a wisdom different from our own. But if contact with extra-terrestials often turns out to be nothing more than the product of an overly-fertile imagination, the sighting of UFO's has become an almost daily occurence.

Discounting the obvious cases of hallucinations, hoaxes and weather phenomena, there is something in all this which scientists and governmental organizations all over the world seem to take very seriously. Why would governments spend millions of dollars to listen for "messages from the stars" with the aid of giant radio-telescopes? Or why would they engrave messages on plates of precious metal to carry spatial probes into the vastness of the universe, if they did not believe that there was something out there?

Flying Saucers

The term "flying saucer" first appeared in the newspapers in 1949, when it was reported that an American businessman claimed to have seen a formation of seven mysterious saucer-shaped objects flying above the Rocky Mountains.

In reality, the sighting of Unidentified Flying Objects (UFO's) goes back much further. Certain experts on the question have even found evidence of sightings going back to ancient times (the famous visions of the prophet Ezekial in the Bible, for example).

It is difficult, most of the time, to separate the question of UFO's from the question of the existence of extra-terrestials. For a long time, these beings were referred to as "Martians," because the planet Mars is so close to the Earth.

Joint Services Advanced Vertical Aircraft - JVX

JVX, built jointly by Bell and Boeing, should fly for the first time in early 1988, and have its first production delivery to the US Marine Corps in mid-1991.

JVX will be the first operational tilt rotor and is being designed from the outset to be a multi-service, multi-mission aircraft. On January 15, 1985, Secretary John Lehman officially named the JVX—Osprey.

The JVX is taking advantage of recently demonstrated advances in structures, control systems and engines, and combining them with the tilt rotor concept. The result is a versatile, high performance VTOL utility transport that combines the hover efficientcy, controllability and low downwash of a helicopter with the 300-knot cruise speed and efficiency of a turboprop airplane.

Orbital Aircraft

An orbital aircraft could take off from any airport, climb to an orbit 200 miles above the earth, patrol there for several hours and then re-

Renderings of proposed orbital space stations.

The Lockheed Solar HAPP.

turn to land at its point of departure, crossing and recrossing the atmosphere. A dream perhaps, but a realistic one, according to the American military experts.

The Pentagon has two revolutionary projects in the drawers of its laboratories: One of them is the rocket-propelled plane of DARPA (Defense Advanced Projects Agency), which would weigh no more than five tons. The other is a SAC (Strategic Air Command) model, which would carry about ten tons on board.

The SAC plane could place small emergency satellites in orbit whose purpose would be to secure communication, to carry out surveillance in case of conflicts, or even to hold anti-satellite weapons.

The DARPA plane would resemble an ordinary rocket head and, since the interior would not be pressurized, the pilot would have to wear a space suit, like an astronaut.

The Pentagon hopes to see these

projects completely developed by 1990.

Solar HAPP

Lockheed Missiles and Space Co. (Sunnyvale, Ca.) is now working on a new version of a Solar High-Altitude Powered Platform.

It is a flying wing with two booms, and underslung payload pod, a pusher propeller, and solar cells on vertical wing stabilizers and on tops and bottoms of the wingtips. Wingtips would hinge up during daytime to catch maximum sun, returning to the horizontal at night to improve aerodynamics.

The engine would be powered by the solar cells during the day, and at night by fuel cells which had been charged by the sun during daylight hours. It would have a top speed of 92 mph. Its weight would be held to 2000 pounds.

It would be launced in still air, spi-

raling upward in four hours to an operational atitude of 12.4 miles (20 km). It could stay aloft for a year. First flight is possible by 1993.

One of its first missions could be sending sharp images to earth to help farmers make crop decisions, probably in conjunction with Department of Agriculture project in southern Arizona.

Solar Power Station in Space

For a long time now, especially since the energy crisis of 1973, Americans have foreseen the possibility of setting up solar power stations; not only in the sunniest regions of the earth, but also in space.

NASA has worked out full-scale projects in this direction. A solar power station in space could function around the clock without stopping, and the procedure would be completely non-pollutant.

But recent re-evaluations seem to

show that this project is not yet feasible, for economic and technical reasons. It would be necessary to have a space-shuttle capable of putting 400 tons of equipment in orbit, at a distance of 150 miles during each voyage. The existing space-shuttle can transport only 30 tons. In addition, it would take 1500 workers to keep the station functioning permanently in space. As for expenses, they would be astronomical.

Stealth Airplane

Lockheed and Northrop are now working on a type of airplane designed under the code name of "Stealth." Their mission is to go into the defense zone of the enemy without being detected.

Lockheed is studying a reconnaissance and combat plane, while Northrop works on the A.T.B. Bomber. This bomber should be put into service in 1988. It should fly at 3000 km/h, at low and high altitude. It should carry cruise missiles and be nearly invisible to radars. Materials, based on carbon, are not detectable. The engines will be buried in the

stream-lining in order to upset most air-air missiles which search for heat sources.

Submarine tankers

Arctic Energies (Severna Park, Md.), a marine technology company, has proposed an ambitious scheme for exploiting the large deposits of methane in the Arctic.

Methane would be converted at the wellhead into methanol, a liquid fuel, and then put into submarine tankers to transport it beneath the Pole to Norway. The cost would be higher than transporting fuel to Europe from the Middle East, but there would undeniably by a strategic advantage to Arctic Energies proposal.

Transatmospheric Vehicle — TAV

The US Air Force is now studying the plans of various projects of TAVs designed by a few aerospace contactors, among which McDonnell Douglas, Rockwell International and Lockheed.

The general characteristics of the

TAV are already outlined in air Force specifications. It would take off from a conventional airfield, accelerate to the fringes of outer space, and return to the landing site. It would be somewhere between a plane and a spaceship. It would operate at altitudes between 100,000 and 500,000 feet (commercial passenger jets cruise at about 35,000 feet).

TAV missions in the military field could go from strategic reconaissance to destroying approaching enemy bombers. The TAV proposed by McDonnell Douglas could reach any point on the globe in less than two hours. The one designed by Lockheed would operate up to 30 times the speed of sound.

Tsukuba, The Exhibition for the Year 2000

The third millenium may have begun in Tsukuba, Japan. From March 17th to September 16th, this little city on the outskirts of Tokyo had the most fabulous showcase ever dedicated to new technologies: "Expo 85". One of the big events of the year, with twenty million visitors. The theme chosen

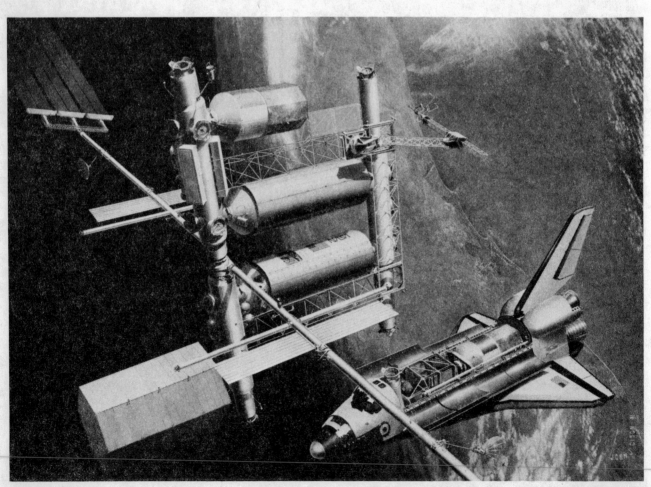

Boeing's study of an Orbiting Spaceport, a small community in space. It would serve as a staging point for space operations. The illustration shows a space shuttle in dock.

was, "Science and technique, at the service of man." Overall, Expo 1985 represented an investment of the order of 2.5 billion dollars. And the entire site was demolished as soon as the last visitor had left.

For Japan, the stakes were high. It was first necessary to convince the entire world about the quality of its research and techniques in the most advanced areas. This was a prestigious operation, on a world-wide scale.

But maybe, more than that, Japan wanted to captivate, astonish, win over the public. This gave rise to the frequent use of films, general public demonstrations, accessible to a broad spectrum of visitors, and the very commonplace presentation of numerous exhibits.

Nonetheless, scientific innovation and technical prowess were well presented in Tsukuba; from Jumbotron by Sony, to the tree bearing 10,000 tomatoes, in addition to the robot organist. It's a fact: in Tsukuba, the 21st century opened before our very eyes.

Some Tsukuba Presentations

Image Cascade

This was no doubt the most original Tsukuba screen. Designed by the Mitsui firm, it was made up of a waterfall on which the images were projected. In order to cope with the transparency of the water, minuscule air particles were injected. This aquatic screen measured 40 x 7 meters.

Large Scale Cinema

Alongside the traditional 16mm and 35mm formats, numerous pavillions had adopted the 70mm format; this allows the use of very large sized screens and generally provides a more or less pronounced dimension. Several 70mm procedures were demonstrated.

- Omnimax, with 5 perforations - the same is used at La Villette - with a spherical screen.
- Dynavision (or Japax), with 8 perforations.
- Imax (or Cinemax-U), with 15 perforations. In this case, the flat screen is used.
- Showscan, or 70mm with 5 perforations, has a speed of 60 images /second, and the six-track Dolby. This procedure was developed by Douglas Trumbull, the creator of the special effects for "Close Encounters of the Third Kind", and "2001, a Space Odyssey". The result is very simply one of prodigious realism.

Machine which recognizes bills, or the "bank note sensor"

Presented at the Toshiba pavillion, a machine permits the sorting of bills and bank notes, for example, at considerable speed (one specimen every 1/25th of a second).

This is made possible by the use of photo-sensors which allow it to recognize colors and shapes with precision, then to track the movement of the bills by means of feelers. All this is done with a greater speed than is possible with the human eye. Deteriorated or dirty bills are then set apart. The machine is also equipped with a vocal command device.

Matsushita Television

For the first time, Matsushita presented its prototype for a flat screen television, 9.9 cm thick (for 25 cm of diagonal). However, the apparatus, which makes use of liquid crystal technology, offers images of rather limited quality. This is, however, a decisive advance in the field of flat screens.

Matsushita also exhibited a new 3-dimensional television procedure which does not require the wearing of filtering glasses. This procedure is used for shooting with 5 cameras, and the images are synthesized using a VCR. The screen measures only 35 cm in diagonal for the moment.

Multilingual automatic translator

Presented by Fujitsu, an automatic translation machine can work in four different languages. It is equipped with a keyboard-operated screen, on which letters or ideograms are designated. In five minutes, the machine translates a simple sentence in French, English, German or Japanese, and then says it by means of speech synthesis.

The principle is based on the universality of mental schemes, regardless of the language. By 1990, this prototype should give birth to a high performance machine which will be capable of operating in "real time."

The Sony Jumbotron

With a height of 42 meters, the Jumbotron dominated the fairgrounds of Expo 85. This 1000 square meter television, built by Sony, was one of the major Tsukuba attractions.

The Jumbotron does not have a cathode tube, as do traditional televisions. Its flat screen is, in fact, made up of 150,000 high luminosity cells called "Trini-lite". Each of them is composed of three elements—red, green, blue—which correspond to the three fundamental colors. Perfected electronic circuits control the light.

As for the signal, it is transmitted in digital form by optic fibers. Finally, in front of the screen, a remote control camera allows the control of the function of the Jumbotron.

This giant television is always visible, even in bright sunlight. It cost the huge sum of 20 million dollars.

The Tree bearing 10,000 tomatoes

This curious tomato plant, approximately 10 meters high, astonished the world. Sown in October 1984, it produced more than 10,000 fruits before the end of the exhibition; the fruits were picked day by day by the inventor, professor Kei Mori, from the University of Tokyo. Better still, it grew without soil or sun!!

This "tree of 10,000 tomatoes" made use of a sophisticated lighting system. Solar captors, commanded by computer, were arranged on the roof of the building. The light thereby "collects" and was filtered and freed of its damaging or useless rays, then carefully dosed before being transmitted to the plant, via optic fibers. This ensured optimal growth.

Further, the plant was fed with a nutritive liquid enriched with mineral salts and fertilizer, according to a very jealously guarded secret.

This procedure might have numerous applications, notably for greenhouse cultivation.

Video Answer Service

Developed by INS, this apparatus allows the receiving of graphic images coming from a data bank, color slides or films at home or by a simple telephone call. All these images, transmitted by optic fiber, are displayed on a traditional TV screen.

In short, this is a real made-to-order domestic programmation scheme. This video response service (VRS) is still in the experimental phase.

Robots of Tsukuba

Fanuc Man, the Hercules of robots

Fanuc can lift a 500 pound load using its two jointed arms. But it is also capable of performing tasks requiring great precision, such as handling a brush in order to do calligraphy.

This ambidextrous automaton is used to assemble robots. It usually works in a nearly deserted factory. Manufactured by the Fanuc company, it weighs more than 25 tons.

The Matsushita robot-designer

Only a few seconds of posing are required. This robot, designed by Matsushita, begins by taking a "photo" of his model. Then it analyzes the photo and converts it into digital signals using a 16 byte integrated computer. Finally, using the jointed arm, it designs the memorized portrait on a large sheet of white paper in two minutes.

At the end of the session, it also asks the model, "Does this look like you?" The question, of course, is possible by means of a system of vocal synthesis.

Robot for the Handicapped

It is called Spartacus. This robot, designed to equip the room of a motorized handicapped person, responds to the voice and even to whistling. It can take care of various tasks, such as giving a drink to the patient, turning the television on or off, opening a door or a window, turning on the heat, etc.

Sculptor robots

These two twin robots possess undeniable artistic talent. Their favorite occupation is sculpting an animal in a block of ice brought to them by an automaton. For this they use high speed scissors attached to the arm.

All the phases of the job to be carried out have been memorized by a powerful computer which subsequently gives instructions to the control system of the robots. This team was developed by the Hitachi firm.

Wasubot, the organist-robot. Through this video camera, it can read a sheet of music in thirty seconds.

Titan III

This four-footed robot, baptized Titan III, is capable of moving over uneven ground and can even climb a staircase. Each of its jointed limbs can move in three directions and has a feeler at its extremity.

Titan III is controlled by a computer. It weighs about 200 pounds and can be used in industry, notably for the exploration of the inside of nuclear reactors.

Wasubot, the organist-robot

On March 17th, it played the Tsukuba inauguration anthem. This astonishing and energetic robot can play the electronic keyboard of an organ and can even move the pedals with its feet. Each of its hands comprises fourteen joints.

It can read a sheet of music in thirty seconds by means of its electronic eye, made up of a video camera. Its repertory ranges from Bach to the Beatles. Better yet, Wasubot—that's its name—obeys vocal orders and can talk!

It was designed in 1984 by a research group at Waseda University in Tokyo, with the help of the Sumitomo firm. Wasubot is equipped with about 50 joints, each controlled by a minicomputer and activated by a motor. This makes Wasubot one of the most complex robots in the world.

WHL-II, the bi-ped

This two-footed robot imitates human walking and slowly waddles over a flat surface—no more than 150 meters per hour. Its hydraulic joints resemble those in the human. Its feet are equipped with sensors and transmit date to the "brain" (the computer), which controls the movement to be carried out.

The difficulty for WHL-II, of course, is to keep its balance. This robot was designed in the laboratory of Professor Ichiro Kato and was built by the Hitachi firm.

medicine

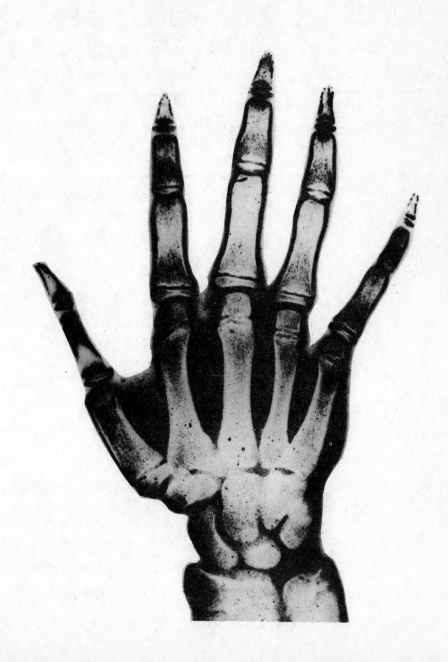

1. Assistance to the Sick

Hospitals (372 A.D.)

The first public hospital was founded in 372 by **Saint Basile** at Caesarea, in Cappadocia. A few years later, a second hospital was built in Rome by Saint Fabiola, and hospitals proliferated thereafter. The first real hospitals were Roman sanitary establishments built behind the "lines," or the fortified zone which protected the Roman Empire against barbarian invasions.

Ambulances (1792)

The first ambulance was created in 1792 by **Baron Dominique Jean Larrey** (1766—1842), private surgeon to Napoleon who wanted to provide care to the wounded right on the battlefields. He conceived of a sort of "flying ambulance," using horse-drawn caissons which would transport the personnel and equipment necessary for the initial treatments.

Assisted by the surgeon P. Percy, Larrey formed an ambulance corps made up of surgeons and stretcher bearers. The corps was active for the first time during Bonaparte's Italian campaign in 1796—1797.

The first motorized ambulance, equipped with a Daimler motor, was presented at the Paris Cycle Show in December 1895 by Panhard and Levassor. The first ambulance of this kind was acquired by the 9th French Army Corps in July 1900. The first civil ambulance appeared in Alencon, France that same year.

Nursing (1854)

In 1854, during the Crimean War, an English woman, **Florence Nightingale**, created the first nursing corps. The scandalous lack of decent health services led her to unite nuns and nurses and to make use of their services in the surgical theater. Her initiative bore fruitful results: Mortality decreased from 42% to 2.2% among the wounded.

Encouraged by Queen Victoria and aided by a national grant, Nightingale went on to found St. Thomas Hospital in London, which became a nursing school. Florence Nightingale, named the "Lady of the Lamp" in remembrance of the time she roamed battlefields searching for the wounded, is universally acknowledged as the foundress of assistance to the wounded.

The first professional school of nursing opened in Paris in April 1878,

A room in the Paris Hotel-Dieu hospital, around 1400.

and soon after there were four of them: Bicetre, la Salpetriere, La Pitie, Lariboisiere (all in Paris).

The Red Cross (1863)

In 1863, four Genevese, General **Guillaume-Henri Dufour**, **Gustave Moynier**, Doctors **Luis Appia** and **Theodore Maunoir**, joined forces with **Henri Dunant** to form the International Committee for Aid to the Wounded, subsequently known as the International Committee of the Red Cross (ICRC).

Henri Dunant was born in Geneva on May 8th, 1828, and was 31 years old when, during a business trip to Italy, he witnessed the Battle of Solferino on June 24, 1859. He became outraged upon seeing so many men die due to incompetent medical care and the virtual non-existence of health services.

At the invitation of the Swiss Confederate Council, a diplomatic conference took place in Geneva, where authorities of twelve states signed a short treaty containing ten articles: "The Geneva Convention of August 22, 1864, for the improvement in the treatment of military casualties wounded in battle." The symbol of the red cross, shown against a white background, was chosen as the unique and distinctive sign for army medical staff as well as for military hospitals and ambulances.

It seems that the choice of the white background color for the badge should be credited to Doctor Appia, and that General Dufour subsequently suggested the addition of a red cross.

In November 1876, Turkey, at war with Russia for six months, declared to the Swiss Confederate Council, which acted as trustee of the Geneva Convention, that the Red Cross emblem offended the religious beliefs of its troops. As a consequence, Turkey adopted the Red Crescent as the symbol for its medical field services.

Robot patient (1970)

This robot, developed by the Japanese firm Koken in 1980 (after six years of effort), simulates a patient. Medical students can save its life when it simulates a cardiac arrest: If the cardiac massages, mouth to mouth respiration and injections have been correctly carried out, it blinks its eyes.

The robot serves as a guinea pig for medical students. It is connected to a computer which continually indicates its clinical condition and the quality of care given it. Its eyes are made of quartz, its heart is metal, its blood an electron flux; its brain is a microcomputer. If it is ill, its lips become purple; they become red again when health is restored.

2. Investigations

Auscultation

Hippocrates of Cos (460—377 B.C.) was the first great physician of antiquity, and he was also the first to place his ear against a patient's chest in order to perform auscultation.

The first real auscultation was

Ascultation in 1345.

practiced by the Frenchman Rene Theophile Hyacinth Laennec (1781–1826). It is said that one day he watched a young boy who, with his ear against one end of a beam, heard signals transmitted by a playmate who struck the other end of the beam with a nail. Laennec immediately drew a conclusion from his observation, and applied a piece of paper, rolled into a cylindrical shape, to the chest of the next patient he visited; he was thereby able to hear the heart sounds clearly.

The stethoscope was born when he replaced this sheet of paper with a wooden cylinder. The apparatus was subsequently modified and improved (in particular by the Austrian Skoda), to become the bi-auricular stethoscope which is generally used today. The electronic stethoscope was invented in 1980 by the Americans Groom and Boone.

Microscope (16th century)

The microscope was invented in Holland at the end of the 16th century, perhaps by the optician **J. Jansen** with the assistance of his son, Zacharias.

The system was based on the use of a convex piece of glass which considerably enlarged the object under observation when examined with an eyepiece serving as a magnifying glass. Described by Galileo in 1609, this microscope gave little detail and inconsequential enlargement.

The Dutchman Antonie Van Leeuwenhoeck (1632–1723) was the first to observe spermatozoids, muscular striations, and some buccal bacteria under a microscope, although enlargement was still less than 200 times.

Modern microscopes have existed since 1880. The electronic microscope was designed by Buesch in 1926, but Zworykin may also claim authorship (1939).

Electronic microscope (1926)

The German Hans Busch drew up the theoretical basis for the electronic microscope. On the basis of his research, two other Germans, Max Knoll and Ernst Ruska, from the Technische Hochschule, Berlin, undertook the initial experimental work in 1928. This resulted, in 1933, in the first operational electronic microscope, developed by Ernst Ruska. In 1943, Knoll and Ruska were able to observe objects the size of one millionth of a millimeter.

Even more recent is the ion emission microscope, which allows observations thought impossible only a dozen or so years ago.

Thermometer (1612)

The first known thermometer was developed by the famous astronomer and physicist **Galileo** (1564–1642) in about 1612, but it was not a medical thermometer.

The first medical thermometer was created in 1626 by the Italian physician Santoria, called Sanctorius. This water thermometer, placed in the armpit, was not practical enough to warrant general use. The model for current medical thermometers, glass tubes with a mercury column and a calibrated stem, was developed by the English physician Allbutt in 1867.

In 1869, the use of temperature sheets was described in Jaccoud's "Treatise on Internal Pathology." Thereafter the establishment of a thermal curve, with graphic interpretation, came into general use.

Disposable oral thermometer (1976)

The disposable oral thermometer was invented by an American, Weinstein, and manufactured by Bio-Medical

Sciences Inc., Fairfield, New Jersey. A plastic tab is marked at one end with 50 thermosensitive points at intervals of 1/10 degree ranging from 35.5° to 40.5°. These white points become green in the mouth. The temperature degree corresponds to the last green point.

Percussion (1761)

Percussion was invented in 1761 by the Austrian physician **L. Auenbrugger**. This clinical mode of exploration allows the condition of certain organs to be known according to the noise obtained when they are tapped with the fingers.

The system was improved by Baron Jean Corvisart (1755–1821), Napoleon's personal physician, and then by the Austrian Skoda, who perfected it considerably.

Measurement of blood pressure (1819)

The hemanometer, a mercury manometer, was invented in 1819 by the French physician and physicist **Jean-Louis Poiseuille** (1799–1869) to measure arterial tension (named "pressure" by the physiologists).

It was not until the invention of the sphygmometer in 1881 (a word forged from the Greek word "sphygmos," meaning "pulse") that the apparatus became truly functional. It was developed by the Austrian Von Bash in order to examine arterial pulsations.

Francois Potain's sphygmographs recorded these pulsations beginning in 1889. In 1896, the Italian Riva-Rocci substituted a large cuff for Potain's sensor, thus permitting a homogeneous and definite compression. The Russian Korotkov perfected the system in 1905 when he developed an apparatus which was able to perform auscultation of the arteries without palpation.

Electrocardiography (1887)

The first human electrocardiogram was recorded in 1887 by Augustus Desire Waller (1856–1922), a Paris-born physiologist working at the University of London.

During the course of his experiments, Waller recorded the electrocardiogram of his favorite bulldog, Jimmy. The initial tracings were of mediocre quality, because they were obtained using the mercury "capillary electrometer" developed by Gabriel Lippmann (1854–1921), professor of experimental physics at the Sorbonne in Paris.

After working with Lippmann from

1894 to 1900, William Einthoven, professor of physiology at the University of Leiden (Holland), developed the string galvanometer in 1901, and thereby became the real inventor of electrocardiography. The equipment designed by Einthoven weighed 300 kg; it took five people to run it.

Radiology (1895)

Following their discovery in 1895, X-rays were first used to diagnose fractures, detect foreign bodies, and examine the damage caused by bone disease.

The initial radiological apparatuses were immediately installed in hospitals. They emitted very weak radiation and subjects had to be exposed to the rays for nearly one hour in order to obtain a clear outline. The danger of such a procedure was very quickly detected (radiologists were afflicted with cutaneous, sometimes cancerous, lesions), and the equipment was soon perfected to shorten exposure time.

Major breakthroughs:

• 1896: First cranial radiography. Welker exposed his patient for one hour, with an interruption every minute for one minute of rest, to prevent the X-ray tube from melting due to heat.

• 1918: Walter Dandy (USA): first ventriculography, first encephalography.

• 1927: Moniz: first cerebral arteriography.

• 1928: Andre Bocage (France): first tomography.

• 1930: Image amplification, invented by Irving Langmuir (General Electric Research Laboratory, USA).

• 1948: Professor Coltman improved the Langmuir tube.

• 1959: The French company Thomson created a tube with a field of 4 inches (11 cm), thus increasing brilliancy by 3000 and allowing the image to be taken using a television camera. This was a revolution in radiological imagery.

• 1967: Scintigraphy (artificial radioelements)

• 1972: Nuclear magnetic resonance (invented by Zavoiski in 1944) was applied in the USA by Lauterbur.

X-rays (1895)

On the evening of November 8, 1895, the German physicist **Wilhelm Konrad Rontgen** (1845—1923) was working with a cathode tube in his laboratory and saw a piece of paper covered with a fluorescent substance begin to shine. He had an intuition that he had discovered a new form of radiation, imperceptible to the human retina and different from cathode rays (known since 1879). This new radiation was an emission of X-rays, so named by the experimenter because he himself did not know what they were.

X-rays, or Rontgen rays, move in a straight line and can cross dense bodies which are otherwise impenetrable to ordinary light. They are electromagnetic radiations which have a very short wave (10^{-8} to 2.10^{-11}) and are able to impress an image on a fluorescent screen or a photographic plate.

Pursuing his research, Rontgen produced the first radiography on December 22, 1895; he X-rayed the hand of his wife who was wearing a ring. Rontgen revealed his discovery on December 28, at a meeting of the Medical Society of Wurzburg. Within two months, the news of this discovery created a sensation throughout the world.

Scanner

In 1971, the field of radiology was turned upside down by G. Hounsfield, who developed the scanner while working in the English company EMI.

Tomography, invented in 1928 by A. Bocage, already permitted radiography in successive steps. The first apparatus was made possible by the work of the Dutchman Ziedses-des-Plantes, in 1924.

The scanner, or tomodensitometer, allows, for example, the exploration

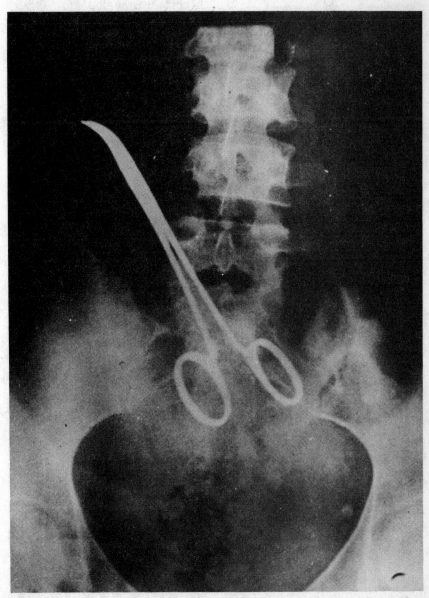

A pair of sissors forgotten by the surgeon are discovered by a radiography.

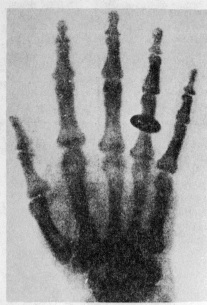

The first radiography in history. On December 22, 1895, Rontgen succeeded in taking a radio of his wife's hand. She was wearing her wedding ring.

of the brain without subjecting the patient to any prior preparation. It performs a large number of very precise radiographic cross-sections of an organ (or of an area in the organism). The data are immediately transmitted to a computer, which instantaneously assembles them and provides an image of the organ under investigation. The surgeon can thus very precisely localize the tumor (for example) on which he must operate.

Neuroradiology was the first area to benefit from such applications. Exploration was subsequently extended to include the entire body.

Nuclear Magnetic Resonance (NMR) (1972)

Water makes up approximately 70% of our body, but its concentration varies depending on the organs. There is much less water in bone than in muscle, for example. The NMR, the principles of which were presented in 1972 by P.C. Lauterbur, professor of chemistry at the State University of New York at Stony Brook, uses magnetic fields to compose a map of the distribution of these water molecules in the organ under examination, and projects the map onto the screen of a computer. Since tumors and malformations, for instance, do not have the same density as the organ in which they are found, they appear more clearly on the screen than they do with a scanner, indeed almost like an anatomical plate. It is even possible to obtain three-dimensional images, thus allowing them to be localized within a millimeter of exactitude.

However, as of 1984, the NMR has still not entered into current usage, mainly because of its high cost; the price of an apparatus is over $1 million.

Portable radiographic apparatus (1977)

Originally invented in 1977 by Doctor Lo I. Yin for the National Aeronautics and Space Administration (USA) to map the "black holes" in space, the Lixiscope is a portable radioscopic apparatus for examination of the hands, arms, feet or other small parts of the body. When the operator pulls on the button to "trigger" the radioactive source, the image appears at the opposite extremity of the Lixiscope, on a screen 5 cm in diameter, which can be photographed.

Electroencephalogram (1929)

In Germany in 1929, Hans Berger, professor of neuropsychiatry at Iena, was the first to record the spontaneous electrical activity of the brain. However, due to the weak amplitude of the signal emitted, his recording met with skepticism.

It was not until the English physician Edgard Adrian (Nobel Prize winner in 1932) published his own results (1934) and became Berger's defender that the latter scientist finally received the approval of professional circles.

Echography (1955)

The American Leskell first used ultrasound to observe the heart in 1955. Echoencephalography followed echocardiography and the technique was then perfected, due notably to the use of microprocessors, to become echography (used for diagnostic purposes) as of the 1970s.

The principle of echography is that of sonar. Emitted from a radio tube, ultrasounds from 1 to 16 MHz are sent to the organ to be investigated. The organ reflects them and sends them back to the receiver. The emissions are intermittent, approximately 200 per second.

In 1956, J. Donald (Great Britain) performed an echography of the gravid uterus, and for the first time it was possible to see the head of a fetus. Since 1978, the spread in the use of ultrasound for pregnancy tests has led not only to the very early (from the 8th/10th week) diagnosis of malformations or chromosomal anomalies and the determination of gender, but also to the care of or operation on the fetus.

Radio-immunology (1977)

Invented by Dr. Salomon Berson and Rosalyn Yalow (which won for her the Nobel Prize in Medicine in 1977), radio-immunology combines two techniques: the biological technique uses the specificity of immune reactions to identify a given organic substance; the other technique, which is physical, marks these substances by introducing radioactive atoms into their molecules.

The sputnik of the intestines (1977)

Designed and created by two researchers working at INSERM (a French national medical research institute), Alain Lambert, a mechanical engineer, and Francis Crenner, an electronics engineer, with the technical collaboration of Sylvain Schmidt, at the request of Professor J.-P. Grenier, this sputnik is used to study the functioning of the intestine.

It is a capsule 2.5 cm long and 1 cm in diameter, weighing less than 5 grams. It has three palpating arms which are equipped with electrodes; the arms spread out or fold in by remote control instructions. Inside the capsule, a recorder gathers information received by the palpating arms. An FM band micro-emitter transmits this information to a receiver at the patient's bedside.

When the sputnik (which has been swallowed by the patient) clears the pylorus, its arms are deployed and folded, still operating by remote control, before being eliminated through the anus.

Rigidimeter (1977—1985)

An apparatus designed to measure the rigidity of the penis in cases of impotency, it is the result of work begun in 1977 by a team from the Center for Study and Research on Impotency (Dr. Ronald Virag) and the firm of Sofimec.

Rythmostat (control of heartbeat) (1979)

The Rythmostat, developed in 1979 by the American company Respironics Inc. and the French company Pragmat, comprises a thoracic belt containing the electrodes and the tachymeter on which the number of beats per minute appears on a liquid crystal dial. Francesco Moser used it in training for his record time on January 21 in Mexico, and Jean-Michel Branchereau for the 100 km track record in 1982.

Stress control (1983)

Invented by Professor **Alfred Barrios** and his team (University of California, Los Angeles), this stress control card is just like a credit card, in the center of which there is a black rectangle containing liquid crystals which is pressed with the thumb about every ten seconds. If the crystals remain black: stress. If they become red or green: the stress is a little less intense; and finally, blue: complete relaxation.

3. Vaccinations

Vaccination

The work of Jenner

On May 14, 1796, **Edward Jenner** (1749—1823), having done considerable work on cowpox (a disease of the cow udder called "vaccine" in French), took a sample of the material from a pustule on the hand of a farm servant, contaminated by her cows, and innoculated it into the arm of a young boy named James Phipps. Ten days later, a pustule appeared on the boy and healed quite normally.

In a second experimental phase, Jenner innoculated the boy with smallpox; there was no effect. The experiment was a complete success, and

in 1798 Jenner published his results. In 1799, he perfected his idea and his technique.

The method spread widely in Europe, in the East, and in the United States. After having subjected his army to the vaccination in 1805, Napolean issued a decree in 1809 in favor of Jenner's method, and had his own son, the king of Rome, vaccinated. Some sixty years later, Pasteur would make a discovery of still greater general interest in the area of disease prevention. However, although the biological principle was different, Pasteur kept the term "vaccination," as a posthumous hommage to Jenner.

The work of Pasteur

Assisted by his students, Roux and Chamberland, Pasteur first isolated a number of microbes which cause disease in man.

Pasteur made his first attempt at vaccination to fight the viral disease of rabies. On July 6, 1885, Pasteur injected young Joseph Meister, who had been bitten by a rabid dog, with dried spinal marrow taken from rabbits he had innoculated with the virus. The result was conclusive.

The method invented by Pasteur was based on a more general principle than that used by Jenner: An infectious agent, the virulence of which has been artificially attenuated, loses

Experimenting for the first time, Jenner administers his vaccine to the young Phipps in 1796.

DISCOVERY OF VACCINES—CHRONOLOGY			
1771	Introduction of Variolation		Great Britain
1798	Publication on the vaccine	Jenner	Great Britain
1885	Rabies vaccine	Pasteur	France
1892	Cholera vaccine	Hapfkine	Russia
1898	Typhoid vaccine	Wright	Great Britain
1913	Diphtheria immunization toxin/antitoxin	Behring	Germany
1921	BCG tuberculosis vaccine	Calmette & Guerin	France
1923	Diphtheria anatoxin	Ramon & Glenny	France
1923	Whooping cough vaccine	Madsen	Great Britain
1927	Tetanus anatoxin	Ramon & Zoeller	France
	Viral Vaccines		
1937	First anti-flu vaccine	Salk	USA
1937	17 D Amaril vaccine	Theiler	
1949	Mumps vaccine		
1949	Development of cellular culture methods	Engers, Robbin, Weller	USA
1954	Inactivated polio vaccine	Salk	USA
1957	Attenuated live oral polio vaccine	Sabin	USA
1960	Measles vaccine	Engers	USA
1962	German measles vaccine	Weller	USA
	Polysaccharidic Vaccines		
1968	C meningococcus vaccine	Gotschlich	USA
1971	A meningococcus vaccine	Gotschlich	USA
1976	First use of hepatitis B vaccine	Maupas	France
1978	Pneumococcus vaccine	Austrian	USA
1979	Hemophilus influenza vaccine		USA
1979	Elimination of smallpox vaccine		

A doctor administers an innoculation to his patient.

Pasteur preparing his antirabic vaccine in 1885.

its toxic potency but keeps its property of stimulating the resistance of the organism which receives it by the development of antibodies.

Pasteur and Roux subsequently noted that some pathological signs are due less to the microbes themselves than to the toxins which they secrete. In 1922, a French veterinarian, Gaston Ramon (1886—1963), managed to isolate a diphtheria toxin and weakened it in formol. He thus paved the way for the vaccines, which have the advantage of not provoking any disorder whatsoever in the subject. The anti-tetanus and anti-diphtheria vaccines belong to this group.

Sera

Sera, obtained by taking samples of blood serum from a diseased or vaccinated patient (serum which thus contains the desired antibodies), allow either preventive or curative action to be taken against numerous diseases and also against bites and stings from venomous animals, by providing the contaminated individual with protective antibodies.

The principal preventative sera were discovered before 1900:

• Anti-diphtheria, discovered in 1890 by the German E. Von Behring, the Japanese S. Kitasato, the Frenchmen E. Roux, L. Martin and A. Chaillou (anatoxin).

• Anti-tetanus, discovered in 1890 by Behring, Kitasato, Roux and Vaillard (anatoxin).

• Anti-plague, discovered in 1894 by the Swiss A. Yersin.

• Anti-anthrax, discovered in 1895 by Italian A. Sclavo and the Frenchman E.E. Marchoux.

• Anti-cholera, discovered in 1896 by the Frenchman Roux, the Russian I. Metchnikoff, and the Italian A. Salimbeni.

Bacteriology

In 1878, Sedillot, the Physician General and civil servant in the French Department of Health, proposed the name "microbes" for all microscopic agents which carry disease. Following a consultation with Littre (who was responsible for the preparation of a dictionary of the French language), and the latter's approval, the term was adopted by Pasteur.

The German R. Koch (1843—1910) shares with Pasteur the title of founder of microbiology. Koch became famous when, in 1882, he discovered the tuberculosis bacillus, and in 1883, the cholera vibrion, which he

MICROBES

The discovery of an incalculable number of microbes, as of the late 1900s, has been the work of scholars and physicians of all nationalities. The list of the most important infectious diseases was essentially established between 1880 and 1900.

1878	Staphylococcus	Pasteur	France
1879	Gonococcus	Neisseur	Germany
1880	Typhoid	Koch	Germany
1880	Leprosy	Hansen	Norway
1882	Tuberculosis	Koch	Germany
1883	Pneumococcus	Talamon	France
1883	Cholera	Koch	Germany
1883	Streptococcus	Fehleisen	Germany
1884	Diphtheria	Loffler	Germany
1886	Tetanus	Nicolaier	Russia
1887	Meningococcus	Weichselbaum	Austria
1894	Plague	Yersin	Switzerland
1898	Dysentery	Shiga	Japan
1906	Whooping cough	Bordet & Gengou	Belgium

detected in less than one month during an epidemic which was rampant in Alexandria, Egypt. Following Koch's lead, the discoveries came in rapid succession (see list below), notably in the Germanic countries. The first microbes were discovered by M.A. Plenciz (1705—1786), an Austrian physician who published his "Medico-physical studies" in Vienna, in 1762.

In his report, written in Latin, Plenciz stated that contagious diseases were probably caused by "very small pathogenic animalcules which move about freely in air and invade man under the influence of certain climatic conditions."

Viruses

The first virus was discovered in 1892 by the Russian D. Ivanovsky; it is the agent of a disease in various plants called the tobacco mosaic, so named because spots form a sort of mosaic on the leaves of the plant.

At first the existence of viruses was simply presumed, since they are not visible under an optical microscope. With the perfection of the electron microscope as of 1939, the list of identified viruses has continued to grow. (See the list below.)

Further, some viruses are now identified before the diseases which they cause have appeared; these are the orphan viruses, or viruses in search of a disease, defined by Werner in 1956.

Variolation

For quite some time, evacuation or flight was the only response possible to smallpox epidemics. Nonetheless, variolation, which was the first known of all immunization methods, was discovered in India or China many epochs ago.

Based on the fact that smallpox, once contracted and cured, definitively prevents a renewed attack of the disease, the concept of voluntarily innoculating and thus immunizing subjects arose, in the hope that the consequences would be benign.

Thus, preventive variolation consisting in the application of dried crusts of smallpox lesions to nasal mucous membranes was used regularly in the Far East almost 800 years before Jenner's discovery. It was not known in Europe until 1717. At this time Lady Mary Wortley Montague, the wife of the English ambassador to Constantinople, revealed that she had had her 3-year-old son variolated.

Parasites

Besides bacteria and viruses, a number of disease-causing micro-organisms exist. These microscopic animals are called parasites.

In 1880, the Frenchman A. Laveran identified the hematozoon, the protozoon responsible for malaria.

In 1881, the Englishman R. Ross and the Cuban C. Finlay discovered the role of the filariae (parasitic worms found in hot climates) in the transmission of malaria and yellow fever. In 1883, the Englishman P. Manson completed his studies by investigating the role of mosquitoes in the transmission of these filariae.

In 1895, the Australian D. Bruce investigated the role of tse-tse flies in the transmission of sleeping sickness.

All these investigations and discoveries form the basis of tropical medicine.

4. Respiration and Blood

Blood circulation

The Englishman **W. Harvey** (1578—1657) is credited with the discovery of blood circulation. In 1628, inspired by

VIRUSES

1898	Foot and mouth disease	F. Loffler	Germany
1903	Yellow fever	W. Reed	America
1909	Poliomyelitis	Landsteiner & Popper	America
1913		Levaditi	France
1913	Herpes	W. Cruter	Germany
1919	Influenza	C. Nicolle and C. Lebailly	France
1930	Measles	Plotz	America
1947	Zona (Shingles)	H. Henle	France
1954	Chicken pox	T. Weller	America

the ideas of the Frenchman Marcel Servet (1511—1553) and the Italian Andrea Cesalpino (1519—1603), he published the "Exercitatio anatomica de motu cordis et sanguinis in animalibus," which constituted a veritable revolution in the field of physiology.

Accompanying each of his affirmations with physical demonstrations, Harvey had the ingenious idea of considering the heart as a pump which operates by means of muscular force. When he died, only the proof of the existence of capillary vessels which link the arterial to the venal system was left out of the diagram he had drawn. That proof was provided in 1661, by the Italian anatomist Marcello Malpighi (1624—1694).

Respiration

The chemical principle of respiration was discovered in 1785 by the French chemist **A.L. Lavoisier**, who, having analyzed a red oxide obtained by heating mercury, demonstrated that air is composed of nitrogen and oxygen.

Working with the French mathematician and astronomer P. Laplace, Lavoisier deduced that a part of the oxygen mixed with the blood in the lungs in order to ensure the combustion of carbon, while the other part of the same gas combined with hydrogen to form water, which is expelled upon expiration.

His only error, the localization of the respiratory phenomenon in the lungs only, was rectified by the Frenchman J. Hassenfratz, who discovered, in 1791, that the carbon gas found in the blood really comes from the tissues.

Blood transfusion (1667)

There seems to be some question as to who should be given credit for the development and use of blood transfusion techniques. Early work in the field of blood transfusion was carried out in England by R. Lower, in France by J. Denis, in Germany by Mayor, and in Italy by F. Folli. Nevertheless, it is practically certain that Lower was the promoter of experimental transfusion in animals, and that Denis was the first to use it for humans. In 1667, Denis injected 1 liter of arterial blood taken from a lamb into a young man whom he had previously bled.

By virtue of the principle involved and because of the severe dangers inherent in its use, the method was immediately condemned and forbidden. In 1821, the study of transfusion in animals was again taken up. The method was defined in 1875, but interhuman transfusion was developed only as of 1900, when the work of the Austrian K. Landsteiner demonstrated the existence of four large blood groups. In 1910, the Czech serologist Jansky designated these groups by the letters A, B, AB and O. It is thus essential that, prior to any transfusion, the compatibility of the donor and the recipient be determined.

On October 16, 1914, the first blood transfusion of World War I took place in the Biarritz hospital. Isodore Colas, a Breton convalescing from a leg wound, made a blood donation to, and thereby saved the life of, Corporal Henri Legrain, 45th Infantry, who arrived from the front suffering from extreme blood loss.

Storage of blood

The problem of storing and transporting blood formed the subject of investigations carried out by Artus, Pages and Peckelharing at the beginning of the century. As of 1914, Hustin was using the anticoagulating properties of sodium hydroxide citrate.

At the beginning of 1917, Hedon, a physician from Montpellier, France, demonstrated that citrated transfusion was possible. On May 13th and 15th of that year, Jeanbrau successfully performed the first three transfusions of stored blood. It was a breakthrough. It should also be noted that the Frenchmen Richet, Brodin and Saint-Girons were experimentally demonstrating the usefulness of plasma injections as early as 1918.

Plasmapheresis

Plasmapheresis is a method which consists of taking blood from a donor and immediately separating the plasma from the corpuscles in order to return the latter to the donor at once. It was invented in 1957 by Stokes and Smolens, professors at the University of Pennsylvania. In effect, plasma is very quickly regenerated by the human body (less than 24 hours), so that the donor can be asked to donate again if necessary.

Rhesus factor

The crowning achievement in the career of Landsteiner came in 1940 when, together with Wiener and Levine, he identified the Rhesus factor and thus provided an explanation for very severe accidents which are specific to newborn infants.

Hemolytic disease of the newborn (HDN)

If the infant to be born is RH+, the RH- mother, during successive pregnancies, will become sensitized to this positive Rhesus factor and will destroy the blood cells of the child she is carrying. To combat this disease, which can cause in utero death of the fetus, the following have been successively used:

● Exchange transfusion. This technique consists of gradually replacing 90% of the diseased person's blood with fresh blood; it was developed in 1947 by Bessis (France), then perfected by Professors Georges David, Yves Buhot and Therese Boreau (Baudelocque Maternity Unit- France).

● Intrauterine transfusion in the fetus, 1963: A.W. Liley at the National Women's Hospital, Auckland (New Zealand).

● Anti-D vaccine (anti-Rhesus), 1965: first trial of this vaccine in 10 Rhesus-negative women, by J.C. Woodrow, C.A. Clarke (Liverpool, Great Britain). July 1976: first application in 35 pregnant women, by F. Pinon, R. Cregut (Paris).

Given systematically to women who require it, the vaccine should make the disease disappear altogether.

Plasmapheresis, which prevents the mother's blood from producing antibodies that destroy the blood of the infant she is carrying. 1968: First attempt by Powell. 1976: In Paris, F. Pinon, R. Cregut, Y. Brossard and P. Maigret use this technique on 12 future mothers; 10 give birth successfully.

Blood group analysis

In 1970, the medical division of the French company Marcel Dassault developed the "Groupomatic," a blood analysis apparatus which can perform up to 6 tests per minute.

This invention was designed by Professors Mathe and Soulier of the National Blood Transfusion Center. The first apparatus was installed in Paris, and the second in Reese Hospital in Chicago.

Artificial blood

In February 1979, the Japanese physician Ryochi Naito injected 200 ml of artificial blood (fluosol DA) for the first time. Fluosol is a completely synthetic, milky white product which allows the replacement of hemoglobin in the transport of oxygen (from the lungs to the rest of the body).

Initial studies began as early as 1933, but interest in such research diminished after 1940. In 1966, two Americans working at the University of Cincinnati, Professors Glark and Gollan, demonstrated that a mouse immersed in liquid perfluorocarbons survived; it was able to take the oxygen necessary to avoid death by asphyxiation from the liquid. However, these fluorocarbons could not be mixed with blood.

In 1967, the American professor Henry A. Slaviter, working at the University of Pennsylvania, emulsified the perfluorocarbons by adding albumin. The emulsion nevertheless risked agglomeration and blockage of some capillary vessels.

Ryochi Naito was the first to succeed with the experiment; in February 1979, a dose of fluosol DA was injected in a man who had undergone emergency surgery at the Fukushima Medical Center, and whose blood type was so rare that transfusion was not possible. This first experiment was performed before the Japanese Health Ministry had given its approval for the use of fluosol DA. The result was an overwhelming success.

Iron lung

Philip Drinken, an American professor at Harvard University, designed the iron lung in 1927; it was tested on a young girl at Boston Hospital on October 12, 1928.

The first model was made of diverse materials: Two household vacuum cleaners alternatively produced a positive and a negative pressure on the patient's thorax.

5. Surgery

Origins of surgery

The first known surgical operation, an amputation, goes back to the Neolithic period (5000—2500 B.C.), but a Neanderthal man, whose skeleton was found in the Zagros mountains in Iraq, and who lived about 45,000 years ago, had also undergone an amputation—he was missing his right arm. This was due neither to chance nor to accident.

Sorcerers, physicians and surgeons of the day were also the first to practice trepanation on human beings, some of whom survived this dreadful operation. In effect, some proofs of scarring have been found on a few trepanated skulls. The prehistoric surgeon used instruments made of stone and bone: knives, blades, triangular points, etc.

Autopsy

The study of anatomy, which appeared in the 14th century, and the ensuing development of surgery are due to autopsy. At the beginning of the Renaissance, man was badly educated as regards the organs of his body. At that time, one did not speak of researchers but of profaners. The formal ban on dissecting the human body, both in Christian and Islamic countries, made almost any development in the field of anatomy impossible.

The first human dissection took place in Bologna, Italy, in 1281. Then, in 1316, the Italian anatomist Mondino di Luzzi wrote in his "Anatomy" of the dissections which he had performed.

Appendectomy (18th century)

An appendectomy is an operation which consists of removing the appendix when it has become infected. The English military surgeon Claudius Amyan (member of the Royal Society who died in 1745) performed the first successful appendectomy.

On April 27, 1887, in Philadelphia, George Thomas Morton (the son of William Morton, the pioneer in anesthesia) operated on a young man, 26 years of age, who was suffering from acute appendicitis, thereby saving his life. In July 1888, von Volkmann, a Swiss surgeon (Halle), operated on two patients following the examinations performed by his assistant, Charles Krafft, and cured them.

Krafft proposed the principle that once a diagnosis of appendicitis has been made, it is never too soon to operate. Another American surgeon, John Benjamin Murphy from Chicago, also had the idea that it was essential to intervene before the infection had a chance to develop. A young

Larrey, Napoleon's surgeon, during the Egypt campaign, 1798.

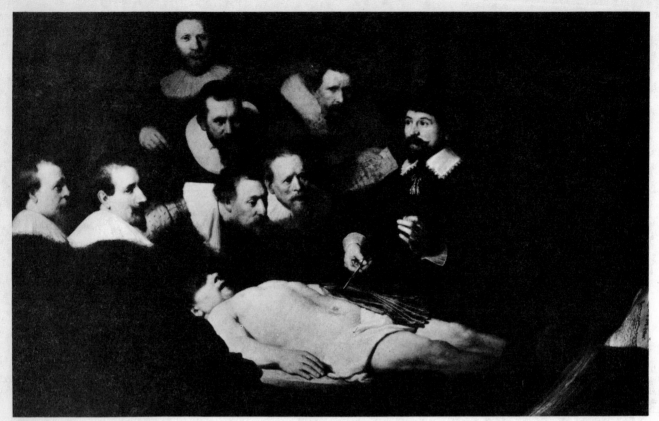

Rembrandt's Anatomy Lesson.

worker by the name of Monham was the first patient to be operated on by Murphy. The surgery took place on March 2, 1889, a few hours after the patient had complained of abdominal pains.

Neurosurgery

In 1879, Macewen (Scotland) performed the ablation of a cranial meningioma.

In 1884, Godlee (London, England) performed the first ablation of a brain tumor.

In 1887, Sir Victor Horslay was the first to successfully perform an ablation of a tumor in the spinal marrow.

Modern neurosurgery was born in the United States around 1918 due to the efforts of Harvey Cushing and the techniques of Walter Dandy.

In 1929, Berger (Germany) invented the electroencephalogram.

In 1936, Thierry de Martel created French neurosurgery, in cooperation with Clovis Vincent, and invented the technique which allows brain surgery while the patient is in a seated position. The same year, Egaz Moniz (Portugal) invented the arteriograph, and received the Nobel Prize for his discovery.

In 1937, Fiambert invented the lobotomy, an operation which consists in cutting part of the frontal lobe and is used to treat some behavioral disorders of a predominantly anxious or obsessive nature. It is now virtually unheard of, its use having been abandoned.

In 1950, Talairach (France) invented stereotaxy, a method for detecting a point inside the brain which allows surgery to be performed via a simple hole drilled in the skull. One of the first applications was made by Fenelon (France) to operate on Parkinson's disease. In 1952, Raymond Houdart, a French neurosurgeon, operated on a patient suffering from Parkinson's disease on live television.

In 1960, Guiot (France) operated on a tumor of the pituitary gland via nasal route.

In 1962, Djindjian (a radiologist) and Houdart performed the first arteriography of spinal marrow, which permitted the operation of angiomas in the marrow. In 1970, Serbedinko (USSR) eliminated the need for the operation by introducing a balloon catheter into the artery and directing it to a given site in order to block the artery (for the purpose of embolizing it).

Cosmetic surgery (1891)

Although the Aztecs knew how to deform heads for reasons of tribal identification, true modern cosmetic surgery was born at the same time on both sides of the Atlantic. In 1891, the American **Roe** invented cosmetic rhinoplasty. This was a surgical modification of the shape of the nose, for purely cosmetic reasons, in a normal individual who was simply dissatisfied with his/her image.

In France, in 1907, H. Morestin described his method of resection for mammary hypertrophy.

In 1925, a woman surgeon, Suzanne Noel, operating in her own home and using local anesthesia, performed face liftings. Finally, it was in Vienna that H. Biesenberger, in 1928, and E. Schwarzmann, in 1930, made a fundamental contribution to mammary reduction plasty.

Cosmetic surgery has undergone broad development in the United States and France, where Morestin, Bourguet, Bassot, Claou and Helene Marc paved the way for numerous operatory techniques between the two World Wars.

The demonstration of the Superficial Musculo-Aponevrotic System of the Face (SMAS), by the French surgeon Vladimir Mitz in 1973, has led to an improvement in the surgical technique for face lifting by permitting facial skin, and also facial muscles, to be stretched.

Cold surgery

Cryosurgery involves the destruction of pathological tissues using extreme cold (temperatures below -104°F (-40°C). It has two advantages: It is painless and it eliminates the risk of hemorrhaging. I.S. Cooper, with the help of Union Carbide, developed a cryoprobe using liquid nitrogen able to reach a temperature of -356°F (-180°C). Other apparatuses have been used for cataract surgery as well as for the treatment of laryngeal and hemorroidal tumors.

In 1964, M.J. Gonder and W.A. Soanes treated prostate adenomas using cryosurgery.

In 1975, a German surgeon, A.J. Keller, invented cryocauterization. After having operated on an adenoma at a temperature of -374°F (-190°C), he raised the temperature of the probe to +392°F (+200°C), in order to pulverize the tissues destroyed by the cold, and thus allow them to be more quickly and easily eliminated via the urethra.

Transplants

Origins

According to legend, the first to practice transplants were the brothers Saints Como and Damian, who transplanted the leg of a black man on a priest. It is perhaps due to this operation that the two were named saints and the patrons of surgeons.

In 1903, Hopfer attempted to reimplant the amputated paws of a dog.

The discovery of tissue groups, and the invention of the HLA system by Professor Jean Dausset, who won the Nobel Prize in 1958, made the development of transplants possible.

Teeth

The first local reimplantation of teeth accidentally fallen out of the odontobothrion (socket) was performed in China in 3216 B.C., during the reign of the Chinnoug emperors.

Nose

In India, in 750 B.C., Sushruta performed the first nose transplants. Adulterous women were punished by having their noses amputated. The physicians who treated them realized that the skin grafted to the mutilated nose had to come from the same person, and had to be irrigated by the same blood, so they transplanted a skin fragment taken from the forehead.

Skin

In December 1869, the Genevese surgeon Jacques-Louis Reverdin performed the first human skin graft on a wound.

Ear

In 1904, Doctor Louis Sauve, of Paris, France, grafted an external ear.

• Tympanum: In 1953, the Germans Zolker and Walstein invented the tympanic transplant (or tympanoplasty). This involves the repair or the replacement of the tympanum by means of a transplant, sometimes associated with the repair of the ossicles performed under microscope. This operation is successful in more than 80% of the cases, with subsequent restoration of hearing.

• Ossicles: Transplant of the stirrup in 1959 by Portmann, of the anvil in 1961 by Beck, and of the hammer in 1967 by Jean Marquet are noteworthy.

Nerve

On July 7, 1909, Doctor Walter Jacoby (Munich, Germany) transplanted a 1.7 inch (4.5 cm) nerve segment, preserved by means of lyophilization, in the right hand of a 35-year-old manual laborer, Helmut Mitschke. The operation was a success.

Cornea

The first corneal transplant was performed by Doctor Elschwig in 1914, in Prague, Czechoslovakia. He based his work on that of Kissam (end of the 19th century) and Zirm (1906).

Testicle

In 1920, Serge Voronoff, in Nice, France, transplanted the interstitial portion of the testicle of a chimpanzee in a human. In 1977, at Saint Luke's Hospital in Sacramento, California, Doctor Sherman Silver transplanted one of Terry Twomey's two testicles in his brother, Timothy, who was born without testicles.

Endocrine glands

• Surrenal glands: In 1935, Bailey and Keele transplanted the surrenal glands of a still-born infant in a woman suffering from Addison's disease.

• Thyroid: In 1935, May and Turpin attempted the operation.

• Pituitary gland: In 1942, Rochat transplanted the pituitary gland of a still-born infant in a young woman, 23 years of age, who was suffering from Simmond's disease.

• Pancreas: In 1966, this transplant was attempted for the first time by Richard Lillehei and William D. Kelly (Minneapolis, Minnesota).

In 1968, Professor Peter (Prague, Czechoslovakia) transplanted the pituitary gland of a deceased adult in a 30-year-old man who had undergone surgery for cancer of the hypophysis.

Kidney

On June 17, 1950, Ruth Tucker, a 44-year-old American whose kidneys were afflicted with cysts, underwent the ablation and replacement of one kidney with one taken from a cadaver. This medical first was the work of the surgeon Richard Lawler from Chicago. The patient lived for five years, after which complications and finally the ablation of the transplant caused death.

The first success was the iso-transplant performed on December 23, 1954 at Peter Brent Hospital in Boston. Doctor J.P. Merrill (and his team) transplanted a kidney in a 23-year-old man, Richard Herrick; the kidney was taken from his monozygote twin, Ronald. The patient lived nine years with this kidney. Since then, numerous successes have been recorded, and the main problem is now the lack of transplantable kidneys.

Bone

Robert and Jean Judet (France) performed human homograft of bone in 1950.

Bone marrow

In 1957, the American physician Thomas attempted the first transplant of bone marrow in the human, in order to combat leukemia.

On October 17, 1958, six Yugoslavian atomic scientists, working at the Nuclear Institute of Vinca, were accidentally irradiated. They were transferred to the Pierre Curie Hospital in Paris, France. Doctors Georges Mathe, Raymond Latarget and Henri Jammet saved five of the scientists (the sixth died after one month), by transplanting bone marrow taken from foreign donors (one of whom was Professor Leon Schwarzenberg). Within four months, the atomic scientists were able to resume activities.

Hair

In 1959, N. Orenteich from New York invented the hair transplant method. Hairs are taken from the back of the scalp and transplanted in the areas where they are lacking.

Limb or segment of a limb

On May 23, 1962, at the Massachusetts General Hospital in Boston, Doctors Donald A. Malt and J. McKhann reimplanted the right arm of a 12-year-old boy which had been severed from the shoulder.

Spleen

On January 19, 1963, Thomas Starlz transplanted a spleen in a hemophiliac child, in Denver, Colorado.

Liver

On May 5, 1963, Doctor Tom E. Starlz of Denver, Colorado performed the first liver transplant in a 48-year-old patient. Survival: 22 days.

Lung

On June 11, 1963, Professor James D. Hardy at Jackson (Mississippi) University Hospital transplanted a human lung in a patient suffering from cancer. The patient lived for 17 days.

In 1969, a 24-year-old Belgian miner, Lois Vereecken, lived for 10 months after having been operated on by Doctor Fitz Derom, of Gand, Belgium. This technique has since been abandoned.

Heart

The first transplant attempt took place on January 24, 1964. Doctor James D. Hardy, at the Jackson (Mississippi) University Hospital, transplanted the heart of a chimpanzee in Boyd Rush (58 years old). He lived for about 3 hours.

On December 3, 1967, at Groote Schurr (Large Grange hospital) in Capetown (South Africa), Doctor Christian Barnard transplanted the first human heart in Louis Washkansky (54 years old). He lived for 18 days.

The first successful heart transplant took place on November 27, 1968 when Emmanuel Vitria, a Frenchman, received a new heart. He is still alive.

Heart-lung

In September 1968, at the Texas Children's Hospital in Houston, Doctor Denton A. Cooley transplanted a "heart-lung" block taken from a one-day-old anacephalic infant (born without a brain), kept alive by artificial respiration, in a young girl, aged 2½ months. The girl was suffering from auriculo-ventricular deficiency and pulmonary hypertension (a nearly fatal disease). She died 11 hours after the operation.

Fallopian tubes

In October 1978, at Bensham Hospital (Great Britain), Dr. Peter Silverstone transplanted the tubes from a woman who wished to be made sterile to a 37-year-old woman who wished to become pregnant but who had been sterilized five years previously, by means of excision of the tubes. This was the first time transplants between two living persons had been performed. Other transplants were successfully performed, but none of the women became pregnant.

Aortic valve

In 1969, the Frenchman Carpentier and the Englishman Edwards transplanted a pig aortic valve in a human. The Americans Ionescu and Shiley performed the same operation in 1971, using a calf pericardial valve.

Fetal cells

In 1974, Professor Jean-Louis Touraine from Lyon (France) transplanted fetal thymal cells in children afflicted with isolated thymic hypoplasia. In 1976, he transplanted liver and thymal cells in "bubble babies," who were suffering from severe immunological deficiencies or partial enzymatic deficiency.

Fingernail

On March 29, 1980, at the SOS Hand Clinic in Strasbourg, France, Doctor Guy Foucher transferred a toenail to the thumb of Christophe Kempf, 12 years of age, after his thumbnail had been lost as a result of a badly treated case of hangnail. One year later, the graft had taken, due to the microsurgical suture of the vessels in the matrix of the nail.

Brain cells

In January 1982, the teams of Doctor Erik Olaf Backlund (Washington) and Doctor Aka Seiger (Denver) successfully injected medullo-surrenal cells into the rat brain. This led the way for the treatment of Parkinson's disease.

The Ethics Committee of the Swedish Parliament authorized Doctor Backlund to attempt human transplants in the most severe cases of the disease.

Anesthesia

Prior to the discovery of anesthesia, the patient's pain was the greatest obstacle to the surgeon's task. The surgeons of the past attempted to avoid suffering by intervening as quickly as possible. The only anesthesia known at the time consisted in giving the patient a mixture of alcohol and gun powder to swallow. Held down by his friends, the surgical patient grit his teeth against his pipe (often made of clay), to prevent him from screaming. He did not let go of the pipe unless the operation ended badly.

Origins

In 1799, the English chemist Sir Humphrey Davy (1778—1829) described the analgesic and mirth-provoking effects of nitrogen protoxide. To demonstrate these effects, he himself inhaled it to ease the pain brought on by an abscessed tooth.

Some dentists, notably the American H. Wells in 1844, used this chemical compound when extracting teeth. Wells stated: "A new era is beginning in dental surgery. It hurts no more than a pin prick." He died in 1848, from a wound to the femoral artery, and as a final recourse inhaled the gas while dying.

General anesthesia

General anesthesia was first used in surgery by the American C.W. Long. He was also among the first to use ether (1842).

Two Americans, W. Morton and J. Warren, at Massachusetts General Hospital, operated on a neck tumor on October 14, 1846, after having placed the patient under general anesthesia with ether. Thereafter, the use of ether quickly spread to operating rooms in the United States and Europe.

After ether, chloroform came into use. J. Simpson, professor of obstetrics in Edinburgh, Scotland, used the product, available since 1831, and in 1834 he started producing much higher quality specimens of this anesthetic. After Queen Victoria had been chloroformed during the delivery of her seventh child, anesthesia was adopted in all hospitals. It is known under the name of "Anesthesia to the Queen."

Local anesthesia

Local anesthesia came later (in 1884) with cocaine (discovered by K. Koller, an Austrian ophthalmologist). Cocaine was subsequently improved upon by the addition of adrenalin, in 1902, to be replaced, in 1904, by novocaine, which is less toxic.

Intravenous anesthesia

Intravenous anesthesia began to spread after 1902, when the German biochemist E. Fischer synthesized Veronal. Other barbiturates, such as amytal, nembutal, and pentothal, were developed after 1930.

Electric anesthesia

This was invented by Professor Aime Limoge. The first obstetrical electroanalgesia was performed on April 23, 1970, at the Rothschild Hospital in Paris, unit of Professor Guy le Lorier.

The first electro-medicamentous anesthesia was performed on March 23, 1972, by Professors Cara and Debras at the urological clinic of Profes-

PROTHESES AND ARTIFICIAL ORGANS

ORGAN	DATE	PLACE	INVENTOR	COMMENTARY
Kidney	1945	Holland	Wilhelm Kolff	
Hip	1946	France	Robert & Jean Judet	Acrylic.
	1965	U.S.A.	Austin Moor	
		G.B.	Sir John Charnley	First total hip prosthesis.
	1968	France	Lagrange & Letournel, Boutin, R. Judet & Lord	Total prostheses, without cement.
Blood vessel (and arteries)	1951		A. B. Voor Hees B. Jaretki & A.H. Blakemore	
Artificial heart	1953	Hos. Laennec (France)	Dr. Fred Zacouto	First heart transplanted in a dog.
	1966			First temporary left ventricle.
	1969	St. Luke's Hos. (Houston Texas)	Denton & Cooley	Artificial heart transplant, developed by Domingo Liutta, performed on Haskell Karo, while waiting for the opportunity to transplant a human heart. It took place on April 7; the patient died within 24 hours.
	1982	Salt Lake City	William Devries	First human transplant: Barney Clark, a 61 year old dentist, received the heart invented by Dr. Robert Jarvik on February 12, 1982. Barney Clark died on Thursday March 24, 1983, of "natural" causes, after 122 days of survival with his artificial heart, which was then stopped by the physicians.
Knee	1954		Shiers & Waldius	Technique abandoned.
	1968	France	Lagrange and Letournel	Hinged joint (+ rotation in 1976).
	1970	France	Guepar	Hinged joint, without rotation.
		G.B.	Freeman	Partial or sliding prosthesis.
Fingers (joints)	1956	G.B.	Flatt	Made of metal.
	1962	G.B.	Swanson	Silicone implants.
Ear	1957	France	A. Dijourno, Ch. Eyries & Vallancien	Implantation of electrodes on the cochlea, to treat deafness.
	1977	France	Profs. Chouard & Mac-Leod	
Pacemaker	1958	Stockholm (Sweden)	Ake Senning	
Cardiac valves	1960	Univ. of Oregon	Prof. Albert Starr	
Arm	1970	Japan		Responds to wearer's voice.
Elbow	1970	G.B.	Dr. Roger Lee	
Wrist	1970			Replacement of the wrist bones with silastic transplants.
Foot	1971	France	Keller-Lelievre	Metatarso-phalangian joint of the big toe.
Ankle	1973	Prof. Lord		
Liver	1973	Berkeley, Ca.	Dr. Kenneth Matsumara	Disposable appartus placed outside the body, containing animal liver cells. After 6 to 10 hours of use, it must be replaced.
Shoulder	1973	U.S.A. France	Dowson & Uright Dr. J. Y. de la Caffiniere	
Pancreas	1974	Paris (France)		External appartus.
	1981	Montpellier (France)	Prof. Mirouze	First insulin pump implanted beneath the skin.
Lower jaw	1975			Vitallium stalk and bone transplant taken from the patient.
Leg	1976	Montpellier (France)	Pierre Rabischong	
Penis	1976	Montpellier (France)	Subrini	
Skin	1982	Boston (USA)	Ionnis Yannas of MIT & John Burk at Harvard Hosp.	Graft of artificial skin in a man, 80% of the skin had been burned following an explosion. Synthetic skin was used on the neck, chest and arms.

The fiberglass 3M plaster does not dissolve in water. It is light and transparent to X-Rays.

sor Couvelaire (Necker Hospital, Paris), for a polar nephrectomy.

Fractures

Techniques used prior to plaster casts:

The Egyptians were the first to use retention methods; small bands, soaked in mud, were recommended by Athotis, in 3000 B.C.

Arab surgeons, such as Rhazes (850—932), Avicenne (980—1032) and Averroes (1126—1198), used mainly cotton strips covered with egg whites.

Ambroise Pare (1509—1590), the author of a treatise on the means of "joining bones," used splints made of wax, cardboard, cloth and parchment, placed while damp on the fractured limb; the product hardened as it dried.

Plasters

In 1798, William Eton, an English consular officer stationed in Persia, noticed that plaster was used to hold fractured bones in place; the limb was covered with plaster, and a wooden mold was made to hold the prosthesis in place.

In 1850, a Dutch military physician at the Royal Hospital, Doctor Antonius Mathijsen, developed strips of linen powdered with dry plaster; they were prepared in advance, and then dampened at the time of use.

Glass fiber and resin plasters

In 1982, the American company 3M developed Scotchcast, a band made of glass fibers impregnated with a polyurethane base resin. Once applied, it sets in about 15 minutes.

This resin-plaster is extremely re-

sistant, is not damaged by water, and weighs only one-third of a traditional plaster cast. The skin can breathe because the material is porous, and this eliminates maceration and itching. Because it does not block X-rays, monitoring of the course of the fracture is made much easier. The only drawback is the cost; Scotchcast is 5 times more expensive than ordinary plaster cast.

Dressings and bandages
(7th century B.C.)

The first traces go back to the 7th century B.C. On the clay tablets discovered in the archeological ruins of Assur and Niniva, in Assyria, written by one of the best known practitioners of the time, Arad-Manai, one can read: "Hail to the small child whose eye has caused him suffering; I placed a dressing on his face, and towards evening I removed the band I had applied, also taking away the compress which was beneath."

In 1825, the surgeon Antoine Labarraque was the first to chemically disinfect wounds.

In 1840, the Englishmen Astley-Cooper, Liston, Syme and Matcarnay introduced a new kind of dressing; this was a cotton cloth soaked in water and covered with sticking plasters.

In 1864, the Englishman Lister used dressings soaked in an aqueous solution of phenic acid for antiseptic purposes.

Soluble dressings (1947)

The soluble dressing was invented simultaneously in 1947 by Jenkins, in

the United States, and Robert Monod, in France. It consists of small gelatin sponges which can absorb 20 to 50 times their weight in blood, before gradually dissolving in the body of the patient.

Transparent dressings (1953)

In 1953, a transparent dressing was invented in the United States. A liquid plastic kept under pressure in a metallic container is sprayed onto the wound, solidifying around it, forming a transparent film, 1/10th of a millimeter thick.

Stitches (1820)

The French surgeon Pierre-Francois Percy (1754—1825), invented metallic stitches around 1820. Catgut, made from cat intestine, appeared one century later, in 1920. Tergal stitches were invented in 1950.

In 1964, the American company 3M invented the Steristrip, a kind of dressing which joins the edges of wounds and eliminates the need for stitches.

Asepsis (1883)

Asepsis was first carried out by means of a dry-heat sterilizer, and was due to the work of the French surgeons Terillon and Terrier, in 1883.

In 1878, Pasteur stated: "If I had the honor of being a surgeon, I would use only perfectly clean instruments ... I would use only shredded linen, small bands, and sponges which had previously been exposed to air heated to a temperature of 266°—322°F (130°—150°C). I would use only water which had been heated to a temperature of 248°F (120°C)." Thus, asepsis is another application of Pasteur's work.

In 1889, the American Halsted introduced the use of rubber gloves.

Antisepsis (1847)

In 1847, the Austrian obstetrician I. Semmelweis attempted to enforce procedures concerning cleanliness and disinfection in areas where the sick were hospitalized. This was the beginning of antisepsis. Semmelweis also made very important observations concerning puerperal fever (that is, "newborn fever"), which afflicted a large number of mothers. He was convinced that the infection was transmitted by the physicians and students returning from the anatomy unit to women in confinement following labor.

The English surgeon J. Lister (1827—1912) concentrated his efforts on antiseptics. Prior to all operations, he sprayed phenic acid in the operating room and disinfected the instruments and the area of the patient's skin where the incision was to be made. His practices were very soon imitated throughout Europe. The first wide-scale application of antisepsis was due to the work of the surgeon E. Von Bergmann during the Russian-Turkish War of 1877—1878.

Pacemaker (1958)

The cardiac stimulator was invented in 1958 by the Swede Ake Senning. The first implantations took place at the beginning of the 1960s.

In addition to the heart, the pacemaker may also stimulate the bladder, the rectum, and respiration, and it is thought that it will be useful for auditive and visual functions, as well as in certain cases of facial paralysis.

Medical laser (1960)

The first ruby medical laser was developed by Maiman in 1960. It has been used by Freeman since 1964 to treat lesions of the retina.

In 1967, research scientists from Bell Laboratories, notably D.R. Herriott, E.I. Gordon, D.A.S. Hale and W. Gromnos, developed a Light Knife which lances and cauterizes the wound at the same time.

As regards the CO_2 laser, discovered in 1964, it was only in 1970 that it was used surgically via endoscopic route (trachea, bronchia, colon, genital organs of the woman), and for the destruction of tumors of the skin or of tatoos. The latest laser in the line is the Nd-Yag (Neodyme Yag), which has been experimented upon in animals since 1975 and has been used for humans since 1976.

Low power laser (1979)

The principle of this laser was invented by an Italian, Dr. Tarrantini, who began to develop it in 1979 in collaboration with a Frenchman, Dr. Bernard Sillam.

It is used in sports medicine and rheumatology because of its anti-inflammatory and analgesic effects. Further, it has an anti-cellulitis action by virtue of the development of local microcirculation.

Laser for tenia (1983)

For the first time ever, the physician Philippe Raimbert (Laser Center at the Hartmann clinic in Neuilly-sur-Seine, France) killed a tenia using a laser. After having located the head of the worm, clinging to the fourth portion of the duodenum, using a special fibroscope, it was destroyed by two applications of the Yag laser.

Transluminal angioplasty

Punction balloon (1964)

The idea of dislodging an obstacle, such as an atheroma (lesion on the inside surface of an artery) from an arterial pathway by means of an inflatable punction balloon brought into contact with the constricting portion—an idea which was launched in 1954 by Dr. Charles Dotter of the University of Oregon—was applied for the first time in Zurich, Switzerland, in 1964 by Dr. Andreas Gruntzig.

Laser (1983)

Laser vaporization of an atheroma which obstructs an artery was performed for the first time ever in October 1983, in Toulouse, France, by Professors Eschapasse, Fournial, Marco, all from Toulouse, and Drs. Choay (San Francisco), Stentzer and Myler (St. Luke's Hospital, New York).

Extraction of kidney stones across the kidney wall (1976)

Invented by a Swedish radiologist, Dr. Fernstroem, in 1976, and subsequently improved upon by Dr. Alken from Mayence and Professor Wickham from London in 1978, it was introduced to France in March 1973 by Professor Alain Le Duc (Lariboisiere Hospital, Paris). It consists of creating a tunnel between the skin and the kidney through which the calcula are either extracted or broken down, or pulverized using ultrasound, or split apart using hydro-electrolytic shock. Hospitalization is reduced to 3 or 4 days, in uncomplicated cases.

Shock waves for kidney stones (1982)

In 1982, Professors Christian Chaussy, Egbert Schmiedt and Walter Brendel of Munich, Germany, together with the Dornier Company (the famous manufacturer of fighter planes) developed an apparatus able to disintegrate a kidney stone of less than 1.2 inches (3 cm) in diameter using shock waves. The treatment lasts 1/2 to 3/4 hour, with cardiac monitoring. During this period, the patient receives approximately 500 shock waves, felt as a sort of slap; peridural anesthesia is required. The stone is broken down into granules which are then eliminated in the urine.

Technique for combatting infarctus (1982)

In 1982, Jean-Christian Farcot (Ambroise-Pare Hospital in Boulogne, unit of Professor J. P. Bourdarias) developed a technique, called synchronized diastolic retroperfusion of the coronary sinus with arterial blood, for use in decreasing the volume of an infarctus of the myocardium.

In order to limit damage to a minimum, once an infarctus has been revealed, Dr. Farcot had the idea of taking blood from an artery and of leading it, counter-current, to a coronary vein via a small pump and a catheter; this oxygenates the area of the cardiac muscle in which the infarctus is located and, as a consequence, decreases its volume.

Further, the technique allows the direct administration of medicaments to the diseased area.

6. Gynecology

Caesarian section

The caesarian is an operation which was first mentioned in a law set forth by Numa Pompilius (715—653 B.C.), the legendary second king of Rome. According to this law, no woman who had died during her confinement could be buried until her infant had been removed, via abdominal incision. The name of the first of the Roman emperors, Caesar, is derived from the Latin word "caedere," meaning "to cut," because one of his forebearers had given birth by means of a caesarian section.

The first caesarian (in modern

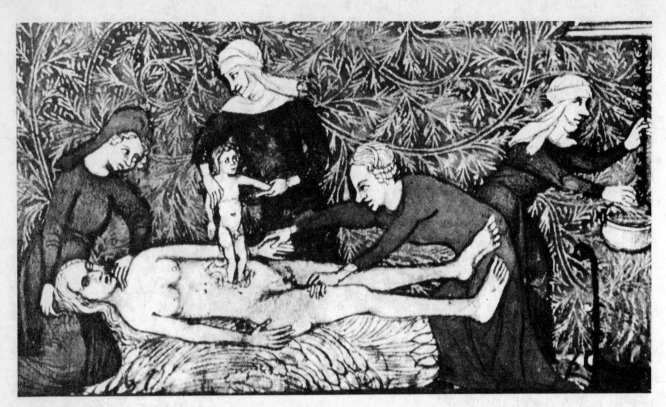

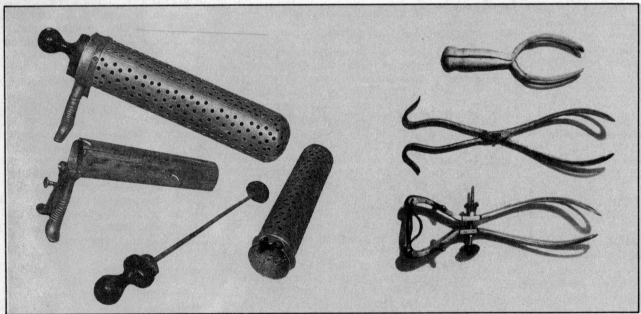

Top: First painting of a Caesarian section around 1350. Bottom left: Speculum. Bottom right: Forceps.

times) seems to have been performed in 1610, by the German surgeon Trautmann.

Speculum

The vaginal speculum was invented by the Romans. An example of it was found in the ruins of Pompeii. Called the *dioptre* in the Middle Ages, it was subsequently abandoned, until the French surgeon Recamier "reinvented" it in 1812 and led to its widespread use. (Recamier also developed the technique of scraping the uterus; the method was abandoned after his death, and was put into use again in Germany.)

In the 19th century, the speculum underwent successive modifications, due notably to the Frenchmen A. Dubois and G. Dupuytren.

Feeding bottle

Feeding bottles have been used since antiquity. At first they were urns with two openings, one of which was used to introduce the liquid, and the other, in the shape of a beak, to feed the infant.

In the 16th century, the design used resembled a duck. Until the end of the 18th century, the nipple was made of a small piece of rolled linen, one end of which soaked in the container with the other emerging from the beak. Nipples were also made of sponge, leather (softened by means of hydrochloric acid), or dried cow udder.

Rubber was discovered in the second half of the 19th century and it soon became the material of choice for feeding bottlés because it was so hygienic. Two holes were punched in the rubber, one for aeration, the other for aspiration. Today, only synthetic materials are used.

Forceps (17th century)

Forceps were invented in the 17th century by the Englishman P. Chamberlen. Previously, hooks called "head grips" had been used.

The Frenchman A. Levret developed the technique for the instrument, and in particular the curve of the spatulas, in the 18th century. In 1838, 144 kinds of forceps were available.

Subsequent to new improvements, the famous forceps of the French obstetrician S. Tarnier became the instrument of choice.

Forceps—leech cup

This permits extraction of the infant by applying the leech cup to the top of the skull. It was designed in 1962 by a Danish physician, Doctor Land. The instrument eliminates the strong intracranial pressure which is characteristic of the tonged forceps.

Artificial insemination (1785)

Doctor Thouret, Dean of the Paris Medical School, was the first to experiment with human artificial insemination in 1785. He injected sperm into his wife's vagina, using a syringe. Nine months later, a healthy baby was born.

The first successful storage of sperm by means of freezing took place in 1950. Since 1953, such preservation has been successfully applied to human sperm. These results, obtained in the United States, are based on the work of the Frenchman Jean Rostand. In 1946 he discovered the protective effect of glycerol against cellular lesions caused by the cold. As of 1963, an important improvement in the technique has been achieved by the use of liquid nitrogen, which increases storage possibilities. And now the rapid development of sperm banks is under way.

Incubators (1880)

Incubators were invented by the Frenchman Budin in 1880. They were made of wood, and pans of hot water were placed in drawers beneath the incubators.

The incubators of today are made of metal, plastic and Plexiglas, with apparatuses which permit control of temperature and humidity. They have permitted a decrease in infant mortality, from 86% mortality in infants weighing less than 2.2 lbs (1000 g) in 1950, to 40% in 1980.

Hygienic napkins (1921)

In 1921, the American company Kimberley-Clark, of Neenah, Wisconsin, launched an improved bandage. Marketed under the name of Kotex, this was the first commercial hygienic napkin.

Ernest Mahler, a German chemist, working in the United States, had invented a cotton substitute made from wood to compensate for the lack of dressings. The nurses acquired the habit of using these cellulose padded dressings as hygienic menstrual napkins. When the Kimberley-Clark Company, which was already manufacturing cotton-wool bandages, learned of this, it began marketing them as napkins.

Tampons (1937)

In the 1930s, the American **Earl Hass** thought of a way to modify the surgical tampon. He wanted to eliminate the inconvenience and the embarrassment caused by the hygienic napkin. In 1937, he applied for a patent and founded the Tampax Company. After some improvements, the use of the tampon spread throughout the world after WW II.

Colposcopy (1925)

Colposcopy is an examination of the neck of the uterus, using an enlarging binocular magnifying lens which permits the detection of lesions which are invisible to the naked eye and their sampling for laboratory examination (biopsy). It was invented by Dr. Hirselman in Germany in 1925.

Since then, the efficiency of the apparatus has increased. Colpomicroscopy was developed in 1948, by Tassiglio Antoine, in Vienna, then perfected by the Frenchman Fernand

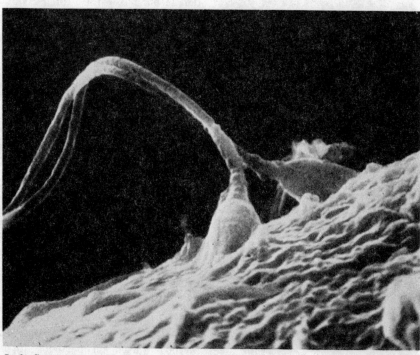

Left: Spermatozoids. Right: Woody Allen in the part of a spermatozoid.

Coupez in 1955. In 1979, the Frenchman Jacques Hamou created a new colpomicroscope, which is lighter and more accurate. In 1973, it became possible to destroy lesions in the neck of the uterus by means of lasers.

Test-tube baby (1979)

Louise Brown was born on July 25, 1979, at 11:47 p.m., in the Oldham General Hospital (Great Britain). She was the first test-tube baby, or the first baby conceived outside her mother's body.

This "first" was the work of the British physicians Patrick Steptoe and Robert Edwards, who, in 1965,

provided the proof that *in vitro* fecundation was possible. In particular, they described the precise process of maturation of male and female gametes.

However, it was not until 1978 that these two research scientists succeeded in producing an *in vitro* fecundation (with Mrs. Brown, another Englishwoman, and an Indian woman).

The egg was taken from Mrs. Brown, 31 years of age, who was incapable of fecundation because of her impaired tubes. The ovum was placed in a test-tube in contact with the sperm of her husband, a 38-year-old truck driver. The fertilized egg was then placed in Mrs. Brown's uterus.

The invention of Tampax in 1937 made life easier for women.

Hormone pump (1982)

Some women are sterile because their hypothalamus (an area of the brain) does not secrete hormones correctly, consequently causing a cessation of ovulation.

Professor **Jean-Pierre Bautray** and his team (Hospital of Creteil - France) developed a small pump, hardly larger than a pocket calculator, which is connected to a vein by means of a catheter and which sends the lacking hormones every 90 minutes, as the hypothalamus would do if properly functioning.

The woman must wear this "pulsatile" pump during the entire duration of the menstrual cycle. The first baby born due to this pump, in September 1982, was a boy. His 27-year-old mother had been sterile for numerous years, despite all kinds of treatment.

Contraception

The oldest contraceptive methods are probably coitus interruptus (the man withdraws to ejaculate outside the woman's body—this is called the sin of Onan in the Bible), and reserved coitus, or carreza (the man does not ejaculate).

Male contraception

Preventives: The invention of the preventive is attributed to the Italian **Gabriele Fallopia** (1523—1562), professor of anatomy at the University of Padua from 1551 to 1562. The preventive was made of cloth, and was to be placed over the glans penis and tucked under the foreskin. Fallopia claimed to have experimented on 1,100 men. His invention was essentially a means to combat venereal disease; its contraceptive value was only secondary. Condom, a physician who resided at the court of Charles II of England (1630-1685), was the inventor of the modern contraceptive named after him.

Pill for men: The work on hormonal contraception for the male is recent (1976) and is due to Professor Salat-Baroux, in Paris, France. Gossypol, a cotton seed extract, was studied in China (Jiang-Tsu province) beginning in 1957, because of its effects on spermatozoa. Daily absorption of the product, over several months, causes sterility which then recedes once ingestion has ceased. However, the side effects and toxicity, which vary among men, have not allowed it to be marketed yet.

Female contraception

Methods based on periodic abstinence from sexual relations:

The Ogino-Knaus method (1932), named after two gynecologists, one Japanese, the other Viennese, is based on the theory that ovulation occurs on the 15th day before the next menstrual cycle in a well-regulated woman. Abstaining from intercourse from the 12th to the 18th day of the cycle (it is necessary to allow for the life duration of spermatozoa) should eliminate the possibility of fertilization. Following a period of great enthusiasm, this technique has been virtually abandoned, due to the number of "Ogino-babies" born because of its failure.

"Temperature" method: This was developed in 1947 by Ferin, following the studies made by the Dutch obstetrician Van de Velde, who in 1904 described the biphasic aspect of the temperature curve during the menstrual cycle. From the time of menstruation until ovulation, the temperature taken in the morning, upon waking, is below 98.6°F (37°C); it drops slightly and rises above 98.6°F (37°C) at the time of ovulation, and remains there until the following menstruation, when it again drops below 98.6°F (37°C), or else remains at the plateau in the case of pregnancy. With this method, the dates of periodic abstinence are based on individual temperature variations.

Local contraception: An old Egyptian papyrus mentions the use of vaginal tampons composed of shredded linen and crushed acacia branch powder. In other civilizations, women used small sponges soaked in products which were reputed to be spermicidal, such as perfumed vinegar water. Syrian sponges were the most sought after, because of their delicacy.

Modern spermicides date from the 1970s; in the form of an ovule, a unidose or a cream, their reliability is about 97%.

Return to the source: A synthetic sponge soaked in spermicide is now being marketed.

Diaphragms and coils: In 1881, the first birth control clinic was opened in Holland under the direction of Doctors Rutgers and Aletta Jacobs; they advised women to use the diaphragm, developed by the German Mensiga. The need to employ spermicidal products in order to ensure their safety led to a considerable decrease in their use, since today spermicides alone are sufficient.

Intrauterine contraception, or the intrauterine device (IUD): The Arabs inserted a small round stone into the uterus of their camels, via a hollow tube, to prevent conception. Ever since, various objects have been inserted in women's uteruses (modern practice began between 1920 and 1930). Used to prevent pregnancy, the objects have ranged from glass to metal to ebony, which in the form of buttons were attached to horsehair or to a silver thread.

However, the first coil worthy of its name was the silver loop designed by the German Ernst Frafenberg in 1928. Measuring 3/5 inch (1.5 cm) in diameter, it was made of a silver spiral having good elasticity. Since then, the shapes and materials used have changed considerably. An IUD soaked in hormones is soon to be marketed.

Oral contraception: The pill was invented in 1954 by the Americans Gregory Pincus of the Worcester Foundation for Experimental Biology, in Shrewsbury (Massachusetts), and John Rock. These physicians worked for five years to develop a definitive contraceptive, which would be "without danger, safe, simple, practical, applicable for all women, and ethically acceptable for the couple." The result was an association of ovulation-blocking hormones.

The initial clinical tests were performed in 1954, and the first large-scale experimentation took place in 1956, in San Juan (Puerto Rico) with 1308 volunteer women.

The first pill to be marketed was Enovid 10, manufactured by G.D. Searle Inc. of Skokie (Illinois). Today, the number of pills produced and marketed is quite considerable.

R.U. 486 (1982): Contragestif was developed by Professor Etienne-Emile Baulieu and the Roussel-Uclaf Pharmaceutical Group, who presented it to the Academy of Sciences on April 19, 1982. Absorbed at the end of the cycle, this "anti-progesterone pill," associated with prostaglandins using a prescribed dose determined under the direction of Professor Bygdeman (Sweden), brings about the onset of menstruation whether or not pregnancy has begun. The name "Contragestif" indicates "against gestation."

This is not a "contraceptive" pill in the strict sense of the term. The pill has not yet been marketed.

7. Drugs and Treatments

Tablets, pills and preparations

Tablets were prepared for the first time in 1843, in England. However, most medications were available in the form of powders, pastes or lozenges, and were made by hand.

In 1873, the French pharmacist Limousin designed the tablet. A mold for suppositories was invented in 1897. The manufacture of capsules (envelopes in the form of tiny eggs) made of glycerin began in 1893. The preparation of ampoules was developed by Limousin and Pasteur, and subsequently by Doctors Duflocq and Berlioz, who, in 1907, improved the technique by using a vacuum to fill them.

Aspirin (1853)

Aspirin, or acetylsalicylic acid, the first medication to be prepared directly in the laboratory, was synthesized by the Frenchman **Charles Gerhardt** at the University of Montpellier, in 1853, but he was not particularly interested in its practical use. In 1829, Henri Leroux, a French pharmacist, extracted salicyn (salicylic alcohol) from the bark of the willow tree. From then, salicylic aldehyde, salicylic acid and finally acetylsalicylic acid were investigated and brought into use.

Why the willow tree? For centuries the willow tree, along with meadow sweet, had been used to treat fever and rheumatism, since both are rich in salicylic aldehyde.

In 1893, Felix Hoffman, a chemist at Bayer, "rediscovered" aspirin and used it to treat his rheumatic father. Bayer began to market it in 1899, under the name "Aspirin": "a" was for *acetyl*, "spir" was for the Latin name for meadowsweet (*Spiraea ulmaria*), and "in" because it was a fashionable ending for medicaments.

In the Treaty of Versailles (1919), Germany surrendered the brand name of "Aspirin" to its conquerors, France, England and the United States, as part of its war reparations.

It was not until the 1960s that aspirin's mode of action and its properties were understood (it blocks prostaglandins, prevents platelet aggregation, etc.). However, even today, not everything about this truly miraculous product is known.

Antibiotics

During the course of his work, Pasteur noted the vital competition which makes some bacteria fight against others. Later, this fact was repeatedly confirmed, and was attributed to the action of an "antibiote" by Vuillemin in 1889. This is the origin of the term antibiotic. Chronologically, penicillin is the first antibiotic. It is still the first choice of doctors, because of its effects and its almost total lack of toxicity. After penicillin, numerous discoveries have added to the knowledge acquired in this field. The United States is largely in the lead, thanks to its continuing efforts in research and development.

• In 1944, S. Waksman discovered streptomycin, whose action against tuberculosis is remarkable.

• In 1947, J. Ehrlich, P. Burkhoder and D. Gottlieb isolated chloramphenicol which, when synthesized, has a powerful effect on a number of infections.

• In 1948, H. Duggar discovered aureomycin.

• In 1950, G. Findlay discovered terramycin.

Since then, antibiotics have continued to make striking progress. Antibiotics are active only against bacteria.

Penicillin (1928)

It was by chance that the English bacteriologist Sir **Alexander Fleming** (1881—1955) discovered penicillin, the first known antibiotic, in 1928. This is how Fleming related the event to the French Academy of Medicine: "In September 1928, I was working on staphylococcus, when a mold contaminated one of my cultures. This event was not without precedent, but what was extraordinary was that all around the molds, the staphylococci

Sir Alexander Fleming who discovered penicillin.

were undergoing lysis [destruction]. I was not investigating contamination by molds; it was pure chance."

Thus, through a stroke of luck, Fleming discovered penicillin (so named because it is present in an ascomycete mushroom, the scientific name of which is *Penicillium*). Dilut-

ed 800 times, it stops microbial development. A team of researchers at Oxford, H. Florey, B. Chain and N. Heatley, continued the work begun by Fleming and concentrated extracts of *Penicillium notatum* in order to obtain purified penicillin, which is more active. These trials continued from 1939 to 1942.

Given the increased demand, industrial production of the drug began in 1943. Fleming's discovery (for which he won the Nobel Prize in 1945) is one of the most striking successes in the history of contemporary medicine. The revolution it started in the area of curative treatment is comparable to the one vaccination brought in preventive medicine.

Hormones

These chemical compounds, secreted by numerous organs, especially the glands, have many variable and complex functions which have only recently been brought to light. The very existence of hormones was completely ignored for centuries.

Bloodletting

Bloodletting was prescribed in antiquity. Little by little it replaced the use of leeches. Practiced with a lance, this treatment developed extensively in the Middle Ages and sometimes posed hygiene problems.

Thus, in 1370, an edict from Charles

DISCOVERY OF THE HORMONES			
GLAND	DATE	HORMONE	INVENTOR
Thyroid	1926	Throxine	Harington
Medullo-surrenal	1856	Adrenalin	Vulpian
	1901	Adrenalin	Takamine
	1942	Noradrenalin	Blaschko
Isles of Langerhans (Pancreas)	1921	Insulin	Banting & Best
Testicles	1935	Testosterone	Butenandt
Ovaries	1924	Folliculin	Courrier
	1929	Oestradiol	Doisy
	1929	Progesterone	Corner
	1938	Synthetic artificial oestrogen	Dodds
A Ovarian destination	1925	Folliculo-stimulating hormone	Ascheim & Zondek
	1965	Luteinising hormone	Reichert
	1972	Prolactin	Wang et al.
Parathroid (Pancreas)	1925	Parathormone	Collip
Cortico-surrenal	1936	Corticosterone and desoxycorticosterone	Kendall
	1938	Cortisone and hydrocortisone	Kendall
	1952	Aldosterone	Simpsan, Tait
Pituitary gland	1966	STH or Growth hormone	Li
	1963	ACTH (Corticotrope) synthesis	Li & Evas
	1930	Thyreostimulin	Loeb & Aron

V decreed that barbers who bled their clients in the morning were required to dispose of the blood before 1 p.m. at the latest. For those who were bled in the afternoon, a two-hour deadline for disposing the blood was imposed; if not respected, a fine of 5 sous was applied.

Bloodletting did not decrease until the 18th century, when it was realized that, although useful in some cases, more often than not it needlessly weakened the patient.

Lavage

Lavage was practiced by the Egyptians, who left samples of lavage horns; the Assyrians and the Babylonians also made use of the method to cure numerous afflictions such as indigestion and intestinal pains.

The enema was invented in 1532.

Injection syringe (15th century)

The principle of the syringe comes from the 15th century Italian Gattinara, but it was not until the 17th century that the first practical trials were carried out, mainly by the Englishmen C. Wren and R. Boyle, in 1657.

Anel, a surgeon in the armies of Louis XIV, had the idea for a piston apparatus. There followed the syringes designed by the German anatomist Neuner, in 1827, then those of the Frenchman C. Pravaz, the first of which dates from 1853. His syringes were made of silver and could hold 1 cubic centimeter of liquid; the rod of the piston was threaded.

The Englishman Fergusson was the first to use glass, whose transparency allowed the injection to be monitored.

However, it was the Frenchman Luer who, in 1869, produced the first all-glass syringe. The technical advance was considerable; risk of infection due to injections was greatly reduced.

The Irishman Rynd, and especially the Scotsman Wood, working respectively in 1845 and 1853, deserve the credit for the development of the method of subcutaneous injection.

Intravenous injection (16th century)

Elshots was the first to inject medicinal products into human veins, in the middle of the 16th century; however, it is known that "intravenous infusions" had already been tested in animals by that time.

In 1665, Schmidt was the first to treat syphilis using intravenous infusion.

Electrotherapy (1786)

A stroke of chance in the laboratory is the origin of the discovery of what was to become electrostimulation.

In 1786, in Bologna (Italy), the anatomist L. Galvani noted that upon contact with two different metals the muscles of a frog reacted convulsively. He understood that this contraction was due to the passage of an electric current.

Following the success obtained in 1795 by the Frenchman J. Halle in a case of facial paralysis, the laws of nervous stimulation have formed the subject of numerous investigations, such as those of Magendie, Faraday and Du Bois-Reymond. They have permitted the development of electrodiagnosis and electrotherapy. Electrostimulation began to enjoy widespread application as of 1960 with the devel-

opment of a cardiac stimulator (pacemaker) with incorporated transistor and batteries in America.

Radiotherapy (1895)

The use of radiation in therapy is the descendant of the discovery of X-rays (1895) and of radioactivity. It became an independent radiodiagnostic specialty in 1934 when **Irene Joliot-Curie** (1897—1956), the daughter of Pierre and Marie Curie, discovered, in collaboration with her husband, **Frederic Joliot** (1900—1958), artificial radioactivity. The first apparatus able to emit radiation which was likely to attack relatively deep tumors was supplied by a 250,000 volt current.

In 1936, a one million volt apparatus was put into service in England. However, this method has been progressively abandoned.

Psychoanalysis (19th century)

Psychoanalysis is everything at once: It is a method for understanding psychological and psychopathological phenomena and for treating mental illness. It was developed at the end of the 19th century by an Austrian physician, **Sigmund Freud** (1856—1939), a Viennese practitioner.

At first received with considerable resistance by the medical community due to its novel propositions regarding sexuality, psychoanalysis has acquired a more and more important place in medicine, in psychology and in the human sciences.

It stresses that it is necessary to understand the pathological processes at work in neurosis, psychosis, or group psychology, artistic production, as well as in the normal development of the child.

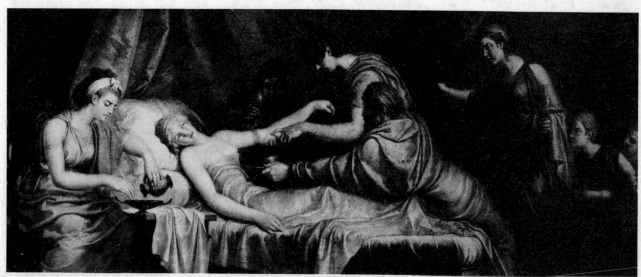

For quite a long time, bloodletting was the only prescribed medicine.

Oxygen tent (around 1900)

The French physician **Charles Michel** (1850—1935) must be credited with the development of the oxygen chamber which, once perfected and adapted to the patient's bed, became the oxygen tent.

Sense isolation chamber (1953)

In 1953, at the Bethesda Institute of Mental Health (near Washington), Doctor **John Lilly**, an American neurophysiologist, had himself placed in a chamber in order to prove that the brain functions without any sensorial stimulation.

Current chambers contain water at 34°F (6°C), saturated with magnesium sulfate, in which the body floats in the dark and in absolute silence.

The undeniably relaxing role of the chamber is attributed, by some, more to its psychological than its physiological action.

Cobalt bomb (1956)

In 1956, cobalt-radium bombardments appeared for the first time at the Hartmann clinic in Paris; artificial radioactivity was used. These bombs multiply the effect of radiation by 4.

Since the 1960s, machines have been creating very powerful electron beams; they are the betatrons and particle accelerators. Their potency is 20 to 25 times stronger than that of the cobalt bomb: the betatrons were discovered in 1940 by Kerst (USA). The synchrotrons date from 1946 and are due to the work of McMillan (USA) and Oliphant (Great Britain).

Interferon (1957)

Interferon was discovered by the Englishman Isaacs and the Swiss Lindenmann in 1957. It consists of a family of proteins secreted naturally by the organism and equipped with great biological potency. Today it is known that 15 to 20 interferons exist, and that much longer investigation will be necessary to clarify all the properties. Further, it is very difficult to obtain interferon. Research in the field of genetics, like that undertaken by Roussel-Uclaf on gamma interferon, seems to be most promising.

Anti-mosquito pill (1959)

The pill has a vitamin B1 base. It was invented in 1959, by the German physician Muting, who noted that mosquitoes less frequently attacked subjects who had just absorbed at least 10 mg of this vitamin.

Aciclovir, against herpes virus (1974)

An antiviral agent which acts against herpes group viruses, it was discovered by **H.J. Scheaffer** and was patented in the United States in 1974.

Electrophoresis in Space (1975—1984)

The first space medicament was isolated in 1975 by EOS (Electrophoresis in Space) during the Apollo-Soyouz mission; it is an enzyme, *Urokinase* (an anticoagulating agent).

Charles D. Walker, the first lay astronaut, manufactured a compound (kept secret) by means of EOS during the flight of the space shuttle Discovery (August 30, 1984—September 6, 1984), but experimentation on humans was forbidden due to a microbial contamination.

Monoclonal antibodies (1975)

The first investigations on monoclonal antibodies were published in May 1975 by two immunologists, George Kohler (Germany) and Cesar Milstein (Argentina), both working at Cambridge (Great Britain). They won the Nobel Prize in 1984 for this work.

Their research consisted in fusing, *in vitro,* mouse antibody-producing lymphocytes with myeloma cells (cancer). They thus obtained hybrid cells (hybridomes) which were able to survive indefinitely and to continually produce a single antibody which was specific for a disease against which the animal had been immunized.

Antibodies are currently making an important contribution to the diagnosis and treatment of numerous infectious diseases and of certain forms of cancer.

Cavitron (1980)

This apparatus, produced by Cooper Lasersonics (California) in 1980, uses ultrasound to destroy a tumor without having to remove it, or without touching its casing; this is particularly useful in neurosurgery. In the United States, there are about 120 Cavitrons; there are about 100 in Japan, 14 in Germany, 4 in Italy, 3 in England, 1 in Belgium and 1 in France.

3M Tenzcare (1980)

A transcutaneous neurostimulator developed in 1979 by **Dr. Eric King-Smith** and marketed by 3M in the United States since 1980, it permits the easing of traumatic pains as well as those associated with childbirth.

A woman who is in the process of giving birth may modulate its intensity by pushing the pre-selected buttons.

Cold box (1982)

The cold box, patented under the name of Rhinotherm, was developed in May 1982 after 8 years of research carried out at the Pasteur Institute in Paris, working in collaboration with the Weizmann Institute in Tel-Aviv.

Presented by the Frenchman Andre Lwoff, who won the Nobel Prize for medicine in 1965, the apparatus produces moisture-saturated air at a temperature of 109.4°F (43°C). The patient inhales this hot air; three half-hour sessions (taken the same day) are sufficient to "kill" the cold.

The results of the experiments conducted by the head of the Beer-Yaakov Allergy Clinic (Israel) are positive: 73% of the patients are cured after the treatment. In the case of chronic rhinitis, the percentage is slightly lower: 65%.

In March 1983, the scientific explanation for the box's action on allergic rhinitis was given before the French Academy of Medicine by Doctor G. Chevance of the Pasteur Institute.

Growth hormone (1982)

Isolated in 1982 by Professor Guillemin, a Frenchman working in the United States (Nobel Prize in 1977), this substance, which orders the secretion of the growth hormone, might soon be used to treat dwarfism of hypothalamic origin, fractures and severe burns.

However, the most futuristic aspect of the GRF (Growth Hormone Releasing Factor) concerns aging. In effect, Professor Guillemin noted that the production of growth hormones as a response to GRF disappeared after 40 years of age. From this, it may be hoped to understand the phenomenon of aging more fully, and perhaps to delay it.

Energy-pack (1983)

Invented by **Doctor Kokoshinegg**, of the Institute of Biophysics and Radiation Investigation of Vienna, and **Professor Bischko**, director of the Ludwig-Boltzmann Acupuncture Institute of Vienna, in 1983, this treatment for muscular or inflammatory pain (cramps, sprained neck, slight sprain, lumbago) using an alternating magnetic field is presented in the form of a 6 cm x 10 cm flexible plate in the middle of which magnetic bands alternate between north and south ori-

MEDICAMENTS			
Diabetes		1673	Thomas Willis noted the sugary taste of diabetic urine, and Frederique Dekkers proposed the use of urinary reagents for the purpose of prognosis.
		1848	Investigation of glycosuria by Hermann von Fehling (who gave his name to the test and the reagent).
	Insulin	1921	Discovered by Pauletto (Romania). Later, in 1921, it was isolated at the University of Toronto by Charles Herbert Best (USA) and F.G. Banting (Canada). Synthetic insulin was obtained in 1964 by Panatotis and Kastoyannis at the University of Pittsburgh (USA). It is not exactly identical to the pancreatic hormone, but it has a similar structure, with 21 amino acids.
	Synthetic hypoglycemic agents	1942	Discovery of hypoglycemic sulfamides by Loubatieres (France).
		1957	Discovery of the biguanides.
		1964	Treatment using hypoglycemic sulfamides: Franck and Fuchs (Germany).
Malaria	Quinine	17th cen.	Cinchona (or quinaquina in quichoo,' the official language of the Incas) was discovered by a Jesuit missionary in South America, which explains its common name of "Jesuit bark." Used to relieve fever, it quickly became a universal remedy. In 1820, the French pharmacists, Pelletier and Caventou, isolated the alcoloid contained in the cinchona bark: quinine. In 1836, at the beginning of the conquest of Algeria, the French military physician, Maillot, used quinine sulfate as a remedy for malaria. Partial synthesis of the product was performed in 1931 by Rabe, but it was not until 1944 that Doerning performed the complete synthesis of the drug.
	Amoebas	1960	Discovery of metronidazole.
Neurology/Psychiatry	Morphine	1804	Alcaloid was isolated from opium by two Frenchmen, Derosne and Seguin, and a German, Frederic Sertuner. They baptized this potent narcotic "Morphium," after the mythological Greek god of sleep, Morphius. Morphine acts essentially on the central nervous system (brain): following a brief period of excitation, there is a paralysis of the nerve centers. Subjects no longer perceive pain. Morphine is thus a central hypnotic and analgesic drug. It is extremely dangerous for irresponsible users, as it is very quickly addictive.
	Cocaine	1844	Isolated by the German Heinrich Emmanuel Merck, who began the industrial production of cocaine in 1862. An Austrian military physician, Theodore Aschenbrandt, gave it to Bavarian soldiers during their maneuvers. He noted that cocaine delayed fatigue. In addition, cocaine suppresses the sensitivity of the areas to which it is applied. Its use should be limited, as it can be addicting.
	Barbituric acid	1864	Adolf von Baeyer (1835—1917) succeeded in condensing urea and malonic acid. He obtained malonyl-urea, or barbituric acid, an inactive substance which did not seem to him to be of any interest.
		1903	Two German researchers, Emil Fischer, a former assistant to Baeyer, and J. von Mering, joined two ethyl radicals to barbituric acid, thereby synthesizing veronal. This was the first of an important group of medications having hypnotic properties and various actions on the central nervous system.
	Amphetamines	1887	The first synthetic amphetamine was prepared in 1887, but the first amphetamines were not marketed until 1937. The American physicians, Smith, Kline and French launched, that same year, the Benzedrex inhaler, which contained an amphetamine: benzedrine. It had been discovered that benzedrine had the property of raising blood pressure, dilating the respiratory tract and stimulating the brain. Thus, patients suffering from sinusitis of common cold felt much better after having inhaled a dose of benzedrex.
	LSD	1938	The Swiss physician Albert Hoffmann isolated lysergic diethylamide acid and tested it on animals. This did not reveal any noteworthy reaction. The acid was virtually forgotten. Then, on April 16, 1943, while working in the Sandoz laboratories, Hoffmann inadvertently swallowed a small dose of LSD 25. In 1962, the American government forbid the sale of LSD to persons other than physicians, psychiatrists and researchers.

Neurology/Psychiatry	Psychotropic drug (Librium and Valium)	1950	Leo Sternbach, the chemist, was studying the chemical class of "Benzodiazepines" at the Roche Laboratories in Nutley, New Jersey. The pharmacologist Lowe Randall observed that one of the substances synthesized by Sternback had a marked sedation effect. The molecule was marketed in 1960 under the name of "Librium," and was followed, in 1963, by "Valium," which has become the trunk of a large tree of benzodiazepines, since it combines four effects: tranquilization, release from anxiety, muscular relaxation and anti-convulsive effects.
	Largactil	1952	Discovery of the neuro-leptic properties of largactil is due to the work of Delay and Deniker.
	Anti-depressive drugs	1957	The MAO inhibitors (monoamino oxydase inhibitors).
	Endomorphines	1977	Discovered by an American, Roger Guillemin —Nobel Prize winner.
Rheumatism		1848	Discovery of hyperuricemia in gout patients by A.D. Garrod.
	Cortisone and Hydrocortisone	1938	Kendall.
		1948-1949	Synthesis and therapeutic use of Cortisone. Teichstein-Kendall.
	Sulfamides (Antibiotics)	1932	Discovery of parasulfa midochrysoidine by Klerer and Miezch.
		1935	Gerold Domagk discovered that several sulfamide compounds have a direct antagonism for bacteria. He obtained Protonsil or Rubiasol.
		1935	Similar work carried out in France by Trefouel and Fourneau.
Allergy	Antihistamines	1906	Discovery of allergy by Clement von Picquet (Austria).
		1932	Large quantities of histamine were identified in the blood of dogs in which an anaphylactic shock had been provoked.
		1937	First active antihistamine, discovered by Bouet and Staub.
		1942	First synthetic antihistamine: Halpern (France).
Cancer	Chemotherapy	1927	I. Berenblum discovered the properties of nitrogen mustard, a derivative of mustard gas.
		1948	Farber obtained the first remission of a case of acute lymphoid leukemia treated with a folic acid antagonist compound; folic acid is essential for cellular division.
		1957	Isaacs and Linderman discovered Interferon (see listing).
Myocoses	Anti-fungicidal drugs	1950	Nystatin, invented by Hazen.
		1955	Amphotericin B, invented by B. Gold.
Tuberculosis	Rimifon	1952	Discovered by Fox (USA).
Hypertension	Anti-hypertension drugs	1957	The diuretics.
		1964	The Beta-blockers. The Englishman Prichard treated coronary insufficiency in his patients with a beta-blocker, Pronethalol; he noted that the medication also acts on arterial hypertension. Pronethalol was the first of a wide range of drugs used to treat the disease.
Liver	Chenodesoxy-cholic acid	1973	Dr. Dowling, England. At the Mayo Clinic in the USA, Hofmann developed the dissolution of gallstones by means of medication.
		1977	The Houde laboratories marketed the discovery in France.
		1980	Marketing of another medication: ursodesoxycholic acid.

entation. A non-magnetic area is placed between each inverse direction band.

Placed on the painful area, it takes effect within a few hours, or three days maximum. It may be reused. Counter-indications: pacemaker, pregnancy, mental disorders, spasmophilia.

Alka-Seltzer (1931)

Alka-Seltzer was developed by the American company Miles Laboratories in Elkart, Indiana.

They are effervescent tablets used for headaches and nausea, usually due to the excesses of the night before. They contain bicarbonate of soda, monocalcic phosphate, aspirin and citric acid.

8. Optics and Acoustics

Eyeglasses

Who invented eyeglasses? The Roman Emperor Nero, who watched the gladiatorial games through an emerald? The Chinese? The Arab scholar Al Hazem (965—1039)? The English philosopher Roger Bacon (1214—1294)? The Neapolitan J.B. Porta (1540—1615)?

All that is known is that very early man looked for ways to remedy eye disorders. However, it was not until the 11th century that the magnifying glass was invented. The Florentine physicist Salvino degli Armati (1245—1317) impaired his vision while performing light refraction experiments, and subsequently sought to improve it. In 1280, he found a way to enlarge objects by using two pieces of glass having a specific thickness and curve. Thus, he should get credit for inventing eyeglasses.

However, he told his secret to one of his friends, Alexander della Spiga, a Dominican at the convent of Saint Catherine of Pisa, who revealed the knowledge. They used convex lenses to treat farsightedness. Concave lenses, for nearsightedness appeared at the end of the 15th century.

Framed eyeglasses (1746)

The frame, which allows the eyeglasses to be placed over the ears and nose, was invented in 1746 by a French optician, Thomin.

Corrective lenses, with variable strength (1959)

The Varilux was invented in 1959 by the French Society of Opticians. These new lenses can replace bifocals or trifocals.

Variable sunglasses

Dr. Edwin Land, the founder of the Polaroid Company, invented polarizing sunglasses which eliminate reflection.

In 1965, the American company Corning Glass invented photochromic lenses which darken upon contact with the light.

Unbreakable lenses (1955)

Organic lenses appeared in 1955. They were made with a thermo-hardening material, Orma 1000, and were lightweight and practically unbreakable.

In 1971, the American company Gentex invented the Troiplus, a new unbreakable lens. It combined the properties of organic glass (plastic) with those of mineral glass.

Eyeglasses for insomniacs (1967)

Invented in 1967 by the German company Bosch, they create electrical impulses sent to the brain by means of electrodes placed on the eyelids and behind the ears. The impulses relax the muscles, slow respiration and induce sleep.

Cataract surgery (1748)

Cataract surgery was mentioned as early as the Code of Hammurabi, the Babylonian king who reigned about 2000 B.C., but it was not until 1748 that the Frenchman **J. Daviel** developed the operation. The first cataract operation consisted in extracting the lens. In his day, Daviel's reputation was without equal, except for the English-

man W. Cheselden, who restored the sight of a man who had been born blind.

Surgical techniques have progressed considerably since then. One of the most ingenious methods, but also the most controversial, is phacoemulsification of the lens. In 1976, the American Kelman invented an apparatus which permits removal of the cataract, after it has been fragmented by means of ultrasound, via a minuscule incision.

Artificial lens

Invented in 1952 by the English physician Harold Ridley, a polymethylmethacrylate (Plexiglas) lens is placed behind the iris.

Bifocals (1752)

The American **Benjamin Franklin** invented bifocals for nearsightedness in 1752. At that time, two lenses were joined in a metallic frame.

Bifocal lenses made of one part, sculpted in the mass, were developed in 1910 by Bentron and Emerson from the Carl Zeiss Company. In 1899 J.L. Borsch invented a more economical procedure, allowing the two lenses to be welded.

Contact lenses (1887—1892)

The first to think of a contact lens system was Leonardo da Vinci. In his *Code on the eye*, he described an optical method for correcting refraction defects by immersing the eye in a water-filled tube that was sealed with a lens. In 1686, the Frenchman Descartes performed the experiment and it was performed again at the end of the 18th century by two Englishmen, Thomas Young and John Herschel. This second experiment involved applying a layer of gelatin held in place by a lens in order to correct ametropy and astigmatism.

• 1887—1892: Contact lenses really made their appearance, due to the efforts of Muller in Germany, Fick and Sulzer in Switzerland, and Kalt in France. The only material used at this time, and until the 1930s, was blown or molded glass (Zeiss glass, in particular).

• 1936: The German company IG Farben made the first Plexiglas contact lens. This same material is still used today for the manufacture of the "hard" contact lenses.

• 1945: The American Tuohy invented the corneal lens, which covers only the cornea.

• 1964: The Czech academician Wich-

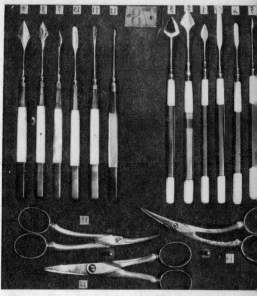

Instruments for operating on cataracts.

terle developed a flexible hydrophile lens.

Electric hearing aid (1901)

The American **Miller Reese Hutchinson** was 26 years old when, on November 15, 1901, he invented the first electroacoustic apparatus designed to amplify sounds for the deaf. His hearing aid was a small case containing batteries and a receiver (similar to the one in a telephone) and it was worn on the ear.

One of the first users of the invention was the English Queen Alexandra, who wore her Acousticon during the entire coronation ceremony. She presented a medal to the young inventor, as a mark of her gratitude for his endeavor.

Cornealize (1949—1983)

Called lamellary refractive keratoplasty (rectification of the curvature of the cornea), the principle of this treatment for some cases of severe myopia or hypermetrophy was devised in 1949 by Professor Barraquer, of Bogota (Colombia).

It consists of taking a very fine corneal sample, freezing it in order to shape it on a cryo-tower, then replacing it, after it has been de-frozen, inside the cornea from which it was taken. This lens has the same optical properties as those of the contact lens, but it is placed inside the cornea. For cataract patients, the lens is obtained from a donor.

Auditive prosthesis (1952)

An auditive prosthesis is an electronic amplifier which compensates for hearing defects. Transistors to aid those who are hard-of-hearing appeared in 1952. The constant progress in electronic miniaturization has permitted smaller and smaller hearing apparatuses; and the latter have been greatly improved during the last twenty years. All cases of deafness which wouldn't benefit from surgery should be helped by an auditive prosthesis, or hearing aid.

Surgery and otospongiosis (1952)

Otospongiosis is a hereditary disorder of the development of some bone structures in the ear; it causes a blockage or ankylosis of the third ossicle (the stirrup). The absence of movement in the area prevents the transmission of sound vibrations to the inner ear. This is a frequent cause of deafness, and is now curable with surgery.

Surgery on the stirrup began in 1952, thanks to the binocular operatory microscope. The operation was perfected between 1960 and 1970. It currently provides spectacular results, with almost complete recovery of hearing. The procedure used was developed by Shea (USA). It consists in removing the blocked stirrup and replacing it with a Teflon piston which is placed in contact with the inner ear.

Radial keratotomy (1955)

The technique consists of radial incisions on the inner side of the cornea in order to correct myopia (nearsightedness). It was perfected in 1955 by T. Sato (Japan), then abandoned, to be taken up again in 1979 by Sviatoslav Feodorov (Moscow).

Deep implants for the deaf (1961)

In 1961, **Georg von Bekesy**, a Hungarian working in the United States, obtained the Nobel Prize for his discoveries relating to the mechanisms of stimulation of the internal ear (cochlea).

3M cochleal implant (1973)

This partially replaces the work carried out by the cocheal cells, the role of which is to transform sound into an electric current. This implant does not restore normal hearing but it allows the totally deaf to hear and to interpret most sounds.

It was implanted for the first time in 1973, and 750 implantations have been performed since then throughout the world.

Surgery with ultrasound (1976)

In 1976, the American **Kelman** invented an apparatus which permits, via a minuscule incision, the removal of a cataract after it has been fragmented by means of ultrasound; local anesthesia is given.

Bertin implant (1977)

Developed by the French physicians **Pialoux**, **Chouard** and **MacLeod** for the Bertin Company, this surgically implanted prosthesis was designed for the totally deaf who still show a positive response to the auditive nerve stimulation test.

Australian bionic ear (1978)

Developed by Professor **Graeme Clark** of the University of Melbourne, this implant has been marketed by the Nucleus Company since 1981. Surveys carried out by the company have shown that there are approximately 4 million totally deaf people in the world.

Surgery with laser (1979)

In 1979, Professor **Daniele Aron-Rosa** developed a surgical technique which makes use of an ultra-rapid pulsated Yag laser; this permits surgery without opening of the eye.

Electronic glasses for the blind (1983)

The apparatus, developed by Doctor **Claude Veraart**, from the Catholic University of Louvain (Belgium), looks like a pair of ordinary glasses equipped with small ear plugs and emits ultrasounds which detect obstacles. These echoes are recorded and transmitted to the listener who then thereby locates the obstacle with precision. It allows the blind to move about much more safely.

9. Dentistry

Toothpastes

The use of plants and other products to combat halitosis (bad breath) has been recommended since antiquity. Formulas for water and tooth powder mixtures have been handed down from the earliest times. Scribonius Largus mentioned those of:

- Messaline (15—48 B.C.): burned deer antler, Chio mastic, ammoniac salt.

- Octabius (70—11 B.C.): sundried

rose powder, white sand, Indian nard (Indian nardostacky, or Valerianaceae).

The countless known formulas for these powders, opiates, pastes and cleansing waters often contained the most unexpected and most revolting vegetable and animal products. Roman society women paid dearly for Portuguese urine (the most prized because it was the strongest, having come all the way from Portugal!). Spanish urine was also prized since it had multiple virtues: It whitened teeth and strengthened gums. This use of urine in mouthwashes continued into modern times. All the 18th century authors, for example, recommended its properties! Our ancestors were unwittingly making use of the ammonia ion which was later used in a modern dental paste (1950).

Fluoride and cavities

Some time ago (1802), in the region surrounding Naples (Italy), spots were noticed on teeth whose enamel had numerous imperfections. However, these teeth, although not attractive, were free of cavities. From this it was inferred that the presence of fluorides decreases the incidence of cavities. Desirabode (1728—1793), a dentist at the Court of France, recommended "Hunter's lozenges" in 1843. They contained fluorides which children and adolescents were recommended to absorb.

The first test trials with fluoridated drinking water took place in 1915, in the United States, thanks to the efforts of MacKay and Black. Today, fluoride toothpastes are very widespread.

Material for filling of cavities

In antiquity, fillings were made up of widely varying mixtures: Nubian soil, hydrated copper silicate, stone chips, turpentine resin. Arculanus (Giovanni d'Arcoli, 1484) discussed filling cavities with solid gold leaf.

In 1853, gold leaf was replaced by spongial gold (Makins), by non-cohesive soft gold, and then by cohesive or adhesive gold (Arthur, 1855). In 1847, gutta-percha was first used to fill cavities. Hill patented a mixture of gutta-percha, chalk, quartz and feldspar.

Amalgams

In 1819, Bell conceived of a mixture of mercury and silver to fill cavities.

In 1850, Regnart proposed adding mercury to Darcet's metal in order to lower the melting point. Fillings using cast metal appeared:

• for inlays in 1884, attributed to Aguilhon de Sarran;

• for onlays in 1886, attributed to Litch.

Extraction of teeth

In China, the tooth pullers' favorite instruments were their fingers. They practiced pulling out nails hammered into thick planks for five to six hours a day in order to acquire the necessary skill.

The heavy-toothed forceps was invented by Thomaseus in 1525. Around 1720, a Frenchman, J.C. de Garengeot, lent his name to a "wrench" used for extracting teeth, although he was not its inventor. The first so-called anatomical dental forceps were made according to the shape of the various teeth for which they were destined, and were invented by the Frenchman J.M. Evrard (1800—1882). Later, as a political refugee in London, he made and polished drill bits according to the indications provided by John Tomes (1815—1895).

Dentures

The first dentures for upper and lower jaws appeared in Switzerland around 1560.

L. Heister (1683—1758) was the first to speak about a removable prosthesis; he recommended that the apparatus be kept very clean, and that it should not remain in the mouth while sleeping. Numerous improvements in the various materials used were tried before the appearance of vulcanized rubber. Developed by C. Goodyear in 1840, it was already being used by a large number of dentists by 1854.

Gold has been used for prostheses for a very long time. Beginning in 1906, cast gold, embossed steel, various alloys and porcelain were used. Today, synthetic resins, which appeared in Germany in 1935, are used. One of the first of these was Paladon.

Dental equipment

In his *Treatise on Medicine and Surgery*, Abulcasis, the Arab surgeon from Cordova, Spain (who died in 1013), described a considerable dental arsenal: pincers, elevator chairs, ligature thread, etc.; this influenced dental surgeons for some time.

Dentist's chair

For centuries, the dentist had his patients sit between his legs, on the ground. In 1810, a chair with a re-

The beginning of a bad tooth.

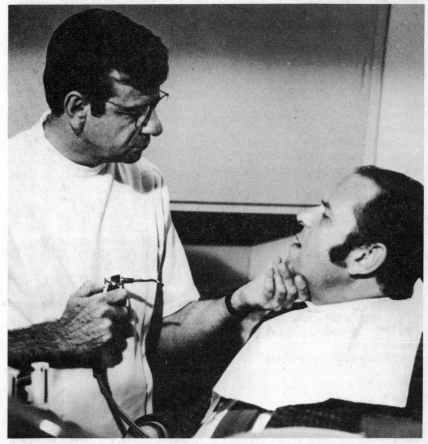

At the dentist.

Dentures.

tractable dentist's stool was built. A basin and an instrument table were attached to the chair. The headrest was a great advance (it appeared in 1848), but there remained the essential problem of placing the patient at the desired height.

Ball invented the first hydraulic chair in 1855.

Harris manufactured the first spiral conveyor chair in 1871.

In 1872, Alexander Morrison invented an iron chair which was raised by means of a manual jack and could be tilted for drug administration.

In 1877, the Johnston brothers made the first hydraulic pump chair, activated by the foot.

The electrically operated chair appeared in 1950.

Dental basin

In 1866, Whitcomb invented the first dental basin; it was cleansed by a continuous stream of water.

In 1881, the Johnston basin, with a faucet and saliva pump (using a water pump system), appeared.

Instrument table

A two-level round table, rotating on a tripod, equipped with drawers and compartments with sliding covers, was built by Sir John Tomes (1815—1895) in 1872. The idea was taken up again in 1928 by Brochere and was named the Girator table.

The dental drill

Cleaning a cavity filled with decayed dentine was first done with enamel scissors. The first rotary instruments (drills) followed; they were turned in both directions, between the thumb and index finger, and thus had to have two cutting edges. Attempts were made to increase rotation speed with a single-edged instrument for removal and elimination of decay. In 1728, Fauchard described a drill whose rotation was ensured by the two-directional displacement of catgut threaded around a cylinder or a watchmaker's bow activating a pulley support.

In 1838, a crank-driven drill was invented (John Lewis).

In 1864, Harrington, of Portsmouth, England, designed a drill activated by a clock movement whose spring was raised and then released. This was the first "mechanization."

In 1868, in the United States, Green presented a drill which was activated by a pedal bellows. On February 7, 1871, another American, J.B. Morrison, patented the pedal bur drill; a flexible arm held the drill in a sort of "hand" part. This was a considerable advance, allowing a speed of 600 to 800 turns per minute.

Green built an electric bur drill in 1874; it was manufactured in the United States by the SS White Company, and in Germany, in 1891, by a company in Erlangen, which added an electric mirror and a galvano-cauterizer. The first "Equipment Unit" was born. Using a foot-driven pedal, rotation speed reached 1200 to 4000 turns per minute.

The rotation speed of drills, whose sizes and shapes, as well as cutting angles, have multiplied since then, increased to 25,000 turns in 1922 (Huet, Brussels, Belgium). In 1958, air-driven turbine models began to appear; they provide rotation speeds of 300,000 to 400,000 turns per minute.

Dental apparatus (1728)

The treatment of dental malpositions (orthodontics) was first foreseen by Celse (using repeated pressure applied by the fingers), and by Pliny and Abulcasis (using filing, often a cause of cavities, and extraction of the poorly placed tooth).

In 1728, Fauchard invented orthodontia. He used silk, linen and metal threads, ligature of the teeth using springs, or metallic blades (more or less extensive, and which prefigured modern arcs).

The true "father of modern orthodontics" is Angle, since, in his first Treatise (1899), he established a classification which is still used today.

False teeth

Until the beginning of the 19th century, false teeth were made from common or exotic animal bone; hippopotamus bone was the most widely used because it was the most resistant. However, such teeth wore out quickly, became brown, and gave off a nauseous odor, which meant that they had to be changed after eighteen months to two years of use. Non-animal materials were then investigated.

As early as 1770, the pharmacist A. Duchateau was offended by the odor and color of his denture, made of hippopotamus bone, and he attempted to make an entire denture from mineral paste.

In 1788, Dubois de Chemant, a Parisian dentist, produced durable dentures using impressions from which a plaster cast was made.

In the United States, the first porcelain teeth appeared in 1817. Made by artisans at first, it was not long before they were industrially produced.

Crowns and bridges

Gold crowns serving as supports for false teeth or as protective jackets to consolidate the teeth have been found in Etruscan tombs (700—500 B.C.).

The crowns often support "aprons" replacing one or several missing teeth. Soldered together they make a bridge, or bridgework. Despite its American name, the apparatus was invented by Pierre Fauchard (1678—1761), a Frenchman. W.H. Dwinelle promoted it in 1856. Toward the end of the 19th century, a permanent prosthesis was developed in the United States. Until 1906, bridges were rather fragile, but later they benefitted from the technique of casting gold.

Industrial porcelain or resin crowns, called jackets (between 1890 and 1900), are placed on the remaining stump of a tooth or on a metallic stump and reconstitute the shape of the tooth.

Solid ceramic (1979)

Developed by Dr. Sozio (USA), this new ceramic permits the reconstruction of artificial teeth which are superior to the natural element. It is so completely solid that it no longer requires any reinforced metallic support.

10. Parallel Medicine

Phytotherapy (3rd millenium, B.C.)

Phytotherapy is one of the oldest forms of medicine. In effect, the Sumerians noted formulas for medicaments having a plant base on their slate tablets as early as 4000 B.C. Closer to our time, Hippocrates, in about 400 B.C., described nearly 250 medicinal plants.

The fantastic medical discoveries made during the last century have rather overshadowed phytotherapy; however, it has been making a forceful comeback over the past ten years.

Acupuncture

The Chinese began practicing acupuncture, one of the branches of their traditional medicine, 2000 years before our era. Sharing the general Chinese philosophy of the balance between two forces, "yin" and "yang," the traditional theory holds that the life force circulates along 14 cutaneous meridians (kings) which usually correspond to an organ.

All pathology is explained by a disturbance of energy flux and can be cured by acting on one or several points located on these meridians. The treatment most frequently involves pin pricks, but may also take

the form of heating (moxas) or massages. Today, an electric current or even a laser ray may be used. One of the diagnostic stages in establishing an energy "profile" is the study of the pulse.

Homeopathy

The German C.S. Hahnemann (1755—1843) created homeopathy in 1796.

In 1790, he had been struck by the description of the properties of cinchona and the incoherence of the explanations given for them, and decided to test its action on himself. He took strong doses of cinchona for a period of several days and soon began to experience the intermittent febrile condition which was identical to the very fevers which quinine cured. Hahnemann then extended his experiment to belladona, digitalis, mercury, and verified the law of similitude: Any substance capable of inducing certain symptoms in the healthy man is also able to make the analogous symptoms in a sick man disappear. This was a breakthrough of some consequence.

Hahnemann based his theory on his extensive experimentation with

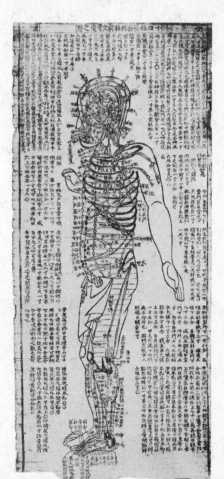

Acupuncture treatise.

healthy subjects, and this led him to pronounce the second fundamental principle of homeopathy: The remedy acts not by virtue of its quantity but in proportion to its dilution, which exerts a potentialization effect. The method subsequently spread very rapidly.

The other broad principles of homeopathy prescribe treating each patient individually, determining the cause of the disease, and adapting the remedy to the disease as it manifests itself in the individual.

One of Hahnemann's principal disciples, C. Hering of New York, created isopathy, which requires that the patient be treated with his own natural or pathological products. The auto-vaccines, used today, are one of the most typical applications of this idea.

Osteotherapy (1892)

The founder of osteotherapy was a physician from Missouri, Andrew Taylor Still. In 1874, he had an intuition which was to lead him to found this science: "Structure governs function." In 1892, he founded the American School of Osteopathy; this marked the entry of the discipline into medical history.

Osteopathy consists in using manual manipulations on the osteo-muscular system, and more particularly on the spinal column, for therapeutic purposes. Today there are approximately 19,000 osteopaths in the United States.

Chiropractic (1897)

After having cured patients by placing pressure on the back, an American hypnotist, Daniel David Palmer, founded the first school of chiropractic in Iowa in 1897.

The method makes use of treatment which is similar to that used in osteopathy. The profession has been recognized in the United States since 1977. Since that date, physicians have been authorized to send patients to a chiropractor. There are about 23,000 licensed chiropractors in the United States.

Mesotherapy (1950)

Mesotherapy was born in the 1950s, due to the work of Dr. Pistor (France).

The technique consists of injecting very weak doses of medicaments, via the use of special needles, into an area to be treated. The principal disorders treated are neuralgia, tendonitis, and trauma resulting from participation in sports.

VITAMINS		
VITAMIN	DATE	DISCOVERY OR INVENTION
Bl (Thiamin)	2600 B.C.	Description of "beri-beri" in China.
	1885	Nutritional cause noted by the Japanese Admiral Takaki.
	1910	Isolated in the rice sheath by Furk, who first gave the name "vitamin" to this life-essential "amine."
	1936	Synthesis by Williams.
B2 (Riboflavin)	1920	Identification of a thermostable growth factor present in yeast by Emmut. The substance is identical to a yellow pigment isolated in milk in 1879 and called lactoflavin.
	1933—1935	Isolation.
	1935	Synthesized by Kuhn and Karrer.
B5 (Pantothenic acid)	1931	Description of a skin disease in chickens subjected to a deficient diet.
	1933	Williams gave the name of pantothenic acid to a growth factor isolated in yeast.
	1940	Isolation and synthesis.
B6 (Pyridoxin)	1935	Gyogyi revealed the presence in yeast of a factor which differs from vitamins Bl and B2 and which cured dermatitis in the rat.
	1938—1939	Isolation, determination of the structure and synthesis.
B12	1925	Anti-anemic action of calves' liver, demonstrated by Whipple.
	1948	Isolation, in the liver, of a red, crystalline substance which cured Biermer's pernicious anemia.
	1955	Determination of the chemical structure.
C (Ascorbic acid)		Antiquity Allusion was made to scurvy in the Old Testament and in the writings of Pliny the Elder.
	1535	Jacques Cartier described the symptoms of the disease observed in sailors.
	16th cen.	Discovery of the specific virtues of fruits of the citrus family.
	1753	Publication by James Lind, a physician in the British Navy, of the first scientific study on the subject.
	1907	Scurvy was experimentally induced in the guinea pig.
	1928	Isolated by S. Gyorgyi from lemon juice.
	1933	Determination of the structure and synthesis carried out by Reichstein.
D (Cholecalciferol)	1645	First description of rickets.
	18th & 19th cen.	The efficacy of cod liver oil in combatting rickets was recognized by various physicians, among them, Bretonneau and Trousseau.
	1890	Protective role of sun rays was established by Palm.
	1919	The curing effect of ultraviolet rays was ascertained.
	1931—1935	Isolation and synthesis.
	1967	Development of more and more active derivatives by Luca.
E (Tocopherol)	1920	Cessation of reproduction noted in the rat subjected to an experimentally deficient diet.
	1922	Discovery by Evans and Bishop of a liposoluble dietary factor present in green leaves and wheat germ; a deficiency of the substance in the gestating rat resulted in the death of the fetus.
	1936	Extraction of vitamin E from wheat germ oil.
	1938	Synthesis carried out by Karrer.
Folic acid	1935—1939	Discovery of vitamin substances in liver and yeast; absence of these substances provoked anemia in the animal.
	1938—1943	Determination of a structure common to all the substances which are abundantly found in leafy vegetables (especially spinach); thus the term "folic acid" (folium in Latin means "leaf").
	1945	Synthesized by Angier.
H (Biotin)	1916	Observation of toxicity caused by a diet containing a high proportion of raw egg white.
	1931	Hypothesis set forth by Gyorgyi that the cause was vitamin deficiency.
	1936	Isolation of a growth, called biotin, from the egg yolk.
K	1929	Demonstration of hemorrhagic accidents in chickens subjected to a fat-deficient diet by Dam.
	1935	Dam gave the name of vitamin K to various substances, of animal or vegetable origin, which are required for blood coagulation.
	1939—1940	Isolation, determination of the structure and synthesis.

VITAMINS (Continued)		
VITAMIN	DATE	DISCOVERY OR INVENTION
PP	1912	Awareness of a disease called pellagra had existed for many centuries. The existence of a vitamin factor which prevents pellagra was ascertained (Funk).
	1926	Experimental induction of the disease and cure by means of yeast absorption.
	1937	Identification of vitamin PP with nicotinic acid, known since 1867 and isolated in yeast as early as 1912.

11. Addendum

Anti-headache glasses (1984)

Invented in 1984 by a 48-year-old Californian, **Glenne Whitten**, Dzidra glasses, equipped with liquid crystal diodes and activated by a battery, carry out a photo-stimulation by means of slow waves and by projecting a shadow over open or closed eyes at regular intervals for 5 minutes. They are designed to relieve ordinary headaches. They were successfully tested by Dr. G.D. Solomon, of the United States Air Force.

Ballistivet: Vaccination from a distance (1984)

This rifle, developed by Ballistivet Inc. (Minnesota), permits vaccination, particularly against rabies, of restive animals: cows, dogs, buffaloes, from a distance (about 30 feet). The lyophilized vaccine is sanitarily placed in small bullets which are crimped under vacuum; the bullets are made of biodegradable material which dissolves after injection under the animal's skin, thereby freeing the vaccine.

Chinese male contraception (1984)

Doctors at the Hospital of Taiwan inject a liquid polymer into the *vas deferens*; the polymer solidifies without adhering to the canal wall and obstructs it, thus providing contraception which, according to them, will be 100% sure and reversible. This technique must be further tested before it can be developed on a broad scale.

Corrective lenses with variable strength (1984)

The Essilor Company (France) developed in 1984 a glass lens for nearsightedness in which the strength decreases progressively beginning with the base, and becomes neutral in the upper part. With these lenses, the nearsighted person is no longer obliged to remove his glasses to see at a distance.

Cryogel (1985)

This is a supple cooling cushion, developed by 3M Health, which accumulates cold. Its use is particularly indicated for hematomas, contusions, insect bites, sprains, etc.

Dental laser (1984)

Developed by S.A. Telecommunications, the compact laser became an odontological laser in 1984, thanks to the Research and Technology Fund. Very small, very manageable and extremely powerful, it has been found to be an excellent tool for small dental work.

Diagnostic test for primitive cancer of the liver (1984)

This test (1984) was developed conjointly in France (Gustave Roussy Institute-Villejuif) by Doctors Claude Bohuon and Dominique Bellet and in the United States (Massachusetts General Hospital-Boston) by Doctors Jack R. Wands and Kurt J. Isselbacher. It permits very rapid detection (in less than one hour) of primitive cancer of the liver, which is very frequent in Africa and Asia and which kills in 3 to 6 months if diagnosed too late.

Ektachem DT 60 (1983)

This portable laboratory, the size of a typewriter, allows seven of the most currently used quantity determinations of blood to be made in only a few minutes on a drop of blood, taken from the fingertip. The Ektachem does not require any special hookup, since it uses the Kodak technique for instantaneous development. It will be sold for less than $6,000 in the United States.

Embryonic transplantation (1984)

On February 3rd, 1983, the world learned with stupefaction that a 30-year-old American (her name was not revealed—let us call her Mrs. X) from Long Beach, California, had just given birth to a small boy who was not her own.

Another woman was artificially fertilized with the sperm of this American woman's husband, and the resulting embryo was removed from her uterus after 5 days and transferred to that of Mrs. X, where it continued to develop. This was the first case of embryonic transplantation in the human species (*see* Genetics).

Heat-Pak (1984)

Invented by a Japanese laboratory in 1980 and marketed in 1984, the Heat-Pak soothes painful joints by means of heat diffusion. Once taken out of its plastic case and shaken, the Pak reaches a temperature of approximately 45°C. It then simply needs to be put in place. It stays warm for 24 hours and is reusable.

Instantaneous counting of bacteria (1984)

Developed by the SII company in Nice (France), the Microcompta apparatus permits the bacteriological control of a substance in real time.

Its applications: research on blood and malaria, as well as the control of the quality of fresh milk, automatic activation of purification stations, or even research on pollution of swimming areas.

Nitrogen baby (1984)

Zoe was the first infant to be born from a frozen embryo; this was in Australia in the spring of 1984.

Before taking up position in her mother's uterus, Zoe's embryo, frozen by Drs. Linda Mohr and Alan Frounson, biologists at Queen Victoria Hospital in Melbourne (team of Professor Carl Wood), had spent two months at -196°C in liquid nitrogen. This was the first attempt to implant a frozen embryo, and it was a huge success.

Operatory microscope with vocal command (1984)

Invented in 1984 by a young Alsatian, **Martine Kempf**, the first microscope to obey the surgeon's voice was used by Doctor Daniele Aron-Rosa for an eye operation. It is a computer weighing only about 11 ounces, and it has the particularity of responding in a few milliseconds (instead of at least one second by other computers) to the order given. The order is recorded twice by the surgeon, so that the computer unambiguously recognizes the voice and the command.

Robot for manufacturing dental crowns (1985—1986)

Invented by the Frenchman Francois Duret, a dental surgeon, this combines a video camera which makes a three-dimensional measurement of the stump of the tooth to be crowned and computer software which calculates the shape of the crown, then directly instructs a machine-tool to form it. This revolutionary procedure, to be marketed in 1985—1986, allows the manufacture of a crown which has a precision of the order of one micron (one thousandth of a millimeter) much more quickly than using the artisanal techniques which exist today.

Solar lenses (1984)

In 1984, the Essilor Company (France) developed the first soft solar brown-tinted contact lenses, under the brand name of Lunelle; they exist in two versions: corrective for ametropia, and neutral for simple solar protection. They are especially designed for sportsmen (regatta sailors, wind-surfers, etc.) and may be worn for an extended period of time.

Sterikit (1985)

In February 1985, 3M Health launched the Sterikit, a family version of the Steristrip. After the wound is cleansed, Sterikit allows it to be closed painlessly, without stitching.

Test for the diagnosis of AIDS (1985)

Blood transfusion centers will have this test available for the diagnosis of AIDS (Acquired Immune-Deficiency-Syndrome), a mortal disease which has continued to spread since its appearance in 1980 and which may be transmitted via contaminated blood.

This test is a world's first. It is the result of the research of the Pasteur Foundation (team of Professor Montagnier, who was the first to isolate the virus responsible for AIDS in January 1983).

The detection kit uses the sandwich-type immuno-enzymatic technique, called ELISA. Worldwide exportation will be possible due to collaboration with the American firm Genetic Systems.

Topographic images of the body (1984)

This is a technique invented in 1984 by a medical photographer, **Robin Williams**, of the Charing Cross Hospital Medical School, London, to measure the precise volumes of some parts of the body. Grids made up of parallel bands which are made to coincide are projected onto the body; this provides a topographic image which is then placed on a photographic film.

TPA: Enzyme which destroys blood clots (1984)

TPA (Tissue Plasimogen Activator) is an enzyme which has a thrombolytic action (destruction of blood clot). It was synthesized in 1984 by Genentech (USA) on the basis of the work carried out by Doctor Collen (Louvain-Belgium), and is currently being tested.

Injected intravenously, TPA can dissolve a blood clot which obstructs a blood vessel within 10 to 45 minutes. This will thus be a valuable tool in the emergency treatment of infarctus.

Video-endoscopy (1984)

The video-endoscope was invented by research scientists working at the American company of Welch Allyn in 1984; it combines endoscopy and video by means of an association of data-processing and microelectronics.

It is thus possible to record the images on video in order to study them at a later time. There is a disadvantage: The endoscope tube measures 14 mm in diameter, while the traditional tubes measure only 8 or 9 mm. It is thus more difficult to insert the tube. Another video-endoscope model has been introduced by the Japanese firm of Fujinon.

Voice coding (1985)

The Voiscope, the invention of Professor **Adrian J. Fourcin** (University of London), on which Professor Pierre Elbaz and Dr. Gerald Pain (Paris) are working, permits the extraction of thousands of signals recorded during the reading of a text or a song, the mean vibration per second of the vocal cords, and therefore the fundamental bass frequency perceived even by the totally deaf. This "coding" of the human voice indeed represents enormous progress.

science

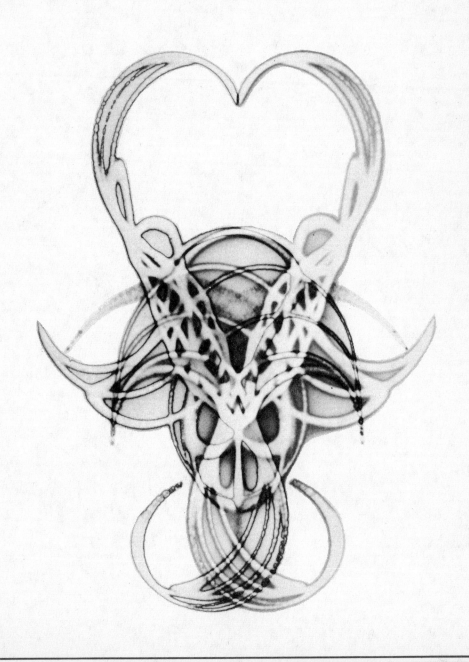

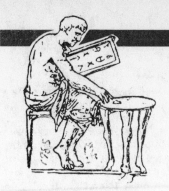

MATHEMATICS

1. Arithmetic

Origin (6th century B.C.)

A form of sign language, still practiced by certain primitive Oceanic tribes, preceded oral and written numbers (*See* Numerals, p. 242). From that "practical" arithmetic, the school of **Pythagoras** founded a true science of numbers at Crotona in the 6th century B.C. For the philosopher-mathematicians of the time, all things were numbers. As for whole numbers, they were imbued with quasi-magical properties.

In the 3rd century B.C., Diophantus of Alexandria summarized and classified the numerical knowledge of the time in his *Arithmetic* (only six volumes of which have survived), which proposed some 200 problems. Unfor-

tunately, Diophantus' work was forgotten, as the Romans took little interest in mathematics. Luckily, however, Arabic science continued the Greek tradition and made considerable advances in mathematics. Revised by the Arabs, the work of Diophantus did not reappear in the West until the 16th century. His *Arithmetic* was published in 1621. A second annotated edition appeared in 1670, with comments by the brilliant French mathematician Pierre de Fermat: (1601—1665). This opened the way to a considerable amount of research and numerous works that were to edify classical arithmetic in the 17th and 18th centuries. The names generally associated with this period are: P. de Fermat; the Swiss Leonhard Euler (1707—1783); the

Frenchman Joseph-Louis de Lagrange (1736—1813); the Frenchman Adrien-Marie Legendre (1752—1833), author of a *Theory of Numbers*; and the German Karl Friedrich Gauss (1777—1855), author of *Disquisitiones Arithmeticae (1801)*.

Zero (4th century B.C.)

The oldest zero in human history is Babylonian in origin and dates back to the 4th century B.C. Mathematical tablets found in Uruk (Iraq) attest to the use of the zero, mainly to denote the absence of units with a number.

But zero had not yet become an operational entity. For example, instead of using zero as the result of a completed distribution of grain, a bookkeeper of the time, not knowing quite how to express the situation, would write in full something like "the grain supply is exhausted."

As early as the 3rd century A.D. zero was known to Mayan astronomer-priests, although they didn't use it as a figure. Astronomers and mathematicians in India were the first to use zero as a true number, represented by 0, well before 628. This was the year that the astronomer-mathematician Brahmagupta wrote a work on the manner of performing the six operations on gains (positive numbers), losses (negative numbers) and zero (nonentity). Multiplying by 0 cancels out a quantity while adding or subtracting 0 makes no change in that quantity.

Etymologically, the word *zero* evolved as follows: From the Sanskrit word *sunya* meaning "nothing," it became *sifr* in Arabic (from which the English word *cipher* is derived). Leonardo Fibonacci Latinized it to *zePhirum*, and finally, a Florentine treatise, *De Arithmetica Opusculum*, established the word *zero* once and for all in 1491.

2. Geometry

Origin (3rd century B.C.)

With the puflication of Euclid's *Elements* (Greek mathematician of Alexandria, 3rd century B.C.), geometry, defined as the study of shapes (figures and volumes), ceased to be empirical. Until the 19th century, the *Elements* dominated elementary mathematics. In the thirteen books comprising his treatise, Euclid not only collected and classified the work of his predecessors, but in addition, founded a geometry that was based on strict definitions starting from assumptions he termed "common notions."

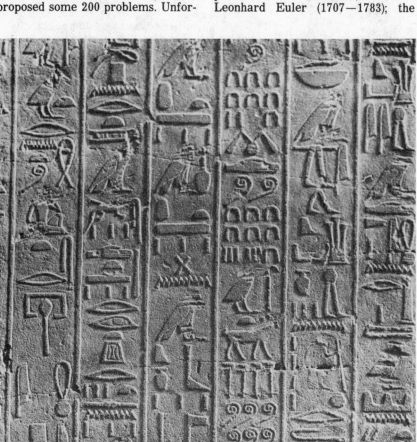

Egyptian numbers.

Euclid's postulate
(3rd century B.C.)

The most famous of the postulates contained in the *Elements* by the Greek mathematician **Euclid** (3rd century B.C.) is the parallel postulate. This postulate, known as the fifth postulate, or Euclid's postulate, states in substance that through any given point there exists only one line parallel to a given line.

Up until the 18th century, Euclid's postulate was the *sine qua non* condition for the application of mathematical reasoning to geometry. This postulate was demonstrated over and over again. But one of its demonstrators, the Italian Jesuit mathematician Gerolamo Saccheri (1667—1733), inadvertently opened the way to a coherent theory based on denial of the postulate. The *reductio ad absur-*

dum (i.e., disproving a proposition by logically deducing an absurd consequence) practiced by Saccheri marked the beginnings of non-Euclidean geometry.

Conics (3rd century B.C.)

Appolonius of Perga (ca. 260—200 B.C.), a Greek mathematician of the Alexandrian school, owes his reknown to the eight volumes he devoted to conic sections. In his work, Apollonius studied the curves obtained from right and oblique cones by varying the angle of inclination of the intersecting planes. He introduced the names of *parabola, hyperbola* and *ellipse,* borrowed from the Pythagoreans, and devoted his research to the basic characteristics of these three types of conics. In 1639, the young Blaise Pascal (1623—1662) rediscovered compu-

tations that were 2000 years old, recording them in his *Essay on Conics.*

Numerals (3000 B.C.)

Some 3500 years ago, counting was performed pragmatically using a soft clay ball inside of which were placed as many clay beads as there were items to be counted. Written numbering dates back to 3000 years before Christ, as attested by the clay tablets found at Susa and Uruk (modern-day Waruk, in Iraq).

Whether expressed in hieroglyphics, cuneiform, or written words, numbering was an additive process. It attributed a distinct sign to a unit, to 10, and to powers of 10 by repeating the symbol as many times as necessary. Decimal notation leading up to our present numbering system first appeared some 15 centuries ago in northern India, where the numbers 1-9, as well as 0, were used. The Arabs were the first to become aware of their system. In 829 A.D. the Arab mathematician Mohammad ibn-Musa al-Khwarizmi (780—850) published in Baghdad, the intellectual center of the Muslim world at the time, a treatise on algebra in which he urged bookkeepers and businessmen to use this new system of notation and corresponding methods of calculation.

It took "Arabic" numerals two centuries to reach Spain and the city of Cordoba. They were finally spread throughout Europe by Leonardo Fibonacci, known as Leonardo of Pisa, by means of his *Liber Abaci* (1201). Arabic numerals reached England at the same period.

Thales' theorem (ca. 600 B.C.)

The Greek Thales of Miletus (ca. 624—ca. 548 B.C.), was one of the seven sages of antiquity. Mathematician, physicist, astronomer, and philosopher, he studied notably the angles of triangles and proved that the angles at the base of an isosceles triangle are equal.

Thales' theorem can be stated as follows: Any line parallel to a given side of a triangle, and intersecting the two other sides of that triangle, will divide those two adjacent sides into proportional segments, thereby creating a second triangle similar to the original.

Ancient wise men like Thales knew how to measure the height of objects like the Pyramids from their shadow.

Infinity

For the Greek philosopher Aristotle (384—322 B.C.), student of Plato, founder of the lyceum at Athens, and tutor of Alexander the Great, infinity

Top: Numbers have been written in many ways since their invention. Here, two illustrations showing a 1900 design. Bottom: Different ways to write the same numbers.

In 600 B.C., the Greek Thales knew how to calculate the height of a pyramid based upon the shadow it cast.

existed only as a potentiality, because a limited quantity can be divided an infinite number of times without ever reaching an end to the process. It was not until the 12th century that one of the great Indian mathematicians, Bhaskara, approached the modern concept of this notion, where infinity appears explicitly as the result of the division of a finite number by zero.

The symbol of infinity in the form of a horizontal eight was first introduced in the treatise *Arithmetica In-nitorum* (1655) by the Englishman John Wallis, chaplain to Charles II.

It should be noted that since the 19th century, mathematicians have distinguished between several different orders of infinity. For example, the infinite number of points existing between any two given points is a much larger infinity than the infinite number of natural whole numbers (0, +1, +2, +3 . . .).

Pi (π)

For the Babylonians and Hebrews, the ratio between a square whose side measures *a* and a circle with a radius of *a*, was 1:3.

Archimedes of Syracuse (287—212 B.C.), the greatest scientific mind of antiquity, invented the number *pi*. Starting from the regular hexagon inscribed in a circle, whose sides he doubled four times in succession, Archimedes demonstrated that *pi* is comprised between $3 + 10/71$ and $3 + 1/7$ (two exact fractions).

Ptolemy, a mathematician of the Alexandrian school (ca. 85—165 A.D.) attributed to *pi* the value now familiar to all schoolchildren, that is, 3.1416. In 1874, the Englishman William Schanks extended the calculation of *pi* up to 707 decimal places. This record has since been surpassed by computers.

The use of the symbol , which is the first letter of the Greek word perifereia, for "circumference," appears to date back to the 18th century. It is found in a treatise by the Englishman William Jones, published in 1706. It did not come into common usage until the publication of Euler's *Introductio in Analysi Infinitorum* in 1748.

Equations (9th century A.D.)

As early as the 9th century, **Mohammed ibn-Musa al-Khwarizmi** (780—850 A.D.), librarian to the Caliph of Baghdad, found solutions to a great many equations, including those of the second degree. Khwarizmi's methods were derived from Greek and Indian sources. However, it was not until the 16th century that third and fourth-degree equations were first solved by Italian mathematicians. Girolamo Cardano (1501—1576), Ludovico Ferrari, and Niccolo Fontana, known as Tartaglia (ca. 1500—1557), engaged in mathematical competitions that often ended up as drunken brawls! But this did not prevent them from contributing significantly to the advance of algebra.

Trigonometry (modern, 1464)

The precursors of trigonometry were the Greeks Aristarcus of Samos (ca. 300—230 B.C.), Hipparchus of Nicea (ca. 180—ca. 135 B.C.), who introduced the division of the circle into 360 degrees according to the Babylonian system; and Claudius Ptolemy of Alexandria (80—160 A.D.), author of the monumental treatise on trigonometry, the *Almagest*.

In 1464, the German mathematician **Johan Muller** (1436—1476), known as Regiomantus, wrote *De Triangulis Omnimodis Libri Quinque*. Not published until 1553, after Muller's death, this work laid the foundation for the development of modern trigonometry, making it a science independent of astronomy.

Georg Rhaeticus (1514—1576), Prussian mathematicin and disciple ofCopernicus, gave trigonometric relationships their modern definition. Rhaeticus established tables giving, with at least ten decimal places, the values of the different functions for angles at ten-second intervals. Unfortunately, he died before completing his tangent tables.

In its primary meaning, trigonometry signifies the measurement of triangles. However, as a calculating tool trigonometry has been extended far beyond its original scope. For example, sine and cosine functions, which are periodic functions, provide a mathematical means of studying electrical circuits as well as the most complex vibratory phenomena.

Negative numbers
(in Europe, 1484)

The first European mathematician to understand the concept of negative numbers was the Frenchman **Nicolas Chuquet** (1445—1500) in 1484.

How can you take ten tangerines from a basket containing only five? This question remained unanswered for centuries, and the result of the operation 5 minus 10 minus 5 was meaningless for the mathematicians of antiquity. Toward the 5th century, negative numbers were used in India to express debt in bookkeeping operations.

Chuquet's ideas were not developed until 1545, when Michael Stifel described all of the properties of negative numbers, which he referred to as *absurd numbers*.

Complex numbers (1572)

The Italian mathematician **Raffaele Bombelli** (1526—1573) invented complex numbers, of imaginary roots, toward the end of his life in 1572, the year in which his work *Algebra* was published. The difficulty to be surmounted was as follows: The product of a number multiplied by itself (i.e., a square) necessarily being a positive number, how can you extract the square root of a negative number? This is where Bombelli proposed to employ complex numbers.

A complex number is any notation in the form $a + ib$, where a and b are real numbers and i represents a new element such that $i^2 = -1$; in other words, $i = -1$. The imaginary quantity is ib. Complex numbers can be used to deal with all second degree equations having no real root. The use of i to designate the square root of minus 1 was first introduced by Euler in 1797 and taken up again by Gauss in 1801, in his *Disquisitiones Arithmeticae*, which contributed significantly to the general adoption of i.

3. Algebra

Origin (late 16th century)

Algebra in its modern form was invented by the Frenchman **Francois Viete** (1540—1603).

Algebra is generally said to have originated with Diophantus of Alexandria (3rd century B.C.), author of the famous work *Arithmetic*. In that work can be found the premises later known to the Indians (Brahmagupta in the 7th century, Bhaskara in the 12th century) and the Arabs. (The librarian of Baghdad, Al-Khwarizmi, published a treatise in 829 containing the Arabic word *hisab al-jabr*, from which our word *algebra* is derived.)

In 1858, the young Scotsman Henry Rhind, on a sojourn to Egypt for health reasons, acquired at Luxor a papyrus covered with symbols. The translation of these symbols revealed close to a hundred different problems, complete with their solutions. One example read: "Haha, its seventh, its whole number . . . makes 19." This seeming gibberish, which demonstrates the perplexity of the scribe in the face of an unknown quantity, along with the symbols used for addition (a pair of legs moving forward), can be considered as algebra's "birth certificate," no less than 3600 years ago.

The aim of algebra is to solve problems involving unknown quantities. These unknown quantities are represented by symbols that are manipulated until their numeric values are determined. We refer to this process as solving equations.

Algebraic symbols (1591)

In his treatise In *Artem Analyticum Isagoge*, published in Tours, France in 1591, **Francois Viete** (1540—1603) adopted symbols similar to our present algebraic notation. For example, he used the German plus (+) and minus (-) signs. In addition, he represented known quantities by consonants and unknown quantities by vowels. Rene Descartes (1596—1650) used modern algebraic notation in his *Discourse on Method*, published in 1637. Descartes notably employed numbers as exponents to indicate powers.

In the 14th century, to represent

Archimedes.

x^4/x^3, they would write: "first square of the square divided by the cube."

Plus and minus signs (1489)

It was the great German mathematician Michael Stifel (1487—1567) who popularized the use of the plus (+) and minus (-) signs as totally abstract algebraic symbols, as a result in his treatise on algebra *Arithmetica Integra*, published in 1544. However, as early as the Renaissance, these symbols had already replaced the notations *P* for addition and *M* for subtraction. They appeared in print for the first time in an arithmetic book for tradesmen published in 1489 on the initiative of the German **Johann Widman** of Leipzig. But until Stifel, these signs remained linked to the notion of concrete quantities and designated credits and debits in inventory and bookkeeping operations.

Square root (1526)

The symbol, used to express a square root, is thought to be a distortion of the lower case *r*, originating in *Die Coss (The Unknown)*, published in 1526 by the German **Cristoff Rudolff** (ca. 1500—1545).

Equal sign (1557)

The symbol = was first used to express equality in *The Whetstone of Witte*, by the English mathematician and Oxford professor **Robert Recorde** (1510—1558), published in 1557, just a year before the author's death. This work dealt with the extraction of roots as well as calculations involving irrational numbers (i.e. certain roots).

Logarithms (1614)

Starting from the ancient idea of comparing arithmetic and geometric progressions, the Scotsman **John Nepier** or Neper (1550—1617) invented logarithms in 1614. Neper's theory of logarithms is contained in his work *Mirifici Logarithmorum Canonis Descriptio* published in Edinburgh that same year.

The logarithm of a number *a* is the power to which a constant number, called the base, must be raised in order to obtain *a*. Logarithms can be used to replace, in a sense, all multiplications and divisions by additions and subtractions. Archimedes had already studied the series of whole numbers $1, 2, 3 \ldots, n$ and the series of the powers of a number $a, a^1, a^2, a^3, a^4 \ldots, a^n$.

In 1484, the Frenchman Nicolas Chuquet observed in his *Tripartite on the Science of Numbers* that there is a correspondence between

The Arabs of the 7th century were remarkable scientists.

the sum of the terms belonging to the first series and the product of the terms of the same rank belonging to the second series, e.g. $2 + 5$ corresponds to $a^2 \times a^5$.

Neper introduced enough intermediate numbers to make it possible to proceed from one series to another. His theory led him to construct tables with four decimal places or more. In the Neperian logarithm, *e* is the base, that is, the real number such that $\log e = 1$, hence $e = 2.718281828459045235 \ldots$

Base-10 logarithms (1624)

In 1624, the Englishman Henry Briggs (1561—1631), Neper's friend and colleague, published his *Arithmetica Logarithmica*, which contained the first decimal logarithm tables. Taking the number 10 as base, numeric calculation was simplified. Briggs gave the logarithms of the first 31,000 whole numbers computed to 14 figures and indicated how to go from base-10 logarithms to Neperian logarithms.

Coordinates (1636)

The concept of coordinates was not clearly expressed until the 17th century, by the French philosopher Rene Descartes (1596—1650) and the French mathematician Pierre de

Chilias 1.			Chilias 1.			Chilias 1.		
Nu.	Logarithmi	Differ.	Nu.	Logarithmi	Differ.	Nu.	Logarithmi	Differ.
1	0,00000,00000		51	1,70757,01761		101	2,00432,13738	
2	0,30102,99957	17609,12590	52	1,71600,33436	843,31675	102	2,00860,01718	427,87980
3	0,47712,12547	12493,87366	53	1,72427,58696	827,25260	103	2,01283,72247	423,70529
4	0,60205,99913	9691,00130	54	1,73239,37598	811,78902	104	2,01703,33393	419,61146
5	0,69897,00043	7918,12461	55	1,74036,26895	796,89297	105	2,02118,92991	415,59598
6	0,77815,12504	6694,67896	56	1,74818,80270	782,53375	106	2,02530,58653	411,65662
7	0,84509,80400	5799,19470	57	1,75587,48557	768,68287	107	2,02938,37777	407,79124
8	0,90308,99870	5115,25224	58	1,76342,71236	755,31379	108	2,03342,37555	403,99778
9	0,95424,25094	4575,74906	59	1,77085,20116	742,40180	109	2,03742,64979	400,27424
10	1,00000,00000	4139,26852	60	1,77815,12504	729,92388	110	2,04139,26852	396,61873
11	1,04139,26852	3778,85608	61	1,78532,98350	717,85846	111	2,04532,29788	393,02936
12	1,07918,12460	3476,21063	62	1,79239,16895	706,18545	112	2,04921,80227	389,50439
13	1,11394,33523	3218,46834	63	1,79934,05495	694,88600	113	2,05307,84435	386,04208
14	1,14612,80357	2996,32234	64	1,80617,99740	683,94245	114	2,05690,48513	382,64078
15	1,17609,12591	2802,87236	65	1,81291,33566	673,33826	115	2,06069,78404	379,29891
16	1,20411,99827	2632,?387	66	1,81954,39355	663,05789	116	2,06445,79892	376,01488
17	1,23044,89214	2?32,35837	67	1,82607,48027	653,08672	117	2,06818,58617	372,78725
18	1,25527,25051	2348,10959	68	1,83250,89127	643,41100	118	2,07188,20073	369,61456
19	1,27875,36010	2227,63947	69	1,83884,90907	634,01780	119	2,07554,69614	366,49541
20	1,30102,99957	2118,92990	70	1,84509,80400	624,89493	120	2,07918,12460	363,42846
21	1,32221,92947	2020,33861	71	1,85125,83487	616,03087	121	2,08278,53703	360,41243
22	1,34242,26808	1930,51552	72	1,85733,24964	607,41477	122	2,08635,98307	357,44604
23	1,36172,78360	1848,34057	73	1,86332,28601	599,03637	123	2,08990,51114	354,52807
24	1,38021,12417	1772,87670	74	1,86923,17197	590,88596	124	2,09342,16852	351,65738
25	1,39794,00087	1703,33393	75	1,87506,12634	582,95437	125	2,09691,00130	348,83278
26	1,41497,33480	1639,04162	76	1,88081,35923	575,23289	126	2,10037,05451	346,05321
27	1,43136,37642		77	1,88649,07252	567,71329	127	2,10380,37210	343,31759

The Scot John Napier invented logarithms in 1614.

Fermat (1601—1665). Descartes wrote his Geometry at Leyden, Holland, in 1637. But he was in fact preceded by Fermat, who had published his Ad Locos Planos et Solidos Isagoge in 1636. Coordinates nevertheless came to be termed as Cartesian.

Coordinates are numbers that, according to a set of conventions, determine one and only one element in a given geometric system (e.g. a point in a plane). The notion of coordinates, more or less implicit in the work of Archimedes and Appolonius of Perga, was nonetheless foreign to Greek geometric thought. Some precedent is found, however, in the solutions provided to certain geometry problems by Arab mathematicians and by Italian mathematicians of the Renaissance period.

Probability theory (1656)

In 1656, the Dutch scientist Christian Huygens (1629—1695) published the first complete work on probability theory under the title De Ratiociniis in Ludo Aleae. Based on the correspondence between Fermat and Pascal on the mathematical theory of games of chance, this work included several unsolved game problems and added those for which Fermat and Pascal had already found solutions. The aim of probability theory is to determine the mathematical chances an event has of occurring.

Differential and integral calculus (1665)

Between 1665 and 1667, Sir Isaac Newton (1642—1727), one of the greatest scientific intellects of all times, developed differential calculus, which he referred to as the "theory of fluxions." While a student at Trinity College, Cambridge, the young Isaac Newton was forced to return home because of an outbreak of plague. He learned of Galileo's discovery that the acceleration of a falling body increases uniformly with time. To calculate what trajectory the body would follow, Newton invented his method of fluxions. This was published as an appendix to his Principia Mathematica in 1687, nearly twenty years after it had been devised. Three years earlier, the German philosopher and mathematician Gottfried Wilhelm Leibniz (1646—1716) published his work on the calculation of maximums and minimums. Since Leibniz used the same mathematical methods as Newton, the situation gave rise to a long and bitter controversy between the two men as to who had first developed differential and integral calculus, also known as infinitesimal calculus or analysis.

Functions (1748)

The Swiss mathematician **Leonhard Euler** (1807—1783), student of Jean Bernoulli and subsequently professor at the University of Saint Petersburg before becoming attache at the Academy of Berlin, developed a theory of functions totally independent of geometry. In his *Introduction in Analysi Infinitorum* (1748), Euler considered the algebraic function as an *analytical expression* comprising one or more variables and one or more constants. He distinguished between rational and irrational algebraic functions on the one hand, and implicit and explicit functions on the other.

The concept of functions arose empirically from the study of relationships that can be observed between two variables, for example, time and space, the area and side of a square. The simplest function is the *linear function*, which expresses proportionality. Linear functions are represented graphically by a straight line. *Homographic functions*, which express inverse proportionality, are represented graphically by a hyperbola.

Statistics (1748)

The German Gottfried Achenwall (1719—1772), professor at the University of Gottingen, introduced the word statistics in 1748. Achenwall's work was confined to the purely descriptive operations of collecting and tabulating information. It was the Frenchman Pierre Simon de Laplace (1749—1827) who made statistics into a mathematical tool in his *Analytical Theory of Probabilities*.

Vectors (1799)

In 1799, the Danish surveyor-geometrician, **Caspar Wessel** used the notion of vectors for the first time to represent complex numbers graphically. However, the publication of Wessel's research in Danish received little attention from mathematicians of the time, and it is the name Jean Robert Argand (1768—1822) that is usually associated with the invention of vectors.

Non-Euclidean geometry (18th century)

After Saccheri (*see* Euclid's Postulate, p. 332), a large number of 18th century mathematicians began questioning the validity of the parallel postulate, thereby opening the way to the development of non-Euclidean geometrics. This movement was begun by the Frenchmen **Jean-Henri Lambert** (1728—1777) and **Adrien-Marie Legendre** (1752—1833). Adapting Euclid's *Elements,* Legendre wrote a treatise on geometry in 1794. Finally, as early as 1799, the German Karl Friedrich Gauss (1777—1855) realized the necessity for enlarging the concept of "geometry," and developed a spatial geometry based on the denial of Euclid's postulate.

Hyperbolic geometry (1826)

Two young mathematicians, the Hungarian **Janos Bolyai** (1802—1860) and the Russian **Nicolai Lobatchevski** (1793—1856), independently systemized non-Euclidean geometry around the same time (1826—1829). This new geometry, which had been anticipated by the German mathematician Gauss nearly thirty years before, in 1799, was based on a surface having constant negative curvature. It is known as *hyperbolic geometry* as opposed to *plane geometry*.

Elliptical geometry (1854)

The founder of elliptical geometry was the German **Bernhard Riemann** (1826—1866), a student of Gauss, who presented his doctoral thesis at Gottingen in 1854. Although Reimann had originally come to study theology, he soon placed his intuition and imagination in the service of mathematics. Not published until 1868, Riemann's thesis constituted the starting point for the revision of classical geometry.

The non-Euclidean hypothesis according to which the sum of the angles of a triangle is greater than the sum of two right angles is the basis for Riemann's elliptical geometry, which starts from a surface of constant positive curvature.

Erlangen Program (1872)

The "Erlangen Program" was the inaugural paper given in the town of Erlangen in 1872 by the German Felix Klein (1849—1925), the leading figure of the German school of mathematics at the time. In this program, all existing geometries converged for the first time in an overall conception that brought to light their common nature. Klein applied group theory to geometry. As a result, geometry was no longer confined to the study of shapes in three dimensional space, but became, in a more general framework, the study of groups operating on sets.

Matrices (1838)

In 1838, the English mathematician **Arthur Cayley** (1821—1895) anticipated the importance of the concept of matrices. He defined their nature and formulated the matrix theory in 1843.

A matrix is a table consisting of rows and columns forming compartments occupied by numbers that are associated in a given manner. The simplest matrix is the addition table. The Pythagorean or multiplication table is another example. A matrix can be infinite.

Cayley introduced the notion of matrices in order to simplify the resolution of simultaneous linear equations and worked on the rules of matrix calculations up until 1858.

Symbolic logic (1854)

In 1854, the self-educated Englishman **George Boole** (1815—1864) published *An Investigation of the Laws of Thought*. Bertrand Russell was later to consider this work as the basis for symbolic logic and pure mathematics. Boole defined classes, the operations between classes, the properties of these operations, and the correspondence existing between these operations and logical operations. For example, union, represented symbolically by the + sign, corresponds to the logical operation *or,* while intersection (.) corresponds to the logical operation *and.* As a result, Boole announced the laws governing logical calculus. Boole gave his name to the branch of algebra that studies the properties of operations occurring between sets.

Set theory (1874)

In 1874, the Russian mathematician **Georg Cantor** (1845—1918) formulated a general theory of infinite sets. It took Cantor 20 years to convince his contemporaries of the validity of his work, and for his influence to spread. But even in antiquity, philosopher-mathematicians had investigated particular sets of numbers and points.

The fundamental notion of set is based on the concept of belonging, a characteristic property enabling one to determine whether a given element does or does not belong in a particular grouping. By studying the properties of sets, both finite and infinite, as well as the operations that can be performed on them, set theory provides a stringent code and, above all, a precise language to all branches of mathematics.

PHYSICS

1. Standards

Weights and measures

The first units of measure were most often derived from the human body: the thumb, the forearm (cubit), the distance between the tip of the king's nose and the tip of his middle finger for the medieval English (yard); or they had some relation to physical activity: the pace, the league, which equaled an hour's walking, etc. The Celts used the capacity of two cupped hands united as a measure, while the ancient Egyptians derived their measure for liquids from the mouthful or draught.

For centuries units of measure spread throughout Europe in haphazard fashion, varying to a large extent from one region to another. Thus, the English *ell* or French *aue,* a measure used chiefly for cloth, ranged from 0.513 meters to 2.322 meters depending on the locality and sometimes even on the goods. In Paris, it was 1.588 meters, in Metz 0.68 meters, while in Laval it was 1.43 meters. All attempts to unify weights and measures had to contend with a wide variety of local customs and usages, until the time when scientific and technological developments necessitated the introduction of a single coherent system. In France, weights and measures become standardized with the institution of the decimal metric system.

In the English-speaking countries, the International System of Units was not adopted until the second half of the 20th century; Great Britain's conversion to the metric system was only completed in 1980.

Illustrating the uses of the metric system, 1799.

Metric system (1795)

The principle of mandatory units for weights and measures was established in France, after the Revolution, by decree of the 18th Germinal of the Year III of the Republic (April 7, 1795). This decree instituted the metric system, established the nomenclature of the units and, for the first time, legally defined the meter—as a fraction of the distance between the North Pole and the Equator, measured on the meridian from Barcelona, Spain, to Dunkirk in northern France by the Frenchmen Jean Delambre and Pierre Mechain in 1791. The unit of weight is the kilogram, which is equal to the weight of a cubic decimeter of water.

On June 22, 1799, the first prototype of the meter and kilogram were deposited at the National Archives in Paris. That same year, in France, the law of 19th Frimaire of the year VII (December 10, 1799) made the metric system mandatory.

The use of the metric system spread slowly. On January 1, 1840, it was established as the legal system in France. In 1875, the Metric Convention, signed by a large number of countries on both sides of the Atlantic, created the International Bureau of Weights and Measures. In 1876, France gave the Bureau its headquarters at the Breteuil pavilion in Sevres, near Paris. Since that date, this domain has become an international enclave, enjoying the privileges of extraterritoriality.

Meter

In 1899, the meter was redefined by the International Bureau of Weights and Measures as follows: The meter is equal to the distance at 0°C between two lines on a platinum-iridium bar deposited in a vault at the International Bureau of Weights and Measures.

New Definition of the Meter (1983)

The metric system will be definitively and permanently adapted for scientific and technical progress. Today, the most precise measurements of length make use of laser interferometers, and the previous definition has become insufficiently precise.

A new definition was adopted on October 20, 1983: "The meter is the length of the path covered by light in a vacuum in 1/299,792,458's of a second.

International System of Units (1960)

The International System of Units was created in 1960. It is the result of an international investigation culminating in the Eleventh General Conference on Weights and Measures attended by representatives of the governments of a large number of countries. The system retains the main elements of the metric system established under the French Revolution. It defines seven base units from which the other units are derived:

- Length (meter)
- Mass (kilogram)
- Time (second)
- Electric current (ampere)

    Thermodynamic temperature (Kelvin degree, which is equal to the Celsius degree—but the Kelvin scale starts at absolute zero instead of 0° C)

- Quantity of matter (mole)
- Light intensity (candela).

    The United States initially accepted the use of the metric system parallel to its own units, then in 1893 decided to derive its units from metric standards. In 1975, a law known as the Metric Conversion Act was passed for the purpose of fostering the country's conversion to the metric system.

Standard unit of time

In 1967, the Thirteenth General Conference on Weights and Measures defined the legal second as follows: A second is the duration of 9,192,631,770 periods of radiation corresponding to the transition between two hyperfine levels of the fundamental state of the cesium atom.

The Bureau International de l'Heure (International Time Bureau), an international organization having its headquarters in Paris, established an international atomic time (AT) based on the mean of the indications in cesium seconds provided by atomic clocks dispersed throughout the world. The International Time Bureau publishes on a regular basis the deviations of each transmitter with respect to the AT, enabling all users to correct the hourly signals broadcast.

The second was originally defined as a division of the mean solar day and subsequently as a subdivision of the tropical year.

2. Statics—Mechanics

Statics (16th century)

The Flemish scientist **Simon Stevin** (1548—1620), known as Simon of Bruges, is considered as the founder of modern statics. Statics is a branch of science dealing with the equilibrium of material bodies. The brilliant precursor of Stevins was the Greek Archimedes of Syracuse (287—212 B.C.).

Centers of gravity (3rd century B.C.)

The greatest scientific mind of all antiquity, the Greek **Archimedes** of Syracuse (287—212 B.C.), was the first to determine the center of gravity of such geometrically defined homogeneous solids as cylinders, spheres and conoids. Archimedes developed this notion in his work *On the Equilibrium of Plenes,* also containing a rigorous theory of levers.

This inventive genius, who had to be reminded by his servants to eat and drink (absorbed in his thoughts, he would completely forget to nourish himself), knew how to put his theories into practice. During the siege of Syracuse, the Roman soldiers in Marcellus' command were given the opportunity to experience the quality and ingenuity of Archimedes' war machines: long and short-range catapults, mobile devices that hurled enormous missiles at approaching ships, etc. After three years, the Romans finally succeeded in taking Syracuse by surprise. Archimedes was stupidly killed by a soldier. Thus came the end of great man who is purported to have said, "Give me a place to stand on and I will move the earth."

Balance scales (5000 B.C.)

Seven thousand years ago, the Egyptians were already weighing their wheat using a two-pan balance suspended from a beam.

Roman balance (1000 B.C.)

This balance scale with two unequal arms was, in reality, invented by the Chinese. It is thought to date back to 1000 years before the Christian era and to have been brought to the West by nomadic horsemen shortly before the birth of Christ. The Roman balance is still used today under the name spring balance.

Roberval balance (1670)

In 1670, the French mathematician, physician, and mechanic **Gilles Personne de Roberval** (1602—1675) presented one of his inventions to the Academy of Science of Paris. It was a balance with two pans supported by a beam and connected to a counter-beam by two rigid guide bars.

Hydrostatics

Archimedes' principle (3rd century B.C.)

The first law of floating bodies was formulated by **Archimedes** (287—212 B.C.) and bears his name. It states that a body immersed in a fluid (liquid or gas) in equilibrium is subject to an upward force equal to the weight of the fluid displaced by the body and applied to the center of gravity of that part of the fluid displaced.

The circumstances of this discovery are a much debated question. People like to imagine Archimedes running naked through Syracuse crying "Eureka! Eureka!" ("I found it!") shortly after discovering the principle of floating bodies. In fact, things probably did not occur that way. The following story is more likely closer to what actually happened.

Hieron, tyrant of Syracuse and a naturally distrustful man, is supposed to have given some pure gold to a goldsmith one day, with which to make him a crown. After each manipulation, the story goes, he weighed the water that overflowed. It turned out that the finished crown weighed less than the pure gold and more than the silver, proving that it had been fraudulently mixed with silver. The goldsmith, who had stolen some of the pure gold, was obliged to make amends.

Hydrostatic paradox (1586)

The Flemish physicist and mathematician **Simon Stevin** (1548—1620), known as Simon of Bruges, inspector of dikes for the States of Holland and in that capacity directly interested in the internal forces of liquids, was the first to make a truly scientific study of these forces.

In 1586, Stevin's three books on mechanics were published. These contained his famous hydrostatic paradox: The pressure of a liquid at the bottom of a receptacle depends solely on the height of the liquid above the bottom and not on the shape of the receptacle. Conversely, the weight of the liquid therein contained depends on the shape of the receptacle.

Fundamental relationship (1651)

In 1663, the Frenchman **Blaise Pascal's** (1625—1662) *Treatise on the Equilibrium of Fluids* was published. In that work, written between 1651 and 1654, Pascal restated the paradox formulated by Simon Stevin and enunciated the fundamental relationship of hydrostatics: The difference in pressure between two points of a fluid in equilibrium is equal to the weight of a column of liquid (of which the section is taken as unit), the height of which is equal to the difference in level between the two points. At a later date, Pascal deduced the *principle of communicating vessels:* In a fluid at rest, all points on a horizontal plane are subject to the same pressure. In particular, all points on the free surface of the fluid are affected by atmospheric pressure, whatever the shape of the vessel and whatever the form of the fluid.

Fundamental principle

Taking the fundamental relationship as a starting point, Pascal also deduced his fundamental principle: Pressure applied to any one point of an incompressible fluid at rest is transmitted without loss to all other parts of the fluid.

Falling bodies (1602)

The Italian physicist and astronomer **Galileo Galilei** (1564—1642) demonstrated that solid bodies fall at a velocity independent of their mass, if the resistance of the air is discounted. In *Discourses and Mathematical Demonstrations Concerning Two New Sciences,* published in 1638, Galileo Galilei (1564—1642) demonstrated that solid bodies fall at a velocity independent of their mass, if the resistance of the air is discounted. In *De Motu,* published in 1602, he demonstrated that the velocity of the fall is proportional to its duration and that the distance covered is proportional to the time squared.

In *Discourses and Mathematical Demonstrations Concerning Two New Sciences,* published in 1638, Galileo formulated the principle of inertia. He also stated the laws concerning constant acceleration, which are the following:

• All bodies fall with a constant acceleration.

• The acceleration of a body moving on an inclined plane increases or decreases in proportion to that incline.

Around 1600, Simon Stevin built a wind chariot that could attain speeds of 75km/h.

• On a horizontal plane, a body moves at constant speed. Once at the edge of the plane, the body describes a parabola, whose trajectory results from the constant velicity and the constant acceleration of the fall.

Fundamental laws of mechanics (1687)

In 1687, the English physicist, mathematician, and astronomer **Sir Isaac Newton** (1642—1727), taking inspiration from Galileo's work, stated in his *Principia Mathematica* the three fundamental laws of mechanics that he had formulated several years before:

• A body remains in its state of rest, or uniform motion, except when acted upon by an external force.

• If a force acts on a body, the rate of change of the velocity (i.e., the acceleration) of that body is directly proportional to the force. In fact, the rate of change of momentum (momentum being the product of velocity and mass) is proportional to the force. Newton assumes that the mass is constant, which is why he establishes a relation between the force and the change of velocity. Consequently, if there is no force, the body's momentum is constant.

• Every action generates an equal and opposite reaction.

Principia Mathematica ends with the statement of the principle of universal gravitation (see Universal gravitation, p.251).

Fluid mechanics-- characteristic numbers

Classical equations can theoretically solve any problem in physics. In practice, however, this system is too complicated. Therefore, experimentation is often preferred, especially in fluid mechanics. For example, scientists use scale models to study the behavior of an automobile or airplane in a flow of turbulent, viscous fluid. The results obtained on these models are then transposed to life-size models known as prototypes.

In order to obtain an accurate transposition from scale model to prototype, the variables characterizing the phenomena studied must satisfy in both cases similar geometric, kinetic, and dynamic conditions. These conditions to the phenomena studied are equal.

Reynolds' number

The English physicist and engineer **Osborne Reynolds** (1842—1912) car-

In 1934, a California woman defied the laws of gravity.

ried out research in the field of hydrodynamics, studying notably the flow of viscous fluids. Reynolds' number is a dimensionless number expressing the ratio of the inertial forces to the viscous forces.

Froude number

The English physicist and engineer **William Froude** (1810—1879) was the first to study by experimental means the resistance of a fluid to motion. To carry out his experiments, he devised the first model study tank. The Froude number (F) is the ratio between the forces of inertia and the square root of the forces of gravity. If F is less than 1, the force is fluvial. If F is greater than 1, the flow is torrential.

The Austrian physicist and philosopher **Ernst Mach** (1838—1916) was the first to recognize the role of velocity in aerodynamic flows. The Mach number (M) is the ratio of the inertial forces to the square root of the forces of pressure. If M is greater than 1, the flow is supersonic (or greater than the speed of sound in air, which is 34 meters per second). If M is less than 1, the flow is subsonic.

Gravitation

For the philosopher-astronomers of antiquity, the stars, being of divine essence, described perfect and natural movements. Most of the explanations advanced were far from coherent: The universe was described as a vault, the Earth as flat, a sort of cake floating on an ocean at the limits of which was a staggering drop. The sun was said to follow the Earth in a barque.

The Pythagoreans (5th century B.C.)

In the 5th century B.C., the Pythagoreans (followers of the famous Greek philosopher and mathematician Pythagoras) believed, with the Greek Eudoxus of Cnidus, in a system of concentric spheres of which the rotational axes, each in a different plane, all passed through a common center: The Earth. In this conception of the universe, the Earth was the center, followed by the moon, Mercury, Venus, the sun, Mars, Jupiter, Saturn, and the celestial vault, in that order. This cosmogony was systematized by the Greek Claudius Ptolemy of Alexandria (ca. 85—165 A.D.), the last noted astronomer of antiquity.

A precursor of Copernicus (3rd century B.C.)

The Greek **Aristarcus of Samos** (ca. 310—230 B.C.) was the first to conceive a heliocentric representation of the universe, 17 centuries before Copernicus. Aristarcus believed that the Earth and the other planets revolved around the sun, and not the opposite, as advocates of the geocentric system claimed (for whom the Earth was the center of the universe). Furthermore, Aristarcus also remarked that the Earth rotated on its own axis.

Copernican system (1500)

The Polish doctor of law, canon, and ardent astronomer **Nicolaus Copernicus** (1473—1543) apppears to have developed as early as the beginning of the 16th century the cosmic system that was to make him famous: The Earth rotates on its own axis and, like the other planets, revolves around the sun.

This tranquil man was well acquainted with the Church. He undoubtedly could foresee the violent reaction that his theory would provoke among the theologians, who firmly believed that the Earth and hence man, "God's image," was the center of the universe; and so he was in no hurry to have his work published. Instead, he entrusted it to his friend and disciple Georg Rhaeticus (1514—1576), who did not send it to the printers until 1543. *De Revolutionibus Orbium Coelestium* was published just a few days before Copernicus's death on May 24, 1543.

Kepler's laws (1609)

The German astronomer **Johannes Kepler** (1571—1630) undertook a systematic study of planetary motions, in particular of the planet Mars. Kepler discovered three laws, which have been named in his honor. These laws had the enormous merit of providing calculations that could be empirically verified.

The first two laws appeared in *Astronomia Nova,* published in 1609, and the third in *Harmonices Mundi,* published in 1619.

• Every planet moves in an ellipse of which one focus is the sun.

• The radius vector from the sun to a planet sweeps out in equal areas in equal times.

• The squares of the periodic times that the various planets take to describe their respective orbits are proportional to the cubes of their mean distances from the sun.

These three laws described the relation between the various components of a trajectory, without, however, proposing an explanation or a cause for planetary motion.

Galileo's system (1632)

The Italian **Galileo Galilei** (1564—1642), a native of Pisa and ardent lover of science, constructed a telescope in 1609 which he turned to the heavens to observe the moon, sunspots, the ring around Saturn, the moons of Jupiter. This instrument also enabled him to discover that the planet Venus had crescent phases like the moon, as anticipated by Copernicus.

Mathematician to Cosimo II, Grand Duke of Tuscany, Galileo supported the Copernican view of the universe, which he substantiated with proofs. But he was investigated and condemned by the Inquisition. Thus, Galileo acted prudently until 1632, when his famous *Dialogue Concerning the Two Chief World Systems* was printed. Accessible to nonscholars, this work demolished Ptolemy's system and vigorously defended the ideas of Copernicus: Immobile, the sun occupies the center of the universe, while the Earth revolves around the sun. The protection enjoyed by Galileo and the success resulting from his book did not prevent the Inquisitors from intervening. In 1633, during a dramatic trial, they forced him to recant, before condemning him . . . to silence. "Eppur', si muove!" ("And yet it moves!") he is said to have exclaimed.

Universal gravitation (1687)

In 1687, the Englishman **Sir Isaac Newton** (1642—1727), one of the greatest scientific minds of all times, formulated in his work *Philosophiae Naturalis Principia Mathematica* the general law governing planetary motion. In expounding his theory of universal gravitation, Newton accomplished the unification of celestial mechanics and terrestial physics, which he is thought to have developed between 1665 and 1667.

Newton considered the elliptical motion of the planets to be caused by a force, the attraction of the sun, which modifies their rectilinear inertial movement. In 1684, he communicated to the Royal Society the totality of his results, contained in his treatise *De Motu.* The following year, he expressed the force of gravity by a new formula that was absolutely universal: Two bodies attract each other with a force that is inversely proportional to the square of the distance between them. The proportionality factor thus obtained is the gravitational constant.

Between 1670 and 1685, the Dutchman Christian Huygens (1629—1695) measured the acceleration due to gravity, and the French priest Jean Picard (1620—1682) measured the Earth's radius. These measurements allowed Newton to verify his hypothesis mathematically, leading him to the conclusion that gravity and astral attraction obey the same law. They are simply different aspects of the same phenomenon.

According to Voltaire's famous anecdote, Newton's discovery was inspired by the observation of an apple falling from a tree. Why doesn't the moon fall to Earth like the apple from that apple tree? This incident is supposed to have given Newton the idea of extending the concept of gravitation to the moon and sun: The Earth attracts the moon because of the moon's low specific weight, while the sun attracts the planets because of its high specific weight.

The quantum theory of gravitational fields (1905)

In 1905, the German physicist **Albert Einstein** (1879—1955), naturalized as

Galileo's system as depicted by an 18th century artist.

a Swiss citizen in 1940, published a fundamental treatise on the theory of limited relativity that modified the laws of Newtonian mechanics by posing the equivalence between mass and energy.

How can the Newtonian forces of gravitation act both at a distance and instantaneously? That problem was resolved by the Einsteinian model, in which gravitation is propagated by waves having a velocity equal to that of electro-magnetic waves, and in a finite, curved space-time continuum (pseudoriemannian).

In 1919, this concept was upheld in the *general theory of relativity,* a fundamental theory of gravitational fields in a four-dimensional (time being the fourth dimension) curved and finite universe.

3. Changes of state— thermodynamics

Gas (17th century)

The Flemish physician and chemist **Jan Baptist Van Helmont** (1577–1644), persecuted by the Church in much the same manner as his contemporary Galileo, was the first to recognize the existence of different gases such as carbon monoxide, carbon dioxide, and oxygen, which would not be identified until much later. The experiment performed by Van Helmont that led him to coin the word *gas,* proceeded as follows: Charcoal was burned in a closed container. Al-though the charcoal was reduced to ashes, Van Helmont observed that the weight of the container was the same as at the beginning of the experiment. Indeed, part of the charcoal had been transformed into an "invisible" (gas), which had not succeeded in escaping from the container.

The Greeks referred to that "huge and murky space that existed before anything came into being" as *khaos,* which Van Helmont phonetically transposed into the word *gas*.

Until the 17th century, knowledge of this state of matter was purely empirical. Air had been considered as a separate element since antiquity, the other elements being earth, water, and fire. Vacuums simply didn't exist, as "nature abhors a vacuum" (Aristotle).

Air (17th century)

In antiquity, air was thought to be one of the four natural elements, along with earth, water, and fire. The first person to assert that air is a mixture was the English physiologist and chemist **John Mayow** (1640–1679), a student of Boyle. Mayow contended that just one of the components of air maintained breathing: the *spiritus nitroaerus* (1668).

However, a century later nothing new was known on the subject, and the only theory that attempted an explanation of air's chemical interactions did not meet with general acceptance, which is not surprising. According to its author, Georg Ernst Stahl (1660–1734), air contained *phlogiston,* an invisible and mysterious substance that acted by some unknown process and, depending on the circumstances, could vary in weight, become weightless, or even, incredible as it might sound, have negative weight.

Oxygen and nitrogen (1777)

In 1772, **Antoine Laurent de Lavoisier** (1743–1794), the founder of modern chemistry, contested the phlogiston theory (*see* above) and conducted a series of fundamental experiments to disprove it. Lavoisier concluded that when a body is heated in the presence of air, part of that air is either given off or absorbed and the heated body undergoes some change (increase or decrease in weight).

In 1774, the English clergyman and chemist **Joseph Priestley** (1733–1804), a proponent of the phlogiston theory, prepared a *dephlogisticated* gas by heating mercuric oxide. He observed that this gas activated combustion. It appears, in fact, that even before Priestly—as early as 1771—a Swedish apothecary, Carl Wilhelm Scheele (1742–1786), had described dephlogisticated air, calling it "fire air."

In that same year, 1774, Lavoisier met Priestley. As a result of their discussions, Lavoisier concluded that air is a mixture of two types of substances: "vital air" (i.e., Priestley's dephlogisticated air), which is absorbed in combustion. It appears, in fact, that even before Priestly—as early as 1771—a Swedish apothecary, Carl Wilhelm Scheele (1742–1786),

In 1772, Lavoisier discovered oxygen. Here, in an illustration by his wife, Lavoisier is pictured in his laboratory.

had described dephlogisticated air, calling it "fire air." In that same year, 1774, Lavoisier met Priestley. As a result of their discussions, Lavoisier concluded that air is a mixture of two types of substances: "vital air" (i.e., Priestley's dephlogisticated air), which is absorbed in combustion and maintains breathing, and "nonvital air." In a paper written in 1777 but not published until 1782, Lavoisier gave the name *oxygen* to vital air literally "that which produces an acid"). According to Lavoisier, during combustion of a body, oxygen combines with the body to form an oxide. In no event does one witness the liberation of phlogiston during this chemical process. Lavoisier gave the name *nitrogen* to non-vital air (literally, "that which does not maintain life").

As early as 1772, the physicist—physician Daniel Rutherford (1748—1819) had, in his doctoral thesis entitled *De Aere Mephitico*, recorded the discovery of nitrogen.

Rare gas (1783)

In 1783, the English chemist **Henry Cavendish** (1731—1801) made the first relatively precise analysis of air. Cavendish found air to contain 20.8% oxygen and 79.2% nitrogen and reported the presence of a "bubble" representing about 1% of the gaseous volume analyzed. The bubble appeared to present remarkable inertness.

In 1894, two English researchers, Sir William Ramsay (1852—1916) and Lord John William Rayleigh (1842—1919), detected by spectroscopic analysis the presence of an inert gas in air. The two researchers decided to name their discovery argon, which in Greek, means lazy. In 1895, Ramsay and the Swedish chemist Per Theodor Cleve (1840—1905) identified the presence of helium in the mineral clevite. Helium had been observed in the atmosphere during the solar eclipse of August 18, 1868, by the French astronomer Jules Cesar Janssen (1824—1907). In 1898, Ramsay and the English chemist Morris William Travers (1872—1966) isolated the other rare gases contained in air: neon, krypton, and xenon. In 1900, the last inert gas, radon, was discovered in the radioactive waste of radium by the German Ernst Dorn.

Water (1781)

In 1781, the English chemist **Henry Cavendish** (1731—1810) got the idea of burning oxygen and hydrogen together. He measured the quantities of both gases carefully and observed

Bottom: Neon was discovered in 1898. The neon tube was invented in 1909.
Top: Preparation of a hydrogen balloon.

Torricelli at work trying to solve the problems of atmospheric pressure.

pump. An unwieldy device, it took at least three men to operate. A piston sucked out in the air of the vessel in which the vacuum was to be created. When this operation was completed, the vessel was hermetically sealed.

In 1654 at Magdeburg, von Guericke presented his vacuum pump to the nobility of the Holy Empire. He got the idea of placing together two hollow bronze hemispheres, each 50 centimeters in diameter. Creating a vacuum in the sphere thus formed, he then had two horses harnessed to each hemisphere. The horses pulled: Nothing happened. He had two more horses added: still nothing. To the general stupefaction, it took eight horse harnessed to each hemisphere to pull the hemispheres apart.

A man of science, von Guericke was also a showman. During his experiments he demonstrated that burning candles were snuffed out in a vacuum, animals quickly expired, and bells no longer rang. These "miracles," which made his fellow countrymen take him for a magician, were nothing more than new scientific results.

that they had been transformed into a quantity of water having a weight equal to the sum of the weights of the two gases. Water is thus composed of oxygen and hydrogen, more specifically, of one volume of oxygen and two volumes of hydrogen for every two volumes of water in the gaseous state. In nature, water is composed of a mixture of isotopes of oxygen and hydrogen. (Oxygen was discovered by the Englishman Joseph Priestley; the Italian Paracelsus (1493—1541) was the first to report the existence of hydrogen. In 1765, Cavendish first isolated hydrogen by reacting sulfuric acid with iron.)

Lavoisier repeated and perfected Cavendish's experiments. He decomposed water into its two constituent parts. He had the idea of vaporizing water and then of separating the water vapor into its two components, which he then recombined to form water. This series of experiments led him to formulate the famous law on the conservation of matter in a chemical reaction.

Atmospheric pressure (1643)

The Italian physicist and mathematician **Evangelista Torricelli** (1608—1647) was the first to measure air pressure. A student of Galileo, Torricelli came to Florence in 1640 to work as his old master's assistant. Three years later the officials in charge of the city's fountains asked him to solve

the following problem: Why can't a suction pump raise water more than 10 meters?

Torricelli understood that this phenomenon was due to atmospheric pressure and set out to prove it. In order to work on more readily measurable heights, he used mercury, which has a density 13 times greater than water. Filling a vessel with mercury, he then inverted over it a glass tube sealed at one end which also contained mercury. The level of mercury in the tube began descending and stabilized when it reached a height of about 76 centimeters from the exposed surface of the vessel. From this experiment Torricelli deduced that air exerted on that surface a pressure that was balanced by the hydrostatic pressure exerted in turn by the 76 centimeters of mercury in the tube.

Torricelli's tube was the first barometer. The term *barometer* was coined by the Frenchman Edme Mariotte (*see* Expansion of gases, below). In 1647, the famous experiments of the Frenchman Blaise Pascal (1623—1662) at the Puy—de—Dome confirmed and completed Torricelli's work. These experiments showed that atmospheric pressure decreases as altitude increases.

Vacuum (1654)

The German engineer **Otto von Guericke** (1602—1686), a native of Magdeburg, developed the first vacuum

Expansion of gases (1661)

In 1661, the Irish self-educated scientist **Robert Boyle** (1627—1691) demonstrated that the variation in the volume of a gas is inversely proportional to the variation in its pressure. He was interested in the experiments of Van Helmont (see Gas, p.242) and Pascal (see Atmospheric pressure, p. 244) and oriented his research to the compressibility of gases. Using a simple graduated J tube and mercury, Boyle proved that the volume of the air imprisoned in the tube can be reduced by half, by doubling the pressure exerted by the mercury.

Laws of Charles and Gay-Lussac (1798, 1804)

In 1798, **Jacques Charles** (1746—1823), a French physicist passionately interested in hot air balloons (he was the first to have the idea of filling them with hydrogen), stated the following law: The pressure of a gas whose volume remains constant varies as a function of temperature.

In 1804, the French physicist and chemist **Louis Joseph Gay—Lussac** (1778—1850) discovered that the volume of a gas at a given temperature and its volume at 0° C are in a relationship of the same type as the one discovered by Charles.

Ideal gas

The French physicist **Emile Clapeyron** (1799—1864) was the first to voice the notion of ideal gas. Ideal gas is a purely theoretical perfect fluid representing the ultimate bounds of a real gas whose temperature would tend toward absolute zero and whose pressure would become extremely high.

Gas liquefaction

Origins *(1818)*

In 1818, the Englishman **Michael Faraday** (1791—1867) discovered a means of liquefying gas (i.e., transforming it from the gaseous state to the liquid state) by increasing the pressure when cooling the gas. Son of a blacksmith, Faraday started out as a simple errand boy in a book shop and later became a bookbinder. An avid reader, he taught himself physics. To perfect his scientific knowledge, he enrolled in an evening course taught by Sir Humphrey Davy and became Davy's assistant.

In 1818, Faraday oriented his research in a direction that was entirely new for the day: studying the effects of pressure and cold on gases. Faraday successfuly liquefied hydrogen sulphide and sulphuric anhydride but did not succeed with oxygen, hydrogen, or nitrogen.

Liquid oxygen *(1877)*

In 1877, the Frenchman **Louis—Paul Cailletet** (1832—1913), ironmaster in Burgundy, invented a pump that could achieve and maintain pressures of up to a few hundred atmospheres. Cailletet succeeded in liquefying oxygen by causing a sudden expansion of the gas contained in a capillary tube in which he decreased the pressure from 300 to 1 atmosphere, thereby lowering the temperature to minus 118.9° C. A few days after this successful experiment, **Raoul-Pierre Pictet** (1846—1929), professor of physics at the University of Geneva, published the results of similar research. Using the same process, both men were able to liquefy nitrogen as well as other gases considered up to that time as being in a permanent state.

Liquid air *(1895)*

In 1895, the German inventor and industrialist **Karl von Linde** (1842—1934) succeeded in liquefying air by compression and expansion with intermediate cooling. He thus prepared a liquid oxygen that was almost pure.

In 1902, the French scientist **Georges Claude** (1870—1960) invented another process for liquefying air, by reducing the pressure and applying an outside force. From the liquid air,

Left: George Claude liquifies air in 1902. Below: Otto von Guericke's experiments in 1650 led to his development of the vacuum pump. Bottom: A liquid air delivery truck ca. 1924.

he isolated liquid oxygen, nitrogen, and argon by fractional distillation, thus developing the first industrial process for gas liquefaction.

Liquid hydrogen *(1889)*

In 1889, the Englishman **Sir James Dewar** (1842—1923) obtained boiling liquid hydrogen (critical point minus 240° C), under atmospheric pressure and at a temperature of minus 252.77° C, using the air-liquefaction process invented by Linde.

Liquid helium *(1908)*

In 1908, the Dutch physicist **Heike Kamerlingh Onnes** (1853—1926) liquefied helium in his famous cryogenic laboratory at Leyden. The critical point of helium is minus 268.9° C or 4° C below absolute zero, the temperature at which the molecules "freeze."

Although helium was the last gas to be liquefied, research has continued in this field. In 1971, when liquefying helium 3 (an isotope of ordinary helium, which is helium 4) at less than 2.7 millikelvins, it was discovered that helium 3 also exhibits the phenomenon of superfluidity (previously discovered for helium 4), that is, it loses virtually all its viscosity and behaves in an acrobatic manner.

Thermodynamics

In 1849, the Englishman **Sir William Thomson**, he future **Lord Kelvin**, first used the word *thermodynamics* to designate the study of the relations between thermal and mechanical phenomena, but it can be considered that this fundamental discipline was founded by the Frenchman **Sadi Carnot** (1796—1832). Thermodynamics represents mathematically the changes occurring in the parameters defining a system when there is a transformation of energy. These parameters are temperature, pressure, and volume.

Carnot principle (1824)

After studying the steam engine invented in 1703 by the Englishman Thomas Newcomen (1663—1729), **Sadi Carnot** stated in his only published work (*Reflections on the Motive Power of Fire*, 1824) that mechanical energy could be produced by simple transfer of heat.

Work-heat equivalent (1842)

Following Carnot, the German physicist and physician **Julius Robert von Mayer** (1814—1874), also studied gases and the motive power of heat. In 1842 Von Mayer published the results of his research in *Annalen der Chemie und Pharmacie*. He stated intuitively what was later to become known as the *principle of work—heat equivalence*: When a system returns to its initial state after having exchanged mechanical work and heat with the external environment, it furnishes work if it receives heat and furnishes heat if it receives work. The ratio between work and heat is constant.

The principle of work—heat equivalence was to be confirmed by the large number of precise measurements carried out from 1840 to 1845 by the Englishman James Prescott Jcule (1818—1889) in his famous paddle wheel experiments. In these experiments Joule turned a wheel placed inside an insulated container that was filled with a liquid, whose rise in temperature he measured; he analyzed the quantity of heat output produced during a given mechanical work input (cf. *Philosophical Magazine*, 1845).

Work-heat equivalence means heat and energy can be measured using the same units, that is, either the calorie (a measure of heat no longer favored by modern-day physicists) or the joule (another unit of energy; 1 calorie = 4.184 joules).

Three Laws of Thermodynamics

In 1852, **Sir William Thomson (Lord Kelvin)** stated the first two laws of thermodynamics based on the work of Carnot, von Mayer, and Joule.

● *First law* (1852). When a system

The ENGINE for Raising Water (with a power made) by Fire.

To arrive at his theory, Carnot studied Newcomen's engine.

passes from an initial state to a final state by exchange of heat and work with the external environment, the algebraic sum work + heat is constant, irrespective of the transformation involved.

● *Second law* (1852). Heat always passes from a hotter body to a colder body, in which it excites the motion of the molecules.

The *internal disorder* of a cold body increases. This notion was introduced in 1864 by the German physicist Rudolf Clausius (1822—1888), who named this disorder *entropy* in 1865.

● *Third law* (1906). The third law of thermodynamics was formulated in 1906 by the German physicist and chemist **Walter Nerst** (1864—1941). This law states that it is impossible to cool a system to absolute zero, the temperature at which the gas molecules are immobile and have zero energy.

Temperature

In 1848, **William Thomson (Lord Kelvin)** formulated the zero principle of dynamics. This principle allows one to define thermodynamic temperature and to establish an objective

method for measuring it. When two systems are in thermal equilibrium with a third, they are in thermal equilibrium with each other. This equilibrium is represented by the equality of their temperatures. If we assign a value to the temperature of a system in a given physical state, the other temperatures can be determined by what is known as *thermodynamic measurement*.

The General Conference on Weights and Measures of 1961 designated the Kelvin (K) as the legal unit of thermodynamic temperature. The Kelvin is defined as the degree on the thermodynamic scale of absolute temperatures in which the triple point of water is 273.16 K (equal to 0° C). At this temperature, the solid, liquid, and vapor phases of water are in equilibrium.

International Practical Temperature Scale (1968)

Thermodynamic measurements are only exact with an ideal gas (*see* Ideal gas, p. 246). This means that in practice some amount of distortion will always be found. In order to get around this difficulty, an international temperature scale was adopted as early as 1925 by the General Conference on Weights and Measures. Revised in 1948 and subsequently in 1968, this scale is a convenient reference tool enabling one to improve the accuracy of all temperatures used.

The eleven fixed points defined are:

—Freezing point of gold   1337.58 K

—Freezing point of silver   1235.08 K

—Freezing point of zinc   692.73 K

—Boiling point of water   373.15 K

—Triple point of water   273.16 K

—Boiling point of oxygen   90.188 K

—Triple point of oxygen   54.361 K

—Boiling point of neon   27.102 K

—Boiling point of hydrogen   20.28 K

—Equilibrium between the
 liquid and vapor phases
 of hydrogen   17.042 K

—Triple point of hydrogen   13.81 K

In practice, the temperature scale used is the Celsius scale, on which the unit of measurement, the Celsius degree (°C), equals the Kelvin. The equivalence to the international scale is expressed as the relation $0° \text{ C} = 273.15 \text{ K}$ (see Temperature, p. 256).

4. Electricity

Electrostatics

Origins

Electrostatics is a branch of physics that deals with phenomena of static electric charges in equilibrium.

The Greek Thales of Miletus (ca. 624—548 B.C.) was the first to observe that amber (elektron in Greek) attracts light bodies when rubbed. Aristotle himself used the electric charge from a skate fish to treat a bad case of gout.

The English physicist and physician **William Gilbert** (1544—1603) made the first scientific study of electrical attraction in his treatise *De Magnete*. Gilbert made a clear distinction between magnetic attraction and electrical attraction. He reported, among other things, that metals do not become electrically charged when rubbed.

Guericke's electrostatic machine (1650)

The German physicist and engineer **Otto von Guericke** (1602—1686) (*see also* Vacuum, p. 244) was the first to imagine a machine for producing electricity. The first electrostatic ma-

Top: Otto von Guericke experiments with the first machine to produce electricity, 1650. Above: An 18th century scientist studying natural electricity.

On July 2, 1729, Stephen Gray proved that electricity would travel from one place to another.

chine (1650) consisted of a sulfur sphere mounted on a spindle that was rotated with one hand and rubbed with the other, using a piece of woolen cloth. This machine, which could produce electricity on demand, gave scientists the opportunity to perform all sorts of experiments.

Guericke's electrostatic machine was later perfected, the sulfur sphere being first replaced by a glass one, then by a disk rotated by a crank. The disk machine, invented in 1768 by the English physicist Jesse Ramsden (1735—1800), was widely used for over a century.

Electrical conductivity (1729)

In 1729, the English physicist **Stephen Gray** (1670—1736), after observing that metals, which conduct electricity, can be electrically charged if they are insulated, discovered electrical conductivity. Conductivity is a specific property of particular substances known as conductors.

In 1733, the French physicist Francois du Fay (1698—1739) observed that a very thin gold leaf is attracted by an electrically charged glass rod, then repulsed by another glass rod, then attracted again by a resin rod. He deduced the existence of two types of electricity; one, specific to transparent bodies, known as *vitreous*, the other specific to resinous bodies and known as *resinous*. Little by little, the first type came to be called *positive* and the second type *negative*.

Leyden jar or electric condenser (1745)

The German **J. von Kleist** (1700—1748) and the Dutchman **Petrus Van Musschenbroeck** (1692—1761), working independently, discovered that the electricity produced by an electrostatic machine can be accumulated. In 1745, in his laboratory, Musschenbroeck and two assistants were working on an electrostatic machine. A brass wire connected the machine to a glass flask filled with water in whibh the wire was immersed. After having his assistants turn the machine and then removing the flask, Musschenbroeck touched the wire. To his great surprise, he received an intense electric shock. The brass wire and flask device had accumulated the electricity produced by the machine and discharged it all at once to the experimenter, who was grounded.

This was the first electrical condenser. It was termed the *Leyden jar*, after the town where its inventor was born. In the 18th century, it was extremely fashionable to go to physics laboratories to experience electrical shocks. In these labs, scientists indulged in simple but spectacular experiments, using Leyden jars.

Coulomb's Law (1785)

In 1785, the French engineer **Charles de Coulomb** (1736—1806) performed several series of precise measurements concerning electrical attrac-

tion and formulated the following law: The force of attraction between two electrically charged bodies is directly proportional to the two electrical charges and inversely proportional to the square of the distance between them. Known as Coulomb's Law, this law constitutes the quantitative basis for electrostatics, which until that time was only known qualitatively.

In 1767, the Englishman Joseph Priestly (1733—1804) hypothesized for the first time that an analogy exists between gravitational attraction (*see* Universal Gravitation, p. 241) and electrical attraction.

In 1772, the English physicist Henry Cavendish (1731—1810) determined the capacity of a conductor—the ratio existing between the quantity of electricity stored (the charge) and the degree of electrification of that conductor (electric potential). Cavendish established that two bodies of different shapes connected by a conducting wire have the same electrical potential without carrying the same charge. The capacity of a conductor is therefore dependent on the geometry of that conductor.

Electrodynamics

Electric currents (1800)

The first electric battery was invented in 1800 by the Italian **Alessandro Volta** (1745—1827) (*see* Volta's battery, p. 318). This led to the discovery that electricity is not necessarily static but can also by dynamic—that is, involve moving charges that give rise to an electric current. In addition, Volta's battery was used to measure electric potential (this was the voltameter), hence to make quantitative studies of electric currents.

Nature of currents

In a conducting metal, currents are produced by the displacement of free electrons. There are a large number of these electrons in oood conductors.

In an insulator, the electrons are firmly held by the nuclei of the atoms.

In a gas or liquid, a current can be produced by displacement of atoms or molecules that are electrically charged—that is, by atoms having one or more additional electrons. These atoms are negative ions (electrons being particles with a negative charge).

In a semiconductor (i.e., conducting crystal to which impurities have been added), there is movement of *holes*. The holes are positively charged because they are no longer filled by electrons (*see* Semiconductors, p. 000).

An external force (magnetic, chemical, etc.) is necessary in order to release and maintain the movement of electrons in a conductor. This force, known as the electromotive force, creates a difference of potential (or voltage) at both extremities of a circuit, causing the electrons to move from one extremity to the other.

Ohm's Law (1827)

In 1827, the German physicist **Georg Simon Ohm** (1789—1854) used a hydaulic analogy to formulate a precise definition of the quantity of electricity, the electromotive force, and the intensity of a current, thereby formulating the law that bears his name. Ohm likened electric current to a liquid flow and the electric potential created by an electromotive force to a difference in level.

Joule effect, fuse (1882)

When an electric current passes through a homogeneous conductor, heat is produced. Since 1882, this phenomenon has been known as the Joule effect, after the English physicist **James Prescott Joule** (1818—1889) who in 1841 formulated the law stating that the heat produced in a given time by passage of electricity through a conductor is jointly proportional to the square of the current and the resistance of the conductor.

The heat generated by the passage of the current turns the filament of incandescent lamps red (*see* Incandescent lamps, p. 170). Resistances,

which generate heat, are still used today in electric stoves, electric radiators, etc. Also based on the Joule effect is the fuse: a metallic wire with low melting temperature inserted in an electrical circuit. If the current exceeds a certain intensity, the wire melts and interrupts the circuit.

Glow discharge

Glow discharge is produced by the passage of an electric current through an insulating or nonconducting medium. In a tube filled with gas and fitted at both ends with electrodes, a luminous discharge will be produced if a certain voltage is applied. This voltage is known as *breakdown voltage*.

The electrons move in the direction of the anode (positive electrode) and the positively charged ions in the direction of the cathode (negative electrode). These movements are acclerated if the gas pressure is decreased. In order for the current to be self-maintained, the positive ions must have enough energy to pull the electrons away from the cathode. In addition, the electrons must be sufficiently accelerated to ionize the atoms of the gas with which they collide when moving inthe direction of the anode. When the gas/electron collision takes place, the atoms absorb energy, which they restore in the form of a luminous discharge.

Paschen's Law

Two physicists, one German, **Friedrich Paschen** (1865—1947), and the other English, **Sir John Townsend**

(1868—1957), demonstrated that the breakdown voltage is a function of the product of the gas pressure and the distance between the electrodes. This is the minimal voltage for a given value of the product in question.

Paschen discovered this law empirically, while Townsend expressed it scientifically.

Neon tube (1909)

In 1909, the French scientist **Georges Claude** (1870—1960) invented the neon tube. This is a tube containing neon at low pressure, in which the neon becomes luminous when a certain voltage is applied to the electrodes. The neon tube is one of the best-known applications of glow discharge.

Electromagnetism (1819)

In 1819, the Danish physicist Hans Christian Oersted (1777—1851) demonstrated that the passage of an electric current creates a magnetic field in the space surrounding the conductor. Oersted placed a magentized needle on a table and moved a conductor back and forth above it. When the current passed, the needle was deflected. When he reversed the direction of the current, the needle moved in the opposite direction, by the same amount. The closer he placed the conductor to the needle, the more the needle was deflected.

Oersted concluded that the passage of electric current created a magnetic field in the space surrounding the conductor. This phenomenon is known

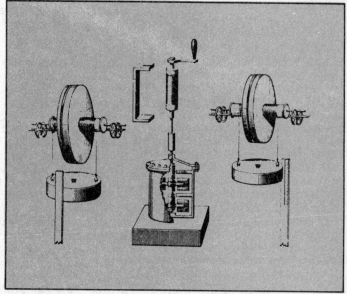

Left: Oersted discovering electro-magnetism. Above: The mechanism by Joule to measure the mechanic equivalent of the calorie.

as *induction*: The passage of a current *induces* a magnetic field. (*See* Induction, below.)

In 1820, the French scientist Francois Arago (1802—1892) constructed the first electromagnet. Around a magnetized soft iron bar, he wrapped a wire coil through which he passed an electric current. This electric coil had the properties of a magnet.

In the same year, 1820, the French physicist Andre-Marie Ampere (1775—1836) developed a theory of the magnetic effects of electric currents. Ampere studied the mutual interaction between currents and magnets. He then stated the rule known as the *Bonhomme d'Ampere* (Ampere's little man), that is, illustrating the direction of magnetic deflection by a hypothetical little man lying on a conducting wire. In 1827, Ampere exposed the results of his research in a paper entitled *On the Mathematical Theory of Electrodynamic Phenomena Deduced Entirely from Experiment.*

Induction (1831)

In 1831, while studying the results obtained by Oersted and by Ampere, the Englishman **Michael Faraday** (1791—1867) discovered the principle of electromagnetic induction. Faraday's discovery had enormous technological ramifications. Among other things, it served as the basis for electricity- producing machines known as generators: the *magneto*, as early as 1832, the *dynamo*, for producing direct current, the *alternator*, for alternating current (*see* these words in the Energy section).

Foucault currents

The self-educated French physicist Leon Foucault (1819—1868) was the first to demonstrate the existence of electric currents inducted by an alternating magnetic field in a massive conductor. Named in honor of their discoverer, these currents create a magentic induction whose flow opposes that of the alternating magnetic field that produced them. The mechanical force resulting from the passage of these currents is therefore always a resisting force.

Foucault currents are used in many applications, including induction heating and electromagnetic braking systems.

Electromagnetic waves (1864)

A student of Faraday (*see* Induction, above), James Clark Maxwell (1831—1879), one of the most illustrious physicists of the 19th century, was the first to suppose the existence of electromagnetic waves, in 1864. Although Maxwell was not able to prove his theory experimentally, he hypothesized that light has an electromagnetic nature.

Hertzian waves (1887)

In 1887, the German physicist **Heinrich Hertz** (1857—1894) demonstrated the existence and properties of electromagnetic waves for the first time. These waves, named in his honor, are a remarkable verification of the theory formulated by Maxwell (*see* Electromagnetic waves, above).

Hertz observed that electrical oscillations produced in a circuit gave rise to wave-type emissions. He determined their wave lengths (approximately 0.60 centimeters) and proved experimentally that these waves had most of the properties characteristic of light waves—reflection, interferences, and stationary waves. He also demonstrated that they confirmed the laws of optics. Hertz's discovery made it possible to transmit waves over long distances (*telegraphy*) without a conductor (*wireless*) and opened up a new age in communications.

Electronics

Semiconductors

A semiconductor is a special crystal having electrical properties in the range between an insulator and a conductor. In an *intrinsic* semiconductor, these properties exist in the pure state. An *extrinsic*, or *doped*, semiconductor acquires them by the addition of impurities such as antimony, arsenic, etc. A low-temperature insulator, the semiconductor becomes a conductor by the effect of heat or light.

Curious phenomena concerning electrical conductivity had already been observed, although not explained, by Faraday as early as 1839 in his experiment on silver sulphide, and later by Ferdinand Braun (1874, galena). However, the first coherent theory of conductivity in solids (covering the properties of insulators, conductors and semiconductors) is attributable to Felix Bloch (1905—), a Swiss-born physicist who became a naturalized United States citizen in 1939. In 1929 Bloch proposed the theory of bands. During the period 1925—1935 Bloch and a number of other scientists from various countries developed the theory of semiconductors.

In a crystal the bands of conduction and bands of valence are separated by a "forbidden band." In a semiconductor the energy gap between the bands of conduction (empty) and the bands of valence (filled with electrons) is small. Excitation by heat or light suffices to cause the valence band electrons to jump into the "forbidden band," producing a current. The spaces left vacant in the valence band behave like particles identical to electrons, only positively charged (electronic *holes*). In a semiconductor, electrical conductivity is produced by electrons or holes.

The understanding of the mechanism of semiconductors led to the development of transistors (*see* that word), which were to revolutionize electronics.

Cathode rays (1850)

Around 1850, the German **Heinrich Geissler** (1815—1879), who had been a glass blower before becoming a manufacturer of laboratory equipment, constructed glass containers equipped with electrodes, in which he created a vacuum. He thus obtained very attractive lighting effects that varied with the shape of the container and the type of gas used. This phenomenon was further observed by two Germans, Julius Plucker (1801—1868) in 1854, and Johann Hittorf (1824—1914) in 1869.

In 1879, the Englishman William Crookes (1832— 1912) carried out experiments proving once and for all that rays were emitted by cathodes of such glass tubes. Indeed, a small very sensitive wheel starts revolving when placed in the path of cathode rays.

In 1895, the Frenchman Jean Perrin (1870—1942) demonstrated that cathode rays are charged with negative electricity and that they are deflected by electric or magnetic fields.

In 1897, the Englishman Joseph Thomson (1846—1940) calculated the ratio between the charge and the mass of the particles emitted. It then became apparent that these particles were electrons. The existence of electrons was first hypothesized in 1874, by the Irishman George Stoney (1826—1911) who gave them that name in 1891.

Cathode-ray oscillograph (1897)

The cathode-ray oscillograph was developed in 1897 by the German physicist **Karl Ferdinand Braun** (1850—1918).

Cathode rays travel in a straight line, although in an electric field they undergo deflections that are proportional to the voltage applied. This property enables variable voltage to be represented by a curve.

Cathode tubes are also used to transmit images onto televison screens. The transmitter or the distribution network emits a signal which is received by the viewer's antenna. (*See* Video Camera, p. 118.)

Atomic physics

Atom (5th century B.C.)

The Greek philosophers of the 5th century B.C. were the first to suggest that all matter was made up of invisible particles, or atoms. Democritus of Abdera (ca. 460—370 B.C.), for example, is known for his materialistic theory of atoms (exposed by Aristotle) in which he probably restated, if not to say outright plagiarized, the idea of his predecessors and contemporaries: Pythagoras (6th century B.C.), Anaxagoras (ca. 500—428 B.C.), Leucippus (460—370 B.C.).

In the centuries that followed, numerous thinkers, including the Roman poet Lucretius (ca. 98—55 B.C.), restated the theory of atoms using a philosophical approach that was far from scientific.

It was not until the end of the 18th century that the atomic hypothesis recovered its scientific interpretation.

Dalton's atom (1808)

The Englishman **John Dalton** (1766—1844), a weaver's son, started out as a local schoolmaster at the early age of 12 before going on to become professor of mathematics and natural history. Although his name is generally associated with color-blindness (or Daltonism, *Daltonien* in French), of which he was himself a victim, Dalton is most famous for the atomic theory he developed around 1808, based on ideas of the ancient Greeks.

According to Dalton, all elements are composed of identical atoms of a determined mass, the atom representing the quantitative unit of matter involved in a chemical reaction. Equal volumes of gas contain the same number of atoms.

This concept may have had the merit of being simple, but it soon met with criticism. As early as 1810, the Frenchman Louis-Joseph Gay-Lussac (1778—1850), experimenting on chemical reactions in the gaseous state, established that the volume of a gas's component parts is in simple relationship to the volume of the gas. For example, one volume of gaseous

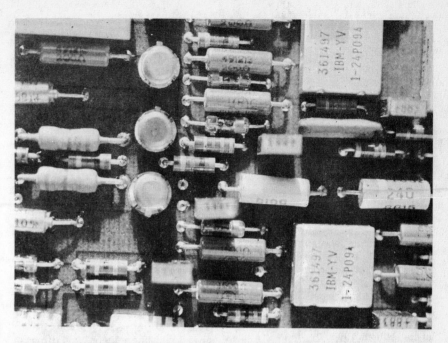

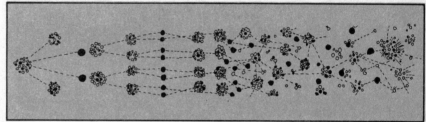

Top: Transistors and diodes on a logic card. Bottom: An illustration of a nuclear chain reaction.

oxygen combines with two volumes of hydrogen to form two volumes of water vapor, thus refuting Dalton's theory.

The explanation for this was provided by the Italian Amedeo di Guaregna e Ceretto—Count Avogadro—(1776—1856), who made a clear distinction between atoms and molecules. Avogadro intuitively supposed that different substances in the gaseous state exerting the same pressure and occupying the same volume at the same temperature, have the same number of molecules. (*See* Avogadro's Number, p. 252.)

Thomson's atom (1897)

Dalton's simplistic model had to be abandoned in 1897 when the Frenchman **Jean Perrin** (1870—1940, Nobel Prize of 1926) and the Englishman **Joseph Thomson** (1856—1940, Nobel Prize of 1906) discovered that the atom is not as indivisible as was previously supposed. It can, in fact, emit smaller negatively charged particles, first known as corpuscules, then electrons. (As early as 1874, the Irishman George Stoney had assumed the existence of electrons.)

Thomson imagined a static model of atomic representation that was very quickly abandoned, as it did not explain either the spontaneous emissions of radiation by certain bodies (see Radioactivity, p. 252), or the changes in direction that the emitted particles undergo as they pass through an obstacle (a thin metal sheet, for example).

Rutherford's atom (1911)

The explosion of a radioactive atom (see Radioactivity, p. 252) propels alpha particles with tremendous force. When a bundle of alpha rays passes through a thin sheet of metal, some of those particles undergo significant deflection. In 1911, to explain this phenomenon, which had been demonstrated by the experiments of the German Hans Geiger (1882—1945) and his research team, the English physicist **Lord Ernest Rutherford** of Nelson (1871—1937, Nobel Prize of 1908), used the hypothesis of the nuclear structure of the atom, proposed as early as 1901 by Jean Perrin: The entire mass and positive charge are concentrated in a small central nucleus, creating an intense field of attraction in which the electrons move like the Earth revolving around the sun.

Rutherford calculated that the positive particles are 1837 times heavier than the electrons, and named those particles *protons*. Thus, the neutral helium atom has a nucleus composed of two protons, around which two electrons revolve.

Bohr's Atom (1913)

An electron moving around a nucleus should theoretically radiate energy continuously and get closer and closer to the nucleus, with the frequency of that radiation increasing.

How then do you explain the presence of well-defined lines in the emission spectra of atoms that undergo an electrical charge?

The Danish physicist **Niels Bohr** (1885—1962, Nobel Prize of 1922), working from the quantum theory proposed in 1900 by Max Planck (1858—1947), suggested, in 1913, a theory founded on two postulates: 1. Any atom can only exist in certain stationary (nonradiating) states; 2. Electromagnetic radiation is emitted or absorbed when an atom passes from one stationary state to another. The frequency of the spectral lines is directly proportional to the difference in energy of the final and initial states. The stationary state corresponds to the circular orbits that the electrons describe around a nucleus.

As early as 1915, Arnold Sommerfeld, a German physicist and mathematician (1868—1951), applied the new quantum mechanics derived from quantum theory to the atom in order to explain the fine structure of the lines in the hydrogen spectrum. Sommerfeld replaced Bohr's circular orbits with elliptical ones.

Atom in wave-mechanics (1925)

For all its richness, the Bohr-Sommerfeld model, in which the positions of electrons are localized in orbits, does not allow one to develop a coherent system of atomic mechanics that takes into account all of the phenomena that must be explained by an atomic theory. It was not until 1925—6 that a satisfactory theory was developed, by two physicists working independently.

The Austrian, **Erwin Schrodinger** (1887—1961, Nobel Prize of 1933), applied the fundamental idea of the Frenchman Louis de Broglie (born in 1892) to the atom: A wave is associated to every particle.

The *uncertainty principle*, formulated by the German **Werner Hei-** senberg (1901—1976, Nobel Prize for 1932), combines Schrodinger's mathematical formalism with a satisfactory physical interpretation of the wave-particle duality: For Heisenberg, it is impossible to measure to any desired degree of accuracy both the momentum and position of a particle. There is always an uncertainty. Electrons do not describe well-defined orbits around a nucleus; all it is possible to calculate with respect to their position is the probability of their being at a given place at a given time.

Avogadro's number

In 1811, the Italian **Amedeo di Guaregna e Ceretto, Count Avogadro** (1776—1856), professor of physics at the University of Turin, established a law that was named after him (*see* Dalton's atom, p. 251). Avogadro assumed that in analogous conditions of temperature and pressure, equal volumes of gas contain the same number of molecules.

This concept may have had the merit of being simple, but it soon met with criticism. As early as 1810, the Frenchman Louis-Joseph Gay-Lussac (1778—1850), experimenting on chemical reactions in the gaseous state, established that the volume of a gas's component parts is in simple relationship to the volume of the gas. For example, one volume of gaseous oxygen combines with two volumes of hydrogen to form two volumes of water vapor, thus refuting Dalton's theory.

The explanation for this was provided by the Italian Amedeo di Guaregna e Ceretto—Count Avogadro—(1776—1856), who made a clear distinction between atoms and molecules. Avogadro intuitively supposed that different substances in the gaseous state exerting the same pressure and occupying the same volume at the same temperature, have the same number of molecules.

Periodic classification of the elements (1869)

The periodic table of the elements, a fundamental step in the development of chemistry as a modern science, was devised in 1869 by the Russian **Dimitri Mendeleyev** (1834—1907). This table established the relationships between the various chemical elements, considered up to that time as independent entities, thereby enabling us to understand why certain elements exhibited similar properties.

A large number of elements were discovered between 1750 and 1850. Many attempts were made to classify them by groups presenting analogous properties. Among these were the triads of the German chemist Dobereiner (1829), the helix of the Frenchman Chancourtois (1858), and the octaves of the Englishman Newlands (1869). The main interest of Mendeleyev's table is that it showed the periodic similarities and trends in the physical and chemical properties of the chemical elements when classified in order of increasing atomic weight (the atomic weight of an element equals the number of protons *and* the number of neutrons; also, in any given atom of an element there are as many protons as electrons).

X-rays (1895)

(*See* also Medicine section, p. 195)
X-rays were discovered in September 1895 at Wurzburg, Germany, by the German physicist **Wilhelm Konrad Rontgen** (1845—1923). Rontgen referred to these rays as "X" because their nature was as yet unknown. And it was only in 1912 that another German physicist, Max von Laue (1879—1960), succeeded in diffracting them with a crystalline blade.

X-rays are electromagnetic waves that pass through matter which is normally opaque to light. They result from the shock of cathode rays (composed of electrons) on the anode of a vacuum tube. These rays are of very short length (0.0001 micrometer, i.e., one ten-millionth of a millimeter). As a comparison, the wave lengths of visible light range from 0.4 micrometers (for violet) to 0.75 micrometers (for red).

The discovery of X-rays immediately aroused worldwide interest. Rontgen became a national hero before the turn of the century and received the Nobel Prize in physics in 1901.

Natural radioactivity (1896)

In nature, a certain number of heavy nuclei present natural radioactivity. The Frenchman **Henri Becquerel** (1852—1908) discovered this phenomenon in Paris, in 1896, while performing experiments on uranium.

In fact, it was the discovery of X-rays (*see* X-rays) that led to the discovery of radioactivity. Radioactivity is present in certain elements whose atomic nuclei are unstable. These elements spontaneously emit heavy particles: helium nuclei (alpha particles); light particles: electrons (beta rays) and photons (gamma rays).

Top: In 1897, customs officials used radiography to trap smugglers. Bottom: A radiation detector used to locate uranium beds, 1953.

nally, after two years of unrelenting and meticulous work, the Curies revealed the existence of not one but two elements that emitted this strange radiation: radium and polonium. The second one was named in honor of Marie Curie, nee Sklodowska, who was Polish by birth. For these discoveries Henri Becquerel shared the 1903 Nobel Prize in physics with Pierre and Marie Curie.

Sir Frederick Soddy, English physicist and chemist (1877—1956, Nobel Prize of 1921 in chemistry),explained the phenomenon of radioactive decay of atomic nuclei, thus paving the way for research in nuclear energy.

Artificial radioactivity (1934)

In 1934, in Paris, **Irene** (1897—1956) and **Frederic** (1900—1958) **Joliot-Curie,** daughter and son-in-law of Pierre and Marie Curie, obtained radioactive phosphorus by bombarding aluminum with alpha particles (helium nuclei).

Certain elements such as radium, neptunium, and actinium exhibit natural radioactivity. In addition, nuclear ractions can be used to obtain radioactive nuclei that do not exist in nature, by inducing the disintegration of atomic nuclei. The elements which have undergone this process are said to have artificial radioactivity.

This discovery, for which the Joliot-Curies won the 1935 Nobel Prize in chemistry, enables manufacturers to produce the isotopes used in medicine, biology, and metallurgy.

Synchrotron Radiation

On October 18, 1984, the installation of the European Laboratory of Synchrotron Radiation in Grenoble was decided upon. This ring, with a circumference of 772 m, and which cost $13 billion will provide an X-ray source is unequalled in the world.

Synchrotron radiation was so named because it was observed for the first time in the Berkeley (California) synchrotron. When electrically charged particles penetrate the curved force lines of a magnetic field at a speed approaching that of light, they emit electromagnetic rays (X-rays, in particular) tangent to their path. Moreover, this only reproduces artificially what occurs naturally in space.

In 1955, the discovery of intense electromagnetic radiation near Jupiter allowed astrophysicists to deduce that an intense magnetic field existed around this planet; this was subsequently verified.

Synchroton radiation will be used in solid physics, chemistry, biology, medicine and industry.

One cloudy day in Paris, Becquerel set up an experiment designed to verify whether a sample of pitchblend exposed to sunlight emitted X-rays. Unable to complete his work because of the weather conditions, Becquerel put his equipment away. He resumed the experiment another day when the weather was more favorable, placing the samples of pitchblend, uranium ore, and potassium on a photographic plate that had not been removed from its wrapping. When later developing the plate, he was surprised to see an image appear whose contours followed the outline of the ore sample perfectly. What could possibly be the origin of the "energy" in the ore that was capable of imprinting a photographic plate?

Becquerel took the matter up with his friends Pierre Curie (1859—1906) and his wife, Marie (1867—1934). Examining the pitchblend more closely, Pierre and Marie Curie discovered that the radiation had been caused by at least one substance that was much more radioactive than uranium. Fi-

Carbon-14 dating (radiocarbon dating) (1947)

In 1947, **William Frank Libby**, an American chemist specializing in the radioactivity of living organisms and recipient of the 1960 Nobel Prize in chemistry, explained the formation of carbon 14 in the atmosphere.

Carbon 14 is an isotope of common carbon, carbon 12. It has two more neutrons than carbon 12.

The dating of archeological specimens with carbon 14 is based on the measurement of the residual activity of isotope 14 of carbon. Living organ-isms generally absorb carbon dioxide a few years after they have come into contact with the atmosphere. At that time, they contain almost all of the carbon 14 produced by the action of cosmic rays on carbon 12. Once the organism dies, the activity of the in-corporated carbon begins to decrease (it is transformed back into carbon 12) according to the half-life period of carbon 14. The period generally as-sumed for carbon 14 is 5,730 years. Measuring the proportions of carbon 12 and carbon 14 therefore provides a reasonably accurate dating.

A number of other radioactive iso-topes contained in specimens from various occurrences are also used for dating purposes, following the same principle (i.e., measuring the residual activity of a radioactive isotope whose half-life is known). One of the most commonly used types is argon-potassium.

For example, this method was em-ployed on coal specimens found in the Lascaux caves, allowing us to deter-mine that these caves were inhabited as early as 13,000 years before the Christian era.

CHEMISTRY

1. Mineral Chemistry

Earthenware (antiquity)

In antiquity, the Babylonians were the first to make earthenware: The brick friezes on their palaces were overlaid with a tin-bearing glaze. Earthenware is clay pottery coated with a tin-oxide base glaze.

In the Middle Ages, the Arabs redis-covered the ancient techniques for making earthenware, and transmit-ted them to Spain, North Africa, and then Sicily. Toward 1400, this clay pottery came to be known as *Faenza*, after the name of one of the principle Italian towns where it was produced.

The celebrated French ceramist Bernard Palissy (1510—1584) is said to have burned all of his furniture in order to continue the experiments that enabled him to discover the se-crets of the Italian craftsmen.

Porcelain (6th century A.D..)

In China, the first porcelain manufac-tories known with certainty date back to the 6th century A.D.

The paste body is composed of a malleable clay (kaolin), a lubrifier (quartz or flint), and a flux (feldspar, talc, lime phosphate), which is a vitri-fying agent.

Feldspar was initially used as an external glaze. However, under the T'ang dynasty, it was discovered that this material could be incorporated into the clay before molding. After firing, the paste body became vitri-fied, hence nonporous.

In 1698, the first European kaolin deposit was discovered in Saxony by Baron Schnorr, a German industrial-ist. During the same period, the Saxon von Schirnahaus (1651—1708) and the German Johan Friedrich Bottger (1682—1719), discovered the secret process for making true porcelain. The same process was undoubtedly also discovered by other chemists, in-cluding the Milanese canon Manfredo, the Englishmen Dwight and Francis Place, and the Dutchman Aelgret de Keiser, to mention but some of the best-known names.

In 1708, Augustus II the Strong, Elector of Saxony and King of Poland, founded the first European porcelain factory at Meissen. From Meissen, the industry spread to Vienna in 1718, then to Hohst, near Frankfurt, in 1750, and finally throughout Europe.

In 1725, the Prince of Conde creat-ed the first French porcelain factory at Chantilly. In 1748, two workers from the Chantilly plant founded the Vincennes porcelain factory, which was moved to Sevres in 1756. Four years later, this establishment be-came France's Royal Porcelain Works.

The discovery, in 1752, of the Saint-Yriex-la-Perche kaolin deposit in cen-tral France made possible the devel-opment of the famous Limoges porcelain factories, founded a few years later.

Metallurgy (10th millennium B.C.)

What metallurgical craftsman made the first piece of gold jewelry, the first copper platter, the first iron hook? No one will ever know. But we do know that as early as the 10th mil-lenium before the Christian era, our prehistoric ancestors were already familiar with metals and employed them even before mastering the use of fire. In the rivers they found gold nuggets and native copper, which they hammered into shape with stones. It was probably by accident that they discovered that these "stones" melted when subjected to heat. As early as the 5th millenium B.C., from China to the Caucasus, men were using heat to fashion jewelry and tools.

Bronze Age (3500 B.C.)

The Bronze Age probably began with another accidental discovery. Copper was melted with tin and, miraculous-ly, a metal infinitely more useful than copper was obtained. It was rigid. Molded by potters, this alloy was used to make agricultural tools, weapons, and statues.

Iron Age

The Egyptians knew how to extract iron ore as early as 3500 B.C. By burn-ing charcoal in rudimentary furnaces and producing a good draught, they succeeded in melting the ore, while at the same time producing the first "steel" without knowing it or intend-ing to do so. (*See* Etruscan furnace, below). Indeed, charcoal diffuses throughout the metallic mass.

Fluxes

Craftsmen discovered the advantage of using fluxes in metal production empirically through trial and error. By adding sand or clay during the melting operation, a better separation of the gangue (i.e., the useless part of the ore) and the metal is obtained.

Tempering

The discovery of this technique, which improves the quality of carburized iron by quenching it in water, was probably accidental. The Odyssey relates that tempering was commonly used by Middle Eastern craftsmen of the time (9th century B.C.).

Etruscan furnace

A mysterious people for archeologists and historians, the Etruscans left splendid vestiges of their metallurgical techniques, especially their furnaces. Such furnaces were common in the ancient world and are still found today among certain African tribes.

Above a hearth about two meters of alternating layers of iron ore and charcoal were stacked, the entire arrangement being covered with a clay-like mud. An opening was made in the top of the roof for gas discharge and holes were perforated in the lower part to improve the air flow. Once the melting operation was completed, the furnace was destroyed. The molten metal recovered from the hearth was then worked by the craftsmen.

Catalan furnace

This furnace, of which several examples have been found in the Pyrenees region of southwest France near Catalonia, from which its name is derived, was widely used in antiquity. It presents two main differences with respect to the Etruscan furnace: First, the hearth is buried in the ground; and second, the furnace is of the forced draught type, this is, the air required for combustion of the charcoal-ore mixture is provided by rudimentary bellows.

Blast furnace

The antecedents of the blast furnace date back to the 13th century. In the 14th century shaft furnaces were used in which the air blast was produced by a hydraulic wheel. The inventor of the modern blast furnace is **Abraham Darby** (1711—1763), an English ironmaster from Coalbrookdale.

In modern blast furnaces, pig iron (an alloy of iron and carbon, at more than 2.5%) can be obtained directly from ore.

Top: Greek pottery. Bottom: The methods of modern day potters bind them to those of antiquity.

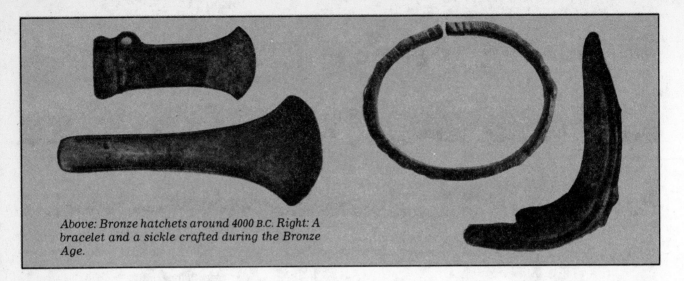

Above: Bronze hatchets around 4000 B.C. Right: A bracelet and a sickle crafted during the Bronze Age.

The second basic improvement after Darby's is the work of Wilhelm Siemens, a German immigrant to England, who in 1857 got the idea of recovering the calories from the combustion gases to heat the air blast.

Coke (1735)

In 1735, the English metallurgist Abraham Darby used coke as fuel in a blast furnace for the first time. Coke is a combustion agent produced by the carbonization or destructive distillation of coal (i.e. heating in a closed container at a temperature between 900° C and 1150° C).

Steel (antiquity)

We have seen that as far back as antiquity, certain craftsmen (the Hittites, for example) were already using iron containing small amounts of carbon to make tools and weapons. In England, starting in the eighteenth century, some steel was produced by ironmasters after Darby.

But it was Bessemer's discovery (*see* below) of a direct conversion process that laid the foundation for the true industrial production of steel.

Bessemer converter (1854)

Son of a typeset manufacturer, the Englishman **Henry Bessemer** (1813–1898) became interested in mechanics and metallurgy at an early age. In 1854, Bessemer designed a converter that was capable of producing steel of acceptable quality from pig iron, by removing the carbon. The steel industry owes its development to Bessemer's discovery, which was made in his own Sheffield cutlery works.

Martin (open hearth) furnace (1865)

A rival of the Bessemer converter from the time of its development, the open hearth, or Martin, furnace was invented in 1865 by the Frenchman **Pierre Martin** (1824–1915), son of an ironmaster. Martin's ingenious idea, the use of recycled steel scrap, improved the efficiency of blast furnaces and enabled them to produce a more refined steel.

Ceramics (7000 B.C.)

Early ceramics were found as far back as the 7th millenium B.C. in Iranian Kurdistan; the 5th millenium B.C. in Anatolia and also in the Aegean Sea area, 4500 B.C. in Uruk (today Waruk in Iraq), the end of the 3rd millcnium B.C. in Central Anatolia, and the beginning of the 2nd millenium B.C., again in the Aegean Sea area.

The pottery of antiquity had one very important defect: It was porous. That is why for centuries attempts were made to develop a water-tight paste body (*see* Porcelain, p. 254), as well as an external glaze.

Chlorine (1774)

In 1774, the Swedish apothecary **Carl Wilhelm Scheele** (1742–1786) discovered chlorine by a reaction of hydrochloric acid on manganese dioxide. Starting from saltpeter, Scheele had thus discovered tartaric acid, oxalic acid, lactic acid, and glycerine.

In 1810, the English chemist Sir Humphrey Davy (1778–1829) identified chlorine as a new element.

A component of all living organisms, chlorine does not exist in nature in the free state. It is found exclusively in the form of chlorides (i.e., salts of hydrochloric acid).

In recent years, the chlorine industry has experienced considerable growth, due to the increasing development of plastics (see Polyvinylchloride, p. 259).

Javel water (1789)

In 1789, the French chemist **Claude-Louis Berthollet** (1748–1822) discovered the bleaching properties of hypochlorites, which he called Javel water after the name of the locality near Paris, on the banks of the Seine, where women went to do their washing.

Hydrogen peroxide (1818)

In 1818, the French baron and chemist **Louis-Jacques Thenard** (1777–1857), prepared the first hydrogen reacting sulfuric acid with barium peroxide at room temperature.

Aluminum (1825)

In 1822, the French mineralogist Pierre Berthier (1782–1861) discovered near the village of Les Baux, France, the first deposit of an ore which he named bauxite. Bauxite is, in fact, aluminum hydroxide.

Aluminum was first obtained in powder form in 1825 by the Danish scientist **Hans Christian Oerstad** (1777–1851), and in ingot form, in 1827, by the German scientist Friedrich Wohler (1800–1882).

For more than twenty years aluminum remained a curiosity. Finally, Napoleon III granted à large subsidy to the French chemist Sainte-Claire Deville (1818–1881) to develop a method for the industrial production of this metal, believing that because of its light weight, aluminum could revolutionize arms manufacture. In 1854, Sainte-Claire Deville obtained the first pure aluminum metal by reduction of aluminum chloride.

The following year, aluminum occupied a place of honor at the Paris Exhibition. Still considered as a precious metal, it was used to manufacture luxury items such as Napoleon

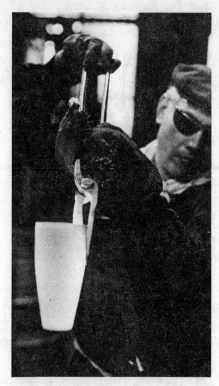

Top: A goblet full of molten gold. Bottom: A catalan furnace around 1400.

III's famous armor and an equally famous table service.

In 1886, the American Charles Martin Hall (1863—1914) and the Frenchman Paul Heroult (1863—1914), working independently, discovered an electrolytic process for producing aluminum from the alumina obtained from bauxite. This process began the phenomenal development of aluminum, which as a result became an inexpensive commodity. In 1887, world production was 26 metric tons. By 1980, it had risen to 14 million metric tons.

Heavy Water (1932)

In 1932, the American chemist **Harold Clayton Urey** discovered deuterium and heavy water. He obtained the Nobel Prize two years later for this discovery.

In 1933, using Urey's method of preparation, the American physicist and chemist Gilbert Newton Lewis successfully prepared a few milliliters of almost pure heavy water by fractional distillation of ordinary water.

The term heavy water, a chemical compound similar to water, is used to describe deuterium oxide (the heavy isotope of hydrogen). This oxide is contained in all waters of the globe in virtually constant porportions, that is, one molecule of heavy water for approximately 1000 molecules of water.

2. Organic chemistry

Organic chemistry is the "chemistry of carbon." On Earth, carbon is of organic origin. Plants produce it by photosynthesis. Carbon and petroleum (hydrocarbon) result from the fossilization of living organisms.

Bakelite (1909)

In 1909, **Hendrick Baekeland** (1863—1944), a Belgian chemist residing in the United States, invented the first thermostatting phenoplastic resin: bakelite.

Rubber

Natural Rubber (1736)

In 1736, **Charles-Marie de la Condamine** (1701—1774), a French naturalist on mission to Peru, sent back to the Academy of Science of Paris the first sample of a blackish resinous product with which the Indians made various everyday objects. In 1751, with his mission completed, La Condamine transmitted his paper *On an Elastic Resin Newly Discovered in Cayenne: Cahuchin (crying wood* in the native language) to the Academy of Science.

Natural rubber, a substance characterized by its elasticity, is obtained by coagulatinq latex. Latex is the milky liquid collected mainly from the *Hevea brasiliensis*, a tree of South American origin (in the family Euphorbiaceae).

Vulcanization of rubber (1839)

The vulcanization of rubber was discovered accidentally in 1839 by the American **Charles Goodyear**.

Natural rubber, which is elastic and viscous at room temperature, becomes stiff under the effects of cold. In addition, it softens and turns black with age. This is why rubber in its natural state has very limited use.

In 1819, the Englishman T. Hancock reported that rubber was amenable to mastication. Gum, when masticated for a long time, gives a malleable paste to which various powdered substances can be added.

In 1839, Charles Goodyear was testing the drying agents of rubber. For one reason or another he left a piece of gum stabilized with sulfur on a hot stove. When he came back, he observed that the rubber appeared to be stabilized. He had just invented vulcanization, which stabilizes the elastic properties of rubber, enabling it to be used industrially. The Englishman T. Hancock made the same discovery very soon afterwards in London.

Synthetic Rubber (1880)

In 1860, the Englishman Greville Williams decomposed rubber by heat, thus isolating a substance called *isoprene*. The chemical name of this substance is dimethyl-butadiene. Butadiene is a hydrocarbon. Twenty years later, the Frenchman **G. Bouchardat** prepared a solid substance from isoprene. This was the first synthetic rubber, although the initial isoprene had been extracted from natural rubber.

In 1884, Tilden synthesized isoprene by decomposing turpentine vapors. This discovery laid the foundation for the production of synthetic rubber, for which researchers in several countries set out to develop an industrial process.

Buna (1926)

In 1910, patents for the industrial production of synthetic rubber were filed, first in England and then in Germany. In 1926, **IG Farben Industrie** succeeded in making *Buna* by placing the polymer of butadiene in the presence of sodium. The word *Buna* is derived from the first two letters of the words butadiene and natrium (in chemistry, sodium).

These elastomers show better resistance to wear than natural rubber.

GR-S (1942)

In 1942, the United States government began industrial production of GR-S (or government rubber + styrene), as the Asian rubber plantations were under Japanese control. GR-S are synthetic rubbers made by polymerizing butadiene and styrene. They bear a strong resemblance to natural rubber.

Collecting natural rubber in Ceylon.

Neoprene (1931)

As early as 1931, the American company **Dupont de Nemours** began marketing, under the name of *duprene*, products obtained by the polymerization of chloroprene (a derivative of butadiene). These products, of which neoprene is the best-known, have excellent heat and chemical resistance.

Cellulose

Cellulose is the chief component of the cell walls of plants. Found in an early pure state in cotton, it is composed of long molecules. Cellulose is used in the manufacture of three main categories of products: paper (*see* Paper, p. 83), explosives (*see* Explosives, p. 84), and plastics. This last category does not employ pure cellulose, but some of its compounds: nitrate, xanthate, and cellulose acetate.

Cellulose nitrate (1833)

Obtained as early as 1833 by two French pharmacist-chemists, **Theophile-Jules Pelouze** and **Henri Braconnot**, cellulose nitrate (or nitrocellulose) was not made industrially until 1847, the year in which the German Christian Friedrich Schonbein, first developed cotton powder and collodion.

Cotton powder is a nitrated cotton having explosive properties (*see* Explosives, p. 44). Collodion is a solution of cotton powder in a mixture of alcohol and ether. It is used as a coating for photographic films.

Nitrocellulose was also used to make the first artificial silk. In addition, it was employed in the manufacture of the first plastics (*see* Celluloid, below), but in this area, as in that of synthetic fibers, it is being replaced more and more by less flammable substances.

Cellulose acetate, rayon acetate (1869)

In 1869, **Paul Schutzenberger**, physician and assistant director of the chemistry laboratory at the Sorbonne in Paris, obtained an acetyl cellulose (cellulose acetate) by the action of acetic anhydride on cellulose. In 1884, the Englishman Charles F. Cross and Edward J. Bevan took out patents for the industrial production of cellulose acetate. In 1921, Bevan patented a process for making rayon acetate. This rayon is known in France under the trade names Rhodia and Seraceta.

Cellulose acetate is also used in the manufacture of plastics, varnish, and photographic films.

Celluloid (1870)

Celluloid was invented in 1870 by two Americans, the **Hyatt brothers.**

In 1863, two American industrialists offered a prize of $10,000 to whoever could develop a substitute for ivory to use in the manufacture of billiard balls. Hyatt, who owned a small printing business in New York state, set out to work on the problem with his brother. After seven years of research, they obtained celluloid by hot mixing of cellulose nitrate, a macro-molecular vegetable substance, and camphor, a plasticizer.

Cellulose xanthate, viscose rayon (1891)

In 1891, the Englishman **Charles F. Cross** filed the first patent for the industrial production of cellulose xanthate. This product is used in the manufacture of rayon.

Cellophane (1908)

On November 14, 1908, at Thaon-les-Vosges, France, **Dr. J. E. Branderberger**, a specialist in the dying and printing of textiles, used viscose to treat cotton. A fine film stuck to the cotton, and cellophane was invented. This cellulosic film is transparent and gives a shiny appearance to fabrics. In 1912, Branderberger deposited the trademark *Cellophane* for the cleaners and dyers of Thaon, at the International Trademark Office at Berne, Switzerland.

Initially, the clothing industry was the only market for cellophane. Today, however, this product is commonly used for the packaging and conditioning of a wide range of consumer goods.

Dyes

Prussian blue (1704—1710)

Prussian blue, a blue mineral pigment, was discovered between 1704 and 1710, by **Diesbach**, a pigment manufacturer in the Prussian city of Berlin, hence the name *Prussian blue*. The process for making this ferrocynide pigment was developed by the English chemist Woodward.

In 1749, the French chemist Macquer demonstrated that Prussian blue could be used for dying and printing fabrics. The French manufacturer Jean-Michel Haussman was the first to employ this product industrially.

Mauveine (1856)

Mauveine, the first synthetic dye, was invented in 1856 by the Englishman **William Henry Perkin**. Industrial production of mauveine began in 1857. In 1859, *aniline red* was developed, followed in 1863 by *aniline black*.

Alizarine (1869)

In 1869, just twenty-four hours before the Englishman Perkin obtained his *aniline red*, the German chemist **Charles Graebe** filed a patent for the production of a synthetic dye that was also red: *synthetic alizarine*. This product is derived from anthracene, itself extracted from coal. For this discovery, Graebe had been assisted by Liebermann and Henrich Caro.

Lingerie is often woven from synthetic fibers.

Indigotin (1880)

Indigotin was first synthesized in 1880 (after mauveine and alizarine) by the German **Adolf von Baeyer**. Up to that time, natural indigo, a dye giving a deep blue color, had been extracted from the leaves of indigo plants. As indigo was a very good dye for cloth, its synthesis represented an important breakthrough for the clothing industry. Nevertheless, an industrial process for making indigotin was not developed until 1895, the work of Heumann.

Polymers

Polyvinylchloride (or PVC) (1913)

The first process for the manufacture of polyvinyl-chloride was patented in 1913 by the German chemist **Professor Klatte**. Between the two World wars, chemists began discovering the tremendous possibilities offered by this substance. Because of its chemical inertness and, above all, its electrical resistance and compatibility with all sorts of plasticizers and additives, PVC can be used successfully in an extremely broad range of applications.

As early as 1931, IG Farben developed the fiber P.C., using the method described by Klatte. In France, Jacques Corbiere launched the fiber Rhovyl at Rhodiaceta in 1941.

Polystyrene (1929)

Styrene was first prepared in 1831. At that time it was already known that when heated or exposed to light, styrene is transformed into a vitreous solid. However, it was not until 1929, nearly a century later, that the German chemist **Hermann Staufinger** (1881—1965) took a closer look at the polymerizing process. Using a high temperature pressure-fed method, Staufinger prepared a series of sty-

rene compounds ranging from dimer (a compound formed from two molecules of styrene) to enormous macromolecules. The properties of these compounds vary depending on their molecular weight.

Polystyrene (a compound, as its name indicates, of a large number of chained molecules) was first marketed in 1938. In foam form, polystyrene is a very light substance used chiefly in the manufacture of packaging and building materials.

Polyamides (1930)

Polyamides were discovered in 1930 in the experimental laboratories of DuPont de Nemours, by the American, **Wallace Hume Carothers**.

• *Nylon* (1938). Nylon was first produced industrially in 1938 by the American company **DuPont de Nemours**. Nylon is a polyamide obtained from adipic acid and hexamethylene. At the end of World War II, the use of

nylon, which became associated with nylon stockings, experienced spectacular growth.

• *Perlon* (1938). In the same year, 1938, the German **P. Schalk** synthesized perlon, another important polyamide. This fiber was not manufactured industrially until after the war.

• *Rilsan* (1951). Rilsan has been produced industrially in France since 1951 by **Organico**, a subsidiary of the large French chemical company Pechiney-Ugine-Kuhlman. Rilsan is a polyamide obtained from ricin. It is widely used by the packaging material and container industry.

Polyethylene (1933)

In 1928, the English company **Imperial Chemical Industries** (ICI) began research aimed at obtaining an ethylene polymer under high pressure (1000 to 2000 bars). In 1933, ICI succeeded in synthesizing a macromolecular compound, and three years later started

Nylon was invented in 1938. Very soon thereafter it was used in the manufacture of stockings.

operation of a continuous machine for producing LDPE (low density polyethylene). These products received immediate recognition as the best materials for cable insulation.

In 1954, the German Karl Ziegler (1896—1973) developed a process for the manufacture of an HDPE (higher density polyethylene). Ziegler's catalysts won him the 1963 Nobel Prize which he shared with the Italian Giulio Natta (1903—1979).

Polytetrafluorethylene (PTFE)—Teflon (1938)

In 1938, **Roy J. Plunkett**, an engineer at the American company, DuPont de Nemours, discovered the polymer of tetrafluorethylene by accident. This fluorocarbon is marketed under the name of Fluon in Britain, and Teflon in France and the United States.

PTFE was patented in 1939 and first exploited commercially in 1954.

PTFE is resistant to all acids and characterized by its exceptional stability and excellent electrical insulat-

ing properties. That is why it is used in making piping for corrosive materials, in insulating devices for radio transmitters, and in pump gaskets. In addition, its nonstick properties make PTFE an ideal material for surface coatings (see Teflon pan, p. 154).

Polyacrylonitrile (1941)

In 1941, chemists from the American company **DuPont de Nemours** (and in 1942, chemists from the German company Bayer) discovered solvents of acrylonitrile. It was several years, however, before a process for making synthetic fibers from this product was discovered.

Since World War II, acrylonitrile has become one of the great intermediate substances of organic chemistry. Among other uses, it is the raw material for two categories of large growth products: acrylic resins and polyacrylic synthetic fibers.

In 1953, DuPont de Nemours placed Orlon on the market. The Bayer com-

pany followed with Dralon in 1954. In France, Crylor was developed in 1949 by Rodiaceta, and placed on the market in 1956. That same year the British firm Courtaulds came out with Courtelle.

Polyester (1941)

The American **Wallace Hume Carothers** (see Polyamides, p. 259) was the first to undertake research on the synthesis of polyesters. After that, two engineers from the British company Calico Printers Association obtained the first polyester fiber from terephthalic acid. This product was patented in 1941.

The first polyester to be marketed was Terylene, which was brought out by the British company Imperial Chemical Industries in 1950. In 1951, Dacron was launched on the marketplace by the American company DuPont de Nemours. In 1956, Tergal, another polyester, appeared on the French market. This fiber was produced by Rhodiaceta.

GENETICS

The science of heredity, genetics, is a branch of biology; its laws were discovered by Gregor Mendel in 1865. Long applied empirically to the breeding of domestic animals and the cultivation of edible plants and decorative flowers, genetics only came into its own beginning in 1919 with the work of the American T. H. Morgan, on the drosophila or vinegar fly. A distinction is made between pure genetics which, in the laboratory, aims at penetrating the internal mechanisms of heredity, and applied genetics, which aims at improving animal and vegetable species, via genetic crossing. The roots of genetics go back to the Greeks of antiquity.

Genetic alphabet

Genetic language has broad analogies to the alphabetic language. Thus, the 26 letters of our alphabet form words that are in turn assembled into phrases are of variable length. The

genetic language comprises only four letters: A, G, C, T, which correspond to the initials of the four bases that constitute DNA (adenin, quanin, cytosin, thymine). Three additional letters define a word that corresponds to a precise amino acid. Then these words, once arranged in a well-determined order, provide a phrase that corresponds to the genetic message of a given protein.

Gametes (5th century B.C..)

The first to suspect the existence of male or female reproductive cells was the father of medicine, the Greek, **Hippocrates** (460—377 B.C.): In the formation of the human embryo, he in effect gave credence to two (male and female) seeds. These two seeds are the spermatozoid and the ovule. The spermatozoid is the characteristic cell of the sperm produced by the testicles. The male produces approximately 200 million daily. The ovule is

produced by the ovary and a young female possesses, at birth, about 400,000 ovules, of which the number subsequently diminishes: 10,000 at puberty, and only 400 of which reach maturity. An embryo is the result of the fertilization of an ovule by a spermatozoid.

Spontaneous generation (4th century B.C.)

This is a theory according to which there exists a spontaneous formation of living beings from mineral matter or from decomposing organic substances. The first to develop this theory was the famous Greek philosopher **Aristotle** (384—322 B.C.). Having remained nearly unchanged for two millennia, it was definitively affirmed by the Frenchman Louis Pasteur (1822—1895).

Hereditary character (4th century B.C.)

The Greek philosopher **Aristotle** (384—322 B.C.) was the first to mention hereditary characteristics when he wrote: "Malformed infants are born of malformed parents, for example, the lame give birth to the lame, the blind to the blind, and, as a general rule, children resemble their parents insofar as anomalies are concerned." Due to the progress made in genetics, it is now known that characteristics are transmitted by the genes, that is, by the deoxyribonucleic acid (DNA of the chromosomes.

Cell (1665)

The Englishman **Robert Hooke** was the first to use this word (from cellula, "small chamber") to describe the minuscule cavities which he observed in 1665, using a rudimentary microscope, in cross-sections made in a piece of cork. The cork was a dead tissue. Hooke in fact only observed the walls which surround the cell. At that same time, the Dutchman **Leeuwenhoek** used a microscope to observe in a more perfected manner many isolated cells, such as blood globules, bacteria, spermatozoids.

It was not until 1824 that the Frenchman Dutrochet declared that a living tissue is made up of a juxtaposition of cells. Then, in 1833, the Scotsman Robert Brown described the nucleus of a cell. Today, a cell is described as a formation of cytoplasm and a nucleus contained inside a membrane.

Species (1700)

The Englishman **J. Ray** (1627—1705) was the first to provide a definition of species (about 1700): "He said that those which belonged to the same species were plants which came, one from the other, by the intermediary of seeds." Today, the species is defined as a group of animal or vegetable individuals having a similar appearance, a particular habitat, fertile among themselves, but usually sterile with regard to individuals of other species.

Chlorophyllic function (1779)

It was the work of the French chemist A. L. de Lavoisier that was at the origin of the discovery of the phenomenon which is specific to green plants and some bacteria. In 1776, he discovered that air was made up of carbon gas and oxygen. The following year, he demonstrated that the combustion of an organic substance produces water and carbon gas. In 1779, the Englishman **J. Priestley** demonstrated that green plants are capable of emitting oxygen, and later the Dutchman G. Ingenhousz (1730—1799) completed this observation by demonstrating that only the green portions of plants release oxygen, and only in light, while they release carbon gas (carbon dioxide) in darkness.

Today the chlorophyllic function is described as the phenomenon by which green plants build up organic materials from mineral elements through the use of carbon dioxide from the air. This is translated by the giving off of oxygen.

Lamarckism (1809)

In his *Zoological Philosophy,* published in 1809, the botanist **Jean-Baptiste de Monet**, chevalier de Lamarck, expounded on his theory of the evolution of human beings by transformism. According to this author, the species are transformed with time under the influence of the milieu, and by the intermediary of habits and needs. Thus, aquatic birds have webbed feet by virtue of their having to hit the water, giraffes have a long neck since they have to graze on high foliage, and pigs have a snout in order to root out things from the ground. These characteristics, once acquired, are transmittable, that is, they are hereditary.

Microbiology (1857)

In 1857, the Frenchman **Louis Pasteur** (1822—1895) discovered yeast (microscopic fungi) and explained the fermentation process. Pasteur extended his research to bacteria. This was the beginning of microbiology. (For further details, *see* the Medicine section, Vaccinations, p. 201).

Darwinism (1859)

In his book entitled *The Origin of the Species by Natural Selection,* published in 1859, the English naturalist and biologist **Charles Darwin** studied, after Lamarck, the problem of the evolution of the species. According to him, the species are not unchangeable entities that are the result of distinct creations, but are entities progressively transformed by selection of the individuals who are best adapted to their environment (survival of the fittest) and the elimination of those who do not adapt to modifications in the environment or who cannot endure competition.

Heredity (1865)

In 1865, the botanist **Gregor Mendel**, born in Moravia in 1822, demonstrated that hereditary characteristics are transmitted by the intermediary of distinct elements, which are today called genes. Two sets of experiments allowed him to reach this conclusion. The first set consisted of crossing peas of stable lines that differed among themselves in terms of a couple of characteristics, for example peas having smooth or wrinkled seeds, with green or yellow cotyledons. The crossing of such plants produced first generation plants in which dominant characteristics were revealed—for example, the hybrids obtained by the crossing of smooth and yellow peas with wrinkled and green peas were smooth and yellow. But when the hybrids were crossed, the parental green and wrinkled characteristics only appeared in a quarter of the second generation hybrids.

From these observations, Mendel deduced that hereditary factors determining traits went in pairs, and that the recessive characteristic was only expressed in the plant when both factors were found together, while one dominant characteristic only suffices for the latter to exist. Such experiments permitted Mendel to determine that in peas the smooth and yellow characteristics are dominant, while the green and wrinkled characteristics are recessive. He also drew up laws that bear his name and which, by revealing the segregation of characteristics, prove that hereditary factors behave independently: They join together and separate across the generations and hybridizations, according to the statistical norms of chance.

Meiosis (1866)

In 1866, the Belgian zoologist **Edouard Van Beneden** observed that the nucleus of sex cells (ovules and spermatozoids) contain a single number of chromosomes (n), while the nucleus of cells that produce sex cells contain the double (2n). This phenomenon is called meiosis and is the result of two consecutive divisions. It involves only those cells which produce sex cells, that is, those cells which are called germinal.

Mitosis (1875)

Means of cell division were observed in 1875 by the German botanist E. **Strasburger** in vegetable cells. Mitosis leads to two daughter cells from one mother cell, by means of the fragmentation of the initial nucleus into two parts and by longitudinal cleavage of the chromosomes present inside this nucleus.

Bacteriology (1870)

Beginning in 1870, Pasteur cultured and identified staphylococci, streptococci, and anthrax-caused bacteria. Pasteur demonstrated that if the environment is favorable, all cultured germs will proliferate.

Since that time, nearly 100 different tissue-culture media have been investigated, resulting in the isolation of many different strains of germs that can be identified today. Each germ has been named after the person who discovered it.

Virology (1880—1885)

In 1880—1885, when working on rabies, an incurable disease at the time, **Pasteur** was confronted with the problem of a disaase for which he was unable to culture the infectious agent. That agent was not a bacteria but a virus. Nevertheless, Pasteur did succeed in developing a rabies vaccine.

In 1898, the Swede **Beijerinck** discovered micro-organisms that were even smaller than bacteria. These tiny bodies were capable of passing through porcelain filters. Beijerinck demonstrated the existence of the tobacco mosaic virus, and virology was born.

In 1935, the American biochemist Wendell Meredith Stanley (1904—1971) crystallized a virus for the first time. In 1959, using the electronic microscope, X-ray diffraction, and biochemical methods, the Frenchman Andre Lwoff (born in 1902) established a definition of the virus that opened the field of virology to further advancement. This definition was based on nucleic acids.

Fecundation (1975)

In 1875, the German **Oscar Hertwig**, professor of anatomy in Berlin, observed in the transparency of a fertilized egg in the sea urchin that the two nuclei of the egg and the spermatozoid joined to form a single nucleus.

The essential significance of fecundation is the reconstitution of the complete number of chromosomes, that is, 46 in the human. Thus, when the nucleus of the egg (containing 23 chromosomes) and that of the spermatozoid (also containing 23 chromosomes) come together, 46 chromosomes (more exactly, 23 pairs of chromosomes) are reconstituted, and the development of a new individual may then begin.

Chromosome (1888)

Chromosomes are short rods, usually curved (measuring 0.005 millimeters in the human) and are found in the nucleus of cells. These rods were named chromosomes by the German anatomist **G. Waldeyer**, in 1888. For each species the number of chromosomes present in a given cell is constant, and chromosomes exist in pairs (2n), except for reproductive cells (ovules and spermatozoids), which have only a single set of chromosomes (n). Thus, in humans, sex cells contain 23 chromosomes, while the other cells of the body contain double that, or 46.

Among these 46 chromosomes, 50% are of paternal origin and 50% are of maternal origin. The essential component of the chromosomes is deoxyribonucleic acid, or DNA.

Mutations (1901)

The first observations of mutations were made in 1901 by the Dutch botanist **Hugo de Vries**. By cultivating and studying different plant species, he observed individuals that did not correspond to what their seed of origin should normally have provided. These individuals showed differences in comparison to their parents, and these differences were hereditary. The name *mutation* was given by de Vries to these hereditary variations. It was subsequently shown that these modifications were the result of alterations of the genes. Mutations are observed both in animals and in plants. To name but a few examples: dwarf guinea pigs, mice with corkscrew tails, rabbits with a fine down coat. Today it is known how to obtain experimental mutations using radiation and chemical products.

Gene (1910)

The outstanding figure of the gene story was an American Southern gentleman, born in Kentucky in 1866: **T. H. Morgan**. In 1910, while working with the vinegar fly, *Drosophila melangogaster*, an insect that usually has red eyes, he observed a male mutant which had white eyes. Morgan crossed the male with a normal female (having red eyes) and noted that all the descendants of the first generation had red eyes. Conversely, among the descendants of this first generation, he found white-eyed flies once again. The "white eye" trait had been transmitted according to Mendel's laws by a chromosome segment to which Morgan gave the name *gene*. Today it is estimated that the genetic heritage of humans contains approximately 500,000 genes distributed over 46 chromosomes. The essential component of the genes is DNA.

In vitro vegetable multiplication (1952)

In 1952, **G. Morel** and **J. Martin**, two French researchers at the French National Institute for Agronomic Research (INRA), obtained an authentic transplantable dwarf plant from a few cells cultured in a synthetic medium.

The extraction of cells from a complex organism and realizing their multiplication in an appropriate nutritive medium was postulated in 1902 by the botanist Haberlandt. Haberlandt had forseen the tremendous importance of *in vitro* multiplication. (*In vitro* culture means culture under glass in sterilized conditions, as opposed to *in vivo* culture, i.e., culture in a natural environment.)

The multiplication process is very rapid, producing thousands of individuals strictly identical to the original, as they all have the same genetic makeup. In 1912, the French physiologist Alexis Carrel (1873—1944), working at the Rockefeller Institute of New York, successfully completed the first cultures of animal cells. Carrel's work represented an important scientific breakthrough and won him the Nobel Prize that same year.

In 1924, Robbins, followed by Philip White in 1931, both Americans, succeeded in getting isolated roots to grow for several months.

In 1952, G. Morel and J. Martin achieved success: Their plant could at last be used as standard stock for producing plants of impeccable quality.

This technique can be used to regenerate viral-infested varieties and to obtain the multiplication of identical plants. As many as 200,000 rose bushes can be produced in a year from a single rosebud, while one bud from an almond-pit peach tree can give 10,000 stock for grafts.

This technique is commonly employed in flower production (orchids, dahlias, hydrangeas, lilies of the valley). It has revolutionized the production of strawberry plants (work of the Japanese Nishi and Oshawa, in 1973, and the American, P. Boxers, in 1974), apple and cherry tree stock for grafts (Jones, 1977), and peach tree stock for grafts (Martin-Zucharelli, 1979). It is currently being studied in connection with reforestation.

Genetic code

The initial experimental arguments in favor of the transmission of hereditary traits from the information provided by DNA go back to 1943, and to the work of the microbiologist S. Luria, of Italian origin. Since then, we have known that each hereditary trait is coded in the form of a message on a small segment of the DNA or gene, and that the translation of this message produces a specific protein producing the hereditary trait. The synthesis of the protein requires the intervention of RNA, on which the gene's message is recopied. The machinery is then used to translate this message in order to produce a protein.

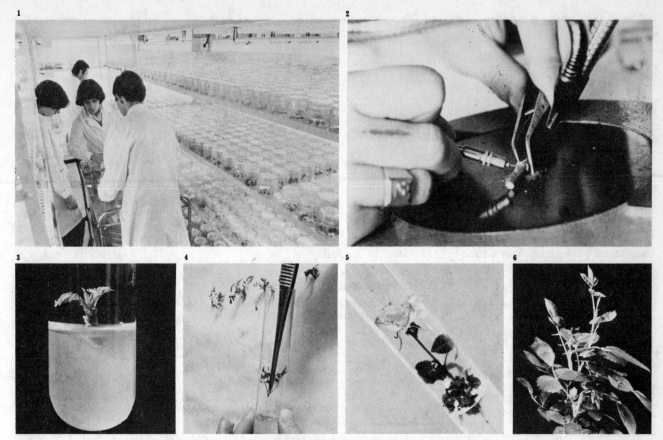

Producing a rose plant in vitro: 1. Preparation of the nutrients and asepsis of the material. 2. A fragment of plant is taken from the mother-plant. 3. The fragment is placed in the appropriate nutritive matter. 4. The tufts are divided and each part is planted individually. 5. After 3 weeks, the plant is grown. 6. The rosetree can then be planted in soil and readied for sale.

Immune system

Louis Pasteur was the first, one hundred years ago, to speculate about the existence of this system when he was working on vaccines. However the description and function of this system only began to be known as of 1950.

The immune system encompasses all those mechanisms that allow the organism to recognize microbes, worn out cells, and abnormal or foreign cells (in the case of grafts, for example), and to destroy them. This system essentially relies on three kinds of white globules: The B lymphocytes attack the aggressors with a sort of grenade, called the antibodies; the T lymphocytes engage in hand-to-hand combat. The macrophages are in charge of digesting the aggressors once they have been killed.

Nucleic acids, deoxyribonucleic acid (DNA) (1944)

In 1944, the American bacteriologist **O. T. Avery**, of the Rockefeller Institute in New York, demonstrated that the transmission of hereditary characteristics from one bacterium to another takes place due to the interme-diary of the DNA molecules. Due to this discovery, it is known that the genes, whose primordial role in the phenomena of hereditary transmission was known, are made up of DNA.

In 1953, using diffraction via X-rays, the electron microscope, and chromatography, the Englishmen J. D. Watson and F. M. Crick described the exact structure of the DNA molecule in the British journal *Nature*. This won for them the Nobel Prize. The DNA molecule exists in the form of a double strand helix made up of a series of units called the nucleotides.

Ribonucleic Acid (RNA)

RNA is also made up of a series of nucleotides each formed from a sugar (ribose), a phosphate, and a base. RNA plays an important role in the synthesis of proteins.

Protein (1953)

A protein may be represented as a linking of 20 kinds of different amino acids. In 1953, the English chemist L. **Sanger** was the first to determine the complete sequence of the amino acids of a protein. The protein was a hormone—insulin.

Proteins form the make-up of the organism: They constitute connective tissue, the skin, the pilous system, ligaments and tendons. Proteins may also intervene in the metabolic processes, in which case they are called enzymes. Some hormones are also proteins. This is true for insulin, which today can be synthetically produced.

In vitro (laboratory) protein synthesis (1954)

The first synthesis of two proteins was performed in a laboratory in the United States, in 1954, by the American biochemist **Du Vigneaud**. These proteins were two hormones normally secreted by the posterior lobe of the pituitary gland: oxytocin and vasopressin, each of which is composed of nine amino acids.

More precisely, proteins are very complex molecules normally synthesized in living cells by the combination, according to the genetic code, of tiny units, the amino acids, of which there are twenty.

In 1969, Merrifiels and Gutte, of New York's Rockefeller Institute, achieved the first synthesis of a large molecule by inducing the successive attachment of 124 molecules of amino acids on a polystyrene support. For this manipulation, which lasted over three weeks, they used electronic equipment enabling them to obtain 369 automatic chemical reactions in 11,931 steps.

The Merrifiels method is widely used today for protein synthesis. In addition, proteins can also be produced by grafting two fragments of cells belonging to different species. All of these "miraculous" operations are cataloged under the heading of *genetic enqineering*.

HLA System (1958)

In 1958, the Frenchman **Jean Dausset** first described the HLA system (for human leucocyte antigen) in white globules. The HLA system is a series of proteins present at the surface of all the cells of an individual. These proteins are analogous to fingerprints. They vary from one individual to another and in some way may be thought of as one's identity card. Today it is known that tolerance to grafts depends on the resemblance of the HLA systems of the donor and the donee.

Caryotype (1960)

Caryotype is the microphotographical representation obtained of the morphological appearance of the set of chromosomes found in the nucleus of a cell. The individual identification of these characteristics is based on numerical classification, developed by a group of researchers in Denver, Colorado, in 1960.

Caryotype is the same for all of the cells of the same species. Thus, the normal caryotype of the human comprises 46 chromosomes, of which 22 pairs are autosomal and 1 pair is heterochromosomal (XY in the male, XX in the female).

Biotechnology (1972)

Biotechnology is a technique that consists of using genetically manipulated bacteria in order to assign to them tasks for which they are not made. Thus, the *Pseudomonas*, a bacterium which digests petroleum, was created in 1972 by a biologist of Indian origin, **Dr. M. Chakrabarty**. This bacterium has formed the subject of a patent registration in the United States.

It should be noted that it took eight years in order to overcome the reservations put forth by the United States Patent and Trademark Office. Interferon, a remedy thought to be useful in combatting cancer, is also industrially manufactured using manipulated bacteria.

Hybridomas (1975)

In 1975, after lengthy investigation, the Frenchman G. Barski and the Englishmen C. Milstein and G. Kohler, working independently created hybrids of an entirely new type. These were cellular hybrids. From two separate cells belonging either to the same organism or to two individuals of the same animals species, or again, to two individuals of different species, a new cell was created. A spectacular example of this was the successful hybridization of a human cell with a cell taken from a mouse!

Embryonic transplants (1975)

Embryonic transplants consist of taking an embryo from the uterus of one female after six or seven days of gestation and implanting it in the uterus of another.

The first experimental transplants on bovines were attempted in England in 1975 in the laboratories of Drs. **Rawson** and **Polch**. A perfectly healthy calf was born. On February 23, 1979, in France, the team of Dr. Plat at France-Embryon attempted the same experiment on the cow Teena, resulting in the birth of more than 40 calves of a race obviously superior to that of their "adoptive" mother.

In 1983, a method was developed for splitting an embryo by microsurgery, yielding two monozygotic (i.e. having the same genetic makeup) calves.

In 1984, in the United States, the first embryonic transplantation in the history of mankind was performed on a 30-year-old woman (*see* Medicine, p. 197).

Cloning

Clones are fragments or regenerated descendants resulting from a single individual. The best-known examples of cloning are cuttings, which are the regeneration of a plant from fragments taken either naturally or artificially. The cuttings take root and produce a new plant.

The Swiss professor **Illmensee** and his research team attempted the same type of experiment, only this time with animals. In 1980, he obtained three perfectly viable baby mice by stimulating cells from a mouse's ovary. These "fatherless" mouselets are exact replicas of their mother, as they have the same genetic makeup.

Genetic engineering

This science of genetic engineering consists of splicing pieces of DNA or of genes and of introducing them into other cells, which thus take on characteristics which they did not initially possess.

In 1983, the team of **Professor R. L. Brinster** of the University of Pennsylvania succeeded in grafting a human gene—the one that contains the synthesis of the growth hormone—into mouse embryos that had previously been taken from the uterus of their mother. The mice developed into giants.

OCEANOGRAPHY

Pisciculture (Antiquity)

The Romans were the first to have the idea of cultivating sea animals, fish or sea food, in a closed medium (fish tanks). Chinese also used artificial spawning.

In the 20th century, the term aquaculture is replacing the traditional word pisciculture. In France, the National Center for the Exploitation of the Oceans has launched operations for the cultivation of bass, turbot, sea bream, lobster, sole, salmon and shrimp in Brittany, on the Riviera and in Tahiti.

First Underwater Photographic Observations (1856)

The first underwater photograph was presented by **William Thompson** to the London Art Society in 1856. The photographic apparatus had been placed at the bottom of the Bay of Weymouth, at a depth of 6 m. The lens, controlled from the surface, stayed open for 10 minutes.

In 1880, the French zoology professor, Louis Boutan, developed the technique of underwater photography by perfecting a complicated underwater system using a magnesium flash. The photographs thus obtained were published in 1900 in an album which was the first of its kind.

In 1935 the Englishman William Beebe, took photographs from the porthole of a bathyscape at a depth of 900 m. Ten years later, an American, Maurice Ewing, developed an automatic underwater camera to photograph underwater geological structures at depths between 400 and 7000 m.

In addition to these purely photographic techniques, other means of looking at the sea depths have come into existence. One of the latest to date, the Sea Beam, from the French company, Thomson, consists in taking radar images (viewed on a TV screen) with an apparatus mounted on a ship. Using this system, Thomson hopes to study the wreck of the Titanic, off the coast of New Foundland.

First Oceanographic Exploration (1872)

On December 21, 1872, the English schooner, *Challenger*, the first oceanographic ship in history, left port. It would not return to port until 1874, having carried out explorations at 362 stations in all the earth's seas. This 2306-ton schooner was manned by members of the Royal Society and the aim was to study everything which was related to the sea.

This was the follow-up of a completely chance discovery: in 1860, the trans-Atlantic telegraphic cable, which had been laid two years previously between the Old and the New World, was brought up from a depth of 1800 m. in order to be repaired. To the surprise of the technicians and scientists, strange forms of plant and animal life, as yet never seen, were found clinging to it!

Challenger's expedition has remained famous in the annals of science; at each station it carried out measurements on the temperature of the water, the speed and direction of the currents, and it lowered dredges to collect samples of rock, flora and fauna.

Polymetallic nodules (1873—76)

Polymetallic nodules were revealed for the first time during the voyage of the famous oceanographic ship, *Challenger*, in 1873—1876. Polymetallic nodules are small black spheres as large as potatoes. They lie at the ocean depths, over large areas (450,000 km^2, in the Pacific).

Since the 1960's they have been of interest to industrialists because of their high nickel, copper, cobalt and dmanganese content. During the past ten years, more than a hundred million dollars has been spent throughout the world by about thirty private or State organizations on the research and evaluation of the economic feasibility of exploiting these fields of nodules.

Wave Energy (1875)

The idea of using the force of waves seems to have been scientifically envisaged for the first time by the Australian, **R. S. Deverell**, in 1875.

In 1889, a very rudimentary installation was made operational off the coast of New York (Ocean Grove).

Gold from the Oceans (1927—28)

In 1927—28, in order to try to pay the debts incurred during the First World War, the German chemist, Fritz Haber, equipped an oceanographic ship, the *Meteor*, to rake the waters of the Atlantic and to attempt to recover part of the gold dissolved in sea water. One km^3 of sea water (one thousand billion liters) in fact contains 6 kilograms of gold. Since all the oceans together comprise 1.5 billion m^3 of water, the total quantity of gold dissolved may be estimated to be 10 million tons.

The procedure used by Fritz Haber was found to be rather fruitless, and he had to abandon his research because of the very small quantity of gold collected.

Aquaculture (1933)

In 1933, the Japanese Professor, **Fuginaga**, was the first to solve the problem of the reproduction of the shrimp, *Panaeus Japonicus*, in captivity.

Between 1890 and 1910, billions of cod and lobster larvae had been placed in the ocean by American biologists in order to seed the sea, in the hope of increasing productivity. It was all in vain. Professor Fuginaga's experiments finally opened the way to modern aquaculture, for which current production (fresh water and salt water) exceeds 25 million tons.

In 1960 the Japanese devised floating cates to raise senoles (a sort of tuna). Today, more than 3000 fish farms exist.

The Ocean in the Laboratory (1953)

In 1953, the American scientist, **Stanley Miller**, created the primordial ocean in his California laboratory by submitting in the action of electric area mixture of steam and chemical elements such as methane, ammonia, and hydrogen.

To replace lightning created by storms or volcanoes, he used electric arcs. In this culture soup, recalling the primitive ocean which covered the planet some 4.5 billion years ago, he found — one week later, amino acids, the basis for all living things.

Stanley Miller was inspired for his experiment by the ideas of the Russian scientist, Alexander Oparine, who made them known in 1925.

In 1961, another American biochemist, Melvin Calvin, demonstrated that not only amino acids but sugars, urea, fatty acids as well as other substances which are of great biological import could be obtained using the same method. These experiments won for him, also in 1961, the Nobel Prize for Chemistry.

Underwater dwellings (1956)

In order to mitigate the difficulties involved in diving from the surface, and to permit working for a longer period of time at the sea bottom, the American, **Edwin A. Link** (inventor of the Link Trainer, a flight simulator), had the idea in 1956 of building an underwater decompression chamber, named the SDC (Submersible Decompression Chamber). Since the atmospheric pressure inside the chamber was equal to that of the water, a diver could come in and rest in dry quarters.

The SDC was tested by the US Navy in 1957. In 1962, the Belgian diver, Robert Stenuit, remained immersed at a depth of 66 meters for 24 hours off the coast of Villefranche, in the Mediterranean. This gave the thrust for the development of larger sized caissons which would be real "sea dwellings".

From September 1963 (Experiment Continent 1, J. Y. Costeau, 7 days at a depth of 17 meters) to October 1977 (Janus IV, Comex, at a depth of 500 meters in the Mediterranean), the American (Sealab I and II, in 19964 and 1965) as well as the French experiments were numerous. Nonetheless, today, underwater dwellings have been abandoned.

Tide-driven power station (1966)

In 1855 the French engineer, **Albert Caquot**, launched the basis for a project of a tide-driven power station in the bay of Grandville englobing the bay of Mont Saint-Michel. His idea was to use the very strong amplitude of the tides (7 to 8 m) to drive generator sets to produce electricity.

In fact, this was not a new idea. A scientific committtee had already been set up in 1919 to establish the principles of operations for an electricity-producing station using the energy of the tides.

In 1943, the Technical Committee for Sea Energy recommended the entire damming up of the bay of Mont Saint-Michel. The concept was judged to be too daring and a decision was made to undertake a more modest project. This resulted in the tide-driven station of La Rance (power of 240 MW), which was inaugurated on November 26, 1966.

In fact, only 3 tide-driven stations exist throughout the world: in France, in the USSR and in Canada, where the Bay of Fundy factory, in Nova Scotia, has just become operational. The cost was 53 billion Canadian dollars.

Glomar Challenger (1968)

Equipped with a 43m high derrick, the oceanographic explorer ship, Glomar Challenger, was launched in 1968. Its mission: to carry out underwater drilling up to a depth of 6000 m, for the American Science Foundation in the context of the "Deep Sea Drilling Project". Its objective: to explore the ocean floors in order to decode the history of our planet. In addition to other important discoveries, two major feats are owed to the Glomar Challenger:
— The Mediterranean, dry 12 million years ago, filled up with water from the Atlantic 7 million years ago via the Straits of Gibraltar.
— The sediments taken from the Indian Ocean reveal that the Himalayas are the result of a collision between India and the Asian content.

In 1985, the highest performance underwater explorer ship, the successor to the Alvin (American) and the Cyana (French), is the new French pocket submarine, the Nautilus, manufactured by IFREMER; it can reach a depth of 6000 m.

Cultivation of Algae (1972)

In 1972, the American, **Howard A. Wilcox**, conceived of the idea of using sea algae to produce methane for use as energy. His calculations revealed that if it were possible to cultivate a "field" of alga, of the *Macrocystis*

Pyrifera species, 856 km in size in the ocean, it would be possible to produce the totality of gas and sea products consumed in one year by the United States.

The idea consists in "harvesting" these algae, which grow very quickly due to the sun's action, and then in having them digested by bacteria which cause the emission of methane.

Geothermal underwater springs (1977)

On February 19, 1977, 320 km to the northwest of the Galapagos Islands, **John B. Corliss** and **John M. Edmond**, aboard the Alvin bathyscape, were the first men in the world to see underwater geothermal springs, at a depth of 3000 m. In areas near the ocean rift, these springs of hot, very strongly mineralized water come out of the ground at a temperature of 250°.

This discovery would in fact not have been very extraordinary in itself if veritable biological communities had not been observed living around these springs: bacteria, long white worms unknown until then, mollusks and crabs. These animals had the particularity of living in a closed ecosystem, completely independent of the sun, contrary to all the rest of life on earth.

Further investigations revealed that they get the energy necessary for their survival from a bacterium which metabolizes hydrogen sulfur present in geothermal waters.

More than twenty underwater geothermal springs have been identified to date in the Atlantic and Pacific Oceans; they constitute one of the major discoveries of the last few years.

Epaulard (1980)

The Epaulard was developed by **IFREMER** (French Institute for Research on Exploitation of the Sea). It is the first uninhabited, completely autonomous submarine and it can reach a depth of 6000 m. It was designed for photographic and bathymetric exploration of the ocean floors.

The longest underwater tunnel (1985)

The longest underwater tunnel in the world was completed in Japan on March 10, 1985. Begun twenty years ago, it joins the Island of Honshu to the Northern Island of Hokkaido, 53.9 km away.

Thirty-four men lost their lives on this Japanese site of the century. The sea entered the gallery four times. The underwater portion is 23.3 km long. Its deepest point is 240 m below the level of the sea. The principal trench is 11 m wide, and it can be equipped with the two Shinkansen tracks (the Japanese high speed train). This project cost 690 billion yen (2.7 billion dollars in 1985 terms), instead of the 200 billion yen originally forecast.

Reactor using wave energy (1985)

In February 1985, the Norwegians undertook the building of the first reactor in the world to use wave energy to produce electricity; this was west of the port of Bergen. It will produce the electrical energy needed to supply 70 average Norwegian dwellings (that is, 1.5 million kw/h.)

Based on the principle of oscilliating columns of water, this installation should furnish electricity at an even lower cost than expected, as was demonstrated by test investigations. The Norwegian government is investing in another project which consists in using wave energy to force considerable quantities of water into a funnel in order to recover it by the intermediary of a conventional hydraulic reactor.

Diamonds from the sea (1985)

Some 90 million years ago, the waters of the Orange, a river which flows in South Africa, began to deposit diamonds collected during its 1500 km of voyage from the heart of the African continent onto the beaches.

Since all the beaches have been combed up to the smallest grains of sand for some time, technicians at the famous diamond company, De Beers, envisage pushing back the Atlantic some 200 m in a clearly laid out area in Nambia. There they will build sand walls, 5 m high, 7 m thick, and 200 m long, to push back the ocean, while constant pumping will dry out the area left behind. Each thus constructed "factory" will be able to last for a few months, leaving the time to prospect the ground before waves destroy the sand walls.

The staff at De Beers estimate that it will be necessary to move 60 million tons of sand and rock to find the equivalent of one million carats of diamonds.

industry

The Pioneer Run Wells oil field in Pennsylvania, 1859.

1. Petroleum

First uses

Petroleum has been known and used since the dawn of civilization. Written records to this effect are among the oldest in existence. The very first recorded uses of hydrocarbons by man were in the Middle East. According to the Bible, Noah caulked his ark with pitch and the mother of Moses used the same substance to coat her son's cradle before launching it on the Nile. Mesopotamian boatmen still use pitch to caulk their small craft. Pitch was also employed by the Egyptians to help preserve their mummies. In fact, the word *mummy* derives from the Persian *mum*, which signifies bitumen or tar.

The Chinese were the first to find a practical home use for petroleum products. They used oil as fuel to heat their houses, to cook their food, and to fire bricks. It was also used for lighting. The crude from tar pits in Venezuela was undoubtedly used for the same purposes by the ancient peoples of Latin America.

In Europe, the earliest mention of petroleum use is for the greasing of wagon wheels and as an ingredient in medicine. Oil or bitumen served in the preparation of ointments for lumbago, contusions, and swellings. This "medicinal" oil was sold and utilized up until the middle of the 19th century in various countries were oil springs were located.

Bitumen products were also used in war — notably as an ingredient of the incendiary "Greek fire."

Birth of an industry

The modern oil industry has its beginning in the 18th century in Alsace (France), Scotland, Galicia (Spain), Rumania, and the Caucasus. Apart from China, where in the region of Sinkiang and Shensi the oldest paraffin lamps known appear to have been fabricated, the use of oil for lighting was late in taking hold. It was only after a colorless, flammable liquid that burned brightly without producing offensive odors was successfully extracted from crude oil that this use was introduced. This liquid, used in lamps, was called lamp oil. In Poland, starting in 1815, the city of Drohobyez employed oil from the Boryslaw region to light the city.

The refining of crude showed that petroleum had an industrial value far beyond that indicated by its initial uses. Even so, the paraffin lamp was not introduced in Europe until 1860.

At about this time, an American, Samuel Kier, refined crude oil and marketed one of the by-products under the name kerosene. In 1850, he discovered oil in Pennsylvania and set up the Seneca Oil Company, which aimed to actively prospect for oil. Kier turned the operations over to an entrepreneur named Edwin L. Drake, who liked to be called "Colonel Drake." August 27, 1859, Drake decided to try drilling near Titusville, Pennsylvania. At a depth of 69½ feet, the well became an oil gusher. Soon after, the region was covered with derricks. The American oil industry was born.

In 1855, the American B. Silliman, making use of studies on the composition of petroleum carried out by European chemists, undertook a series of experiments in refining. He successfully obtained several by-products: tar, lubricating oil, naphtha, solvents for paint, and gasoline (then used as a spot remover and considered a minor by-product).

Seismograph (1914—1918)

Seismography is a method of prospecting for oil that is carried out on the surface. It was developed by the

German, **D. L. Mintrop**. During World War I (1914—1918), he developed a method of seismographic interpretation in order to locate Allied artillery emplacements. After the war, he used his method to study underground geological formations. He set up the first company for seismic exploration in the United States.

The seismographic method consists of artificially provoking slight disturbances at ground level. The resulting waves traverse underground rock layers and are reflected in some cases. Waves returning to the surface are analyzed to determine characteristics and location of the different rock layers.

Data processing (1953)

Seismography has been greatly advanced by the introduction of data processing. The **Geophysical Analysis Group** (GAG), with the aid of the Massachusetts Institute of Technology (MIT), carried out the first experiments from 1953 to 1957. The collected data is stored and analyzed by a computer.

Well-logging, or electrical core sampling (1927)

The first attempt at electrical core sampling was made in 1927. That is to say, the nature of substrata was determined by electrical measurement. The work was done by a team from the French company **Schlumberger**, founded by Conrad **Schlumberger** (born 1878, in Alsace).

Mechanical coring consisted in removing samples of rock by drilling. The drill brought up bits of earth and rock in the form of a cylindrical "core," hence the name.

After World War I, Conrad and his young brother, Marcel, founded the Schlumberger Company. They conducted research into electrical prospecting. In 1927, the Petechelbronn Oil Company in Alsace asked Conrad Schlumberger if his method could help obtain more precise information about the layers traversed by exploratory wells. Conrad gave the job of constructing the appropriate equipment to his son-in-law, Henri Doll. Doll compiled his first electrical log September 5, 1927. The results were

very positive. Correlation between resistance diagrams of rocks, traced during diverse surveys, permitted the obtaining of a clear and precise image of geological structures.

Drilling

Impact drilling

When the presence of petroleum in substrate is suspected, it is necessary to "dig" a hole for purposes of verification. This is done through drilling.

In the beginning, pile-driving (impact) techniques were used. That is to say, a sharp object was driven into the ground by repeated blows. In 1833, the Frenchman J. B. Fauvelle had the idea of running water into the well-shaft in order to make it easier to remove the debris that accumulated around the drill shaft.

Rotary drilling (1844)

The Englishman **Robert Beart** filed the first patent for rotary drilling in 1844. A rotary drill consists of a bit with toothed rollers that turn at the end of a shaft. The shaft is drive on

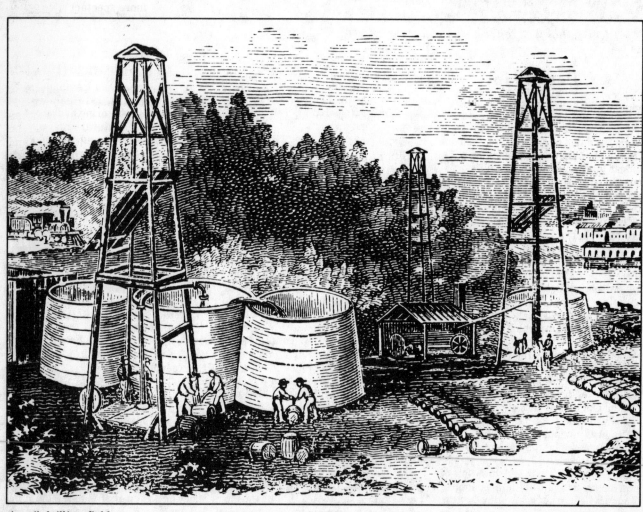

An oil drilling field.

the surface by a motor. Pumps draw off the drilling mud as work progresses. Rotary technology makes it possible to drill much deeper than with impact drilling.

In 1863, the Frenchman Rodolphe Leschot invented a very wear-resistant diamond bit. It was used in Pennsylvania by the American Bullock in 1870. In 1876, a depth of 900 meters was reached using a Leschot rotary drill.

Finally, in 1887, the modern rotary drill was described in a patent by Chapman.

Offshore drilling (1897)

The main problem in offshore drilling is the always moving ocean surface. The drilling platform or ship must be kept as stable as possible. On May 4, 1869, Thomas F. Rowland filed the first patent for a fixed platform. That same year, Samuel Lewis described the principle of a self-elevating, mobile platform. The first working offshore oil well was built off the coast of California in 1897. On August 21, 1928, the Italian-born American Louis Giliasso filed a patent for a submersible barge. It was constructed in 1933 and named after its inventor.

Drilling ships were considered, to increase mobility. The first drilling ship, the *Submarex*, was constructed by the American company Cuss in 1953. The first drilling ship capable of dynamic positioning was the *Cuss 1* from Global Marine; it was deployed in 1961. Dynamically positioned ships are constantly in motion relative to the water around them but maintain a fixed position in relation to the ocean bottom.

The first self-elevating mobile platform was the *Delong-MacDermott No. 1*, built in 1954. As to the *Blue Water 1*, constructed in 1962, it was the first semisubmersible platform. Because of their huge floats, which are mostly underwater, these platforms can withstand very heavy seas.

Turbodrilling (1922)

Turbodrilling was developed by the **Soviets**. The first experiment was conducted in 1922. The bit was powered by circulating drilling mud "driven" by Kapeliushnikov turbines. This way, the drill shaft no longer turned; hence it did not wear.

Electrodrilling (1949)

This method was invented in 1949 by the American company **Electrodrill Corporation**. In short, the motor is attached to a cable rather than to the drill shaft.

Horizontal drilling (1980)

Elf-Aquitaine and the **French Petroleum Institute** (IFP) developed the horizonal drilling technique in 1980. First attempts to radiate outwards from a central well in order to increase .productivity had been made in the United States in 1940, then in the Soviet Union in 1950. In 1980, Elf and IFP completed the first successful horizontal oil well in Lacq. A second well was established at Lacq in 1981. In 1982, these same two companies built the first horizontal offshore well at Rospo Mare, Italy.

The deepest drill barrel in the world (1982)

The French company **Creusot-Loire** and the **French Petroleum Institute** in 1982 developed equipment that made it possible to break the world record for deep-sea oil drilling. A depth of 1714 meters was reached off the coast of Marseilles on December 5, 1982. The drilling equipment is made up of 75 assemblies, each 23 meters long. They are connected by a totally new linkage system. This technology should allow for drilling as deep as 3000 meters.

Refining

It is difficult to fix the exact origin of refining petroleum, since that is lost among petroleum's first uses. Refining permits the breaking down of petroleum into hydrocarbon derivatives. These include propane, gasoline, paint thinner, diesel fuel, and lubricating oil.

Multistep refining (1873)

When the tonnage to be processed increased, it became necessary to invent some means of implementing multistep refining — that is to say, a way to handle all the different steps in refining in the same installation.

Starting in 1873, at Baku inthe Transcaucasus, the Russian **Tawrisow** succeeded in maintaining a constant level of crude in a battery of stills. In 1877, Samuel Van Sycle invented the same process at Titusville, Pennsylvania. In 1880—1881, the Nobel brothers developed a new multistep refining process, using stills in a cascade arrangement. Finally, the refining process column was introduced in the United States in 1926 by the Power Specialty Company.

Cracking (1925)

Catalytic cracking was invented by the Frenchman **Emile Houdry** around 1925. Nonetheless, its industrial implementation was a long process, and it was not until 1936 that Houdry be-

gan to make use of his system (in the United States). The use of cracking greatly increased during World War II.

Cracking increases the efficiency of fuel production from petroleum. It permits transformation of a certain number of heavy residues from refining into light by-products, principally gasoline and diesel fuel.

Catalytic reformation (1949)

Catalytic reformation was invented in 1949 by **V. Haensel**, who was working for Universal Oil Products. This operation permits modification of the composition of gasoline, notably to obtain very high octane ratings.

Oil fire fighters (1861)

On April 17, 1861, the Henry Rouse wells, east of Oil Creek, exploded and caught fire. There were 19 dead and 10 injured. It was three days before the fire could be brought under control by covering the wells with earth. It was around 1915 that K. T. Kinley invented another fire fighting method. Dynamite was exploded to blow out the fire. This method was also used, on a more spectacular scale by his son, Myron Kinley, then by Paul N. "Red" Adair.

Subterranean storage (1916)

It was the German company Deutsche Erdl that in 1916 patented the first process for storing hydrocarbons underground. Natural gas can be stored in underground salt caverns, which are virtually hermetic natural reservoirs.

Methane cargo ship (1959)

The first methane cargo ship, the *Methane Pioneer*, sailed out of the Gulf of Mexico for England on January 31, 1959. Constructed for the North Thames Gas Board, it contained 2200 tons of liquid methane. It is easier and less cumbersome to transport methane in liquid form than in the gaseous state.

2. Building Materials

Cement (Antiquity)

Hydraulic binding agents date back to Antiquity. The Romans used pozzolanic mortars. These were made with lime, which when combined with ash from the volcano Pozzuoli (hence the name *pozzolana*) would harden when mixed with water. In 1775, the Englishman J. Smeaton fabricated a bet-

The Crystal Palace was the symbol of industry at the London Universal Exhibition in 1851.

ter quality lime, which was obtained from limestone containing clay.

Portland cement (1824)

In 1824, an Englishman from Leeds, **J. Aspdin**, invented a new preparation process. A very high roasting temperature produced a cement whose hardness was clearly superior to those used up until that time. Thus, he pointed the way to the preparation of portland cement. It was named "portland" because the mortar prepared with this new cement was like the rock from the Isle of Portland.

Reinforced cement (1845)

Starting in 1845, the Frenchman **J. L. Lambot** built garden furniture out of reinforced cement, and such items as shelves, orange crates, and birdbaths. In 1855 he presented at the World's Fair in Paris a reinforced cement boat he'd built in 1849. (After a century of use on the lake at the Lambot estate near Miraval, the boat was still in good condition.)

Lambot's goal was to replace wood in shipbuilding. However, his patent also mentioned the design of a reinforced cement beam. His work opened the way to materials that would revolutionize the art of construction.

The French gardener Joseph Monier filed a series of patents for reinforcing cement with iron rods, the first of which is dated July 16, 1867. He later applied this patent to railway ties (1877), to reservoirs (1880), to pipes (1885), and to tubing for electrical wires (1891).

Macadam (1815)

The Scotsman **John McAdam** (1756–1836) invented macadam, sometimes called blacktop, in 1815. At the end of the 18th century, the state of roads in Europe was catastrophic. Most had not been resurfaced since the Middle Ages. Meanwhile, commerce was rap-

idly developing. Engineers everywhere began to deal with the problem.

John McAdam was one of these engineers. Born in Scotland, he had made his fortune in the United States. Back in Scotland, he undertook a series of studies, at his own expense, on the paving of roads. In 1815, he was named general administrator of roads for the county of Bristol and was finally able to put his theories into practice.

Macadamization involves a system of resurfacing roads with crushed stone and sand, all of which is compacted by means of steam rollers. Although the name has remained, the macadam road surface has been practically abandoned today. Nowadays, mixtures of cement, cinders, and slag are used. Tar has also been abandoned, giving way to road surfaces made of asphalt combined with the recycled products of previous surfaces. New equipment permits the building of an average of 1 kilometer of highway in one day.

Mechanically formed tile (1835)

The mechanically formed tile was invented in 1835 by two brothers, **Joseph** and **Xavier Gilardoni**, working in their factory in Altkirch, in Alsace, France. Interlocking, mechanically formed tiles can be laid in a single layer, thus providing a savings in roofing material. These tiles are made in a press.

The vivid colors of the first models, as well as the diamond patterns and ribbing used as decorations, were severely criticized. That is why tiles manufactured nowadays are artificially aged, using manganese.

Bondstones (1846)

In order to put up a wall using simple assemblies, the Frenchman **Jean-Aime Balan** developed hollow elements of brick or of plastic mortar.

These are characterized by tongues and grooves and were patented February 2, 1846.

Gutta-percha (1846)

The gum called gutta-percha (from the Malay word for gum, *getah*) was discovered in Singapore. In 1843, an English surgeon reported on it to the London Royal Society. On July 28, 1846, the Frenchman **Alexandre Cabriol**, a rubber manufacturer in Paris, filed the first patent on the applications of gutta percha. He was interested in manufacturing tubes, pipes, and drive belts, among other things.

However, the essential value of gutta-percha lies in its use as an electrical insulator. This property was discovered by the Englishman Michael Faraday (1791–1867). Completely impervious to salt water, gutta-percha is the only material that can be utilized to form the envelope around underwater telegraph cables.

Hollow brick (1848)

Hollow brick is the result of work done by two brothers, **Henri-Jules** and **Paul Borie**, Frenchmen who patented their invention on October 28, 1848. Up until then fired-clay pots were used as hollow filling material in floors and walls. The process of extrusion made it possible to obtain cheap, self-supporting elements. The extrusion technique has been extended, and now diverse materials are used — for example, terracotta (fired earth), used for facing metallic structures on American skyscrapers and for pugging.

Corrugated sheet metal (1851)

Starting in the 18th century, sheet metal and iron were used to roof buildings, notably baroque churches. However, this material easily rusted. In 1837, the Frenchman Sorel developed a process of galvanizing sheet metal. On September 2, 1851, another Frenchman, **Pierre Carpentier**, filed a patent describing a machine capable of ribbing galvanized metal sheets.

Carpentier's invention increased sheet metal's rigidity by corrugating it. It was shown at the World's Fair in Paris in 1855. It permitted construction of lightweight frameworks and roofs spanning wider spaces without intermediary supports.

Prefabricated panels (1888)

On December 12, 1888, the Frenchman **Georges Espitallier** patented the first prefabricated panels. These were of standard sizes, made of varnished, compressed cardboard and waste wool. By combining them in diverse ways, they could be used to construct collapsible sheds. This invention found its first application as part of a rural ambiance presented at the World's Fair in Paris in 1889 by a manufacturer from Pont-a-Mousson.

These first panels can be considered as the model for compatible components, constituted from lightweight, isothermal "sandwich" structures.

Reinforcement brace (1892)

In 1892, **Francois Hennebique** filed a series of patents for a strut linking the cords of traction and compression in support beams, the reinforcement brace.

Reinforced concrete (1892)

Reinforced concrete was invented by the French architect and entrepreneur **Francois Hennebique** (1842—1921) in 1892. He began his research after a fire broke out in a house that he was building for a friend. He was looking for a fireproof construction material. He developed a reinforced concrete slab, than an entire monolithic structure integrated with a wooden framework.

Reinforced concrete as such was not patented. Hennebique continued his work and among other things developed hollow flooring and continuous beams on several supports. However, he did not market his discoveries himself. He turned this task over to a chain of concessionary companies. The success of reinforced concrete was shown by orders that in 1900 amounted to 3 million gold francs.

Plastic wood (1975)

In 1975, faced with the oil crisis, the French company **Creusot-Loire** developed a production process for sections composed of sawdust and plastic. The proportions are 80% to 85% wood fibers and 15% to 20% plastic material.

3. Glass

Origins

The oldest piece of glass known was fabricated under the reign of Pharaoh Amenophis I, between 1557 and 1530 B.C. in Egypt and in Mesopotamia. It was obtained by fusion at a temperature of 1500° C. from a mixture of silicon and carbonate of soda, or potash.

The exact origin of glass remains unknown. The Latin writer Pliny the Elder even contended it was discovered by chance. The story he tells is that glass was discovered by sailors on an Egyptian ship loaded with natron (hydrated native sodium carbonate). After being shipwrecked on a beach in Phoenicia, they used two blocks of natron as stones to support a cooking pot. Heat from the fire caused the natron to combine with sand (silicon) on the beach. Glass resulted from this accidental mixture.

Glassblowing (2nd century B.C.)

Glassblowing techniques were probably developed by the Syrians during the 2nd century B.C.. However it was not until the 1st century A.D. that the technology was seriously pursued by the Romans. Glassblowing replaced molding, a process that involved letting liquified glass harden around a core of sand or ceramic.

In blowing glass, the glassblower collects a bit of softened glass on the end of a hollow tube. He rough forms it with a tool and blows into the tube with his mouth, forcing air into the glass glob. Thanks to this method, glass products became inexpensive to make, so became everyday household items.

Crystal (12th century)

English glassmakers invented crystal at the beginning of the 12th century. Up until then, glassmakers had always tried to imitate natural quartz, which the Egyptians considered a precious stone.

Since English glassmakers used coal in place of wood, they had to add a flux to the melting mass. This was lead oxide (minium). In doing so, they

The body of this car was formed from fiberglass.

accidentally obtained what ever since has been called crystal. Everyone tried to copy the English. In France, glassmaker Saint-Louis was the first to successfully fabricate crystal (in 1782).

Venetian glass (1463)

The invention of Venetian, or crystalline glass, is attributed by many to a master glassmaker from Murano, **Beroverio**, who is said to have perfected this very transparent kind of glass in 1463. However, this claim is contested. Norman (French) glassmakers claim to have intented this technique in 1623. Before that, they say, their ancestors had been practicing similar processes for 250 years in Languedoc.

In fact, crystalline glass probably originated in Syria or Egypt. The Venetians were able to establish a lasting monopoly because of the prestige of their product.

Bohemian glass (15th century)

At the end of the 15th century, Bohemian glass was invented by rock crystal engravers from the city of Prague, situated in the heart of Bohemia. Like the engravers who worked with Nuremburg glass in Germany, the Bohemians needed a glass thicker than Venetian glass in order to practice their art.

Bohemian glass, which is imitation quartz, was at first engraved white glass. As time went on, it was cut in facets and engraved. Finally, master Bohemian glassmakers were able to produce colored, cut, and engraved glass that rivaled English crystal.

Fiberglass (1836)

The Parisian craftsman **Dubus-Bonnel** was granted a patent for the spinning and weaving of glass on November 14, 1836. He described his procedure as weaving of glass strands that are obtained by drawing and maintained malleable:

• At the moment of spooling by executing this operation on a hot shuttle in a vapor bath;

• By carrying out weaving on a Jacquard-type loom in a room heated to 30° to 35° Reaumur.

Dubus-Bonnel wrote out his patent in longhand. There are even spelling mistakes that have been corrected. He enclosed as supporting evidence with his patent application a small square of woven fiberglass.

Tempered glass (1874)

The industrial manufacturing technology for tempered glass was patented, June 13 and July 6, 1874, by the Frenchman **Francois Royer de la Bastie**.

The principle behind tempering glass had been known since 1835 when the formation of "Prince Rupert's drops" was observed in Holland. These were elongated drops of water, produced when molten glass came into contact with water.

Pyrex (1884)

In 1884, **Carl Zeiss**, who was working in a glass factory at Jena, Germany, invented a sort of glass containing boric acid and silicon. This glass was significantly more resistant to heat than other glass. The American company Corning Glass improved on Zeiss's process and created Pyrex.

In-production tempering (1887)

In 1887, at his factory in Choisy-le-Roi, near Paris, **Royer** worked out a process for glassmaking wherein products were directly tempered, without having to pass an intermediary step in a muffle furnace.

Glass is now popularly used in building design.

Wired glass (1893)

In 1893, the Frenchman, **Leon Appert** launched the fabrication of plate wired glass. His process produced industrial sheet glass obtained by lamination of a fluid glass mass into which a metallic mesh had been placed.

The first patents on this process had been taken out in the middle of the 19th century by the Englishman Tenner and the Frenchmen Becoulet and Bellet, but their work remained, purely theoretical.

The first use of wired glass was as casings and windows. The glass roofs in train stations were constructed thanks to this technology.

Automatic glassblowing of bottles (1895)

A bottle-glassblowing machine was patented on February 26, 1895, by the American **Michael J. Owens**, from Toledo, Ohio. This machine permitted molten glass stuck onto the end of a pipe to be molded. A bellows inflated the glass to the desired form. Simultaneously, the bottle would be turned in order to expand evenly. This invention allowed the bottled drinks industry to develop rapidly, because bottled drinks quickly became very reasonably priced.

Laminated glass (1909)

In 1903, the French chemist **Edouard Benedictus** began the research that led to the invention of laminated glass, patented on August 10, 1909. Marketed since 1920 under the name of Triplex, laminated glass was first used for automobile windshields and windows. Today, laminted glass is known under the names Stadip, Tristadip or Multi Stadip. It is used for transportation and more and more in the building industry: for casings, guard rails, windows, doors, etc. Some laminated glass is even resistant to shots from automatic weapons such as machine guns.

Plate glass (1910)

Leon Appert's research into wired glass, which focused on vertical drawing techniques, showed that the coming trend would be away from glassblowing as a means of producing glass for windows.

Two engineers completely unknown to each other simultaneously developed plate glass toward 1910. They were **Emile Fourcault** in Belgium and **Irving Colburn** in the United States.

In 1952, the Englishman **Alistair**

Pilkington invented a much more economical plate glass fabrication method. A layer of molton glass is poured onto molten tin. This produces a perfectly smooth and brilliant glass surface.

Triplex (1920)

Triplex is a kind of safety glass manufactured in France beginning in 1920. It resulted from a discovery by the French chemist **Benedictus**. In 1910, Benedictus noted the adhesive property of nitrocellulose on glass. This led to the production of safety glass. It was constituted of two or three thin sheets of glass glued together by a coat of nitrocellulose. It was marketed under the name Triplex. This bonded glass does not shatter on breaking and is therefore well suited for use in automobile windshields.

Securit glass (1929)

The procedure of thermal quenching, which consists in abruptly cooling a surface heated to 600-650 by means of cold air jets, led to the development of Securit by the French firm **Saint Gobain** in 1929. Chemical quenching procedure, which is more costly, is reserved for products that require exceptional quality, such as the windows for the Concorde aircraft.

Kappafloat (1980)

As a result of his research work, the Englishman **Alistair Pilkington** invented the Kappafloat in 1980. This is an "energetic" glazing that lets external solar heat penetrate at the same time that it prevents internal energy from escaping. The window is transformed into a sort of heat trap, allowing considerable energy savings, even with a greater glass surface.

4. Clocks and Watches

Hourglass (Antiquity)

The Chinese probably invented the hourglass. It then made its way to the West by the intermediary of the Egyptians, who introduced it to the Greeks. The first mention of an hourglass is in a play by the comic poet Baton, who lived during the 3rd century B.C.

Sundial (Antiquity)

The invention of the sundial was attributed by ancient writers to the Greek **Anaximandre of Miletus** (between 610 and 547 B.C.). Modern au-

An example of a sundial.

thorities consider its invention to be much earlier, made in China and in Egypt simultaneously.

A sundial is made up of a style whose shadow is cast on a flat surface. The surface is divided by marks called hour lines, which indicate the hours of the day.

Gnomon (3rd millennium B.C.)

The gnomon was one of the first instruments used by the ancients to measure time during the day. Its invention is attributed by some to the Chinese and by others to the Chaldeans, toward the 3rd millennium B.C. This instrument is composed of a style, which casts a shadow on a horizontal surface. The shadow indicates, as a function of its length, the height of the sun or of the moon above the horizon, and their orientation.

Astrolabe (2nd century B.C.)

The astrolabe was invented by the Nicene Greek **Hipparchus** (2nd century B.C.). It was later perfected by the Arabs.

An astrolabe is composed of two or more rings. These rings have a common center and are inclined in relation to each other. This way, an astronomer can make a number of observations at the same time across the different circles of the sphere.

Clepsydra (1500 B.C.)

The first artificial clock, the clepsydra, or water clock, was known in

Egypt starting in the 15th century B.C. It also existed in China, Greece, Rome, and Gaul.

The working principle behind the clepsydra is the constant flow speed of a liquid.

Destructive timekeeping techniques (ca. 400 A.D.)

During the first centuries A.D., the Byzantines introduced various kinds of clocks based on burn rates. For example, there were fuses with equidistant knots, oil lamps in graduated flasks, and candles that had reference tabs or little embedded weights placed so as to fall out every hour. In spite of improvements intended to increase their utility, the precision of these instruments remained very mediocre, on the order of plus or minus an hour per day.

These kinds of clocks also existed in China. Extremely elaborate, they were often based on the burn rate of sticks of incense and were even able to set off a chime at regular intervals.

Escapement (725)

It seems that the first escapement mechanism was invented by the Chinaman **I Hing** in 725.

The escapement is one of the most important parts of a clock. On the one hand, it regulates, under the control of the oscillator, energy transfer from the motor to the hands. That is to say, it counts the number of oscillations. On the other hand, it imparts to

The Jacquemart of the astronomical clock of Notre-Dame of Dijon (France).

the oscillator, at each period or half-period, energy to compensate for that dissipated through friction. If the escapement does not fulfill these two functions, the regularity of the clock is degraded.

Numerous escapements were successively invented. The peg-wheel escapement, invented by the French clockmaker Amant (1730), was improved by the Frenchmen Lepaute and Pierre Caron, the future Beaumarchais (1753), author of *Le Mariage de Figaro*.

The first escapement utilized was of the swing wheel type. This was characteristic of medieval clocks and was also the kind used in Huygen's clock. It was replaced in watches by a cylinder escapement. Today, the most commonly used escapement is the sprocket-and-lever type, conceived by the Frenchman L. Perron in 1798.

Graham escapement (1715)

In 1715, the English clockmaker G. Graham invented a frictional dead-beat escapement for pendulum clocks. This was called the Graham escapement. It was applied straightaway to the construction of clocks with a second hand, which had pendulums about 1 meter long.

Also in 1715, Graham conceived the mercury-compensated pendulum. Invention of the cylinder escapement for watches is credited to Graham as well, although a first rudimentary version was built by W. Houghton and T. Tompion in 1695.

Anchor escapement (1759)

The English clockmaker **T. Mudge** invented the anchor escapement in 1759. This lever-type mechanism is still used nowadays in many mechanical watches of all kinds.

Weight-driven clock (late 10th century)

The true art of clockmaking was founded the day an escapement and a dead weight acting to drive a wheel were invented. These inventions are credited to the Frenchman **Gerbert D'Aurillac** (toward 938—1003), who was pope under the name Sylvester II in 991. He built the celebrated clock at Magdeburg.

In fact, the weight-driven clock derives from improvements made on clepsydras and hourglasses by the Arabs. They were familiar with the use of weight and counterweight systems.

Automatic striking mechanism (1120)

The oldest record of automatic bell ringing is in the *Usages* of the Order of Citeaux (1120). It stipulated that the sacristan should set the clock so that it would sound the matins.

In 1314, Beaumont constructed in Caen, France, the first public clock that would toll the hour.

Portable clocks (1410)

In 1410, the Florentine architect **Brunelleschi** fabricated a spring-driven pendulum clock. Spring-driven clocks were lighter than the bulky mechanisms that were their forerunners. This meant they could be easily moved. And so began the era of privately owned clocks. Up until then, they were restricted to decorating public monuments.

Pendulum clock (1657)

The first clock regulated by a cyclodial pendulum dates from 1657. It is credited to the Dutch physicist, surveyor, and astonomer **Christian Huygens** (1629—1695).

In 1602, the celebrated Italian scientist Galileo (1564—1642) had discovered the laws and properties of the pendulum. He applied this knowledge to regulating fixed timekeeping instruments. In 1657, the Dutch clockmaker S. Coster, from the Hague, constructed the first pendulum driven clock, based on the drawings and instructions of Huygens. At that time, Huygens was completing work on the theory of pendulums, based on the movement of bodies. This would provide clocks with near chronometric perfection. This clock was small enough to be used in the home. The use of an oscillating system constituted a real improvement in time measurement. It was rapidly adapted to all types of clocks.

Balance wheel and spiral spring oscillator (1675)

In 1675, the Dutchman **Christian Huygens** (1629—1695) invented the balance wheel and spiral spring oscillator. The introduction of the spiral spring into watch design had an effect analogous to that of the application of pendulums to clocks.

Credit for Huygens's invention was claimed by the English scientist R. Hooke (1635—1703). It was later established that the idea of applying a spring to regulate the movement of a balance wheel had, in fact, been set forth by Hooke. However, Hooke had not, like Huygens, indicated the spiral form of the spring. Yet this shape is essential, since it permits large amplitude oscillations of the balance wheel, and places the center of gravity of the spring on the axis of the balance wheel. This spiral shape prevails even to this day.

Chronometer (1736)

The English clockmaker **John Harrison** (1693—1776) gets credit for building a naval chronometer in 1736. This model, for which the inventor received the highest distinction from the Royal Society, had wooden works. Twenty years later, Harrison developed a much smaller chronometer in the shape of a watch and made entirely of metal.

Winding without a key (1755)

In 1755, **P. A. Caron** (a.k.a. Beaumarchais) invented for Louis XV's mistress, Madame de Pompadour, a watch that could be wound without using a key. With her fingernail, she could turn a ring mounted on the watchface.

After several attempts, it was in 1842 that the Swiss A. Philippe succeeded in constructing a winding mechanism with a lug, which permitted both winding of the watch and positioning of its hands. He positioned in the lug a screw with a serrated head. This would later come to be called the winding stem.

Self-winding (1775)

Self-winding watches were invented in 1775 by the French clockmaker **A. L. Perrelet**.

In these watches, an oscillating mass was set into movement by the arm's motion and this would continually wind the spring-motor through the intermediary of a supplementary wheelwork. These watches were called "shaker watches" in France and "weight watches" in Switzerland.

The first self-winding wristwatches were patented by H. Cutte and J. Harwood in 1924.

Electric clock (1840)

In 1840, the Scottish electrotechnologist **Alexander Blain** built the first electric clock.

Distribution of electricity by the hour, regulated by a clock called the master clock, which can be electric or mechanical, was invented by the English physicist C. Wheatstone, also in 1840.

Alarm clock (1847)

The first modern alarm clock was invented by the French clockmaker **Antoine Redier** (1817—1892) in 1847. It was a mechanical device. The electric alarm clock was invented toward 1890.

Alarm clock that responds to the voice (1985)

Developed by **Braun Electric** of France, the Braun Voice Control is the first quartz alarm clock to obey the human voice. In the morning, when the alarm goes off, a single word is sufficient to silence the alarm. The clock repeats its waking message faithfully every 4 minutes for a period of 40 minutes until it either succeeds or is manually turned off.

Diapason (1866)

The idea of using a tuning fork as a resonator for a clock was employed by the French clockmaker **Louis Breguet** (1804—1833) in 1866. Yet, it was not until 1954 that the first electric wristwatch based on a diapason appeared. It was designed by the Swiss engineer Hetzel.

In this watch, the resonator is a diapason that has a specific frequency of 360 hertz. It functions electrically by a coil hooked up to an electrical circuit. This coil interacts, by means of two other coils, with permanent magnets mounted at the ends of the two arms of the diapason.

Time clock (1885)

The time clock was invented by the American **W. L. Bundy** in 1885. Every worker was given a key. When the key was inserted into the clock on arrival at work, the clock would record on a paper tape the number of the key and the time.

The time clock was improved by a traveling salesman from Syracuse, New York, John Dey, in 1894.

Wristwatch (1907)

On November 27, 1907, the aviator Santos-Dumont set the first airborne speed record by traveling 220 meters in 21 seconds. He immediately knew the results by looking at the wristwatch that had been created for him by his friend **Louis Cartier**.

The wristwatch rose in popularity during World War I. French soldiers attached their watches to their arms with watch chains.

The Cartier firm created a "Santos-Dumont" model in 1978. This was in remembrance of the first wristwatch and had the same design.

Waterproof watch (1926)

In 1926, **Hans Wilsdorf** and his Rolex team developed the first absolutely waterproof watch case: the Rolex "Oyster." In 1927, Hans Wilsdorf gave one of his watches to his young Londoner taking part in a cross-Channel swim. The Rolex "Oyster" perfectly lived up to the test, and this unprecedented success was announced on the front page of the *Daily Mail*.

Diving watch (1953)

The first diving watch, the "Submariner," was also developed by the Swiss **Rolex Company** in 1953, followed by the "Sea Dweller" in 1971, the first diving watch in the world, equipped with a helium valve. The Sea Dweller is guaranted up to a depth of 610 meters.

Quartz crystal clock (1929)

Work on the use of quartz as a resonator for clocks goes back to the year

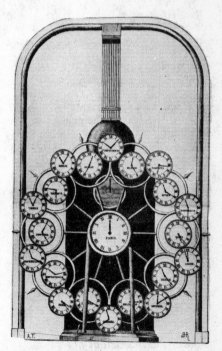

This 1848 clock gave the time of the world's major cities.

Film comedian Harold Lloyd's famous clock scene. This, by the way, is not a special effects shot. Lloyd actually performed this stunt high above street level.

1920. But it was not until 1929 that **Warren Alvin Marrison**, a clockmaker from Orange, New Jersey, fabricated an electric clock with a quartz resonator. His work made use of the piezoelectric properties of quartz, discovered in 1880 by the Frenchmen Pierre and Paul-Jacques Curie.

The master clock is built around a piezoelectric quartz crystal, maintained at a constant temperature and pressure, and mounted in an oscillating circuit. The quartz provides a current whose frequency is extremely stable but too high to run a mechanical device. Therefore, amplifiers and frequency dividers are used, which allow operation of a classical mechanism.

This type of apparatus gives the time with very high precision. The mean functional error of the quartz

crystal clocks used an astronomical observatories is on the order of 0.0001 seconds per day, or less.

These clocks are also often used as simple frequency generators for frequency measurement or to verify the functioning of diapasons in astronomical clocks.

Quartz crystal wristwatches have been marketed since 1970. Their precision is on the order of several tenths of a second per year.

Atomic clock (1948)

It was the American physicist **William F. Libby** who laid down the principle of the atomic clock. The first model was constructed at the National Bureau of Measures in Washington, D.C., in 1948.

In order to avoid the slight frequen-

cy drift observed in quartz oscillators, an "atomic" oscillator was incorporated into a servo-controlled system. Any frequency variation causes a reaction, which makes it possible to reestablish the desired frequency. It is therefore necessary continually to compare the oscillating frequency, which wants to remain constant, with a perfect defined, stable frequency. For this reference frequency, the frequency of a phenomenon occurring in an atom was chosen. Hence, the name *atomic clocks* was given to systems implementing this principle.

The standard frequency is the wave frequency determined in an atom of waves emitted or absorbed at the instant an electron passes from one energy level to another. In order to obtain this result, optical pumping is often used. This technique was de-

A computer watch.

Watches that measure one's pulse beat.

5. Security Devices

Orgins

Apparently the oldest locks come from ancient Egypt. These massive locks, made of hard wood, worked by combinations of cylindrical pins of unequal lengths. The pins were fit to notches cut in removable bits, which served as keys.

Safety locks date from the 18th century. They were improved in the 19th century by Fichet, Haffner, Dorval, and Lhermitte.

Today, locksmithing is considered a part of the hardware trade, which also includes metalworking in industry and the building trades.

Bramah lock (1784)

In 1784, the Englishman **Joseph Bramah** invented a type of lock that is still in use today. Bramah was so sure his lock was burglarproof that he offered a reward of 200 pounds to the first person who could successfully break it open. The reward went to an American, Alfred Hobbs, who succeeded during the London World's Fair, in 1851. It took him 51 hours to open a door secured with this kind of lock.

Drum and pin lock (1851)

Pin locks, sometimes called American locks or Yale locks, were patented May 6, 1851, by the American **Linus Yale**, from New York. Yale was

vised by the French physicist Alfred Kastler (born 1902, Nobel laureate in 1966). The precision is around 21 picoseconds per month over a period greater than one year.

Moreover, an atomic clock is reproducible. Clocks can be constructed that function at exactly the same frequency, something not possible with quartz oscillators alone. Since 1964, a cesium clock has served as the universal standard of time.

Trick clocks (1919)

Starting in 1913, the Parisian jeweler **Louis Cartier** began studying the fabrication of clocks whose works would be entirely invisible. In 1919, he launched the first trick clocks.

Manufactured out of glass or crystal, often objets d'art, these clocks were transparent. From the front, it was impossible to see their works. The hands seemed to move of their own accord.

In 1982, the Frenchman Michel Masson invented an electronic clockwork system that was also invisible. Michel Masson's transparent clocks kept perfect time, thanks to progress made in this technique. They adapt to any environment—aquariums, mirrors, windows, etc.

Swatch (1982-1984)

The Swatch watch is the invention of two Swiss engineers, **Jacques Muller** and **Elmar Mock**. Sold in Switzerland and the United States since 1982, it is waterproof up to 30 meters, shock resistant, lightweight, and guaranteed for one year (three years for the bat-

tery). The Swatch mechanism is mounted in the case. It can't be repaired; it is simply thrown away when it wears out.

Moslem prayer watch (1984)

Invented by the Swiss company **Empire S.A.** in 1984, the watch indicates to Moslems the direction of Mecca at prayer time. The hour section raises and reveals a direction disk that the faithful can consult to know which direction they must face.

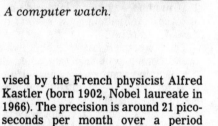

A group of computer watches and the keyboard which is used to input data.

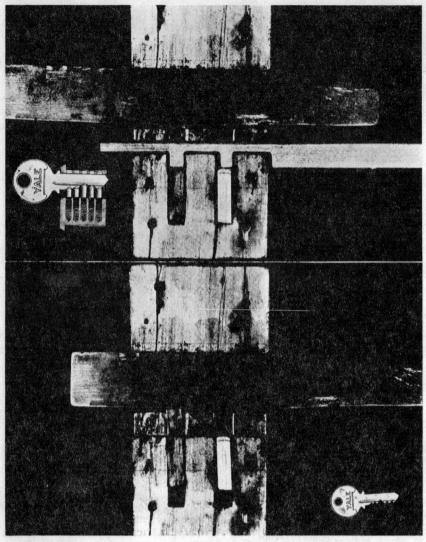

An Egyptian lock that operates on the same basic system used by modern Yale locks.

An intercom in Berlin during the 20's. At that time the intercom was used to summon doctors in the middle of the night.

inspired by the pin-and-bolt locks of Egyptian origin, which were of wood.

Intercom (1920)

It was in Berlin, Germany, in 1920, that the first intercom appeared. Often called door telephones, they were especially used by doctors on night calls in the German capital.

6. Spinning

Origins (Antiquity)

The art of spinning goes back to prehistoric times. A spindle, a sort of stick with a conically shaped point, was first used to spin wool and flax. This method was very slow and only allowed for spinning a meter of thread at a time.

The invention of the spinning wheel, which probably originated in the Middle East, was an important step forward. It permitted spinning great lengths of thread without interruption. The spinning wheel was introduced to the West toward the beginning of the 13th century.

Flying shuttle (1733)

In 1733, the English weaver **John Kay** invented the flying shuttle. This invention was motivated by the impossibility of obtaining cloth beyond a certain width without having to resort to a large, heavy loom worked by two weavers.

At the same time, Kay was looking to save weavers from the fatigue caused by continually having to cast the shuttle, which carried the woof thread across the warp threads. Kay's flying shuttle, instead of being hand cast between the two layers of thread, was propelled by sharply pulling on a cord attached to a driver block. This way, the shuttle "flew" from one side to the other. Thus, the weaver always had a hand free to work the batten.

Spinning jenny (1764)

During the 18th century, increasing demand provoked a serious shortage of thread. A boom in inventions resulted, leading to development of mechanical spinning. This progress was particularly marked in English industry.

The first patent for the mechanical drawing of thread was filed in 1738 by the Englishman, L. P. Lewis. In 1764, again in England, **James Hargreaves** built the first mechanical spinning frame, the spinning jenny.

Sir Richard Arkwright

Linen industrial spinning was invented in 1890.

Water frame (1769)

In 1769, the Englishman **Richard Arkwright** invented a spinning frame that permitted nonstop spinning. The mechanism, whose design was superior to Hargreave's jenny, depended on water power to make it work, hence the name *water frame*.

Arkwright's spinning frame had a fundamental impact on the spinning industry and provided a base for modern, mechanical spinning.

Mule jenny (1779)

Technical difficulties associated with spinning cotton were resolved in 1779 by the Englishman **Samuel Crompton**, who invented the mule jenny. Crompton's spinning machine blended the principles of the water frame (for the drawing mechanism) and the jenny (for alternating operation), thus permitting elimination of the drawbacks of each. Crompton's invention, which has only been refined in a few details, is nowadays known as a winder.

In 1824, the Englishman R. Roberts improved Crompton's machine by introducing a high-speed system for the intermittent movement of the carriage (the self-acting mule).

Ring throstle (1828)

In 1828, the Englishman **J. Thorpe** invented the ring throstle, which W. Mason improved in 1833.

This machine, still known under its original name, is most often used for spinning wool, combed cotton, and synthetic fibers.

Thorpe's machine culminated the series of English spinning inventions. In the space of half a century, spinning had moved out of an archaic state into the industrial era.

Spinning flax (1810)

It is not known whether **Philippe de Girard** (1775–1845) was already preoccupied with spinning flax before the Napoleonic decree of Bois-le-Duc in May 1810. This law promised a compensation of 1 million francs to whoever invented the best and most economical machine system. Whatever the case, on the following 18th of July, he filed his first patent for an entire mechanized industrial complex capable of spinning flax and hemp.

De Girard was left in debt because of the insolvency of the imperial government, which never made good on its promise of a reward, and was hounded without pity by several of his creditors. After Waterloo (1815), he gave in to solicitations by the Austrian government and moved his spinning mill to Hirtenberg, near Vienna.

7. Looms

Origins (Antiquity)

The art of weaving, like that of spinning, goes back to prehistoric times. Nonetheless, even if up until the end of the 18th century handlooms were always made of wood, the idea of mechanizing weaving was still the object of a succession of inventions.

In 1606, the Frenchman Claude Dangon conceived the warping frame. This contrivance made it possible to move the heddles by means of a pulley arrangement. Thanks to this system, more intricate patterns could be woven.

In 1725, the Frenchman B. Bouchon improved Dangon's invention by introducing removable and interchaneable warp protectors. This meant a woven design could be held in reserve, then used again.

H. Falcon further improved this sytem in 1728. He proposed a series of cardboard rectangles in which perforations determined the decorative design.

Machine-made lace (1776)

In 1776, the French engineer **Leturc** presented a lace frame to the Academy of Sciences. The machine was very complicated. It had no less than 700 vertical spools that served as bobbins.

In spite of promises of official subsidies, which Leturc needed to market his invention, these were not forthcoming. A worker even stole his idea. Discouraged, the inventor went to reside in England for about 10 years. He made his experience available to his hosts there and trained numerous workers.

A traditional weaving loom in England around 1890.

The introduction of mechanical looms drove craftsmen to misery.

Mechanical looms (1785)

The first loom to function mechanically was invented by the Englishman **Edmund Cartwright** in 1785. The first attempts at developing such a device had been made by a French naval officer, De Gennes, who in 1678 presented to the Paris Academy of Sciences a machine to make cloth without the aid of any worker.

Jacquard loom (1804)

In 1804, the Frenchman **Joseph-Marie Jacquard** (1752—1834) invented, at Lyon, a loom (known as the Jacquard loom) for brocaded fabric. He also invented various other appliances for the weaving and fabrication of fishnets. On his loom, a weave could be produced mechanically according to the design of the weaver. It was sufficient to keep track of the warp protector cards in order to reproduce any given pattern. Jacquard never took out a patent on his invention.

Automatic loom (1822)

In 1822, the Englishman **R. Roberts** invented an entirely automatic loom, which was soon adopted throughout Europe. It was subsequently improved by different technicians and constructors, who invented various devices to manipulate ever larger numbers of heddles.

In 1890, the American J. H. Northrop conceived an automatic loom, which, with several improvements, is still used by the textile industry.

8. Knitting

Stocking frame (1589)

The first loom for making stockings was invented in 1589 by the Englishman **William Lee**.

In spite of Lee's questioned competence (he received a Master of Arts from Cambridge in 1582), Queen Elizabeth I refused him the right to found a factory in order to exploit his invention. He did not have any better luck in France. The assassination of Henri IV prevented him from exploiting the privilege the French sovereign had just granted him.

Back in England, Lee improved his machine with the aid of Aston. The two men opened a mill in Nottinghampshire, where, for the first time in the world, silk stockings were fabricated industrially. This was proof that machines could work with a material as fine as silk, and ten times faster than man.

In 1657, Cromwell granted a charter to the "masters, directors, assistants, and companies of the knitting trade."

Looms and the hosiery trade

In January 1656, Louis XIV granted the Frenchman Jean Hindret the privilege of founding a mill for the fabrication of stockings, women's undershirts, and other articles of silk. This enterprise was to be located on the Chateau of Madrid in Neuilly. Hindret equipped his mill with Lee-type looms.

Inventors improved these machines in several ways. In 1775, the Englishman Jeremiah Strout launched the side-on system, or stitching at two points. In 1764, Morris and Betts invented the lock-stitch mechanism. In 1781, Morris developed the stilleto loom, or pineapple, which permitted knitting of motifs.

Rip-stop knitting (1775)

Invention of the chain-stitch loom is attributed to the Englishman Crane, in 1775. Even so, the first patent covering the making of rip-stop knitwear was not filed until 16 years later, in 1791, by Crane's fellow countryman, Davson.

Circular loom (1798)

In 1798, the Frenchman **Decroix** patented a circular machine that fabricated tube stockings without seams. This loom was constructed at Lyon by Aubert, who had an eye on the World's Fair in 1802.

At first used in England (Marc Isambard Brunel installed it there in 1806), the circular loom was not widely used in France until 1830

A clockmaker from Troyes, France, Jacquin improved this machine in a number of ways. In 1837, he built a circular loom with a reduced diameter for weaving nightcaps. Jacquin also developed a large diameter loom that produced women's undershirts, men's undershirts, and underpants.

First power loom (1832)

The first attempts at developing automatic looms, operated by mechanical rather than human energy, seem to have been made in England, at Loughborough, in 1828. Nevertheless, it was in the United States in 1832 that **E. Egberts** and **Timothy Baily** opened a mill using the rectilinear loom, which turned out four articles simultaneously and was powered by hydraulic energy. In 1838, an English emigrant to the United States, John Button, set up in Germantown, Pennsylvania, a hosiery factory for children's wear, with the looms powered by steam engines.

Rectilinear loom and spindle whorl (1857)

The first loom to have an incorporated spindle whorl appeared in Troyes, France, toward 1840. It was the work of the mechanic Poivret.

In 1857, the Englishman **Luke Burton** introduced a significant improvement. He built a rectilinear loom with spindle whorl, having a metal framework and automatic movement, which produced socks and fully fashioned stockings.

In 1861, Arthur Paget filed a patent for a further improved model.

In 1863, the Englishman William Cotton filed a patent for a model very similar to Paget's, but it had the advantage of producing from two to a dozen stockings simultaneously, instead of only one.

Automatic fashioning (1834)

In 1834, after years of research, the French hosiery tradesman from Troyes **Joseph-Auguste Delarothiere** filed a patent on the invention of a machine that automatically fashioned textile articles.

Home knitting machines (1860s)

The first home knitting machines appeared during the 1860s. Toward 1865, machines produced by the Englishmen **Herrick** and **John Aiken** were closely modeled on circular loom systems. Operated by a crank, they could be screwed to a tabletop. Another model, built by Lamb and Buxtorf, also appeared in 1865.

9. Refrigerating

Compressed-ether refrigerating machine (1805)

In 1805, **Oliver Evans** presented his design for a compressed ether refrigerating machine, exhibiting a prototype in Philadelphia. He imagined a system for drawing up vapor with a pump, compressing it, and finally condensing it. His particular innovation was the use of a closed cycle for the process.

The American Jacob Perkins, who was born in Massachusetts but who had been living in London since 1819, filed a patent in 1834 on a machine for ether compression in a closed cycle. Since Evans and Perkins had been friends while Perkins had been living in Philadelphia, it is reasonable to consider Evans as the inspiration behind Perkins's invention.

It is James Harrison, a Scottish emigrant to Australia, who takes credit for developing these prototypes into industrial machines. In 1855, he was granted his first Australian patent. His compression refrigerating machine was implemented in the first warehouse for frozen meat storage.

Great Britain granted a patent to Gorrie in 1850. Ironically, his invention caused a scandal in his own country. A major New York newspaper, *The Globe*, wrote: "There is an eccentric, down in Florida, who thinks that he can make ice, with a machine, as well as Almighty God." Gorrie could not get an American patent until 1851.

Expanding-air refrigerating machine (1844)

In the middle of the 18th century, it was already known that expanding air, compressed beforehand, was ac-

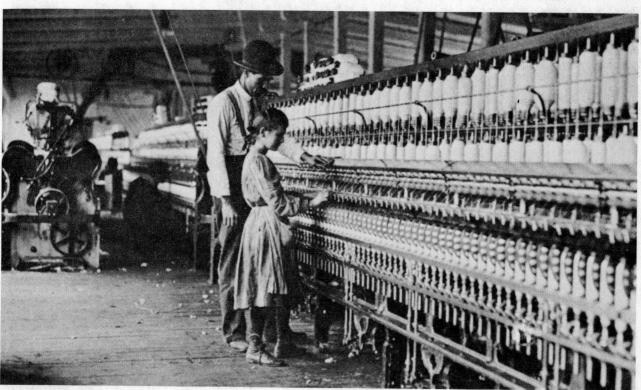

A weaving plant in North Carolina around 1908.

companied by a cooling effect. The American **John Gorrie** merely applied this principle of air expansion when in 1844 he fabricated his expanding-air refrigerating machine. A doctor practicing in Florida, he invented this machine to comfort his patients. This was because natural ice, which came from the far north of the United States, was only available sporadically during the summer.

Absorption refrigerating machine (1859)

The Frenchman **Ferdinand Carre** patented an absorption refrigerating machine in 1859. The new idea was that the refrigerant fluid (which produces the cold effect), after having acted as a coolant, was absorbed by another substance instead of being drawn up by a compressor.

Following Carre's patent, small household machines were constructed that functioned periodically and could produce from 1 to 5 pounds of ice.

A continuously functioning machine employing Carre's principle was the first to be utilized in a significant way in industrial application.

Water vapor refrigerating machine (1866)

In 1755, the Scot William Cullen obtained a small amount of ice by vaporizing water in vacuum under a bell jar. In 1777, another Scot, Gerald Nairne, had the idea of adding sulfuric acid in order to absorb the water and thus accelerate the process. But it was not until 1866 that a water vapor apparatus emerged. **Edmond Carre**, younger brother of Ferdinand (*see* above), developed a practical system utilizing sulfuric acid. Carre's machine was an immediate success in its use as a means of icing pitchers.

SO_2 compression refrigerating machine (1874)

In 1874, a physics professor in Geneva, **Raoul Pictet** (1846—1929), utilized sulfur dioxide (SO_2) as the working fluid in a compression refrigeration system. Cold was produced by vaporizing a fluid.

Sulfur dioxide has the advantage of being self lubricating and extinctive. Its disadvantage is that in contact with humidity, it becomes sulfuric acid, which is very corrosive. Pictet overcame any difficulty by enveloping the motor and compressor in a hermetically sealed enclosure.

Pictet's machine was installed in 1876 at the first artificial skating rink in London.

Even princesses wear knitted apparel.

Ammonia compression refrigerating machine (1876)

The German **Karl von Linde** (1842—1934) is known as the inventor of the ammonia compressor, the first model of which he built in 1876. But it was actually the American **David Boyle**, a Scot who emigrated to Alabama, who obtained the first patent for a compressor utilizing this refrigerant, in 1872.

Nonetheless, Von Linde, a young professor in Munich, made certain of the success of this system by developing two types of ammonia compressor (in 1876 and 1877), which he very quickly marketed. Starting in 1877, machines of this type were installed in the Spatenbrau brewery in Munich, in order to refrigerate wort in the fermentation vats. Later, they were installed at the Westminster brewery in London and at the Carlsberg and Tuborg breweries in Copenhagen.

The winter in 1883—1884 was particularly mild in Germany. The shortage of ice that resulted also contributed to Linde's success.

10. Machine tools

Drilling machine (Antiquity)

The exact date at which the drill was invented is unknown. The need to build machines specifically adapted to drilling seems to have arisen rather recently. Yet drilling machines, along with lathes, are among the oldest machines known. Homer notes their existence in the *Odyssey*.

The horizontal drill press built by the Frenchman Jacques de Vaucanson (1709—1782) belongs to a series of workbench tools of simple design fabricated by 18th-century clockmakers.

One of the best known drilling machines was built by the Englishman J. Withworth in 1835.

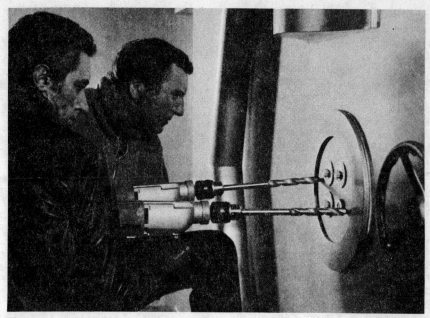

We do not know who invented the drill, but we do have an idea of its various applications.

Welding (4th millennium B.C.)

Heterogenous welding as a means of joining two pieces of different metals dates back to the 4th millennium (approximately 3500 B.C.), at the beginning of the Bronze Age. Autogenic welding is more recent (around 1500 B.C.). The only welding procedure for iron and steel until the end of the 19th century was forge welding, and it was not until 1877 that the American engineer, Elihu Thomson, invented electric welding. Arc welding appeared at the beginning of the 20th century, at about the same time as Charles Picard's oxyacetylene blowtorch in 1904. Today, welding by means of electronic bombardment and laser is more and more used.

Planing machine (1751)

In 1751, the Frenchman N. Focq invented one of the first machines for planing iron in the same way that wood is planed. It was replaced in 1805 by a machine credited to the French locksmith and mechanic Caillon.

The invention of the modern planer is claimed by several English manufacturers: Bentham (1793), Bramah (1802), Murray (1814), Fox (1814), Rob-

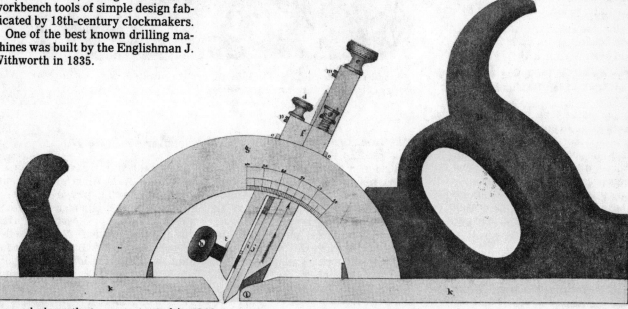

A wood plane that was patented in 1844.

erts (1817), and Clement (1820). The planing machine was improved mainly by the Englishman J. Withworth in 1835.

Slide lathe (1751)

Toward 1751, the oldest known metallic slide lathe was invented by the Frenchman **Jacques de Vaucanson** (1709—1782). Vaucanson most certainly constructed this machine for traversing, that is to say, lathing pieces of metal in a cylindrical fashion to produce cylinders that he used in building his looms.

Reaming machine (1775)

The reaming machine was invented by the Englishman **J. Wilkinson** in 1775. (The term *ream* means to polish and adjust the interior diameter of a cylinder.) Wilkinson was an ironmaster known for the cannons he manufactured. In 1775, he conceived a device that permitted the correct execution of a bore. The brace chuck bar was held at each end by a steady rest. The tool holder worked parallel to the axis of a cutting tool in fix position so that it could generate a cylindrical surface. This device can be considered the first industrial machine tool.

Hydraulic press (1796)

The hydraulic press was invented in 1796 by the English mechanic **Joseph Bramah** (1749—1814).

This machine has the advantage of being able to provide a very strong mechanical thrust, with motion that is simultaneously progressive. The press works in accordance with hydrostatic principles set forth by Pascal.

Screw-cutting lathe (1797)

The threading lathe was invented by the Englishman **H. Maudslay** toward 1797.

The lathe, one of the oldest and most significant of machine tools, was considerably improved and modified throughout the 19th century. Through the addition of various accessories and devices, many different operations can be carried out. One of these is threading, which consists of cutting a helicoidal groove in a cylindrical surface.

Standardization of screw threads (1841)

The standardization of screw threads was the result of work by the Englishman **J. Withworth.**

In his film "Modern Times" (1936), Charlie Chaplin asked the question: Who serves the other, man or machine?

At first, the number of nuts and bolts used in the machine construction industry was small, and each mechanic made up his own as they were needed. The Englishman Maudslay was the first to pursue a threading standard, working between 1800 and 1810. His influence on his brilliant student Withworth was considerable. Withworth's work culminated in 1841 with the adoption of the Withworth thread as a standard. It required a number of years to effect standardization fully.

Circular saw (1799)

The inventor of the circular saw was a Parisian. **A. C. Albert**, a mechanic by trade, filed his patent request in September 1799, although the patent was apparently never granted. One was granted on March 27, 1816, to the Frenchmen Auguste Brunet and Jean-Baptiste Cochot, residing at 65 Bourbon Street, Paris. Their circular saw had tempered steel teeth welded into notches in a sheet-iron disk specially awled and machined for this purpose.

Even though this tool was intended for woodworking, it is the predecessor of segmented metal saws, other bladed tools, including those with set-in parts, which are widely utilized nowadays.

Paper manufacturing in continuous form (1799)

On Fructidor 22, the year VI of the French Revolution, Nicolas Robert, an employee of the book dealer-printer Pierre-Franis Didot, addressed a request to citizen-minister of the interior Francois de Neufchateau to obtain a 15-year patent for "the manufacture of an extraordinary sort of paper, 12 to 15 meters long, without the help of any worker, and by purely mechanical means." On January 18, 1799, he obtained his patent, acknowledging him as the inventor of the continuous form paper machine.

This invention proved very significant to the paper trade. It freed the industry from traditional methods, employed since the Renaissance. It was now no longer necessary to make paper in vats, sheet by sheet.

Milling machine (1818)

The milling machine was invented in 1818 by the American **Eli Whitney**. Whitney, who had already made a name for himself in 1792 by developing the cotton gin, had been director in an arms factory since 1815. The milling machine he designed was to permit machining of rifle parts being made under a government contract.

Some improvements were later made. Toward 1814, two Americans, Gay and Silver, developed a horizontal milling machine similar to Whitney's. However, it was more robust.

Filing machine (1826)

In 1826, the first filing machine was invented by the Scot **James Nasmyth** (1808—1889). Intended to replace cold chisel work but obtain a fine file finish thanks to the regularity of the cut, this device was at first called Nasmyth's steel arm. The most important improvement made on this shaping machine was a rapid tool to return developed by the Englishman J. Withworth.

Grinding machine

There is little precise information on the origin of the artificial grinder. In England, the first sharpening stones, called emery wheels, were made around 1840. And it is known that in 1842 the Frenchman Malbec put together an artificial grinder, bonding quartz with shellac.

Steam power hammer (1836)

Invention of the steam power hammer is generally attributed to the Englishman James Nasmyth, in 1841. However, in 1836, there was already a steam power hammer working in F. Cave's shop, in Paris. He was one of the founders of the French machine tool industry.

Invention of the forging press, a sort of power hammer, had a major influence on the heavy metal industry. This was because of the press's great force, with permitted forging of parts with massive dimensions.

Nasmyth was one of the most productive English inventors. At the age of 19, he made a name for himself by building a steam-powered automobile that he drove around Edinburgh for several months. His steam power hammer made him famous at the World's Fair in 1851.

Paper cutting machine (1844)

The inventor of the paper cutting machine is the French cutler **Guillaume Massicot** (1797—1870). On March 18, 1844, he was granted a patent for a machine with which "it became impossible to cut poorly." This machine was such a success that in France it is known by his name.

Road roller (1859)

The Frenchman **Louis Lemoine** invented the steam roller on May 27, 1859. The invention revolutionized road construction and marked the beginning of industrially constructed highways.

It is the quality of the roadbed that makes for good roads. The French construction system, inspired by Roman roadbuilders and developed by the general inspector of roads and bridges Pierre Tressaguet in 1764, was improved on by the Scot John McAdam (*see* Macadam, page 272), then by Thomas Telford. Each's technique involved packing of the road surface. Previously, this packing was done manually, then later using a roller harnessed to oxen or horses.

Jackhammer (1861)

The pneumatic pick, or jackhammer, was invented by the French engineer **German Sommelier** while digging the Mont Cenis tunnel in 1861. This project, which was to link France with Italy via a tunnel, was begun using traditional methods. Specialists forecast that the work would take 30 years. This motivated Sommelier to look for more efficient methods. Modifying steam-powered drills, he devised drills that worked with compressed air. Work advanced three times faster thanks to his pneumatic drills and was completed in 1871.

11. Printing

Three printing processes were successively invented: relief printing (*typography*), engraving (*heliogravure*), and transfer (*offset printing*).

Typography (1450)

Typography dates back to 1450. This technique was perfected thanks to **Gutenberg**'s past achievements.

Typography, using individual let-

Moveable lead letters were invented by Gutenberg.

Copper-plate engraving in the 17th century.

ters and characters, should not be confused with relief printing, which already existed in the form of wood engravings (xylography) for reproducing combinations of text and illustrations. The Chinese made use of this xylography. Wood engravings dating from the 9th century have been found in the Far East.

Copperplate engraving (15th century)

According to Giorgio Vasari (1550), this intaglio process was discovered when a 15th-century Florentine jeweler, **Masa Finiguerra,** got the idea to ink an engraved silver plate and then apply a sheet of paper to it to obtain a printed impression.

The term *copperplate engraving* covers all intaglio processes from metal plates.

Photogravure (1821)

We owe the invention of photogravure to **N. Niepce,** around 1821. Niepce actually invented photogravure (*see* Chapter 4, "The Arts") before photography. In fact, his invention gave rise to use of a printing form (the engraved copperplate) to produce an existing image by means of a press. Replacing man-powered tools by light, photogravure contributed to the growth of typography and gave rise to heliogravure and offset printing.

Aqua fortis (16th century)

In olden times, copperplates were engraved by means of an etcher's needle. Early in the 16th century, aqua fortis began to be used for etching. Aqua fortis is the old name of nitric acid. It produces depressed lines in copperplates by etching away the parts of the plate free of surface protecting varnish. Jacques Callot (1592—1635) improved this technique by covering copperplates with the hard varnish used by stringed-instrument makers. In this way, the etched lines were thinner. However, the most famous "aquafortist" is the Dutch artist Rembrandt Van Rijn (1606—1669).

Linogravure (1864)

The invention of linoleum in 1864 (*see* Chapter 7, "Everyday Life") gave rise to linogravure, based on the same principle as relief printing. Linogravure, however, has the advantage of being used by everyone, particularly schoolboys.

Rotary printing (1895)

Printing processes are not limited to book printing. The textile industry also uses printing processes. Rotary printing technology began its development in the last century in the textile industry. As early as 1895, the Englishmen **S. H. Sharp** and **L. Marcan** took out a patent for a fabric drawing machine utilizing a stencilplate and an endless band installed on a cylinder. In 1926, an American, **F. Hinneken,** invented a machine designed on the same principle. Then, in 1950, a Portuguese, **A. J. C. de Oliveira Barros,** developed a method to form the

A printing press at the New York Sun around 1850.

stencilplate into a cylinder by joining both ends. However, printing quality and the choice of drawings were not up to expectations. It was not until 1963 that the Dutch company Stork Amsterdam engineered the first color rotary press. Rotary printing is also used for floor and wall covering.

12. High Technology

Photoelectric cells (1893)

Photoelectric cells were invented by the German physicists **Julius Elster** (1854–1920) and **Hans F. Geitel** (1855–1923) in 1893. These devices permit the transformation of a luminous flux into an electric current. Thus, variations in light can be transmitted by an electric current, then reconverted into identical light variations on arrival. This permits transmission of static or luminous imagery.

Geiger counter (1913)

The Geiger counter was invented in 1913 by the English physicist **Ernest Rutherford** and his German assistant **Hans Geiger**. This apparatus detects and "counts" alpha particles, which are part of the radiation emitted by radioactive bodies.

Maser (1955)

The first maser was built by the Americans **J. Gordon, H. Zeiger,** and **C. H. Townes** in 1955. This was an ammonia maser.

A maser is a microwave amplifier (or oscillator). Its working principle is based on the quantum properties of matter. The word *maser* is actually an acronym for "microwave amplification by stimulated emission of radiation."

Laser (1960)

In 1958, the Americans **A. L. Schawlow** and **C. H. Townes** thought of orienting the maser to infrared and optical light frequencies. Soon after, toward the middle of 1960, **Theodore H. Maiman** constructed the first "optical maser," or laser, at the Hughes Research Laboratory (Malibu, California). This was a ruby laser device.

Laser is also an acronym, with the *M* for *microwave* in *maser* replaced by *L* for *light*, used in the broad sense to include ultraviolet, visible, and infrared rays. The terms laser was coined by R. Gordon Gould of Columbia University in 1957.

At the end of 1960, A. Javan built

A laser duel as envisioned for "Star Wars."

the first gas laser; in 1963, semiconductor lasers first appeared. Present technology has introduced new types of lasers—for example, die lasers.

The properties of a laser, among them the intense heat power of the beam, which can be focused to a point a mere several thousandths of a millimeter in diameter, permit various practical applications — in surgery, biology, photography, military technology, telecommunications, etc.

Laser Lancet (1960)

One of the characteristics of a laser is it produces an intense beam that is directional and can be focused. The laser lancet uses this technology. It allows exacting surgery to be performed without harming surrounding tissues. This use was tried the year the laser was invented, in 1960. Laser lancets reduce risk of hemorrhage and are often used in hospitals in combination with endoscopes (*see* Medicine). The laser lancet revolutionized the world of surgery by permitting operations requiring only small incisions in the patient's skin.

Laser disk (1979)

In 1979, **Philips Data Systems**, a subsidiary of the Dutch company Philips, invented a laser-diode optical recording process. This process allows storage capacity on dual-sided disks to reach the equivalent of 500,000 typed pages, recorded in a half second's time. This technique is now used for video disks and mass-storage compact disks.

Metal cutting laser

The focused beam of certain carbon dioxide (infrared) lasers can cut

through a metal plate several inches thick. The French Commission for Atomic Energy (CEA) uses these lasers to cut the complex parts that make up their nuclear power plants. The leather and clothing trades use less powerful versions of the same kind of lasers. They are connected to a computer and programmed to cut several fabric layers at a time.

Laser weaponry (1960)

In 1958, A. L. Schawlow, one of the inventors of the optical laser, declared in a conference that he knew nothing of the military applications of the laser and did not want to know them. However, in 1960, shortly after the first laser was produced the United States government commissioned **Hughes Aircraft Company** to carry out research in the field of laser weaponry. Missile guidance applications were first tried.

Today research is directed toward blaster laser weapons for destroying missiles, aircraft, or satellites. These new weapons are very controversial. However, certain specialists insist that laser weapons are so powerful that they would instantly neutralize any nuclear attack and thus make nuclear war impossible.

The shortest flash (1982)

In 1982, the Bell Telephone Company developed a procedure that permits production of a laser impulse 30 femto-seconds long (30 x 10 - 15 seconds). The capability of producing such short laser impulses is expected to allow for the study of ultrarapid processes—in microelectronics, for example. They should also help biologists to manipulate certain cells.

Uranium enrichment by means of laser (1980s)

This is a new procedure developed by French and American engineers. The first to ensure the industrial development of the procedure will be guaranteed comfortable commercial profits.

Laser for nuclear fusion

This kind of laser is being developed in the Soviet Union, the United States, and France. The purpose of this laser is to provide enough power to generate nuclear fusion among hydrogen isotopes.

The potential power transferred by this kind of laser beam is several billion watts, equivalent to the electricity in use by a whole continent at any given moment. Nuclear fusion lasers are expected to be operational around the year 2000.

Hologram (1963)

The principle of the hologram was set forth in 1947 by the English physicist D. Gabor, who was studying the possibilities of imaging atoms. Yet, it was not until discovery of the laser, which produces "coherent" light, that practical application could be realized.

A hologram (from the Greek *holos*, "entire," and *gramme*, "written thing") is a relief photograph produced by utilizing the interference grating that results from the superimposition of two laser beams.

The first demonstration was made by the Americans E. N Leith, J. U. Upnatjeks and C. W. Stroke in 1963.

Holographic interferometry (1963)

University of Michigan researchers discovered by chance the first industrial application of holography (around 1963) when wide dark bands crossed the image produced in one of their holograms. A quick analysis determined that a minute movement when taking photographs caused the bands to appear. Applying that insight to precision measurement allows form modifications of about one-millionth of a millimeter. The French tire manufacturer Michelin uses this technique to inspect the quality of its products.

Masterpiece holograms (around 1969)

Toward the end of the 1960s, Moscow Professor **Yuri Denisyuk** applied holography to the exact reproduction of masterpieces displayed in Leningrad and Moscow museums. Thanks to this scheme, many Soviets had the opportunity to "view" masterpieces that never left the museum.

A hologram creates a perfect 3D image when properly illuminated.

Impressed holograms (1975)

Today, a technique identical to the electroformed impression of video or phonograph records is used to mass-produce duplicates of a master hologram (as many as several million copies). The Japanese and the Americans first experimented with this in 1975. Reproduction quality is superb. Each hologram copy is as luminous and accurate as the original.

Integral-relief motion pictures (1980)

Many attempts have been made throughout the world to develop holographic motion pictures. In 1980, N. Aebisher and C. Bainier of the University of Franche-Comte, in France, made a color motion picture featuring gulls flying. Unfortunately, only six people at a time can see this film. Researchers are actively investigating film applications further.

Holographic bubble chambers (1982)

A few years ago, the European Center for Nuclear Research (CERN) in Geneva, Switzerland, replaced traditional photography by holography to take pictures in nuclear particle detectors called "bubble chambers." Holography records the three trajectories of colliding particles in three dimensions. Holographic bubble chambers were first used around 1982.

Holographic scanner (1983)

IBM designed the holographic scanner now in common use at cash market registers. It is the main component of the cash register IBM marketed in 1983. The holographic scanner uses holographic optical elements (HOE), which deflect laser beams in a peculiar way.

Holographic instrument navigation system

In France, Thomson-CSF designed a new instrument navigation system for civilian and military aircraft that allows blind flying and landing. A HOE-type hologram is used (see above). It provides the pilot with more visual comfort than do conventional displays. In a few years, instruments for airliners may resemble the cockpits seen in *Star Wars*. Projects are under way.

Optical fibers (1955)

Optical fibers were invented in 1955 by the English professor **Narinder S. Kapany**. He had studied the works of fellow countryman John Tyndall (1820—1893), a physicist who had in 1870 demonstrated certain light transport properties by reflection.

An optical fiber is a filament composed of two types of glass. A light beam directed at one end of the filament is "sealed in" by a continuous series of total reflections along the inside lateral walls of the filament.

Optical fibers have been used in numerous applications, notably in information transmission (telex, cable television) and in medicine (endoscopes). They are also used with laser beams.

Shape-memory alloys (1960)

Shape-memory alloys are made from titanium and nickel. Their exceptional characteristics allow creation of variable-geometry objects. These alloys "remember" the forms they are given under certain conditions of temperature if these conditions are met again.

In 1960, **William Buehler**, an engineer at the Silver Spring research laboratory in Maryland, experimented with "nitinol," which characteristically changes form with variations of temperature (from - C to +100 C), depending on its composition. Today, shape-memory alloys are used in the aircraft and household appliance industries and especially in surgery. Dr. Philip Sawyer of the State University of New York recently taken out a patent for an artificial heart that uses nitinol.

A new alloy tested in Belgium, Great Britain, and the United States does not contain titanium, which is difficult to find outside the Soviet Union.

information systems

1. Data Processing Machines

Hardware

Hardware consists of all the physical elements used for data processing. These physical elements include the electronic and mechanical equipment that make up an information system: a central processing unit, memory, peripheral devices, etc. In other words, hardware designates everything that is a physical part of a machine—as opposed to software (programming), which is the thinking that goes into data processing.

There is also a step in between hardware and software: solid-state logic. These devices, such as compilers, are wired using integrated circuits. This wiring technology is also the basis of microprogramming.

Designers

Pascal's machine (1642)

In Paris, in 1642, the Frenchman **Blaise Pascal** (1623—1662) designed and built the first authentic, numerical calculating machine.

Officially presented in 1645, Pascal's machine carried out addition and subtraction, as well as permitting exchange rate conversions for differing currencies existing at that time.

It is the predecessor of modern calculators. Pascal originally built his machine to ease his father's work. As president of the Board of Customs and Excise at Clermont Ferrand, Pascal's father was commissioned to reorganize finances, especially taxation.

Stepped reckoner (1671)

Gottfried Leibniz (1646—1716), a German mathematician and philosopher, invented in 1671 a mechanical calculator similar to, but more advanced than the "pascaline".

Pascal's arithmetic machine could only count. The stepped reckoner multiplied, divided and extracted square roots as well. Nevertheless, both these calculating machines used the same mechanical technique of "stepping", that is, repeating a process such as a series of additions. Even today, this is still the way many digital computers function.

Babbage's machine

In 1833, the English mathematician, **Charles Babbage**, designed a calculating device that followed instructions written on a perforated tape. Babbage's *analytical machine* was supposed to find the solution for any

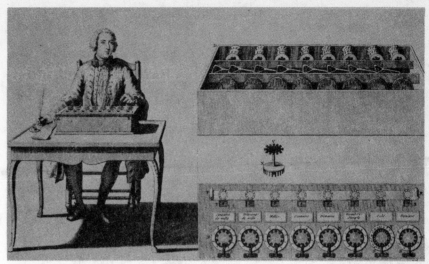

Blaise Pascal and his calculating machine, 1642.

equation, and to carry out all the operations of mathematical analysis.

To make his machine work, Babbage employed two technologies. The first was that of *jaquemarts*, mechanical figurines which sound the hour in belfries. The second was the perforated card invented by Joseph-Marie Jacquard in 1805, for use in textile weaving.

A forerunner of computers, the Babbage machine proved too ambitious for the technology of the period. Not until construction of the difference tabulator by the American company, IBM, in 1930, did the first equipment that approached Babbage's ambitions appear.

Boolean algebra

Boolean algebra is credited to **George Boole** (1815—1864), an English mathematician. In 1847, while a professor at Queen's College in Cork, Ireland, he published "The Mathematical Analysis of Logic." The first in a series of articles, this work gave birth to algebraic logic.

This method of calculation, which uses the binary system, is well adapted to logical elements that work in binary states. That is to say, two states are used, corresponding to two units of information: 1 and 0 (i.e., on and off).

At the time of their publication, the works of Boole were not particularly celebrated. It was not until 1938 that the American Claude Elwood Shannon (born 1916) proposed the first applications of Boolean algebra in information theory. And later, the American G. Stibitz used Boolean algebra in the first binary calculator, which he built at Bell Laboratories in the United States.

Hollerith's machine (1890)

The American **Herman Hollerith** (1860—1929) constructed—at the time of the American census of 1890—a device which was to launch the business machine industry. It was an electromechanical machine, using perforated cards; it consisted of a punching device, a sorter, and a tabulator with mechanical counters.

To this day the coding on perforated cards is often called "Hollerith code" in memory of this innovator. Blending the spirit of enterprise with that of invention, in 1896 he founded the Tabulation Machine Company, which, in 1957, became the International Business Machines Corp. (IBM).

Turing's machine (1936)

In 1936, the Hungarian **Alan Turing** was sent by the University of Manchester (Great Britain) to take classes at Princton, New Jersey. With the collaboration of the Argentinian logician, A. Church, he formalized the notion of calculableness, and adapted the notion of an algorithm to the computation of certain functions. Thus, Turing's machine was defined as being theoretically capable of computing any calculable function.

In 1941, Turing, who had returned to the University of Manchester, participated in the building of Colossus (*see* p. 296), which was intended for British Ciphers.

von Neumann's machine (1948)

In 1944, the Englishman, **J. von Neumann** came to join the team at the Moore School, at the Institute of Advanced Study in Princeton, New Jersey. Also working in this group were the Americans, Mauchley, H. H. Gold-

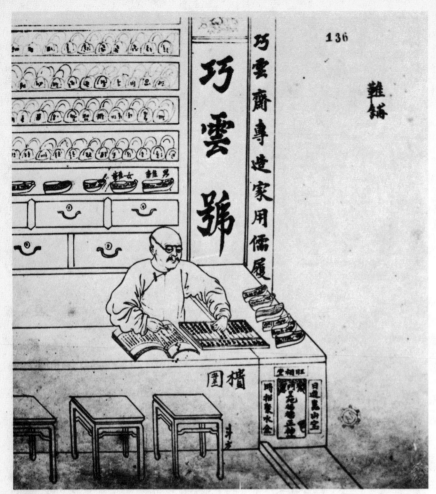

Still in use in some countries, the abacus permits very complicated calculations.

The Mexican Mayas used these strings to calculate.

stine, and J. P. Eckert. It was here that the idea of a machine with a stored program was conceived.

In this type of calculating machine the program, as well as data, are stored in the memory. A programmed calculator is a *computer*. The first machine built by von Neumann was the computer, IBM SSEC (*see* p. 297), between 1945 and 1948.

Only von Neumann's name is remembered, since he was the secretary of the Moore School, which is really an injustice as concerns the other members of the team.

Machine Types

Abacus *(circa. 3,000 B.C.)*

Ancestor of the modern calculating machine and computer, the abacus is probably of Babylonian origin sometime early in the 3rd millenium B.C.

The word *abacus* comes from the Semitic word for *dust*. Indeed, in its primitive form the abacus was a wooden tablet sprinkled with fine sand, the numerals being written with a stylus.

Archeological evidence shows that the Hindus, ancient Greeks, and later on, the Romans of the classical period all used some kind of abacus. By the 10th century, it was in universal use throughout Europe as well.

Today, the abacus is a board with beads sliding in grooves or on wires in a frame. It is widely used in India, China (suan pan), Japan (soroban), and Russia (tschoty). An abacus performs the four arithmetic operations: addition, subtraction, multiplication and division.

Mechanical computing or "Napier's bones" *(1617)*

In 1617, a Scottish inventor, John Napier (1550—1617), wrote a paper explaining division in terms of a series of subtractions, and multiplication by a series of additions. He demonstrated his ingenious method for performing these fundamental operations with the help of "bones" (so called because they were printed on sticks of bone).

Napier's technique opened the way to computation by mechanical means,

since it allowed any calculation to be made simply by repeating one or another process over and over again.

Calculator *(1624)*

The German **Wilhelm Schickard**, professor at the University of Heidelberg, built, in 1624, the first arithmetic machine capable of performing the four basic operations. He called it a *calculator-clock*.

Analog calculator *(19th century)*

At the beginning of the 19th century, analog machines, built around voltmeters and ampmeters, were developed.

An analog calculator embodies a computational scheme that uses operators which function in analogy with the orders of magnitude to be simulated. The analogy used in this type of calculator is an electrical one.

The operating devices interact in such a way as to form an electrical circuit. The circuit corresponds to mathematical equations which are representative of the phenomenon under study.

Binary calculator *(1937)*

The first binary calculator was built by the American mathematician, **George Stibitz** in 1937. He worked in the laboratories of the American firm, Bell Telephone Company. The machine was a logic contrivance whose output was the sum of the input.

Stibitz's calculator used telephone relays which can only be on or off (meaning they only use the digits 1 or 0). He was developing a general-purpose machine. To this end, he worked with a few discarded relays, two light bulbs, and a cigar box. He assembled it in one weekend.

Computer (1948)

On January 24, 1948, the IBM SSEC (Selective Sequence Electronic Calculator) was presented to the public. It carried out extremely complicated calculations, based on stored data. (*see* p. 297).

A computer is an electronic machine that is intended for the automatic processing of information. It is a general-purpose calculator which works from a stored program. The circuits of the computer constitute the hardware, while the different programs are the software.

For the first time, instructions and data coexisted within a machine—the computer was born.

Perceptron (1958)

In 1958, the American physiologist, **Frank Rosenblatt**, built the Perceptron Mark I.

The Perceptron was an *intelligent* machine, constructed on the model of the human brain. The imput devices consisted of photoelectric cells, and the output device was a CRT (Cathode-Ray Tube) screen. Interconnections within the machine were established by chance. Chemical devices, called *neuristors*, functioned as memories.

Radically different from computers, the Perceptron is a novel link in the chain which leads to *artificial intelligence*.

Integrated circuit (1959)

The American, Jack S. Kilby of Texas Instruments, filed a patent in 1959 for the first integrated circuit. The patent was granted in 1964.

An integrated circuit is an electronic device. Its different components are diffused or implanted, then interconnected, directly on a thin semiconductor substrate, such as silicon. This makes it possible to form complex electronic circuits, corresponding to complete functions.

The first commercial application

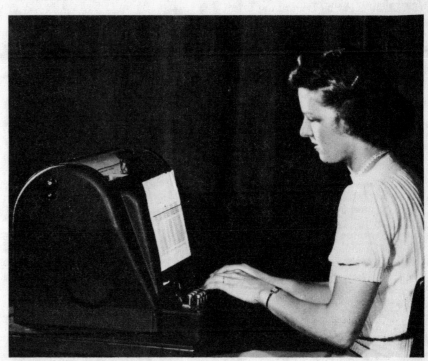

Working on a Stibitz calculator

The calculators room at Columbia University (New York) in 1930.

of the integrated circuit was in 1964. An integrated circuit by Texas Instruments was used in a hearing aid.

In 1969, Kilby was awarded the National Medal of Science.

Minicomputer *(1960)*

In 1959, a team at the American company, Digital Equipment Corporation (DEC), directed by Benjamin Curley, launched the development of the computer PDP 1, which would be marketed in 1960.

A minicomputer is a computer that takes up little space. It is often designed for a specific application, and is well adapted for real-time data processing.

Microprocessor *(1971)*

In 1971, an American company, Intel Corporation, announced the first microprocessor, named Intel 4004.

A microprocessor, commonly called an *electronic chip*, is a whole computer, layed out on a surface of several square millimeters. The circuit in the Intel 4004 microprocessor contained 2,300 transistors. Today, a circuit of the same size, such as the Iapax 432, also manufactured by Intel, contains 225,000 transistors!

In 1969, the Intel team, then headed up by an American, Marcian E. Hoff, was working for the Japanese firm, Busicom, to develop a family of integrated curcuits, intended as components for a calculator.

Microcomputer *(1973)*

The first microcomputer, called MICRAL, was marketed in 1973 by the French company, R2E under the direction of Trong Truong.

A microcomputer is constructed around a microprocessor, to which is added the program and the necessary hardware for the complete handling of information.

Electronic pocket calculator *(1972)*

The first electronic pocket calculator was developed by the Americans J. S. Kilby, J. D. Merryman, and J. H. Van Tassel, of Texas Instruments. The patent was filed in 1972, and granted in 1978. The original model is kept in the permanent collection of the Smithsonian Institute in Washington, D.C.

In 1973, Hewlet-Packard announced handheld calculators, whose functions were preprogrammed for limited domains of application (finance; economics). In 1976, this same American manufacturer marketed the first programmable calculators, actual pocket computers.

The Harvard Mark I weighed 5 tons.

Firsts

Zl *(1931)*

In Germany, in 1931, Konrad Zuse built Zl, the first calculator. Making use of binary technology, this calculator was followed in 1939 by Z2—a mechanical calculator.

In 1949, Zuse built Z3. This calculator employed relays, and was capable of executing a multiplication in 3 or 4 seconds. The Z3 was destroyed during a bombing raid during World War II, but another version, the Z4, was preserved and installed, in 1949, at the Eidgneossische Schuhle of Zurich.

In 1945, Zuse developed an algorithmic programming language which he called "Plankalkul." This made it possible for him to limit the drawbacks of the circuitry of his day.

Colossus *(1941)*

In 1941, a team at the University of Manchester (Great Britain), under the guidance of M. H. A. Neuman, built Colossus, a very powerful computer.

Manufactured at the request of British Ciphers, this computer contained more than 2,000 electronic tubes. Ten machines were assembled.

This machine, which was developed with Alan Turing's collaboration, made it possible (in 1943) to keep track of the German Army's secret codes on a day-to-day basis.

Mark I *(1944)*

The first calculator in the world to have a built-in, stored program was the Mark I, still called the IBM Automatic Sequence Controlled Calculator. It was completed in 1944 in the United States.

Research concerning this project began in 1939, based on the work of H. H. Aiken, with the encouragement of T. J. Watson, then president of IBM. This calculator considerably improved on C. Babbage's dream, offering two innovations: (1) a clock intended to synchronize the diverse sequences of operations; (2) the use of registers, an idea which was picked up by every other manufacturer.

A register is a memory with very little storage space, capable of remembering information of a certain nature. It is generally designated by the nature of this information.

Binac *(1946)*

The Binary Automatic Computer (Binac) was built in 1946 by the Americans W. Eckert and J. Manchly. It was very reliable, using delay-line memories (512 words). In fact, it was made up of two computers, which simultaneously carried out the same calculations, then compared the results.

Eckert and Manchly founded the Electronic Control Company in 1946, and produced the Binac for the Northrop Aircraft Company.

The Binac was the first computer to work in *real time*. Its reliability was phenomenal for the period. In 1949, one of the two processors in Binac operated for 44 hours without a single fault.

Eniac

The Eniac was imagined in 1946 by J. P. Eckert and his colleagues as part of a contract negotiated in 1943. The Ballistic Research Laboratory (BRL), working with the American Army, agreed that the Moore School should construct an electronic calculator able to define ballistic trajectories.

This first general-purpose electronic calculator weighed 30 tons, occupied 160 square meters of floor space, and consumed 150 KW per hour. It blended relatively new electronics technology with the well-known technology of mechanical business machines.

At the official unveiling, calculation of a ballistic trajectory was demonstrated, using a program written by Adele Goldstine, the wife of H. H. Goldstine, author of numerous computer articles.

EDSAC *(1948)*

The Electronic Delay Storage Automatic Calculator (EDSAC) was developed in 1948, at the University of Cambridge (Great Britain), by H. V. Wilkes.

It was the first totally electronic computer and it offered display screens, which served to show the contents of the delay-line memories. Its design was inspired by Ebiac, a machine having 32 delay lines, and using 4,000 vacuum tubes.

This type of computer is the forerunner of conversational graphics terminals, the development of which led to *computer assisted design*.

IBM SSEC (1948)

On January 24, 1948, the first computer in history capable of handling a stored program was presented in New York. It was called the IBM Selective Sequence Electronic Calculator (SSEC).

The IBM SSEC used 13,500 vacuum tubes, 21,000 relays, and was able to add 3,500 numbers of 14 decimal places each, in a single second.

Its construction began in 1945, under the direction of Franck Hamilton and lasted 3 years. Its design was largely inspired by the machine, Harvard-IBM, but it was 100-times faster.

Thus, on January 24, 1948, the first true computer was born.

Univac I (1951)

The first Universal Automatic Computer (Univac) was installed on June 14, 1951, in the United States Office of Census Taking. Thirty units of this model series were produced by the Americans, **Eckert** and **Mauchly**.

Univac I was the first computer that had a magnetic tape unit which functioned as a *buffer memory*, thus providing a constant data input speed to the faster, main memory.

This brought results in terms of output as well. In 1953, the first high-speed printer was connected in a commerical application (600 lines per minute).

The Univac Solid-State Computer, announced in March 1959, was the first, *entirely transistorized* computer.

Bizmac (1952)

The Bizmac was developed in 1952 by the American company, RCA. It was presented as the most powerful machine of its time.

Equipped with one of the first iron-core memories and a magnetic drum, the Bizmac was designed to handle the totality of RCA's information processing. In particular, the central processing unit had direct access to 200 magnetic tape units. Drive speed was 2 meters per second, while the tape had a data density of 499 characters per centimeter. These tape units were controlled by a perforated card reader (150 characters per minute).

Furthermore, the central processing unit had access to its peripherals through an actual network of control unit processors. This design, which is very modern, is often used nowadays.

It is also interesting to note that the Bizmac was used to support the first *data base*.

Whirlwind (1955)

Between 1944 and 1955, the binary computer, Whirlwind, was built by the Massachusetts Institute of Technology (MIT) under the direction of Jay Forrester. It was a machine to control flight simulators, commissioned at the request of the U.S. Air Force and Marine Corps. Whirlwind was operational until 1959.

Using Whirlwind, the first experiments in teleprocessing were carried out. The new technology meant data from remotely located terminals could be processed by a central computer, or results from the main machine could be processed by these distant workstations.

The Whirlwind was also used to hook up the first conversational graphics terminal. This was a visual display that permitted a man-machine dialog for the processing of information.

Atlas (1958)

Three Atlas computers were installed in England between 1961 and 1963. They were installed in 1958 by **Ferranti**, in collaboration with a team, directed by R. M. Kilburn, from the electrical engineering department of the University of Manchester.

This computer offered a paging system which let the user have access to an addressable memory that was larger than the real memory—the first *virtual memory*.

IBM 1401 (1959)

The IBM 1401, a widely marketed computer, was very reliable and relatively inexpensive. Beginning in 1959, it replaced the mechanical business machines used in many companies. This was the first computer to be manufactured in a quantity of more than 10,000 units.

The IBM 1401 was equipped with an iron-core memory. It was mass delivered, with only a reader and card puncher as peripheral devices.

Stretch (1961)

The first model of this supermachine was delivered to Los Alamos in 1961. It had been developed by IBM in response to a call for tenders from the Atomic Energy Commission (AEC).

The Stretch is a transistorized computer. It is a very high performance machine, and has 64-bit data paths. It is particularly well suited to complex scientific calculations, which require large volumes of data.

The IBM-SSEC

One important innovation was to use the *byte*, i.e., eight bits, as the basic unit of information. This idea continues to affect computer design.

Besides a complete secondary memory system on disk storage devices, the Stretch used a magnetic tape drive control scheme, the Tractor, which could manipulate 640 tapes concurrently. It would read or write one and a half million characters per second. The Stretch was operational up until 1971.

CDC 6000 *(1964)*

The CDC 6000 belongs to a family of very powerful computers. This machine was first manufactured in 1964 by the Control Data Corporation (CDC), an American company founded in 1957, in Minnesota.

This line of computers used 60-bit words. The 6600, the most powerful of the family, included a central processor which was associated with a main memory, and to several specialized processors, each equipped with their own memories. This architecture permitted parallel-mode processing, and made possible very high speed calculation.

IBM 360 *(1964)*

The IBM 360, of which six models were offered in 1964, is the result of a study undertaken in 1961 by an IBM team directed by **G. Amdhal** and **M. Blaaum**. The objective was to develop a family of computers totally compatible with each other.

This type of computer, one of the most common at this time, opened the era of modern data processing. Advanced technology was applied to the most ambitious designs.

In 1970, a new series took over, the IBM 370. This was the beginning of the *virtual machine*, and it made virtual memory commonplace.

The designation 360 for this series of computers refers to their capabilities—defined in 1961 by T. J. Watson Jr. as an all-around (360°) family.

Supercomputers *(1976)*

The first commercial supercomputer, Cray 1, was engineered by **Seymour Cray** in 1976. With 200,000 integrated circuits, it was freon-cooled and could perform 150 million operations per second.

Cray 1 was followed by the Cray X/MP, performing 400 million operations per second. A Cray X/MP supercomputer was bought for $12.6 million in 1978 to create motion pictures and animated cartoons (*Tron*, Walt Disney Productions, among others). However, Cray supercomputers have

The Cray 1 supercomputer.

been surpassed by Control Data's Cyber 205, whose output amounts to 700 million operations per second.

Japan's project *(1989)*

Funded by 200 million dollars from the Ministry of Industry and International Technology, six Japanese companies (Fujitsu, Hitachi, NEC, Mitsubishi, Oki, and Toshiba), have teamed up to produce, before 1989, a supercomputer one thousand times more powerful than present models.

2. Implemention of Machines

Software

Software is the package of programs needed to carry out computerized data processing.

Software covers any computer program; from those for the implementation of a machine (operating system, peripheral programs, etc.) to those in a given user's application of the machine (languages, compiler, etc.).

Programming and Operating

Algorithm *(ca. 1800 B.C.)*

As early as 1800 B.C., Babylonian mathematicians of the Hammurabi period set forth algorithms to resolve certain numerical problems.

An algorithm is a finite series of elementary operations, intended to solve a problem. The word "algorithm" itself is much more recent, and has its origin in the Middle East. It comes from the last name of a Persian scientist, Abu Ja'far Mohamed ibn Musa al-Kohwarizmi. This man's work on arithmetic, published toward 825 A.D., continued to influence thinkers for several centuries.

The idea of mechanizing algorithms goes back to the year 1000. The works of the Frenchman, Gerbert d'Aurillac (ca. 938—1003), later Pope Sylvester II, should be noted.

A computer program is the translation of an algorithm into a well-determined language.

Programming *(1833)*

Programming is the field of activities that is oriented toward the design, im-

The Pentagon computers room as seen in the film "War Games."

plementation, testing, and maintenance of computer programs.

The first programmer in history was Lady Ada Byron, countess of Lovelace. She was the daughter of the celebrated English poet of the same name. She wrote numerous programs for Babbage, starting in 1833. (See Babbage's machines, p. 293).

Encoding (1890)

Strictly speaking, the encoding of information goes back to the first perforated-card controlled devices, such as Jacquard's looms or Babbage's machine. The encoding task consists of representing a set of information in a code which is comprehensible by the central processing unit of a computer.

Two engineers have had a particularly pronounced effect on the evolution of encoding. First, an American, Hermann Hollerith, defined a code for perforated cards, which were used in the American census of 1890. Second, a Frenchman, Emile Baudot, invented the code for the original telegraph.

Assembler (1950)

A symbolic assembly language, or more commonly *assembler*, was used for the first time toward 1950. It was employed on EDSAC, at Cambridge (Great Britain), by a team under M. Wilkes' direction.

It is the lowest-level programming language for a computer. It permits description of applications, using a mnemonic code (by symbolic instructions).

In 1951, symbolic instructions were also used by Grace Hopper on Univac I.

The first assembler, delivered by an industrial firm, was the Symbolic Assembly Program (SAP) developed by the United Aircraft Corporation (UASAP), and installed on the IBM 704.

Microprogramming (1951)

In 1951, H. V. Wilkes introduced the concept of microprogramming. The objective was to simplify the functioning of a computer's central processing units.

Microprogramming is a technique used to implement the control program for a central processor. Each instruction is decomposed into a series of microinstructions.

Computer operating system (1954)

In 1954, Gene Amdahl developed the first operating system, on an IBM 704.

Such a system is the set of base programs for a computer. They permit the machine to use all of the capabilities available to it. The operating system also ensures management of work to be done, and of peripheral units.

Program package (1959)

The Computer Science Corporation, is frequently cited as the first to sell packaged programs, in 1959.

A package is a collection of programs, accompanied by their documentation. The package allows users to implement a given data processing application without modifying it. In principle, such a product should be able to run on different machine types.

Multiprogramming (1961)

In 1961, the Stretch was the first computer to use multi-programming; a procedure for operating a computer that permits several programs stored in the computer's memory to run concurrently.

Data processing on a time-sharing basis was first accomplished at the Massachusetts Institute of Technology (MIT), in 1961, under the direction of F. Corbato.

Corbato perfected the Compatible Time Sharing System (CTSS) for operation of IBM 709 and 7090 computers. The first time-sharing system to be marketed was on a PDP 1, in 1962.

Languages

FORTRAN (1954)

The language, Formula Translator (FORTRAN), was created in the United States, in 1954, by **John Backus**. A first version was published in November 1954.

FORTRAN was the first truly advanced programming language. It permitted concise formulation of a problem, and made use of mathematical notation.

In 1955, a compiler, that is to say, the program permitting a transposition from natural language into machine language, was developed for a version of this language, FORTRAN II. It included automatic error detection of syntactical errors made by the user. This compiler was offered to all IBM customers who possessed an IBM 704 computer. It was accompanied by the "Preliminary Operator's Manual," the first work of this type to be published anywhere in the world.

Nowadays, the versions, FORTRAN IV, which dates back to 1962, and FORTRAN V, which was worked out in 1979, are used.

APL (1956)

APL is a generalized programming language. It is founded on dense notation, which is concise and rigorous. It was developed, starting in 1956, by the American mathematician, **Ken Iverson**, at Harvard University.

APT (1956-1957)

The programming language, Automatic Programmed Tool (APT) was roughed out by D. T. Ross in 1956—57.

This language, which is used to implement numerically controlled machines, was developed thanks to work being done on FORTRAN.

Lisp (1958)

The programming language, Lisp, was developed on an IBM 704 in the

United States in 1958. The authors were a team at the Massachusetts Institute of Technology (MIT), directed by J. McCarthy. Very soon thereafter, a second version of Lisp was implemented on an IBM 709 computer. At present, Lisp has become compatible with the majority of commercially-available computers.

This language, designed for handling chains of characters as well as lists, has since gone beyond these first objectives. It has shown itself to be well adapted to data processing, and for handling certain problems in artificial intelligence such as the problem posed in a chess game and those raised by automatic translation.

A Lisp machine is now marketed by Lisp Machine, Inc., a company founded by one of the faculty at MIT.

Algol (1958)

The programming language, Algorithmic Language (Algol) permits the writing of algorithms, that is to say, the description of elementary sequences in a computation.

This programming language, first called International Algebraic Language (IAL), was presented in Zurich (Switzerland), in 1958.

COBOL (1959)

Common Business Oriented Language (COBOL) is a programming language well adapted to management and administration. It was designed, starting in 1959, by a working group, the Conference on Data System Languages (Codasyl). This group was made up of civil servants, university professors, and manufacturers, who, in 1960, published the first report on this language.

COBOL is well suited to data processing, when large files are used. It is also good for manipulating alphanumeric data, and for writing reports from the results.

BASIC (1965)

The programming language, Beginner's All-purpose Symbolic Instruction Code (BASIC), was conceived in 1965 at Dartmouth College, Hanover, New Hampshire. It was the work of a team directed by John Kemeny and Thomas Kurtz, under a grant from the National Science Foundation.

This easy-to-learn programming language was to permit non-data processing specialists to write, on their own, programs for computer applications. At first widely used in teleprocessing through time sharing, BASIC has become important as a programming tool for microcomputers.

PL/1 (1964)

In 1964, a very versatile language was introduced: Programming Language One (PL/1).

It covers the application domains of other languages, such as FORTRAN, COBOL, or Lisp. It is independent of the type of computer used. The semantics and syntax of this language have been defined and standardized.

Pascal (1969)

The first compiler written in Pascal was installed in 1969 on a CDC 6400 computer. Its inventor is N. Wirth.

The programming language, Pascal, was developed in order to teach programming that is easy to install on different types of computers. Its implementation is particularly simple.

Ada (1979)

Following a call for tenders from the United States Department of Defense, a team at CII-Honeywell Bull (France), directed by Jean Ichbiah, developed Ada. It is a programming language which resulted from the Common High Order Language Effort. This project was launched in 1974, and completed in 1979, that is, after 5 years of work.

Ada, which at first was called Green, owes it name to Lady Augusta Ada Byron, countess of Lovelace. (*See* Programming. p. 299). This language is largely based on concepts that were introduced by N. Wirth in the language, Pascal.

Concepts

Automatic control engineering (1913)

Torres Y. Quevedo, a Spanish engineer, deserves credit for having founded in 1913, the science of automatic control engineering. It deals with the study and building of mechanisms, as well as systems, capable of functioning without human intervention.

Quevedo named this science *automatos*, thinking of the implementation of automatons. Unfortunately, since no one needed the computations at that time, he could not find the financing necessary to advance his work.

Remote processing (1940)

In 1940, Bell Laboratories (USA) carried the first experiments in teleprocessing. The telegraph network linking a terminal in Hanover, New Hampshire, with a computer installed in New York was used.

Remote processing is the automat-

ed use of date processing systems, utilizing telecommunications networks.

Data base (1952)

A data base is the complete, nonredundant information set that is necessary to some series of automated applications. The first data base was implemented on the Bizmac, in 1952, by the American company, RCA.

The information is organized into a structure by a special program: the data base management system (DBMS). This makes it possible to retrieve data and to manage its updating.

Cybernetics (1940)

Cybernetics, as a science, was invented by N. **Wiener**, in 1940. But it was not until 1948 that the word itself was coined by N. Wiener and A. Rosenblueth. It comes from the Greek word *Kubernos*, which means pilot.

Cybernetics is the study of communication and control mechanisms in machines as well as in human beings.

Bit

The term *bit* is a contraction of the expression *binary digit*. A bit is a binary unit of information. It designates either of the two elements, 0 and 1, which are used to code information in a computer.

The first use of the word, *bit*, was by professor John Tukey, concerning work on Eniac.

Byte

The word *byte* commonly designates an *8-bit byte*. That is to say, eight binary elements. The byte first appeared as the basic unit of information on the Stretch, a high performance, transistorized computer built by IBM in 1961.

Simulation (1962)

In 1962, the first general-purpose, simulation languages were proposed: (1) SIMSCRIPT by the Rand Corporation and (2) GPSS by IBM.

Simulation is a field of technology which permits studying a system's future behavior through imitative representation. This is usually based on an appropriate mathematical model. The model is programmed in order to study the evolution of different variables which are representative of the phenomenon under analysis.

Access Methods

Plotter (1833)

The first plotter was conceived by the English mathematician, **Charles Bab-**

bage (1792—1871). He thought of this device in 1833, as a way to express the computational results obtained by his analytical machine.

Card sorter *(1890)*

The perforated-card sorter was invented in 1890 by the American **Herman Hollerith**.

This electromechanical machine permits the classification of perforated cards, according to a given reference system, either in ascending or descending order. It also allows sorting by elimination or by selection, as well as counting cards which correspond to different chosen criteria.

Speech synthesis-vocal unit *(1933)*

The first electronic talking machine, the *Voice Demonstrator* (Voder), was built by 1933, by **Dudley**.

After building his Voder, Dudley developed, in 1939, the first Voice Coder (Vocoder).

Vocal units function on two principles: (1) reconstitution of a message by stringing together prerecorded words, and (2) voice imitation, using a multichannel vocoder, that is, a sort of electronic organ.

In the second case, the computer has to calculate the frequencies to be emitted, together with the duration and amplitude of the constitutive phonems in a message. Applications are being found in many industries, such as automobiles (Renault 11), ovens, diverse machines.

A predecessor to modern voice synthesis appeared around the beginning of the 18th century when Baron Van Kampelen introduced his talking machine. Faber, a few years later, perfected the design and it could then produce several intelligible words.

Terminal *(1940)*

The first experiment that involved a terminal connected to a remote computer was conducted in 1940 by Bell Laboratories. The computer was in New York and the terminal in Hanover, New Hampshire.

A terminal, strictly speaking, is a device which allows access to a computer, situated some distance away, and between which there is a data transmission line.

Printer *(1953)*

The first genuinely high-speed printer was developed by Remington Rand (USA) in 1953 for Univac.

A printer is a computer output device which permits printing of predefined characters on paper, in a

A terminal in the form of a cash register.

streaming fashion. The first printer had a print speed of 600 lines per minute, with 120-character lines.

In 1957, IBM introduced matrix printing technology, which enabled print speeds of 1,000 lines per minute.

Thermal printing techniques (Texas Instruments, 1966), and laser printing (IBM 3800), at a print speed of 30 pages per second in a commercial format, are also noteworthy.

The sheets of paper printed in this way constitute computer listings. (*See* Lists p. 302)

Conversational graphics console *(1963)*

In 1963, the conversational system, Sketchpad, was presented to the data processing industry. It was revolutionary, resulting in the birth of computer-assisted design (CAD).

Sketchpad permits an artificial presentation, in the form of an image on a display screen, of numerical data coming from a computer.

Actually, two conversational systems were developed simultaneously: (1) DAC-1, by General Motors, and (2) Sketchpad, by the Lincoln Laboratory of the Massachusetts Institute of Technology (MIT).

It is also worth mentioning that in Europe, beginning in 1968, the Department of Data Processing at the University of Brumel, Uxbridge (Great Britain), played a role promoting this new discipline.

The printing chain of a printer. Printers were 45 times slower than nowadays.

Optical pencil.

The IBM laser printer churns out 30 pages per second.

Light-pen*(1963)*

The first light-pen was presented in 1963. It was a part of the conversational graphics system, Sketchpad, which was developed at the Massachusetts Institute of Technology (MIT) by **I. E. Sutherland**.

A light-pen is a device for graphic interaction with a visual display. It is pencil shaped, and works by detecting the luminous point shown on the display screen. The detected light is transformed into an electrical signal, which is transmitted to the computer. The computer then establishes a correlation with the appropriate information.

Graphic tablet*(1964)*

M. R. Davis and **T. D. Ellis** presented the Rand tablet in 1964; it was the first of its kind, developed by the Rand Corporation (USA).

A graphic tablet is a graphics data-input device for a computer. Davis' and Ellis' tablet was like a drafting board.

In 1966, the Lincoln Laboratory announced the *wand tablet*. This device, using microphone sensors, detected the position, in three dimensions, of an ultrasonic emitter in a pen.

Data Supports

Perforated card*(1805)*

The Frenchman **Joseph-Marie Jacquard** (1752—1834) invented the perforated card, which he used in 1805 on his loom.

The Englishman, Charles Babbage, also used it as an input and output data support for his analytical machine.

However, it was Herman Hollerith who really launched its use, at the time of the American census of 1890, thus creating the business machines industry.

Memory*(1833)*

The first memory, linked to a calculator, was that of **Charles Babbage's** *analytical machine*, in 1833. It was constituted of perforated cards and sprocket wheels.

The first magnetic tapes were connected to the computer Univac I, in 1951.

Iron-core memory was utilized for the first time in 1949, on Forrester's Whirlwind. However, it was not until 1954 that the first computer with iron-core memory was marketed: the Univac 1103.

List*(1886)*

The term "list" was used for the first time by the American **W. S. Burroughs** in 1886. It designated the results obtained from the printer for the Burroughs Adding and Listing Machine.

A list is generated by a printer, and shows the set of results from a processing task, submitted to a computer.

For a long time, lists required a special format. Yet today, certain printers produce them in a commercial format, such as the IBM 3800.

Perforated tape*(1939)*

Starting in 1939, **G. Stibitz** made use of perforated tape as data support during his research at the data processing laboratories of the Bell Telephone Company.

Magnetic drum*(1946)*

Envisaged as early as 1946 as a means of increasing memory capacity, drums were the first magnetic information supports. Towards 1950, they were being manufactured by several firms.

Magnetic tape

The first magnetic tapes appeared very early. They were of steel and very brittle. It was not until 1948 that IBM invented a device which operated in a vacuum.

The first plastic tapes were used in 1951, by Luis Fem, on the Raytheon Faydac.

The first computer that was equipped with a magnetic tape unit was the Univac I (USA). Here, the tape constituted a buffer memory between the computer and its users.

Magnetic disk*(1949)*

The first tests of magnetic disks used as computer secondary memory were carried out in 1949, on the Edvac.

In 1960, *removable disks* (disk magazines) appeared; they were constituted on a pile of disks that could be easily transported.

In 1962, IBM marketed the first removable disks (1311), which had the capacity of 3 million characters.

Towards 1970, floppy disks were introduced. Today, they are used on numerous microcomputers. One should say "re-introduced," because they were invented by the Laboratory for Electronics during the 1950's, but the idea was discarded.

Bubble memory *(1967)*

The bubble memory was designed in 1967 by a team at Bell Laboratories (USA) directed by A. H. Bobeck.

A bubble memory is non-volatile, that is, its contents are never lost, even when power supplies are removed. It was marketed in the late 1970s.

Today, bubble memories have a storage capacity of 92 Kbits. They should reach 1 million Kbits in the next few years.

Smart card *(1974)*

In 1974, the French engineer, **Roland Moreno**, submitted in the name of the company, Inovatron, the first patents for a memory card. It offers a secure reading and writing support.

The electronic card, still called *individual data support*, makes it possible to carry in miniaturized form (tag or card) information relative to the holder, e.g., civil status or bank account number. Logic circuits protect information stored in a read only memory, and prevent anyone but an authorized person from reading it.

Recently, cards even more complex have appeared. Containing a microprocessor, these cards permit most bank transactions, virtually constituting electronic money.

Fields of Application

Art and the computer

A computer can be used as an aid to artistic creation, in music, literature, or painting.

Art and science often overlap. Artists keep their eyes on the evolution of technology as they are always on the lookout for new ways to express themselves.

● Music: In this quest, composers were the first to look to the computer for help in artistic creation. The very first to do this were two Americans, **M. Hiller** and **M. Isaacson**, who composed the *Suite Illiac.*

● Voice synthesizer (1979): Xavier Rodet, a researcher at IRCAM (Paris) has been working since 1979 on a synthesizer, whose capabilities he has improved. This synthesizer can reproduce different vocal tones, notably those of singing voices. Other apparatus of this type can only sound like monotone voices.

The programs are defined from analysis of different voice or instrument sounds. In this way, Rodet has

been able to program an exact sonorous copy of the voice of the celebrated opera singer, Maria Callas.

● Plastic arts: As to plastic arts, they were a little late in making use of computer technology. Starting in 1937, first **Ben F. Laposky**, then C. **Barnet** proposed using Lissajous shapes in applied art. Yet it was only in 1963 that numerous exhibits consecrated to this form of expression began to appear. At this time, the magazine, *Computer and Automation*, organized the first contest for drawings done by a computer.

● Image animation (1951): Image animation was first experimented on a computer at the Massachusetts Institute of Technology (MIT) as early as 1951. But it was not until the early 1960s that the interest in this technique was fully understood. Today, medicine, architecture (with three-dimensional models), space exploration, and chemical engineering make great use of image animation. Simulation allows man to drive a tank on broken ground or land an aircraft safely, methodically and at low cost.

● Movies: Last, but not least, are the movies. Data processing systems to aid film production have existed since 1964. Essentially, they have been used to produce animated films. However, computer generated images have

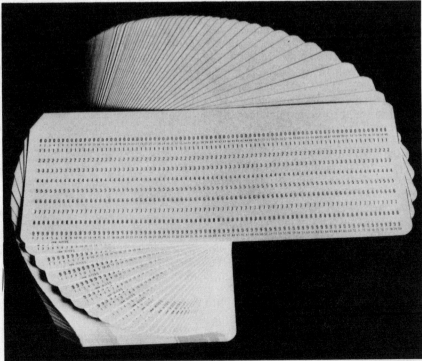

The perforated card was invented in 1805 for the Jacquard weaving machine.

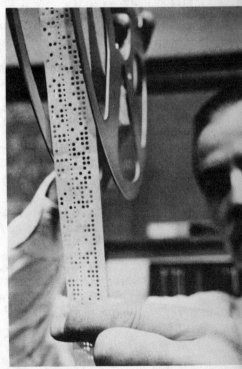

Perforated tape.

been utilized in commercial live-action feature films; Walt Disney Productions' *Tron* and Lorimar Productions' *The Last Starfighter*.

The images created for these films looked much like those filmed more conventionally with models, lights and cameras. Film producers hope that in the future computer generated images may help lower production costs for films requiring exotic locations.

Two American companies at the forefront of this application are Digital (Los Angeles) and ILM (San Francisco). The public is seeing the fruits of their labors in numerous television commercials and network graphics.

Computerized systems for visualizing physical and mathematical phenomena, as well as animated drawing systems in the more common sense, are also worth mentioning. As a reference, the works of Peter Foldes, and particularly his celebrated film, *La Faim*, should be noted.

Artificial intelligence

The source of *artificial intelligence* is found in the article "Intelligent Machinery," published by an Englishman, Alan M. Turing, in September 1947.

Artificial intelligence refers to a collection of technologies that is used in the effort to build automatons which, in a manner of speaking, think like humans. Like all recent sciences, artificial intelligence does not yet have a well-defined domain. Hence, certain definitions are sometimes contrary, or they make sense only in the minds of the scientists who are working in this field.

A series of collected articles, "Machine Intelligence" (which includes Turing's previously cited article), provided the first well-structured approach to artificial intelligence. Volume I is devoted to a seminar held in Edinburgh in 1965. The expression, *artificial intelligence*, was officially used for the first time in 1969, at the first International Joint Conference on Artificial Intelligence, in Washington. Ever since, this large international colloquim has been held every two years. Each time, there have been nearly 500 participants.

Games were an initial field of success. In 1959, A. L. Samuel, one of the best American checkers players, proved the value of his system for the first time. Today, the little machines that play chess are an application of research done in the field of artificial intelligence. Let us also mention Frank Rosenblatt, whose work led to the building of the Perceptron in 1958.

Artificial intelligence programming languages

The earliest programming AI language (1956)

Logical reasoning and the formal calculus have led to entirely new programming methods. Today, more than a hundred languages are used in the field of artificial intelligence.

The first AI programming language, IPL (Information Processing Language), was invented in 1956 by Amercian scientists, **A. Newell, D. Shaw,** and **F. Simon.** It was specifically developed to write the LT (Logic Theorist) program, capable of resolving mathematical logic problems. The major focus in the IPL was list processing.

The programming language of the future (1973)

PROLOG (from the French phrase PROgrammation LOGique) has proved a real phenomenon in the world of artificial intelligence. It has been selected by the Japanese as the programming language of their fifth generation computers.

PROLOG was developed in 1973 by a Frenchman **Alain Colmerauer,** and his team at the University of Marseilles-Luminy, France. The basic concept is that the sentencial (or propositional) calculus is adequate to generate a very high-level programming language.

PROLOG allows the programmer to describe only the problem to be solved, not the algorithm and solution, as with the procedural (or algorithmic) languages mentioned previously.

Automatic translation

Starting in 1946, W. Weaver and A. D. Booth foresaw the possibility of using a calculator to carry out translations. The truth of the matter is that totally automatic translation does not yet exist, and one should say *computer assisted translation*.

In 1946, the technology was still primitive. It was only in 1950 that Weaver and Booth could experiment with their ideas. After 1960, numerous projects were started, aiming at automatically translating from French, or from Russian, into English.

Thus, by the beginning of the 1970's, Doctor P. Thomas had developed a system, Systran, to translate Russian into English. This system has been continually improved. Today, it is capable of translating a dozen languages, but it only produces approximate translations.

Computer assisted design (CAD)

Computer assisted design began in the 1960's. The aeronautics industry produced large programs for the American military.

A range of techniques offers: (1) a means to build up data that describes the object to be designed, (2) a tool for manipulating data conversationally, and (3) a look at the finished form of a design. Besides the military, CAD is now used by civil aeronautics, automobile, and data processing industries.

Starting in 1961, IBM engineers were using CAD for circuit interconnections in Stretch. Univac used the same technology for the Larc. The most significant applications of CAD involve calculations made on evolutive, conversational graphics consoles. Work by I. E. Sutherland, on Sketchpad, in 1963, was instrumental. (*See* Light pen, p. 302)

Computer assisted instruction

The first ideas concerning programmed teaching were defined by the American psychologist, B. F. Skinner, in 1954. At that time, Skinner proposed a linear procedure, which did not permit one to make a distinction of level based on the students' answers. Another American psychologist, Norman Crowder, invented a multiple-choice method.

Towards 1955, work began that applied computers to this field. In 1958, a pioneering experiment was conducted.

The year 1965 saw industry take an interest in computer based eduction. In 1967, a specialized working group was set up by the Association for Computer Machinery (ACM). During this period, several systems were developed: Plato, at the University of Illinois, and Ticcit, of the Mitre Corporation.

The work of Symour Papert at the Massachusetts Institute of Technology (MIT) should also be mentioned. He developed the programming language, Logo, which is well adapted to young children.

Expert systems

An expert system is a computer program that solves problems normally requiring human expertise. Expert systems reason heuristically from a problem raised by the user of a computer.

The system operates from a knowledge base and production rules making up problem-solving engines designed by human experts.

Computer assisted design.

periment data to infer the structures of an unknown compound.

METADENDRAL adds analysis knowledge to DENDRAL by proposing and selecting fragmentation rules for organic structures. METADENDRAL was experimented in 1977 by **M. Mitchell, B.G. Buchanan,** and **E.A. Feigenbaum.**

MYCIN (1976)

MYCIN is an expert system designed to diagnose and treat infectious blood diseases. It was experimented in 1976 by an American, **E. Shortliffe,** at Stanford University, Calif.

Another expert system, TEIRESIAS, was built on top of MYCIN. TEIRESIAS is a knowledge-acquisition system, that is, it helps transfer expertise from the human experts to MYCIN knowledge base. It was developed in 1979 by R. Davis.

BAOBAB is still another system working on top of MYCIN to interface with human experts. BAOBAB was invented in 1979 by a Frenchman, A. Bonnet.

Speech understanding (1976)

HEARSAY-II was developed in 1976 by **Raj Reddy, V.R. Lesser,** and **L.D. Erman** at Carnegie-Mellon University (Pittsburgh, Pennsylvania).

HEARSAY-II can understand connected discourse from a 1,000 word vocabulary. However, HEARSAY-II does not meet the high performance criterion established for expert systems. Other projects have been experimented to improve HEARSAY-II. These are AGE at Stanford, California, and HEARSAY-III.

PROSPECTOR (1979)

The PROSPECTOR expert system was developed by **R.O. Duda. P.E. HART,** and **R. Reboh** at SRI International, Menlo Park, California.

PROSPECTOR was designed to assist geologists in interpreting data and guiding exploration. It uses knowledge of local and regional characteristics of areas favorable for specific ore deposits.

PORSPECTOR helped discover a new extension to an existing molybenum deposit in the state of Washington.

CESSOL (1984)

Developed in 1984 at the University of Savoie, France, in conjunction with Symag Informatique, CESSOL is an expert system used for geophysical studies.

CESSOL assists architects in determining soil characteristics before buildings are erected.

In the early sixties, researchers had already put forward the idea that the laws of reasoning combined with powerful computers should produce systems exceeding human experts' capacities. Expert systems were first used in the early seventies.

The oldest expert system (1961)

In 1961, **J.R. Slagle,** an American scholar at the Massachusetts Institute of Technology (MIT), wrote his doctoral dissertation on a heuristic program for solving problems related to symbolic mathematics.

SAINT (Symbolic Automatic INTegrator) was the beginning of a line of development that culminated in MACSYMA, experimented in 1971 by two MIT reasearchers, W.A. Martin, and R.J. Fateman. Today, an improved version of the MACSYMA surpasses most human experts in performing differential and integral calculus symbolically.

The DENDRAL project (1965)

The DENDRAL project has entered its twenty-first year. This project has been carried out by Stanford University scientists and produced two systems, DENDRAL and METADENDRAL.

DENDRAL was experimented in 1967 by **B.G. Buchanan, E.A. Feigenbaum,** and **J. Lederberg,** who won the Nobel Prize for Medicine in 1958. DENDRAL analyzes chemical ex-

Numerical control for machine tools (NCMT)

Numerical control for machine tools (NCMT) dates back to 1956. At this time, the programming language, APT, was created at the request of the U.S. Air Force.

To a certain point, the principle of numerical control can be traced back to Jacquard's looms. Or, one can go even farther back to in Russia, in 1712. Peter the Great's mechanic, A. C. Nartov, invented the first automatic-drum copier.

Speech recognition

The first machine was built in 1950 by an American, K. H. Davis, who was working at Bell Laboratories, (USA). It was entirely hard wired, and it could distinguish, from a series of acoustic signals, ten spoken numbers.

The Advanced Research Project Agency for Speech Understanding Research (ARPASUR), which coordinates both private and public research in the United States, had, between 1971 and 1976, a total budget of 15 million dollars for this work. The ARPASUR project resulted in the development of ambitious systems. Among others, there are Hearsay II and Harpy, of the Carnegie Mellon University (Reddy and his team), and Hwin of Bolt, Beranek and Newman (Woods, Wolf and their team). These systems have a vocabulary of around 1,000 words and use moderately complex syntax. Sentences are either directly understood, or decoded after an intermediate dialog (one sentence) between the user and the machine.

A special domain of application for this technology is in the education of deaf children. In France, there is the Sirene project at CRI, in Nancy; and there are other projects at the IBM scientific center, all of which are based on a visual aid.

Principles and Components

Register (1944)

The first registers appeared in 1944, at the same time the Harvard-IBM machine was developed.

A register is a device which provides temporary memory for significant information. The Harvard-IBM machine was equipped with 60 memory registers and 72 addition registers.

Electrostatic tube (1947)

In 1947, at the University of Manchester, the Englishman F. C. Williams experimented using electrostatic tubes as computer main memory. IBM used this type of memory on the first computer that it marketed (IBM 701).

As of 1954, the introduction of iron core memories began to replace Williams' tubes for computer memory.

The inventors of the transistor. From left to right; W. Shockley, W. Brattain and J. Bardeen.

The first transistors assembled at the Bell Labs.

Transistor (1948)

In 1948, the American **William Bradford Shockley** invented the transistor while working in collaboration with his fellow-countrymen, John Bardeen and Walter H. Brattain.

A transistor is actually a semiconductor triode. It is the electronic component which characterized second generation computers. A solid-state component that neither needs nor dissipates much energy, it was quickly adopted.

The first transistors were made of germanium, which is very sensitive to temperature variations. From 1960 on, transistors have used silicon, which is much more stable. Ever since, semiconductor technology has continually evolved, leading to microelectronics and integrated circuits.

The first computer to use transistors was the SEAC, built by Standard Eastern Automating Computing (United States Institute of Norms).

Iron cores (1949)

In 1949, **Jay Forester** used iron cores as main storage components in Whirlwind.

Cores are iron rings. They have the properties of a permanent magnet and this allows them to memorize an electrical signal. Iron cores began to go out of use in 1964. They have been replaced by semiconductor memory devices.

Input-output channel (1954)

In the United States, in 1954, it seems that Bob Evans operated the first channel on an IBM 70P4 computer.

An input-output channel on a computer permits the memory to communicate with one or several peripheral units. In the beginning, channels were simply electrical cables that transmitted information in the form of electrical impulses.

The first machines equipped with channels were marketed in 1958. These were IBM 709s. It should also be remembered that in 1959 engineers at Univac built an input-output management system on the Larg.

3. Robotics

Robot ancestors (850 B.C.)

The notion of artificial beings has existed from time immemorial. Homer (8th century B.C.), the celebrated Greek poet, referred to Hephaestus, whose slaves were "animated gold statues that share intelligence, voice, movement, and that received metal-working from the immortals."

Robbie the robot from the film "Forbidden Planet" with co-star Ann Francis in 1955.

Philosophers that built animated machines were many in ancient Times. Egyptian sermonizers impressed their followers with statues of this kind.

Automatic control (1788)

Today's robots could not exist without automatic control. It was invented in 1788 by a Scotsman, **James Watt**, with his ball governor. Eighty years later, another Scotsman, James Clerk Maxwell, laid the mathematical bases of automatic control.

In the 1940s, the works on cybernetics of an American, Norbert Wiener (*See* Cybernetics, p. 300), really gave rise to modern robotics, based on electronics and the use of sensors. Thanks to Norbert Wiener's works, another American, Gray Walter, made several small automatons capable of directing their own course by themselves.

First industrial robot (1962)

The first industrial robot appeared in 1962 with the Unimates. Unimates were mass-produced by Unimation Inc., at Danbury, Connecticut.

These kind of robots offer five to seven degrees of freedom. They can perform almost the same jobs as a worker for a cost three to four times as low.

Lunakhod, first extraterrestrial robot (1970)

The Lunakhod spaceship, which, on November 17, 1970, alighted in the Sea of Rains, on the moon, can be considered the first extraterrestrial robot.

Renault-Acma robots (1974)

In 1974, Acma, a Renault affiliate, designed universal and easily programmable robots for automobile manu-

Car making robots assemble, weld and paint.

facturing. They are a good instance of sophisticated, industrial robots. Particularly suited for image reconstruction, they perform welding, painting and handling jobs.

First definition of a robot (1978)

The first (and only) real definition of a robot was given by the Japanese around 1978.

A robot is a versatile machine (that is, performing different jobs), capable of modifying its environment (manipulation), and adaptable to a changing environment (programmable). There are four types of robots:

• Type A: Man-driven manipulators

• Type B: Fixed sequence robots (sequentially controlled manipulators in which a predetermined sequence cannot be programmed easily). They are

the most widely used. Travelling cranes are an example of this type of robots.

• Type C: Programmable sequence robots (sequentially controlled manipulators in which a predetermined sequence can be programmed). Painting robots adapt to the forms of different automobile bodies, and belong to type C.

• Type D: Learning robots (manipulators that can be programmed by recording manually controlled movements and capable of repeating the learned movements). Examples: arc welding and assembling robots.

The word "robot" (1921)

The word "robot" was coined in 1921 by the Czech dramatist Karel Capek for his play *R.U.R.* (*Rossum's Universal Robot*).

The word robot comes from the Slavonic root *rab*, meaning "slave," which became the Czech or Russian word *robota* (work).

Laws of robotics (1940)

Robots rebelling against their creator (man) have always obsessed people. To keep robots from mastering man, the American science-fiction novelist **Isaac Asimov** invented, in his novel *Robbie* published in 1940, the three famous laws of robotics that govern man/robot relationships:

• Law 1: A robot must not harm a human being, nor must it, by not acting, cause a human being to be harmed.

• Law 2: A robot must obey human beings' commands, except when such commands go counter to Law 1.

• Law 3: A robot must protect itself as long as such protection does not go counter to Law 1 or 2.

Robotized robot making (1983)

The first robotized plant especially designed to turn out robots has been operating since 1983 at the foot of Fuji-Yama, in Japan. The 68 Yamazaki industrial robots do in three days the work that 220 workers would do in a month. Operating 24 hours a day, this plant is equipped with a mainframe computer capable of troubleshooting from 350 references failures.

The mainframe also controls a robot to make the needed repairs. If a failure requires careful consideration, the mainframe discontinues production until men have returned.

This plant is the only one of its kind in the world. It took the 250 people of Yamazaki's engineering department 30,000 hours' work to set up the plant.

Calligrapher robot (1983)

In Japan, a calligraphy contest is held each year for the best schoolboys. In 1983, a robot presented by the Fujitsu Fanuc company surprisingly carried off first prize very easily, thanks to the artistic quality of its handwriting.

The character most highly rated by the jury was *tao* meaning "track," "path," or "road to the future." In fact, this robot reproduced the subtle movements of a human calligrapher that had been stored in its memory.

Nursing robot (1983)

Melkong (Medical Electric King Kong) was created by Professor **Hiroyasu Funakubo** of Japan. It can hold a patient in its arms, wash him, put him back in bed, and tuck in the bedclothes.

Marilyn Monroe (1983)

Marilyn Monroe, the celebrated movie star, is certainly the most humanoid robot of all times. It was created by a Japanese, **Shumeihi Mizuno**.

This robot is a true copy of the movie star. It can sing "River of No Return," playing its own accompaniments on the guitar, with the same intonations and gestures as its renowned model.

A computer synchronizes the voice with the gestures which are generated by 85 cylinders. Mizuno has yet to sell his illustrious robot. Rather, he hires it out for 900 dollars per month to Tokyo department stores.

Guide Dog Robot (1983)

In Japan, the number of guide dogs for the blind is very limited. That is why a Japanese inventor, **Susumu Tachi**, started the development of guide dog robots in 1975. In 1977, the first generation robot, MARK I, was introduced. In 1983, an offshoot of MARK I, MELDOG MARK IV was capable of performing most of the functions of guide dogs.

Diagnosing robots (1984)

Thanks to funds from the Japanese Cancer Institute, a palpating robot is in experimentation at Waseda University, Tokyo. Equipped with fingers covered with artificial skin and connected to a computer, this robot palpates breasts to detect tumors.

First congress of personal robot makers (1984)

The first international congress of personal robot makers was held on April 15, 1984 at Albuquerque, New Mexico.

Several hundreds of amateur robot makers gathered to display their "offspring" and to award the most sophisticated production a gold android. That year, Henry, a household robot created by Bruce Taylor carried off top honors.

Henry is programmed to serve beer in an apartment. It can grasp a glass, carry, and put it down. In addition, it can go "sulky" if orders do not please it. Its maker argues that this is its human side.

Among other robots displayed at Albuquerque, the most suprising were:

● An educational robot equipped with a siren and traffic lights, used by the Rapid City police to teach the Highway Code to schoolboys.

● Rebecca, a household robot which keeps its maker's daughter away from the cakes in the kitchen.

● Ultima, a robot resembling *Star War*'s celebrated R2D2 which obeys voice orders to perform household jobs; it answers the telephone, opens the front door, watches over the house, detects fire hazards and, if need be, calls the police or fire brigade.

Robot sports (1984)

Until recently, robots were mostly designed for work. Now they go in for sports. In April 1984, M. Billingsly, a British researcher at Portsmouth, England Polytechnic, launched the idea of a ping-pong championship for robots. The table will be 0.5m x 2m with a net 24 cm in height. Provision will be made for a frame to keep a robot from hitting the ball outside the field of vision of the movie cameras that control the robots.

For serving, a ball will be released from a box over the serving robot. In order to avoid fastidious exchanges, the following rule will apply: The first of two robot players to hit the ball more than 20 times will win the point. To date, the promoter of this singular contest has already received more than 30 applications.

Sheep-shearing robot (1984)

In 1984, Australian researchers of Western Australia University designed the first robot capable of shearing a sheep. It has been successfully tested on several hundreds of sheep.

The shape of the sheep is stored in the computer's memory, which allows it to control the movements of the arm holding the electrical shear. A detector senses the low electric current on the surface of the skin to monitor the shearing process.

Researchers are improving their robot to shear 95% of skin surface instead of the present 70%. Thanks to this new breed of robots, Australian university researchers hope to compensate for the lack of sheep shearers that affects this traditional sector of the Australian economy.

Sniffing robot (1984)

The first sniffing robot was experimented in 1984 by a team of British scientists directed by Professor **George Dodd**.

An analyzing system allows this robot to determined the molecular structure of atmospheric components. Very soon, Professor Dodd estimates, this kind of robot will be capable of identifying the odor of spoiled meat or gas leaks.

At the present time, the robot knows the fragrance of a clove from that of sage and jasmine. Professor Dodd thinks that his robot will also be able to identify individuals from their odor.

Maintenance Robot (1985)

A MOOTY is its name. It is one of the most useful robots in operation. A MOOTY was introduced in 1985 by a Japanese research team including scientists at the University of Tokyo, and Toshiba Corp. engineers, in conjunction with the Japanese Ministry of International Trade and Industry (MITI).

Basically, A MOOTY was designed as a remote handling equipment for nuclear fuel cycle facilities. It is made up of two parts. Mobility is achieved by a mechanism called TOROVER, which allows stair-climbing, stepping-over, obstacle avoidance, and A MOOTY proper, designed to perform maintenance tasks on machines and electronic equipment.

A MOOTY is typically suited to replace human technicians in hostile and dangerous environment, such as nuclear power plants.

Household Robots

Hero 1 (1982)

This stocky, 50-cm-tall robot was created by the American company Heathkit. It can be programmed from a keyboard installed on one of its sides. An ideal tool for introduction to robotics, it can also walk a dog.

Bob (1984)

The Androbot company at Sunnyvale, California, founded by Noland Buschnel (inventor of Pong, the first home video game), is proud of its first child, BOB (Brains on Board).

This robot can watch over your house and call up the police if need be. It walks at 2 km per hour and can sing and teach foreign languages.

Topo (1984)

Topo is Bob's younger brother. It is a peripheral of the Apple II computer. It is first programmed and then remotely controlled from the computer. Ninety cm tall, it can serve drinks pushing a small trolley and trods away doing antics.

Genus (1984)

Created by Robotics International Corp., Genus is not very attractive but it is very efficient. It has a synthetic voice (he can speak) and a speech recognition system (you can speak to it). Two sensors allow it to identify terrain and pass around obstacles. Eventually, you can use it for vacuum cleaning the house.

Kikuzo (1984)

Kikuzo is a very small Japanese robot (20 cm tall). It can walk and it understands eight oral commands. It is a toy.

Robot paramount

The largest robot manufacturer in the world is the Japanese company Fujitsu Fanuc. It turns out 500 machine-tools and 350 robots every month with only 100 people and 30 robots. Without the latter, more than 500 people would be required.

Robotized robot making (1983)

The first robotized plant especially designed to turn out robots has been operating since 1983 at the foot of Fuji-Yama, in Japan. The 68 Yamazaki industrial robots do in three days the work that 220 workers would do in a month. Operating 24 hours a day, this plant is equipped with a mainframe computer capable of trouble-

energy

1. Hydraulics

Vertical water-raising wheel (5th century B.C.)

The vertical wheel for elevating water, but not for driving (originally), was probably the first waterwheel invented. It was seen in the Middle East between the 5th and 3rd centuries B.C.

The oldest known apparatus for water raising is the *chadouf*, a rudimentary sweep, which appeared toward 2500 B.C. in Egypt. It is still used today.

Archimedes' screw (3rd century B.C.)

One of the greatest scientists of antiquity, a Greek from Syracuse, Archimedes (287—212 B.C.), invented the hydraulic screw. This device is made of an inclined cylinder that encases a broad-threaded screw. It is used to raise water, to serve whatever purpose one wishes. The device is introduced into a body of water, then the screw is rapidly turned, so that the water rises from whorl to whorl.

Undershot waterwheel (lst century B.C.)

The undershot waterwheel was the first hydraulic motor in the history of man. It originated in the same region as the vertical wheel, from which it derives, but the exact date is unknown. Undershot wheels were used to turn grindstones in water mills beginning in the lst century B.C.

Horizontal waterwheel (lst century A.D.)

This appeared during the lst or 2nd century A.D. in the mountainous regions of the Middle East. The horizontal wheel was equipped with paddles driven by the force of flow in mountain streams.

Overshot waterwheel (4th century A.D.)

Toward the end of the Roman Empire (4th—5th century A.D.), someone came up with the idea of replacing the paddles on the undershot waterwheel with buckets. These wheels had little troughs of wood or iron (cast iron beginning in the 1840s) whose bottoms were not perpendicular to the radius of the wheel but tilted back slightly so that when they reached the top of the wheel they were filled by the stream of water than ran over them (hence, the name *overshot wheel*). During the 18th century, this wheel provided the driving force for much of industry.

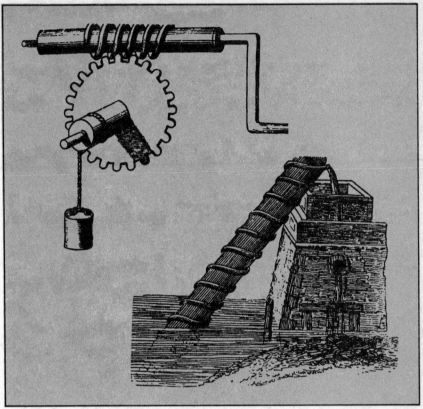

Top: Endless screw. Bottom: Archimedes hydraulic screw.

Breast waterwheel (12th century)

This wheel's paddles were placed on the side of the wheel, perpendicular to the wheel itself, so that the water struck the wheel at about its midpoint. The breast wheel's origins remain obscure; all that is known is that it appeared between the 12th and 13th centuries.

Breast waterwheel with inclined paddles (Sagebien wheel) (1851)

The inclined-paddle wheel (in relation to the water's surface) constituted a considerable improvement to breast waterwheels. It was patented by a Frenchman, **Sagebien**, in 1851. The Sagebien wheel presented the advantage of working as a water-raising machine but had the inconvenience of turning too slowly.

Hydraulic ram (1796)

In 1796, the Frenchmen **Joseph** (1740—1810) and **Etienne** (1745—1799) **de Montgolfier** had the idea of utilizing the kinetic energy of running water in a pipe to force a portion of the liquid mass to a higher level than its source. The energy was transferred through what they called the *ram effect*. The hydraulic ram was improved by Amedee Bollee (1844—1917), notably in connection with use of air compressors.

Undershot waterwheel with curved paddles (Poncelet wheel) (1827)

In 1827, the French general and celebrated mathematician **Jean Victor Poncelet** (1788—1867) published his *Paper on Curved-paddle Waterwheels Driven from Beneath*. This work defined improvements to be made on undershot wheels. The Poncelet wheel foreshadowed turbines.

Outward-flow reaction (Fourneyron) turbine (1832)

In 1832, a Frenchman, **Benoit Fourneyron** (1802—1867), patented the first water turbine, the outward-flow reaction turbine. His invention won him a prize of 6000 francs, awarded by the French Society for the Encouragement of National Industry. This group was anxious to see improvements in hydraulic motors.

Fourneyron was a student of the Frenchman **Claude Bourdin**, who was of Italian origin. Bourdin (or Burdini) was the first researcher to employ the word *turbine* to designate this kind of engine. (The word *turbine* was first used in the 16th century to designate the wheel for spinning wool.) In his work published in 1823, Bourdin drew inspiration from the horizontal waterwheels of antiquity.

The first turbine was installed at

Top: An example of the waterwheel. Bottom: A 1558 water pump based upon the principle of Archimedes screw.

Pont-sur-l'Ognon (Doubs, France). As time went on, higher and higher vertical drops were utilized. In 1838, Saint Blaise Falls in the Black Forest powered a turbine 55 centimeters in diameter, which turned at 300 revolutions per minute and was rated at approximately 40 horsepower with 70% efficiency. By 1843, Fourneyron turbines were in 129 European factories, but they were not used in Great Britain or the United States.

Axial Fontaine turbine (1837)

In 1837, a Frenchman, **Fontaine-Baron**, more commonly known as Fontaine, patented the axial turbine. This machine, an advance over the Fourneyron turbine, was manufactured by Fromont, an industrialist in Chartres, France. Fontaine positioned the distributor horizontally, along the axis of the wheel, hence the name *axial turbine*.

Girard turbine (1853)

In 1853, the Frenchman **Girard** invented an axial turbine from which he and his successors derived numerous models. It was adapted to many different waterfalls and flow rates. Constantly improved throughout the period 1860—1890, Girard turbines were widely employed during this time.

Girard did not see the final version of his turbine. He was killed in 1871 by a Prussian sentry who challenged him as he was returning to Paris from Versailles. Distracted by his thoughts, Girard forgot to respond.

Francis turbine (1855)

In 1855, the book *Lowell Hydraulic Experiments* was published by an American of English origin, **James Francis**. In it he described the invention of a turbine designed for medium and small waterfalls. In fact, the principle of the Francis turbine is actually to be credited to his collaborator, Swain.

Pelton turbine (1870)

An American, **Lester Allen Pelton** (1829—1908), developed the turbine that is most used nowadays with high waterfalls. The design was inspired by the *cup wheel*, a horizontal kind of wheel Pelton saw utilized in the mountainous regions of California. Pelton, a young student at the time of the California gold rush, liked to mingle with prospectors during his vacations. He often suggested new mining techniques for working their claims. But it was only toward 1870, 20 years later, that he built his own turbine.

Kaplan turbine (1912)

In 1912, an Austrian, **Viktor Kaplan** (1876—1934), began work on fitting an axial turbine with a variable-pitch propeller rather than with a wheel. However, industrial construction only began in 1924.

The Kaplan turbine converts the energy of low waterfalls and of wide, slowly flowing waterways. The motor apparatus of a Kaplan turbine is a propeller, similar to those on airplanes. The propeller has four helicoidal blades, the dimensions of which are particularly impressive on heavy-duty units.

2. Steam Engines

Crank and connecting system (ca. 1400)

The advent of the crank and connecting rod system toward the beginning of the 15th century was one of the most important discoveries in mechanics made during the Middle Ages. This system permits transformation of continuous circular motion back into rectilinear, reciprocal (i.e., back and forth piston-type) motion. Absence of a transformation system until then had greatly limited technological development.

In a treatise written around 1400, the German engineer **Conrad Kyeser** illustrated for the first time a handmill operated by a crank and connecting rod system. Applications came quickly for a wide range of technologies. First there were saws, then pumps and spinning wheels, and finally, two centuries later, steam engines.

Huygens' piston machine (1673)

In 1673, the celebrated Dutch scientist **Christiaan Huygens** (1629—1695) demonstrated the first piston motor. His worthy audience included King Louis XIV's chief minister, Colbert, who had called him to France in 1665, and the French Academy of Science. Everyone present was stupified as Huygens' machine lifted "several lackeys" into the air.

Machines made up of a piston moving in a cylinder had been known since antiquity. The Alexandrian Greeks knew of such machines. In the lst century A.D., the mathematician Heron described multiple uses for them. The Romans fabricated great numbers of piston pumps; the Metz museum has a Roman pump with two pistons.

A conclusive experiment was carried out by a German, **Otto von Guer-**icke (1602—1686), who, toward 1650, invented a pneumatic machine. He showed that if the air is emptied from a cylinder with the aid of a pneumatic machine, the piston moves into the vacuum with considerable force. In this case, it was a question of a pneumatic lever and not a motor.

The internal combustion machine invented by Huygens is an actual motor. Gunpowder is placed in the cylinder, closed by the piston, and fired. The gases expand during combustion and contract on cooling. A line links the two ends of the piston, where it runs over pulleys. This way, at the end of each stroke, the piston returns to its initial position.

Huygens had for an assistant a fellow by the name of Denis Papin.

Steam engine (1705)

In France, everyone is convinced that the steam engine was invented by **Denis Papin**, a French physicist, born in Blois in 1647, who died in London in 1714, a pauper. A Protestant, he was condemned to exile when the Edict of Nantes, guaranteeing religious freedom, was repealed in 1685.

In 1679, Papin invented a safety valve for a pressure cooker he had devised. Even then, he was dreaming of building a cylinder engine in which the piston would be powered by steam. Yet it was not until 1707 that he wrote a description of "his" steam engine. That same year, he launched the first steamboat on the Fulda River in Germany.

In 1705, the Englishmen **Thomas Newcomen** (1663—1729) and **Thomas Savery** (about 1650—1715) fabricated the Newcomen steam engine, which precipitated the industrial revolution.

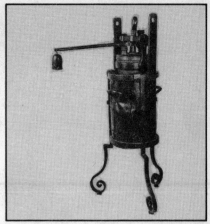

Denis Papin steam machine, 1679.

Savery had patented a machine in 1698, but one lacking a safety valve, so liable to explode at any moment.

Large-capacity steam boilers (1712)

In 1712, the Englishman **Thomas Newcomen** put the finishing touches on the boiler for his steam engine (*see* above). Newcomen's boiler resembled an enormous, spherical pressure cooker, the upper part of which was in the shape of a dome.

At the end of the 18th century, a Scot, James Watt (*see* below), elongated this boiler and flattened the ends. The result was called the *wagon boiler*, because of its shape—also the *Watt boiler*.

Condenser (1765)

In 1765, the Scot **James Watt** (1736—1819) ordered the construction of a Newcomen machine to be equipped with a condenser, a device indispens-

An example of a Pelton turbine.

Top: A steam machine. Right: James Watt.

able to the efficient operation of this kind of steam engine. He had thought of this improvement in 1763 while manufacturing precision instruments in Glasgow, his business since 1757. The idea came following repairs he made on a Newcomen machine, the operating principle of which left him most perplexed.

Double-acting (Watt) steam engine (1783)

In 1783, **James Watt**, who since 1765 had been repairing and refining steam engines (*see* above), transformed the Newcomen machine into a double-acting engine. Watt had conceived of a new machine in 1766; he patented it in 1769. This patent ran until 1800 due to a specially proclaimed law. Naturally, Watt's exclusive rights discouraged other researchers from devising improvements for the steam engine.

In 1775, Watt formed an association with an important Birmingham manufacturer, Matthew Boulton. The following year, two prototype machines were put into service, but these were still only pumps, used in mines to draw off water. Between 1776 and 1800, Watt and Boulton manufactured approximately 500 different machines, constantly improving on them. In 1783, the double-acting engine, designed in 1780, left the factory; in 1785, it was equipped with a centrifugal governor. Later on, Watt fitted the cylinder with a heat jacket.

Watt's engine is a Newcomen machine—that is, a low-pressure machine, since the vapor temperature is only slightly over 100°C. The steam acts in alternating fashion on each of the two faces of the piston (double acting). This way, resistance of the piston to the steam is reduced, since after being pushed, it is pushed back with the same force. Watt considerably increased the power, and therefore the efficiency, of the steam engine. In 1881, physicists gave his name to the international unit of power, i.e., the watt.

Multiple-expansion engine (1803)

In 1803, an Englishman, **Arthur Woolf** (1736—1837), patented the first multiple-expansion engine—that is, one with several cylinders. This machine worked at low pressure, but the principle was later adopted for use with high pressure.

High-pressure engine (1805)

Toward 1805, the Welshman **Richard Trevithick** (1771—1833), a mining engineer, completed his design for the

Left: George Stephenson. Right: Marc Seguin.

first high-pressure steam engine. Trevithick realized that by putting the engine's steam under high pressure, he could do away with the condenser, which was cumbersome and heavy. He began his research in 1798.

Piston with metallic rings (1816)

In 1816, an Englishman, **John Burton**, filed a patent in England for a piston with metallic rings. Nevertheless, this piston was commonly called the *Swedish piston*. It was universally adopted.

Water-tube boiler (1825)

The first truly functional water-tube boilers were developed, starting in 1825, by the Englishman **Goldsworthy Gurney** (1793—1875). He wanted to build steam-powered automobiles and needed a lightweight boiler. Unfortunately, legal complications prevented his building an automobile.

In the 1860s, Gurney's boilers appeared in industry. First, "limited-flow" boilers were fabricated, which were fitted with serpentine tubes. "Free-flow" boilers with practically horizontal tubes were built as well. The most successful models were built by the Englishmen Babcock and Wilcox (1867) and by the Frenchmen

Belleville (1877) and Serpollet (1886). This last manufacturer's product was reputed to be excellent for steam-powered automobiles.

Smoke-tube boiler (1827)

As part of the effort to develop locomotives, several people designed smoke-tube boilers: the Frenchman **Marc Seguin** (1786—1875) in 1827; and an Englishman, **George Stephenson** (1781—1848), who was assisted by his son Robert (1803—1859), in 1829. These boilers permitted increased power at a reduced engine weight. This was indispensable to progress in building locomotives.

Heat extractor (1843)

In 1843, the Frenchman and Alsatian engineer **Gustav Hirn** (1815—1890), a physicist specializing in the study of gases, together with his collaborator, the Alsatian engineer **Schinz**, patented the first heat extractor. This device was intended to increase the efficiency of boilers.

Superheater (1855)

In 1855, the French physicist **Gustav Hirn** (1815—1890) had the idea of superheating water, which was brought to the boiling point in boilers, in order to increase the pressure.

3. Generation of Electricity

(For related principles, *see* Electricity, page 247.)

Origins

According to two German researchers, Professors Wilhelm Konig and Arne Eggebrecht, the **Parthians** invented the first electrical battery in Iraq, toward the year 1 of our epoch. This conclusion is based on the examination of an object unearthed in 1936. It resembles a clay vase and contains a copper cylinder, an iron stem that is covered in lead, and pitch. If an acid or alkaline solution were poured into this vase, say the two professors, it would produce an electric current of ½ volt.

Arched dam (13th century)

In 1843, the French engineer (originally from Venice) Zola, father of the celebrated novelist, built the first modern arched dam, near Aix-en-Provence, France.

The arch had been known by masons and architects since ancient times and was especially utilized during the Roman period, not only for roofing buildings but for bridge construction as well.

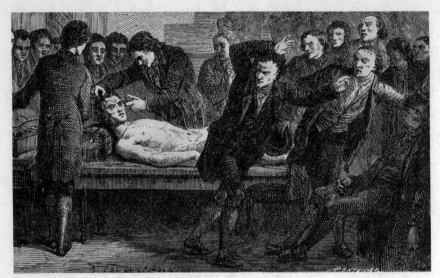

Strange experiments followed the discovery of electricity.

The step from bridge to dam was easy to make. It sufficed to design an analogous construction, the positioning of which was rotated in space.

The oldest arched dams known were constructed in Iran during the 13th century. The technique was not widely practiced. Nevertheless, several dams of this type still exist in Spain.

Voltaic cell, or Volta battery (1800)

The voltaic cell, the first modern electric battery and hence the first modern electrical generating device, was built in 1800 by an Italian, **Alessandro Volta** (1745—1827). This invention earned Volta the favor of his most illustrious correspondent, Napoleon Bonaparte, who made him a noble in 1801.

In 1786, the Italian Luigi Galvani (1737—1798) had noted that two different metals put in contact with a muscle in a dead frog caused the muscle to contract. In a work he published in 1796, he explained this contraction as due to "animal electricity."

Volta, who was known for his work in electrostatics, began studying this phenomenon in 1793. He correctly attributed the muscle response to an interaction with the two metals. This gave him the idea of building the first electric battery. He called it a column battery; it came later to be known as a *Volta battery*, or *voltaic cell*. Volta soldered a copper ring and a zinc ring together. He then piled up, always in the same order, a set number of these soldered copper-zinc rings, separating them with felt rings soaked in acidulated water.

Volta discovered that chemical energy could be converted into electrical energy. For this reason, he is considered the founder of electrochemistry. Moreover, his battery permitted measurement of differences in electrical potential (voltage). The measurement apparatus is called a voltmeter. Volta's name is the source of the word *volt*.

Thermoelectric battery, or thermocouple (1821)

In 1821, in an experimental setup, the German physicist **Thomas Seebeck** (1770—1831) constructed the first thermoelectric battery. The thermocouple is mainly used as a heat measurement instrument.

In 1830, the Italian Leopoldo Nobili (1787—1835) developed a thermocouple that permitted observation of slight temperature variations. To do this, variations in electric current density had to be measured. These variations were produced by heating (either a little or a lot) metals used as *indicators*.

In 1886, the Frenchman Henri Le Chatelier (1850—1936) constructed a pyrometer, which could indicate with precision the temperature in furnaces for the first time. The couple he used was made of platinum (which melts at 1770°C) and platino-iridium. Thus, it could resist high temperatures.

Magneto (1832)

In 1832, a Frenchman, **Hippolyte Pixii** (1808—1835), built the first magneto-electric generator (abbreviated to *magneto*). It worked by induction and was the first apparatus to convert mechanical energy into electrical energy. It produced alternating current.

The magneto was once the most common ignition device for internal combustion engines. It is now mainly used for race car engines.

Daniell cell (1836)

In 1836, an English physicist, **John Frederic Daniell** (1790—1845), invented the first impolarizable battery.

In a Volta cell, hydrogen, which is freed by the electrolyte, migrates to the copper, positive pole. This creates an undesirable barrier at the positive pole. Soon the current no longer flows. This *polarization* renders the battery useless. The phenomenon, known but unexplained for a quarter century, was elucidated in 1826 by the French physicist Antoine Becquerel (1788—1878), the grandfather of Henri Becquerel, who discovered radioactivity. Daniell then developed an *impolarizable* battery, a cell from which the hydrogen would be eliminated.

Photovoltaic cell (1839)

The first photovoltaic cells (or batteries) were constructed in 1839 by **Antoine Becquerel** (1788—1878), the grandfather of Henri Becquerel, who discovered radioactivity. Also called solar cells or photocells, they directly transform light into electricity. Their value lies in their ability to convert sunlight, which falls on the Earth in abundance, into electric energy. Nowadays, they are made of a conductor-semiconductor joined together. They are the basis for numerous solar energy applications.

In 1954, Bell Laboratories researchers G. L. Pearson, C. S. Fuller, and D. M. Chaplin experimented with a solar battery. It was made up of small silicon cells whose capacity, although that tripled over a few years (from 6% in 1955 to 18% in 1978), was very low and cost-ineffective. However, researchers have been trying to reduce the manufacturing cost of silicon cells to allow ground applications where no electric network presently exists—for remote telecommunication relays, lighting, water pumping in dry areas, etc.

Bunsen cell (1843)

In 1843, following up the work of the English physicist Sir William Robert Grove (1811—1896), a German physicist, **Robert Wilhelm Bunsen** (1811—1899), invented a more efficient depolarizing cell than Daniell's.

Lead acid storage battery (1859)

In 1859, a French physicist, **Gaston Plante** (1834—1889), invented the first electric storage cell: the lead acid storage battery. The German physicist Johann Wilhelm Ritter (1776—1810) had earlier observed, in

1803, on a lead-plate voltmeter, the phenomenon utilized in this kind of cell. Storage batteries make it possible to accumulate electricity in order to later distribute it as needed. Indeed, a storage cell is a *secondary battery*. It is commonly called a "rechargeable battery." The "battery" in a car is usually made up of one or more storage batteries.

Gravity dam (1861—1865)

The first gravity dam, the composition of which had been mathematically determined, was constructed on the Furens River between 1861 and 1866 as part of the water supply system for the city of Saint-Etienne, France. The name of the builder is unknown.

Dynamo (1869)

On July 17, 1871, the French Academy of Sciences enthusiastically received the inventor of the dynamo (short for *dynamoelectric generator*), the Belgian **Zenobe Gramme** (1826—1901), who had been living in France since 1856. He was deeply moved. Only a few years earlier, he had been a modest worker taking classes at night school. Two years later, after the World's Fair in Vienna (1873), he became a celebrity known around the globe. In 1878, he invented the alternator.

Gramme's machine was the first generator to produce direct current from mechanical energy. Up until then, only alternating current could be produced from mechanical energy. Gramme, who in 1867 had already improved earlier kinds of generators, blended his dynamo discoveries with devices known since 1832. His invention, strictly speaking, was the *commutator*, built in 1869.

In an automobile, the dynamo furnishes electrical current, which is accumulated in the storage batteries. On a bicycle, a simple sort of dynamo makes the light work.

Penstock (1869)

On September 28, 1869, a French paper manufacturer, **Aristide Berges** (1833—1904), became the first to transform the mechanical energy of a waterfall into electrical energy. He used a cascade in the Alps to produce electricity for operating the machines in his factory. To this end, he utilized penstocks. Hydroelectric power had come of age.

Penstocks direct water from a dam, under pressure, to mechanical generators in a factory. In general, nowadays this is a hydroelectric power plant.

In 1869, Berges was using a waterfall 200 meters high. This produced

1.5 kilowatts of electrical energy. In 1882, he used a waterfall 500 meters high. In 1886—1887, he built the first *hydraulic accumulator*. This was a cavern he ordered dug under Lake Crozet, 25 meters beneath the lake bed, to receive the overflow. It was an extremely risky undertaking. In 1889, Berges had two stands at the World's Fair in Paris. His brochures there employed for the first time the expression *hydroelectric power*.

Dichromate battery (1870)

Toward 1870, the German physicist **Heinrich Ruhmkorff** (1803—1877) developed a Daniell cell containing two fluids. At the time, Ruhmkorff was managing a company in Paris that handled electrical apparatuses.

Leclanche cell (1877)

In 1877, after ten years of study, a French engineer, **Georges Leclanche**, completed the design of a battery with a solid rather than a liquid depolarizant. (The Daniell cell used liquid.) Much later, the physicist Charles Fery (1865—1935) improved this cell, and it is still used today.

Dry cell (1887)

In 1887, the Englishman **Hellesen** invented the first dry cell.

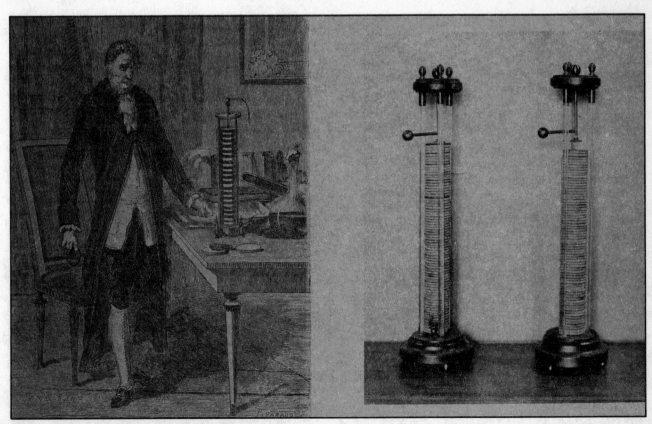

Left: Alessandro Volta. Right: Volta's battery, 1800.

Top left: The thermocouple. Bottom left: The Plante lead-acid storage battery. Right: The Hoover Dam.

Dry cells are batteries in which the electrolyte is in the state of humid paste. Since no liquid can leak out, they are portable. The ingredients have been modified over the years, and the dry cell has become a portable version of the Leclanche cell.

Dam and man-made lock (1912)

The first large-scale dam was the earth dam that was constructed in Panama, 1912.

Since 1918 the technology for building this sort of dam has improved considerably. Among the most spectacular projects are the Hoover Dam on the Colorado River, the series of man-made lakes along the Tennessee Valley, and the Aswan Dam in Egypt, constructed with Soviet aid between 1960 and 1970.

Alkaline storage batteries (1914)

Toward 1914, the ingenious American inventor **Thomas Edison** (1847—1931) developed the first alkaline storage battery. It is called alkaline because the electrolyte is not acid, but basic.

The silver zinc storage battery was developed in 1941 by the Frenchman H. Andre. The positive electrode is made of silver oxide, the negative electrode of zinc, and the electrolyte of a potassium solution. The energy storage efficiency of this kind of battery is two or three times better than that of lead acid batteries.

Fuel cell (1936)

An American named **Bacon** designed the fuel cell in 1936. It was then improved after World War II by a number of American laboratories. The fundamental advantage is that the battery's fuel supply can be renewed. This way, the cell has a lifetime that is much longer than that of ordinary batteries.

Industrial fabrication of fuel cells began in the United States in 1960. Spacecraft use cells whose fuel is hydrogen, with combustion supported by oxygen. The electrodes are of porous roasted nickel.

The biochemical cell is a particular kind of fuel cell. Here, a *primary* fuel (glucose, for example) is transformed into gaseous, *secondary* fuel by bacteria.

Magnetohydrodynamic generator (1959)

In the United States, the **Avco Research Laboratories** (Massachusetts) built the first magnetohydrodynamic (MHD) generator in 1959. This was a purely experimental device, capable of directly transforming caloric energy into electrical energy. It was the end-result of theoretical discoveries made ten years earlier by the Swedish physicist Alfven (born 1908, Nobel laureate in 1970). Alfven had been studying the flow of matter caused by intense magnetic fields—in the solar corona, for example.

4. Electrotechnology

Magnetoelectric motor (1838)

In 1838, a German physicist, **Moritz Hermann von Jacobi** (1801—1872), who worked in Russia from 1832 until his death, outfitted a paddleboat with an electric motor. This motor is the most famous among those invented at the time.

Several researchers made the same important observation soon after the magneto was invented. If alternating current is run to this kind of generator, the machine, although originally designed to transform mechanical energy into electrical energy, does the opposite: It produces mechanical energy, which can be used to turn a wheel.

Direct-current motor (1844)

In 1873, at the Vienna World's Fair, a Frenchman, **Hippolyte Fontaine** (1833 —1917), exhibited the first working, large-scale direct-current motor. He utilized a Gramme dynamo *back-wards*. By powering it electrically, the machine no longer transformed mechanical energy into electrical energy but did the opposite.

Starting in 1885, this type of motor, improved to the point that it was easily distinguishable from a dynamo, was used to run tramways.

It is noteworthy that as early as 1844 the Frenchman **Gustav Froment** (1815—1865) was building direct-current motors to operate calculating machines. They did not need to be very powerful and ran off a simple battery.

Alternator (1878)

Beginning in 1878, the French company Gramme, founded by the Belgian **Zenobe Gramme** and the Frenchman **Hippolyte Fontaine**, industrially manufactured the first alternators. A German company founded by Werner von Siemens (1816—1892) also began building them that same year.

An alternator is an apparatus that transforms direct current into alternating current, i.e., into current that is periodically inversed. Alternating current, which has numerous applica-

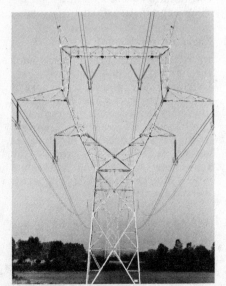

A high-tension line.

tions, is more easily transmitted than direct current.

Alternating-current synchronous motor (1885)

In 1885, an Italian physicist, **Galileo Ferraris** (1847—1897), discovered the rotating magnetic field. Practical application resulted in multiphase, alternating-current synchronous motors. At the same time, an American of Croatian origin, **Nikola Tesla** (1856—1943), developed the multiphase alternator. Very soon thereafter, powerful three-phase current motors appeared on the market. These rotational-field motors are generally called *asynchronous motors*, in reference to their variable speed.

High-tension line (1882)

In 1882, the French engineer Marcel Deprez (1843—1918) transmitted electrical energy for the first time by using high-tension lines.

In order to economically transmit electrical energy over a long distance, a high-tension line is necessary. This is because the current strength is inversely proportional to the electric potential difference (commonly known as tension). Hence, if the tension (i.e., voltage) is increased, the current strength (often measured in amperes) diminishes. Since low current is less susceptible to ohmic dissipation (energy loss), use of high-tension lines is necessary.

On September 25, 1882, Deprez transmitted direct current from Miesbach to Munich. He used an iron cable that was 4.5 millimeters in diameter and had a resistance of 950 ohms. The voltage was 1343 volts and the efficiency 39% (61% of the electrical energy was lost).

On February 6, 1883, a crowd gathered in Paris was astonished when a motor started up at the same time a generator next to it was run, even though the two were not visibly connected. In fact, there was an electrical line connecting the two that ran to Bourget, then back to Paris. Its characteristics were: 160 ohms, 2700 volts, 37% efficiency.

Direct current was used in these tests. Later on, thanks to the application of alternators and transformers, alternating current could be transmitted.

Transformer (1882)

In 1882, a French chemist and physicist, **Lucien Gaulard** (1850—1888), patented a "secondary generator of electric current." This apparatus was the

first industrial transformer (the word *transformer* was not used until later). Gaulard intended it for lighting, but its uses proved diverse.

In 1884, thanks to his transformer, Gaulard showed how electricity could be industrially transmitted from Lanzo to Turin, a distance of 50 kilometers. That same year, he conducted tests in London (applications in the subway), at the Tours fair, and elsewhere.

The issue of credit for the invention of the transformer provoked heated disputes. Various other French and foreign researchers have claimed credit.

5. Engines

Closed system, hot air engine (1816)

In 1816, a Scottish pastor, **Robert Stirling**, invented the closed system, hot air engine. However, the first Stirling engine to be industrially operational was not built until 1844. The brothers Robert and James Stirling built it for an English foundry in Dundee. Then this engine design was abandoned. In recent years, the Dutch company Philips and the American company General Motors took up the design again. They built engines which had an efficiency of 40% and a good power/mass ratio, making them competitive with internal combustion engines. Nowadays, these engines are effectively employed aboard submarines and spacecraft. In the latter case, they can be powered by photovoltaic (solar) cells.

Water-cooled engine (1823)

In 1823, an Englishman, **Samuel Brown**, invented a water cooling system. In Brown's engine, water was circulated around cylinders lined with jackets. Water was driven by a pump and cooled when it came into contact with the ambient air. In 1825, Brown founded a company that manufactured several engines of this kind. One was mounted on a vehicle, another on a boat.

The first operational internal combustion engine, invented by Etienne Lenoir in 1860 (*see* Two-stroke engine, without preliminary combustion, page 332), used a water cooling system.

Ignition by flame transfer (1838)

In 1838, an Englishman, **William Barnett**, designed an ignition system where a steady flame would light an intermittent flame as part of the operating cycle of an engine. Such a system became necessary due to the advent of engines that used preliminary fuel compression. Barnett's intermittent flame system permitted ignition of the precompressed fuel mixture at fixed intervals in the cylinder of an engine. The flame would go out after each lighting.

In 1900, electrical ignition replaced flame ignition. Indeed, increasing rotational speeds in engines made flame ignition impossible.

Induction coil (1841)

The induction coil was invented in 1841 by the French physicists **Antoine Masson** and **Louis Breguet** (1804—1884).

The induction coil applies the electromagnetic induction phenomenon to generate high-voltage alternating current. This coil is standard equipment on the ignition system of all internal combustion engines. As early as 1836, Masson had produced high-voltage currents by causing quick interruptions in battery-generated current. The induction coil that he built in 1841 with Breguet allowed him to produce electrical discharges in rarefied gas.

The induction coil was improved in 1851 by a German physicist, Heinrich Daniel Ruhmkorff, and now bears his name.

Open system, hot air engine (1851)

In 1851, an American of Swedish origin, **John Ericsson**, patented the open system, hot air engine. The first Ericsson engine was constructed for a transatlantic ship that bore his name and sailed in 1858. It was an upright, in-line, four-cylinder engine rated at approximately 250 horsepower at 9 rpm.

Hot tube ignition (1855)

In 1855, the American **Alfred Drake** developed a small, cast iron tube that, when heated red-hot, provoked ignition on contact with the fuel mixture in an internal combustion engine. This ignition system was widely used between 1880 and 1905. Yet, even though it was well adapted to rapid-cycle speeds and high compression, it had a serious drawback. The moment of ignition was imprecise and impossible to regulate. These defects made this system absolutely useless when it came to engines for vehicles, the operating conditions of which are essentially variable.

Two-stroke engine, without preliminary compression (1860)

The first authentically operational internal combustion engine was invented in 1860 by a Frenchman of Belgian

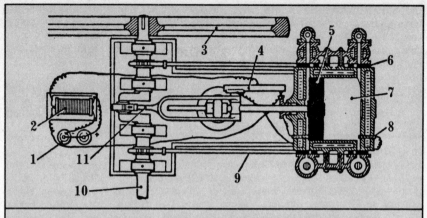

THE TWO-STROKE ENGINE
1. Electric battery. 2. Induction coil. 3. Fly-wheel. 4. Distributor.
5. Piston. 6. Distribution valve (admission and evacuation). 7.
Cylinder. 8. Spark plug. 9. slide-rod. 10. Driving shaft.
11. Driving rod.

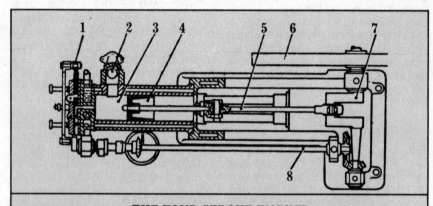

THE FOUR-STROKE ENGINE
1. Inlet and ignition valve. 2. Outlet valve. 3. Cylinder. 4. Piston.
5. Driving rod. 6. Fly-wheel. 7. Driving shaft. 8. Slide-rod.

origin, **Etienne Lenoir**. This engine did not make use of preliminary fuel compression before ignition. It was characterized by a two-stroke operating cycle and the use of coal gas as fuel.

This engine was enthusiastically received by small manufacturers at a time when steam engines, electric motors, and other hot air engines were not entirely satisfactory. In spite of a very low efficiency (on the order of 4.7%), these engines were built over a period of about 10 years. They had power ratings of from 0.5 to 12 horsepower.

"Air-breathing" engine (1867)

In 1867, the German engine builders **Nicolas Otto** and **Eugen Langen** designed the "air-breathing" engine. It became the most widely used internal combusion engine up until the introduction, in 1876, of the four-stroke engine, also developed by Otto.

The "air-breathing" engine was pre-

sented at the World's Fair in Paris in 1867 and was everywhere acclaimed because of its high efficiency. Its fuel consumption was only half that of the other engines on exhibit.

Brayton engine (1873)

In 1873, the American **Brayton** built the first engine that really used preliminary fuel compression as part of its operating cycle. The Brayton engine uses a two-stroke system with combustion at constant pressure, making it a predecessor of the diesel engine. It was mainly used in the United States, up until 1882, at a time when internal combustion engines were still extremely rare.

Air-cooled engine (1875)

In 1875, a Frenchman, **Alexis de Bischop**, for the first time used an air cooling system. His engine included a cylinder covered with vertical ribs. The fuel mixture was uncompressed.

Air cooling systems are widely used on motorcycles and aircraft engines.

Four-stroke internal combustion engine (1876)

Theoretical study of an engine with a four-stroke operating cycle was undertaken in 1862 by the Frenchman Alphonse Beau de Rochas. However, its practical design is credited to the German engineer **Nicolas Otto** (1832–1891), who patented and built the first four-stroke internal combustion engine in 1876. This patent constituted one of the landmark dates in the epoch of the internal combustion engine.

The first engine that worked successfully on a four-stroke cycle was presented by Otto, in 1878, at the World's Fair in Paris. Compared with combustion engines of the time, it was remarkable for its smooth operation, as well as for its reduced size and higher efficiency. It had a single piston that was horizontally oriented, employed flame ignition, and was water cooled. Coal gas served as fuel. The engine produced approximately 3 horsepower at 180 rpm.

This original engine was improved by W. Playbach and was later built in different sizes. By 1890, a power rating of 100 horsepower had already been achieved.

Opposed piston engine (1878)

The first opposed piston engine was built in 1878 at the German firm Hannoversche Maschinenbau A. G., by the engineer **Ferdinand Kindermann**.

Two-stroke internal combustion engine (1879)

In order to get around Otto's patent on the four-stroke engine, a two-stroke internal combustion engine with preliminary fuel compression inside the cylinder was designed in 1879 by the English engineer **Dugald Clerk**. The firm Sterne & Co. in Glasgow handled the manufacturing for engines with power ratings from approximately 2-to-12 horsepower.

Clerk's engine had the advantage over the four-stroke of offering greater motive power for the same cylinder size. However, certain inconveniences made it uncompetitive. As with all two-stroke internal combustion engines, there was a problem with exhausting the burned gases, which also involved some fuel loss. This made it less efficient and more polluting.

Four-stroke engine (1883)

In 1883, the French inventors **Delamare-Debouteville** and **Charles Ma**-landin built the first gasoline-supplied four-stroke engine. It was an 8 horsepower, twin-cylinder engine. The gasoline was supplied through a slide valve and exhausted through an outlet valve. Continuous sparking in the distributor allowed the electric ignition to work. This experimental engine was mounted on a car but was never mass-produced.

Spark plug (1885)

In 1885, the Frenchman **Etienne Lenoir** invented an electrical spark plug similar to those still in use today. In 1777, the Italian inventor Alessandro Volta had already suggested firing a fuel by sparking. The same technique was advocated by Isaac de Rivaz for the internal combustion engine he described in 1807.

Whirlwind engine (1887)

In 1887, a Frenchman named **Millet** developed the first radial, rotating-cylinder engine. It was a five-cylinder engine with electrical ignition and slide-valve timing. At the time, it was used to power a tricycle.

The best-known whirlwind engine was the Gnome engine, built in 1908 by the French engineer Louis-Laurent Seguin.

The merger of the companies Gnome and Rhone resulted in the engine most widely used in aeronautical construction during World War I. This engine had nine cylinders and produced 100 horsepower.

Radial engine (1888)

In 1888, the Frenchman **Fernand Forest** designed a radial engine whose 12 cylinders were laid out in four parallel disks of three cylinders each. This design had been preceded that same year by the first internal combustion engine with its cylinders arranged radially. This was also Forest's idea, working in collaboration with Lalbin.

The radial engine was often employed on automobiles, on ships, and particularly on aircraft because of its compact size.

High power gas engine (1888)

In 1888, the Frenchmen **Delamare-Deboutteville** and **Malandin** were the first to build a gas engine with a single cylinder capable of producing 100 horsepower. They named it Simplex because of its simplicity. It was horizontally oriented. Intake and ignition were regulated by a timing slide-valve system. Exhaust was through a poppet valve.

Toward 1895, Delamare-Deboutteville and Malandin built an engine that produced 300 horsepower. Around 1900, the 1000 horsepower barrier was broken. These engines were intended for industry.

In-line, six-cylinder engine (1888)

The first in-line, six-cylinder engine was designed in 1888 by a Frenchman, **Fernand Forest**.

The six cylinders are parallel to one another and stand in the same (usually vertical) plane. This design excels other types of in-line multiple cylinder engines, as it is better balanced and offers smoother torque. As a result, it was widely used in the automotive industry. Today, the in-line, six-cylinder engine is still in use in the United States, where large-capacity engines are still popular.

V-type engine (1889)

The first V-type engine was built in 1889 by the two German engineers **Gottlieb Daimler** and **Wilhelm Maybach**. Daimler's and Maybach's engine was made up of two cylinders, each at an angle of 17°, joined like the two arms of a V. It produced 1.5 horsepower at 600 rpm. In 1889, they

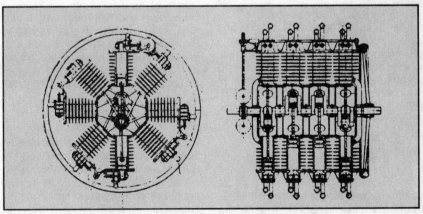

The Forest engine was mainly used for airplanes.

An 8-cylinder Chrysler engine.

installed this engine on an automobile specially designed for it by Maybach. Today, this car is considered to have been the first modern automobile.

The V-type engine was very successful. Most manufacturers used it because of its compact size and well-balanced design. Different models had 2, 4, 8, 12, or 16 cylinders.

Oil engine (1890)

On the basis of a patent he took out in 1890, an English engineer, **Herbert Ackroyd-Stuart**, that same year built an oil engine. Oil was considerably cheaper than gasoline.

This was a horizontally oriented, four-stroke engine. It became well known when its manufacture was turned over to the Hornby firm in Grantham, England. Hornby developed engines that produced from 1 to 6 horsepower at 200 rpm.

The characteristics of this engine were much appreciated It was simple in design and sturdy. Maintenance was easy. Fuel for it was cheap.

As it had a separated combustion chamber (*see* below), this engine was the forerunner of the semi-diesel engine, also known as the hot-bulb engine. In fact, Ackroyd-Stuart is said to have contested the originality of Rudolf Diesel's invention.

Separated combustion chamber (1890)

The first engines to have separated combustion chambers were built in 1890 by the English engineer **Ackroyd-Stuart** (*see* above). This system was later refined on certain hot-bulb

engines (*see* Hot-bulb engine, page 325).

The combustion chamber plays a very important role in all internal combustion engines. This is especially true in diesel engines, where it insures that an adequate fuel mixture is achieved.

In diesel engines, there are two principal categories of combustion chamber: (1) direct injection and (2) injection into a separated chamber.

There are three types of separated combustion chamber: (1) prechambers, (2) turbulent separated chambers, and (3) air-pocket separated chambers.

Carburetor (1893)

A German engineer, **Wilhelm Maybach**, invented the carburetor in 1893.

The carburetor is one of the major components of an internal combustion engine. Inside the carburetor, the fuel is vaporized into a charge of air before being drawn into the engine's cylinder. It was first used with a parallel, twin-cylinder engine known as Phoenix, built by Maybach and Gottlieb Daimler, which was one of the first internal combustion engines mounted on a car. It met with great success.

Throttle valve (1893)

The carburetor was improved by **Karl Benz** the year it was invented (1893). Benz installed a throttle valve on the carburetor to regulate the amount of explosive mixture supplied to the engine. This allowed one to adjust the rpm and power of an engine.

Opposed-cylinder engine (1895)

Developed by a pair of Frenchmen, **De Dion** and **Bouton** in 1895, the flat-twin was the first engine to have opposed cylinders situated in the same plane. This sort of engine could have several pairs of cylinders. The prototype had two, and produced 1.5 horsepower. At the time, its rotational speed of 1000 rpm was a record. It was air-cooled.

In 1896, the German K. Benz began to build these kinds of engines. His had two cylinders and had power ratings from 5 to 14 horsepower at 900 rpm.

In England, William Lanchester, who had built the first English internal combustion engine, one year later constructed an opposed-cylinder engine.

The engine in the famous Volkswagen "beetle," introduced to the German market before World War II, was a four-cylinder opposed engine, air-cooled and situated in the rear of the car. This popular automobile was still being built as late as the 1970s.

Four-stroke diesel engine (1897)

The German **Rudolf Diesel's** (1858—1913) experiments were numerous and

The Earl of Dion and his partner Bouton.

Rudolf Diesel.

strewn with difficulties before he succeeded, in 1897, in building his internal combustion piston engine. His starting point was a study of the Carnot cycle, a fundamental concept in thermodynamics expounded in 1824 by the Frenchman Sadi Carnot (1796—1832).

While trying to develop a thermally rational engine based on a maximally efficient Carnot cycle, Diesel realized that this cycle was impossible to achieve. So he began work on an engine to be characterized by combustion at constant pressure rather than at constant temperature.

A diesel engine compresses air to the point where the resulting rise in temperature is sufficient to automatically ignite the fuel, which is injected at the end of the compression stroke. This kind of engine has the highest efficiency of all heat engines, nearly 45% for supercharged diesels. That is why the diesel engine is still the most widely used heat engine today.

The work cycle and principle of the diesel had already either been advocated or employed by Beau de Rochas, Brayton and Ackroyd-Stuart, to the point at which this last-named inventor contested the originality of Diesel's invention. Yet the diesel engine holds an important place in the history of technology, because it was the first example of a technical achievement which was the result of a theoretical, scientific analysis.

Two-stroke diesel engine (1899)

In 1899, the German engineer **Hugo Guldner** developed the two-stroke diesel engine. This permitted a considerable increase in the specific power of the diesel, from 60% to 80% for the same displacement. Curiously enough, the inventor patented his engine under the name H. Eckhardt and turned over manufacturing to the firm Maschinenfabrik Augsburg.

Today, two-stroke diesels are the most powerful of the internal combustion piston engines. They can produce up to 50,000 horsepower. They are utilized to power naval vessels and as high-power industrial engines.

Distributor (1900)

Around 1900, manufacturers improved high-voltage distribution by perfecting storage batteries. An American company, **Dayton Engineering Laboratories Company** (Ohio), experimented with a directing system that led to the Delco distributor, named after the company.

Prechamber (1909)

The German engineer **Prosper L'Orange**, of the German company Benz, invented the prechamber in 1909 and completed its development in 1919.

In a related development, at the Swedish company Svenska Maskinverken, the engineer Harry Laisner worked on a similar project, starting in 1913.

Hot-bulb engine (1902)

Derived from the Ackroyd-Stuart oil engine, the hot-bulb engine was invented in 1902 by the Swedish engineer **Rundolf**.

The "hot bulb" is a small space in the cylinder head. Since this space is not cooled during operation, it furnishes the extra heat necessary for ignition in each cycle.

Hot-bulb engines are of the diesel type. That is why they are also called semi-diesels.

Turbulent chamber (1919)

In 1919, the English engineer **Henry Ricardo** constructed his celebrated combustion chamber. It had a T-shaped cylinder head.

Around this time, many engine-builders were trying to use turbulence in order to improve air-fuel mixing. This was for diesel engines as well as for other piston-type internal combustion engines.

A combustion chamber typically represents 50% to 80% of the total compression volume. It is generally spherical in shape and situated in the cylinder or piston head. A rudimentary design had been built by Stricland in 1904. In 1920, Taylor introduced the hemispherical shape.

Air-pocket chamber (1926)

This type of chamber was designed by the German engineer **Franz Lang**, in 1926. Air, which has accumulated in the separated chamber, flows into the cylinder during the power stroke, thus improving combustion.

Reciprocating gas generator (1937)

The reciprocating gas generator was developed in 1937 by **Raul de Pasteras Pescara**, an Argentine engineer working in France. It was derived from the reciprocating compressor developed by the German engineer Junkers.

The reciprocating gas generator is nothing more than a simplified internal combustion engine based on the diesel work cycle. It replaces the combustion chamber and the compressor of a classic gas turbine.

6. Turbomachines

Gas turbine

In 1791, the Englishman John Barber patented the design for what could have been the first gas turbine, but it never got past the planning stage.

Combustible-gas turbine (1847)

In 1847, the French engineer **Claude Bourdin**, who coined the word *turbine*, proposed building a combustible-gas turbine. Several related patents followed, taken out by the Frenchman Tournaire (1853), the Swede Laval (1893), the German Stolze (1899), and others. The first attempts at industrially manufacturing gas turbines were made in France, in 1901. This work was done by the Turbo-Motors company, directed by Charles Lemale and the Armendaud brothers.

Steam turbine (1884)

In 1884, an Englishman, **Sir Charles Algernon Parsons** (1854—1931), built the first steam turbine. It was a reaction turbine.

Prior to this, in 1629, an Italian, Giovanni Branca (1571—1645), described in his treatise *The Machine* an apparatus in which steam drove a wheel. And later, progress in steam engines and hydrodynamics led to the industrial development of the gas turbine.

While Laval developed and improved his impulse turbine (*see* below), Parsons was improving his reaction turbine, which was used on ships. In 1894, he undertook construction of the *Turbinia*, which in 1896 broke the world's speed record at sea. The combined power produced by all the Parsons turbine installations in service in 1904 equalled 1 million steam-horsepower, or more than 700,000 kilowatts.

Top: Inside a turbine. Right: Sportscars are now equipped with turbo compressors.

Impulse steam turbine (1889)

In 1889, five years after Parsons, the Swede **Charles-Gustave de Laval** industrially manufactured the first impulse steam turbine.

Pressure-stage steam turbine (1897)

Between 1897 and 1900, a Frenchman, **Rateau**, developed the first multistage impulse turbines. These machines permitted slower rotational speeds, less friction, and higher efficiency.

Gas combustion turbine (1903)

The first patents on combustion turbines were taken out by Peer (Brooklyn, 1890) and by Hordenfeldt and Christophe (Paris, 1894). However, it was a French engineer, **Armengaud**, who in 1903 built the first gas combustion turbine. This combustion turbine was still of the kind that did not use preliminary compression.

Gas combustion turbine with preliminary compression (1909)

The first gas combustion turbine with preliminary compression was built by the German engineer **Holzworth** in 1909. In 1910, Holzworth built a second combustion turbine with preliminary compression and operated it in Mannheim. This was the first gas turbine to be used industrially.

Construction of gas combustion turbines was pursued, but sporadically until just after 1945.

Turbocompressor for supercharging diesel engines (1905)

While building a gas turbine at Carels, Belgium, a young Swiss engineer, **A. Buchi**, envisaged replacing the combustion chamber with an internal combustion engine. This would eliminate inconveniences due to high temperatures and pressures. His research led to supercharging turbines. In 1905, he patented the first turbocompressor.

Turbocharging is a means by which power and efficiency of internal combustion engines can be increased. The engine cylinders are fed air under pressure from a centrifugal compressor.

In 1908, Buchi went to work at Sulzer Freres in Winterthur, Switzerland. This company came out with the first turbocharged engine in 1911. It was a four-stroke, single-cylinder diesel engine with pneumatic fuel injection. Today, this type of engine is used in trucks and automobiles. Gasoline engine turbos work on the same principle.

Turbocompressor for supercharging aircraft engines (1916)

In 1916, a French engineer, **Auguste Rateau**, built the first turbocharged aircraft engine. This engine produced 50 horsepower and weighed 23 kilograms. The turbocharger worked at a speed of 30,000 rpm.

Turbocharged diesel trucks (1953)

In 1953, **Volvo**, a Swedish company, was the first to equip trucks with a turbocharged diesel engine. It was an in-line, six-cylinder direct injection engine.

Mass-produced turbocharged engines (1967)

Mass-produced turbocharged engines have been one of the most outstanding improvements in the automotive industry over the last decades. As early as 1967, SAAB was the first manufacturer to equip mass-produced engines with *turbochargers*. This technique allowed one to supercharge an engine while feeding it less fuel. BMW followed suit in 1974, and in 1983, the 524 series was equipped with a turbocharged diesel engine.

Turbopropeller (1920)

The turbopropeller, which was invented in 1920 by the Englishman A. A. Griffith, is basically a gas turbine whose energy serves essentially to turn one or more propellers. The principal difficulty lies in the high speed of a gas turbine. Powerful speed reducers are indispensable if the propeller is to turn at a speed compatible with its working conditions.

The turboprop was especially used on airplanes flying at 600 to 800 kilometers per hour.

Closed-cycle gas turbine (1940)

The closed-cycle gas turbine was invented in 1940 and developed by the

Swiss company **Escher-Wyss**, in Zurich. It was called the aerodynamic turbine Escher-Wyss A.K. (The initials A and K stand for its inventors, Professor **Ackeret** and Doctor **Keller**.)

In a closed-cycle gas turbine, a certain quantity of gas, which remains constant, is compressed and heated periodically This way, the same gas mass is utilized continually.

Turbojet (1941)

The project of building a turbojet was undertaken by the Englishman **Frank Whittle** in 1928 and brought to completion in 1941.

A turbojet is a jet engine. Exhausting burned gases produce engine thrust by reaction, which provides motive power for the vehicle equipped with this engine. The Whittle turbojet had a centrifugal compressor and developed a thrust of 375 kilograms of force.

The first aircraft equipped with a turbojet was British and went into service in 1941.

In Germany, Hans von Ohain pursued similar research, starting in 1936, and the first Heinkel jet fighter began flight tests in 1939.

The Concorde supersonic airplane is equipped with Olympus turbojets built by the English firm Bristol Siddley.

7. Rocket and Aircraft Engines

Rocket

The first rockets were fabricated in China and in India hundreds of years ago. According to a Chinese legend, a mandarin tried to fly by hanging onto two kites propelled by a large number of rockets. There is clear documentation of their use in the 13th century.

The Tartars used them in 1241 at the battle of Liegnica (Lower Silesia, now in Poland).

Rocket engines propel by reaction the vehicles that carry them. They do this by ejecting hot gases, which the engine produces through burning fuel that the vehicle has on board.

The only rocket engines utilized nowadays are chemical-fuel internal combustion rocket engines. Other types, such as nuclear or electrical rocket engines, are still in the experimental stage.

Solid-fuel rocket engine

Up until the 20th century, the only kind of rocket fuel was gunpowder. Rockets were no longer used in Europe after the 14th century except as fireworks. Their military application was rediscovered by an Englishman, General William Longrave, in India in approximately 1790.

The first rockets were simple cardboard or wood cylinders blocked off at one end. Beginning around 1800, the cylinder was made of metal.

The American Robert Goddard improved the efficiency of rockets 65% by increasing the gas exhaust rate. This was during World War I. Performance was further improved subsequently.

Solid-fuel engines are much simpler than liquid-fuel engines. However, they are less versatile in application. That is why they are used less and less.

Liquid-fuel rocket engine (1897)

The first liquid-fuel rocket engine was built in Paris in 1897 by **Pedro P. Poulet**, a South American living in France.

Four illustrious inventors established the theoretical and practical foundations of liquid-fuel rocket engine technology. They were (1) the Russian Konstantin Tsiolkowski, who, beginning in 1898, developed a complete theory of rockets; (2) the German Hermann Oberth (born 1894), who, beginning in 1923, first recommended using liquid fuels; (3) the American Robert Goddard, who launched liquid-fuel rockets, starting in 1926; and (4) the Frenchman Esnault-Pelterie, who in 1930 published his work entitled *Astronautics*.

Oberth developed German astronautics. A noteworthy project was the V2 (*see* below). Oberth and his student Wernher von Braun are considered the fathers of space technology (*see* Space).

The V2 rocket engine (1942)

Rocket engine research in Hitler's Germany was carried out under the guidance of Hermann Oberth. Later it was directed by Oberth's student **Wernher von Braun** (1912—1977). Work was first done at the military base of Kummersdorf, then transferred to the Peenemunde base in 1937.

The infamous German V2, which used liquid fuel, made its first flight on June 13, 1942. Designed by Wernher von Braun, it was equipped with a rocket engine built by W. Thiel. (Thiel was killed in 1943 during a bombing raid on the Peenemunde base.)

The engine on the V2s that terrorized London in 1944 and 1945 was the last in a series of rocket engines designed initially to power the V1. The rocket attained a velocity of 760 meters per second. Escape velocity for leaving the Earth's gravitational field (for example, in order to put a satellite into orbit) is 11 kilometers per second, or 15 times faster than the V2. After the war, the V2 engine served as

Left: A German V2 rocket in Paris in 1944. Right: The German V1 was propelled by a pulsojet engine. This missile was about 8 meters long and could carry 500kgs of explosives.

a model for both the Americans and the Soviets. W. von Braun immigrated to the United States.

Rocket engine cooling (1933)

In 1933, the Austrian E. Sanger introduced a new cooling system for liquid-fuel rocket engines. Called regenerative cooling, it became the most widely employed system. Sanger also built several rocket engine starters, used during takeoff on the first aircraft equipped with turbojets. However, soon after, he developed a jet aircraft engine that started itself.

Monopropellant rocket engine (1935)

In 1935, the German Helmuth Walter built the first monopropellant rocket engine. A monopropellant rocket engine is a liquid-fuel rocket engine. Contrary to most engines of this type, the fuel of which is carried as two components mixed only at the moment of combustion, a monopropellant engine uses just a single liquid as fuel, hence the prefix *mono*.

The engine that Walter developed used hydrogen peroxide in a concentration of 80% to 85%. It was such an immediate success that Walter was able to found his own company in 1935. In February 1937, a Walter rocket providing 100 kilograms of thrust was installed as a starter engine on the German Heinkel Kadett aircraft. This resulted in the first rocket-assisted takeoff in the history of aviation.

Stratojet (1936)

In 1913, a Frenchman, Lorin, patented the first stratojet, but no industrial application ensued.

Lorin's idea was taken up by the French engineer Rene Leduc (1898—1968) toward 1936, and a first prototype was built.

A stratojet is a very simple jet engine. However, it presented a major difficulty. The vehicle propelled by a stratojet must first be accelerated by an auxiliary engine up to a velocity at which the stratojet can begin to function. It was not until 1949 that the first aircraft powered by a stratojet could be flown. The stratojet was started after the plane carrying it (a Leduc 010) was launched by a mother craft. In 1956, thanks to a turbojet mounted inside the stratojet to unify the air-intake systems, the aircraft Leduc 022 was able to take off under its own power.

This type of engine is rarely used.

Pulse-jet engine (1940)

In 1940, the principle of the pulse-jet was discovered by the German Paul Schmidt. He was trying to develop a stratojet that could start up under its own power. The pulse-jet was immediately put to use on the German flying bomb, the V1 (*Vergeltungswaffe* 1, i.e., "reprisal weapon" no. 1).

Motor-jet engine (1940)

In August 1940, the Italian Caproni company finished building the first aircraft propelled by a motor-jet engine. The plans had been drawn up by the engineer Campini.

A motor-jet engine is a jet engine in which preliminary air compression is achieved through a compressor powered by an internal combustion engine with a piston. However, this type of engine was very quickly replaced by turbojets because of its weight and the vibrations it caused.

Turbofan engine (1940)

In 1940, the American company Metropolitan Vickers developed a turbofan engine. This engine design is between the turboprop and a full-fledged turbojet. It has two advantages over these two other engine types: (1) in-flight fuel consumption is 20% less, and (2) there is less noise at the exhaust nozzle.

The turbofan engine has been widely used. The Snecma company of France has installed its model TF-106 on the vertical takeoff aircraft Mirage III-V. The American firm Pratt & Whitney has built turbojets of this type for the Boeing 707, Boeing 720, Boeing 727, the French Caravelle, and the Douglas DC-8. General Electric and Rolls-Royce have produced turbofan engines as well.

Turbojet with afterburner (1945)

In 1945, the British company Rolls-Royce constructed the first turbojet with an afterburner. Afterburning allows greater thrust to be obtained without substantial increase in engine weight. This increased thrust is particularly useful in takeoff.

The biggest drawback of afterburners is their very high fuel consumption. That is why they are only used very briefly, at takeoff, and to obtain peak speeds with fighter aircraft.

8. Nuclear Energy

Isotopic separation (1922)

In 1922, isotopic separation was performed for the first time in the laboratory by an English physicist, William Francis Aston. To achieve this, Aston utilized a mass spectrograph, which sorted atoms with the aid of a magnetic field.

Isotopic separation consists of isolating a particular isotope from a given substance. In nature, a substance almost never exists in the pure state. Generally it is accompanied in very small proportions by substances with the same atomic number but having a different number of neutrons.

Deuterium is an isotope of hydrogen. It is actually heavier than hydrogen, because it has one neutron, while hydrogen has none.

Industrial sorting methods used nowadays no longer require a mass spectrograph. Two examples of current methods are: (1) heavy water production, the isolation of the deute-

A nuclear plant.

rium contained in light water for use as a *moderator*, and (2) uranium enrichment, which eliminates the undesirable uranium U238, which is abundant but not fissionable, in order to concentrate the valuable uranium U235, which is very rare. The substance U235 has three neutrons less than U238.

Atomic pile (1942)

The first atomic pile was constructed in Chicago under the direction of **Enrico Fermi** (1901—1954), a physicist of Italian origin. It was started up on December 2, 1942. The pile contained 50 tons of uranium and 500 tons of graphite.

The energy produced by atomic fission is given off in the form of heat. This heat is then recovered and transformed—for example, into electrical energy. During fission, there is released not only a great deal of energy but also a neutron bombardment, which in turn induces more fission. A chain reaction results. However, in an atomic pile the reaction is slow enough to be controllable.

Nuclear fuels (1942)

In 1942, inside the Chicago reactor, **Enrico Fermi** employed a uranium oxide, U_3O_8. Since 1960, the oxide UO_2 in a roasted form is the mostly commonly used fuel in reactors. It is the only fuel that can be used in water reactors, which dominate the market and will undoubtedly continue to dominate it for a long time.

Moderator (1942)

In 1942, the Italian physicist **Enrico Fermi** (1901—1954; Nobel prize in 1938) employed graphite as a moderator in the first nuclear reactor, which he built in Chicago.

The probability of a fissionable nucleus undergoing fission depends on the energy of the neutron that collides with it. With the principal fissionable nuclei, those of uranium 235 and plutonium 239, this probability increases when the extremely high kinetic energy of the neutron decreases. In order to moderate this energy, substances (moderators) with very light nuclei are employed. In collisions with the neutrons, they dissipate the neutrons' initial energy, slowing them down but without capturing them too often. The only usable moderators are: carbon, in the form of industrial graphite; hydrogen, in the form of water or certain hydrocarbons; deuterium, in heavy water; and beryllium, either pure or in an oxide state.

In 1932, an American, **Harold Clayton Urey**, was the first to obtain heavy water. He employed isotopic separation (*see* below). Heavy water is used in certain reactors as a moderator.

At the beginning of World War II, a Norwegian factory in Rynkan possessed a supply of heavy water. Thanks to a French initiative in the spring of 1940, this supply was brought to France, from where it was sent to Britain. Norwegian commandos blew up the factory in Rynkan in 1943, and in 1944 destroyed a ferryboat transporting the remaining stock of heavy water to Germany. These operations contributed to the failure of nuclear research in Nazi Germany.

Nuclear reactor (1951)

The different types of reactors and the selection of usable materials were basically determined between 1940 and 1945. But even so, industrial construction, which began in 1951 in the United States, required considerable preliminary study.

Any combination of fissionable nucleus, moderator, and heat conductor (designated as a *line of approach*) can lead to a reactor design. More than 200 combinations have been envisaged, and each one takes into account various possible models. Problems to be resolved have essentially concerned systems design and integration, production drawing of components, and rigorous study of the materials. Only 15 combinations are considered as practical, either for experimental reactors or for prototypes. In the domain of fast-neutron reactors (*see* Breeder reactors), studies have mainly dealt with helium or steam cooling.

For slow-neutron reactors, more classical in design, work done during the years 1965 to 1970 showed three lines definitely to be impractical.

Today, enriched natural uranium reactors are almost the only kind of reactor constructed. They use uranium oxide (UO_2) as fuel and ordinary water both as a moderator and as a heat conductor. These reactors exist in both the boiling water and pressurized water varieties. Pressurized water reactors are the most commonly used in the West.

Less common are reactors that also use UO_2 but are moderated by heavy water and use either heavy water (again) or ordinary water as a heat conductor.

Also noteworthy are two categories of reactors likely to be developed over the long term. First are reactors

Inside a nuclear reactor.

that allow for obtaining temperatures much higher than their predecessors' (800°C as opposed to 350°C). This permits powering of very modern turbo-alternators or of gas turbines. Second are fast-breeder reactors, which can run turbo-alternators with superheated steam. Already, a very small number of operational prototypes exist.

However, there is a contrary trend. Ecologists are mobilizing demonstrators in ever greater numbers against nuclear reactors, which they feel pose unacceptable risks in the event of malfunction. However, it is expected that toward the years 2010 to 2020, controlled fusion will become industrially usable (*see* below).

The first nuclear power plant started operating in 1953, in the United States. The following year, a Soviet power plant became operational. A nuclear power plant is an electrical power plant that uses the heat produced as a result of a nuclear chain reaction (nuclear fission) in a reactor.

The first breeder reactor started operating in the United States in 1963.

A breeder reactor is a reactor that produces more fissionable matter than it consumes. Thus, using up the planet's supply of fissionable uranium (7% of the natural uranium) is no longer a danger.

Controlled fusion

In 1952, when the Americans exploded the first H-bomb (*see* H-bomb, page 47), everyone believed that thermonuclear reaction, known as fusion (of heavy hydrogen nuclei), could eventually be put to peaceful use in the production of energy. But the heat released is so intense that the particles produced cannot be confined in an ordinary reactor. These particles, which constitute a *plasma*, must be made to move along a closed trajectory. To this end, "magnetic traps" were designed during the 1950s. The plasma is contained in a "flux tube" created by an intense magnetic field, which is pinched at each end. One fusion reaction is primarily being studied. This is fusion of a deuterium nucleus (heavy hydrogen with one neutron) with a tritium nucleus (heavy hydrogen with two neutrons). This results in helium while freeing one neutron and a great deal of energy. This reaction has two advantages: (1) deuterium and tritium can be extracted from ordinary water (*see* Heavy water, page 329) and (2) there is theoretically no waste.

European scientists hope soon to set off such a reaction at Culham, England. However, this kind of thermonuclear energy production cannot be expected on an industrial scale before 2010 or 2020.

9. Alternate Energy Sources

Oil drilling (2nd century B.C.)

As early as the 2nd century B.C., the Chinese bored oil wells using bamboo stems or bronze pipes.

"Black gold" (oil) rush (1859)

The first oil rush started at Titusville, Pennsylvania on August 27, 1859. **Edwin L. Drake**, an American industrialist and reputedly an adventurer, had been sent to Titusville by Seneca Oil Company to drill for oil. Oil spurted up when the borehole was 23 meters deep. The news rapidly spread over the surrounding country. Some made a fortune, others went bust. In a few months' time, dozens of derricks were erected around Titusville.

Oil tankers (1886)

The first tanker especially designed for oil transport was a German ship, the *Gluckauf,* launched in 1886. In 1861, barrels of oil extracted at Titusville, Pennsylvania, were conveyed in a sailing ship, the *Elisabeth Witts.*

Modern drilling techniques

In 1922, the Soviets first experimented with turbodrilling, and in 1949, Electrodrill Corporation, a United States company, tested this technology. In 1980, Elf, a French oil company, and the French Institute of Oil (IFP) designed the first horizontal oil-drilling system at Lacq, France, and in 1982, the first horizontal offshore oil drilling system at Rospo Mare, Italy.

Lead-free gasoline (1970s)

Gasoline obtained by the refinement of raw petroleum contains no lead. However, in the 1920s, in order to increase the energy yield provided by combustion engines, chemists decided to add some lead alkyls to it. The spread of this procedure was due to the American automobile manufactuer General Motors.

Conversely, the Americans have been advocating lead-free gasoline since 1975, preceding Japan by two years. The manufacturers were able to produce high performance engines using lead alkyls, that is, engines using little gasoline but providing high power.

The problem is that the exhaust gases from the cars are extremely polluting, causing considerable environmental damage. Ecologists have vociferously lobbied in favor of a decision to eliminate lead from gasoline. This decision will oblige automobile manufacturers to install catalytic converters in all their models in order to alleviate the problem.

The deepest drill in the world

Today, on the Kola Peninsula in the Soviet Union, drilling is being carried out in order to determine the depth of the Earth's crust. During the course of these investigations, one experimental drill managed to reach a depth of 12,000 miles.

Windmill (7th century)

The first windmills were mentioned around the beginning of Islamic civilization, in the second half of the 7th century. However, their existence was not actually documented before the 10th century. The term *windmill* commonly designates any device that serves to capture the kinetic energy of the wind in order to power a machine (notably, mills).

Wind engine, or wind-powered generator (1876)

In 1876, the first modern wind engine was put into service in the middle western United States. (The name

The Rance tide-powered electric plant.

wind engine refers to any machine powered by the wind. It is the modern version of the ancient windmill.)

Since 1975, the National Aeronautics and Space Administration (NASA) has pursued a program for building more and more powerful prototypes. Some vanes are as long as 91 meters, and their power can reach 2.5 megawatts. The most important development is constant computer-corrected orientation of the apparatus to take wind direction into account.

Geothermal energy (1818)

Geothermal energy, which is inexhaustible, comes from the heat of the Earth. The oldest geothermal installation is in Larderello, in Tuscany, Italy. Here, beginning in 1818, the Frenchman F. **de Larderel** (1789—1858) made use of the vapors (*soffioni*) that rose naturally out of the ground. He utilized them as a heat source in order to produce boric acid. Today, the Larderello *soffioni* help run an electric power plant that produces 250 megawatts. The vapor comes out of the ground at a pressure of 40 kilograms per square centimeter and at a temperature of 230°C.

There is a distinction between "low geothermal energy" and "high geothermal energy." The former is practical for domestic heating; the second is used to produce mechanical energy, which in turn is converted into electrical energy. More than 2500 megawatts (geothermic) are produced in the world today from high geothermal energy—mainly in North and South America (United States, Chile, and Mexico) and in New Zealand. Japan, a volcanic country, is only now beginning to exploit its immense geothermal resources.

Tide-powered electric power plant (1966)

In 1966, Electricite de France (EDF) put into service the first tide-powered electric power plant in the world. It is situated in the estuary of the Rance River in Brittany. The mouth of the river is closed off by a dam 750 meters long. The dam contains spillways to allow the tide to come in and go out. There are 24 generating waterwheel sets, specially designed to work in both directions. The alternators can also run as pump motors during off-peak hours. The plant has a total power output of 240 megawatts. Unfortunately, the seawater environment is very corrosive, which makes maintenance expensive.

Only three tidal power stations exist throughout the world: in France; in the Soviet Union, in the Bay of Kis-

A couple of the first solar cars.

laya, on the White Sea; and a third station became operational in Canada, in the Bay of Fundy, Nova Scotia, in 1984.

Water wave energy (1875)

The idea of employing the force of water waves seems to have been studied scientifically for the first time by the Australian R. S. Deverell, in 1875. He attempted to measure water wave energy using an apparatus placed on board a ship. It recorded roll and pitch. He thought this energy could be used to propel ships. However, there were enormous technical difficulties to overcome.

Soon thereafter, the Frenchman de Caligny succeeded in utilizing the oscillating motion of the sea to raise water to a certain level. He was able to drain several coastal lakes this way.

The only mechanical installation that operated reliably was a rudimentary mechanism in Ocean Grove, New Jersey, near New York. It was described in 1889. All that was needed was a number of sluice gates installed between the pilings of a jetty. Their opening and closing motion activated a pump that raised water 12 meters so that it could be used to wash down the city streets. Since then, water wave energy has been the subject of numerous projects, none of which have produced significant results.

Solar energy (1878)

As early as 1878, a Frenchman, **Mouchot**, presented a small solar power plant at the World's Fair in Paris. It furnished enough energy to operate a steam engine. However, this pioneering work was not followed up until similar research was undertaken after 1945. Later, the energy crisis that began in 1973 pushed governments of the principal industrial nations actively to encourage this work.

Solar energy, which consists of the sun's rays striking the Earth's surface, is practically unlimited. The main difficulty in the way of its utilization is the relatively feeble "insolation" at ground level. Under the best of atmospheric conditions, this is only a little more than 1 kilowatt per square meter.

Today, the number of objects that use solar energy in daily life is considerable: televisions (1979); telephones (1981); lighters (1978); water heaters (1981); airplanes (1980); cars (1977); etc.

Solar furnace

The principle of the solar furnace (concentrating the sun's rays) can be traced back to Archimedes (278—212 B.C.). During the defense of Syracuse, he provided his city's soldiers with concave breastplates. This armor concentrated the sun's rays so that beseiging Roman troops believed themselves to be facing mythological "soldiers of fire."

An old solar furnace.

The Frenchman **Antoine Laurent de Lavoisier** (1743—1795), founder of modern chemistry, was the real inventor of the solar furnace. He concentrated the sun's rays in order to burn diamonds in an oxygen atmosphere without using any kind of fuel.

Solar thermal power plant (1960)

The first significant solar thermal power plant was constructed in Ashkhabad, capital of the Soviet Republic of Turkmenistan, in 1960.

In a solar thermal power plant, the sun's rays are concentrated on a steam boiler by a set of mirrors. The steam runs a classical turbo-alternator. At Ashkhabad, solar radiation falls on 1293 flat mirrors arranged in several concentric circles. Each mirror has a surface of 15 square meters, which gives a total surface of about 20,000 square meters for the installation.

There are several solar thermal power plants in the world. The most powerful (10 megawatts) is in Barstow, California. It was put into service in 1982. Japan has built two 1-megawatt installations.

Heliogeothermics (1973)

Since 1973, a group of dwellings in Blagnac, France, have been provided with hot water, using heliogeothermic techniques.

Heliogeothermics, more commonly called solar heat storage, involves storing heat obtained with solar collectors in bodies of water enclosed just below the Earth's surface. At Blagnac, 240 square meters of solar collectors are connected to a hot-water reservoir of 400 cubic meters. This installation serves 26 apartments and makes possible a 50% savings in gas.

Also noteworthy are an experimental installation in Sweden as well as research in Israel.

Solar collectors for a power plant (1980)

In 1980, the French companies **Bertin** and **Creusot-Loire** undertook the construction of an innovative solar power plant. This was the result of research carried out at the Heliophysics Laboratory in Marseilles. Heat obtained with this installation can be transformed into electricity via a steam turbine system.

Solar power plant (1983)

Inaugurated in June 1983, the **Pernod** plant, near Lyons, France, is the first sun-powered industrial complex. Pri-

marily designed to save energy, the plant also perfectly integrates its solar components into the architectural style. Detectors, energy storage, exchangers, and heating systems—the whole plant—is controlled by a computer that senses any variation in sunlight. Energy thus produced accounts for 50% of that used by the complex.

Mobile solar power plant (1983)

The first wheeled solar power plant in the world, *Solar Genny One* (Genny for generator), was invented by an American, **Ty Braswell**, in the early 1980s. With potential use for anything from open-air rock concerts and rodeos to the recording of the first sun-powered disk, the future of this generator looks very promising.

Space-borne solar power plant

For a long time now, particularly since the energy crisis of 1973, Americans have envisaged the possibility of installing solar power plants in space rather than in the sunny regions of the Earth. The National Aeronautics and Space Administration (NASA) has presented complete programs to this effect. For example, stations consisting of two 8 x 8 kilometer panels covered in photocells have been described. Each station would be placed in geostationary orbit at an altitude of 35,680 kilometers. Orbital movement would be matched to the Earth's revolution, and the station would always be over the same point on the ground.

The electrical current produced by this station would be transformed into microwaves, then sent to the Earth by an antenna with a radiating surface of 36 square meters. The microwaves would be received in a desert region by an antenna with a surface of 100 square kilometers. These space-borne power plants could function 24 hours a day and, it is claimed, would be absolutely nonpolluting. However, recent evaluations of this project indicate that it is not yet feasible, for technical and economical reasons. For instance, a space shuttle able to carry 400 tons of cargo to an altitude of 250 kilometers would be required each voyage; the current shuttle can only transport 30 tons. And 1500 space workers would be needed to operate each station on a permanent basis. The expense would be "out of this world."

District heating (1927)

Burning garbage to produce district heating dates back to 1927. However, this technique was not widely used before 1960. In Paris, a 260-kilometer network provides heating to homes,

The Pernod solar plant.

offices, hospitals, and other buildings. Other towns in France (notably Grenoble and Chambery) also have district heating. In West Germany, district heating networks are two or three times as powerful as those in France. In Moscow, the network is thirteen times longer than in France. As for New York's system, although slightly regressing, it remains the world's fourth largest.

Heat pump (1927)

A heat pump is a device that recovers heat from any free source, e.g., outdoor air, exhaust air, ground water, a lake, or a river. It then transfers the heat through use of a conventional system of radiators or heating pipes. The only energy expended is that needed to run the compressor for transferring the heat. More energy is recovered than is consumed, which makes this process exceptionally economical, especially since the energy crisis.

Credit for having thought of this process goes to Lord Kelvin, the famous Irish scientist. (He also invented the absolute scale of temperature that bears his name.) By 1852, he had already set down the principles of a thermodynamic machine that furnished heat as well as cold. Nevertheless, it took 75 years before the first heat pump was fabricated by the Englishman **T. G. N. Haldane**, in 1927. He used it to heat his office in London and his house in Scotland.

Since then, this technology has become very widespread. More than 50,000 heat pumps were put into service in both private homes and heavy industry in 1982.

Thermal energy from the seas (1930)

In the tropics, there always exists a considerable difference in temperature between the ocean surface (26° to 30°C) and its deeper layers (4°to 5°C), which are cooled by currents flowing from polar waters. Jules Verne noted this fact in 1868, when in his book *20,000 Leagues Under the Sea* he posited using this temperature gradient to produce energy. The technical problem is how to reach these deep ocean layers (around 1000 meters). It would require submerging a long tube to permit contact between the cold layer in the depths and the warm layer on the surface.

In 1928, **Georges Claude** carried out tests on the surface in a river near Liege, Belgium, where warm water was being dumped by a steel mill. The next year, he set up an actual installation in Cuba, in the Bay of Matanzas. A tube 2 meters in diameter and 2000 meters long was used. The idea was to reach an ocean layer 600 meters deep. Two attempts failed (in 1929 and in 1930), but a third succeeded, in the autumn of 1930. Claude was able to put a 10 kilowatt turbine into service. Nevertheless, the cost was enormous and the project was abandoned.

Commercial nuclear reactor (1954)

The first civil nuclear reactor went into service in June 1954 in the Soviet Union, after a reactor that could produce 5000 kilowatts was made operational. Prior to that the United States had developed a nuclear engine for military use and designed for a submarine; its power was analogous to that of the Soviet Union's reactor. However, since it was a prototype (and destined for military use), the United States' reactor did not receive international approval.

Biomass (1973)

Different uses of the biomass have been studied since the beginning of the energy crisis in 1973. *Biomass* refers to the totality of the Earth's vegetation. It is an important source of energy, because vegatation naturally stores solar energy. Indeed, 0.5% of the sun's energy absorbed by the Earth is transformed into biomass by photosynthesis. Its various uses are very diverse. The most highly developed forms are discussed in the following paragaphs.

Direct combustion

Today, a significant portion of the biomass is being burned using traditional techniques. In the future, it will be necessary to substitute modern technology. Throughout the world, forests are being used up at an alarming rate. Most combustors only have an efficiency of 5% or 6%.

Biogas (methane) (1776)

It is possible to recover the heat that results from the composting of manure or brush. This can be used to produce hot water cheaply. However, it is even more economically profitable to produce methane (also known as biogas) by composting at 35°C in an airtight pit.

Methane, discovered in 1776 by **Alessandro Volta**, is formed by the decomposition of organic matter via fermentation. The first use of biomethane dates back to 1857. An installation was built in a leper colony near Bombay, in India.

In Lingalsheim (Alsace, France), 350 pigs produce 4 cubic meters of manure per day. The gas that results from composting this manure amounts to the equivalent of 22,500 liters of fuel oil in a year. It is used to heat 15 acres of greenhouses. Furthermore, the manure can still be used as an organic fertilizer after composting and is even of better quality.

Ethanol (antiquity)

This is the scientific term for ethyl alcohol, or simply alcohol. The procedure for obtaining alcohol goes back to antiquity, but its industrial use for energy production took on more importance following the 1973 international oil crisis. Ethanol is obtained in the distillation of weak wines, but can also be produced by the fermentation of beet sugar or potato and grain starch. As a fuel it is rather costly and has only mediocre energy efficiency. Added to gasoline, ethanol improves the performance of automobile engines, but in order to replace just 10% of the world's gasoline by ethanol, 34 million tons of straw would have to be used for distillation.

Methanol (1661)

Methanol, or methylic alcohol, was discovered in wood distillation products by an Irish chemist, **Robert Boyle**, in 1661. In 1812, Taylor compared it to alcohol, but it was the work of two French chemists, Dumas and Peligot, to determine its composition in 1835. Like ethanol, methanol is a liquid fuel, but with a superior energy yield. Since 1973, experiments on the use of methanol for fueling combustion engines have increased. Methanol is also used in the manufacture of solvents, glues, and industrial chemicals.

Right: A model of a wind tower for production of electricity.

Accumulated compressed-air power plant (1978)

An accumulator power plant was presented to the public on December 8, 1978, by the German **Huntorf**. It was built by the Nordwestdeutsche Kraftwerke. Most of the compressors were developed by the Swiss company Sulzer. These stations substitute for hydroelectric power plants in flat regions.

A reservoir, made up of caverns situated about 500 meters underground, is filled with compressed air during off-peak hours. The air is then released to produce extra energy during peak demand. The power plant can produce approximately 220 kilowatts. The installation is not only economical but also offers the advantage of leaving the countryside untouched.

space

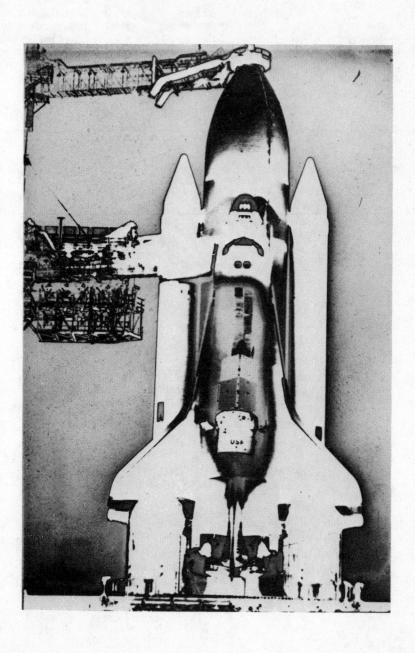

JANUARY   FEBRUARY   MARCH   APRIL   MAY   JUNE

1. Astronomy

Astronomy, together with mathematics, is certainly the oldest science. The first rudimentary astronomical observations were carried out in the great empires of Mesopotamia, Egypt and China several centuries before our epoch. The first astronomers measured time using gnomons and polos, the ancestors of sundials, and hour-glasses and water clocks.

To gaze at the heavens early astronomers used the alidade, a simple rectilinear rod, and the compass, a jointed, double alidade. They were also able to make use of the astrolabe, an instrument for measuring the position of stars above the horizon; Hipparchus, the 2nd century B.C. Greek astronomer, is credited as the inventor of this instrument. It was in this way that the first calendars, charts of the heavens, and astronomical tables were conceived.

Astronomical Satellites

Astronomers have also used rocket-probes, balloons and satellites, to increase their powers of observation of radiation; the latter is difficult, even impossible to receive on the ground because of the earth's atmosphere.

The first observations were carried out in 1946, by Americans, who used V2's recovered in Germany. But they did not allow very extensive observation. The first astronomical satellites were launched by the United States (OAO, OSO, SAS, HEAD, etc.) and by the USSR (Cosmos).

The United States and the USSR also use the Skylab and Salyut space stations (See Skylab and Salyut, p. 350) for astronomical observations.

Space radio-astronomy

The first radio-astronomy satellites were the NASA radio- astronomy Explorer (RAE 1 and 2) satellites launched, respectively, in July 1966 and June 1973. Designed to study the radioelectric waves of stars having long wave lengths, these satellites were equipped with telegraphic antennae (deployable) 230 meters long.

The first space radiotelescope was put into orbit by the USSR in July 1979, in the Salyut 6 space station. The KRT 10 radio telescope, weighing 200 kg, had an antenna 10 meters in diameter; it was carried by a Progress cargo ship and deployed in orbit for centimetric and decimetric wave observations. The Soviets then went on to make similar observations (interferometry of various radiosources) from the Crimean observatory. The KRT 10's antenna was then cast off into space with some difficulty: a cosmonaut had to leave the spaceship to detach it from the space station on which it was coiled.

Calendar

Several thousands of years ago, the Babylonians developed the notions of day, hour and year (the time it took the sun to return to a given point in the heavens). Determining the notion of months, however, was not as easy.

The Babylonians used the lunar calendar in which months began with the full moon. However, the lengths of months were not equal, and it was necessary to add extra months in order to coincide with the seasons. The Moslems still use this calendar.

Solar calendar

The Egyptians invented the solar calendar, which divided the year into 12 months of 30 days each, with 5 or 6 extra days at the end of the year. The names of our months come from the Roman calendar, called the *Romulus*, which had 10 months; there were 304 days in the year.

Under the reign of the King Numa (8th-7th century B.C.), two months were added to the calendar and a 355-day year came into being. In 46 B.C., Julius Caesar, acting on the advice of the Greek astronomer, Alexander Sosigeny, added 10 days and invented leap years. The Julian calendar was used for more than 16 centuries.

Gregorian calendar

The Julian calendar was reformed by Pope Gregory XIII on October 15,

JULY   AUGUST   SEPTEMBER   OCTOBER   NOVEMBER   DECEMBER

1582, because of the time differential which had accumulated since its adoption. This differential (11 minutes 14 seconds per year, or 18 hours 40 minutes per century) amounted to 10 days, and he dropped them in order to base the calendar on the solar system. The year remained set at 365 days, but some leap years were eliminated (3 years for every 400 years, always at the beginning of a century, e.g., 1900). The French adopted the Gregorian calendar in 1582, the English in 1752, the Russians in 1918.

January 1st

Based on a proposal made by the Scythian monk, Denys the Small, the Church decided that, beginning in 532, years would be counted starting with January 1st, a date closely following the date of Christ's birth.

Maps

It was most likely **Anaximander of Milet** who drew the first map of the world, in 575 B.C. It showed the world to be round, and surrounded by a sea.

Ptolemy (90-168), a Greek astronomer, mathematician and geographer from Alexandria, drew 26 maps which were used until the 16th century. They are the basis for Christopher Columbus' error; he thought he had arrived in India when he discovered America.

In the Middle Ages, cartography declined and the only valid maps were the *portolanos* used for navigating from port to port. The oldest of these still preserved is the Pisan map, made in Genoa in 1285.

During the Renaissance, sailors acquired the science of measurement. In 1569, the Flemish scientist, Gerhard Kremer, known as Mercator (1512-1594), drew the entire known world on 18 pages, inventing a projection system which is still used for naval maps. The first maps of countries began to appear at about the same time. Donnus Nicolaus Germanus published the map of Germany; it was called the *Eichstedt*. The Frenchman Oronce Finne (from Briancon 1494-1555), drew a map of France, while the Englishman Mathew Paris drew a map of England based on direct observations.

The first map of the United States dates from 1677; it was the work of John Foster, a Bostonian.

Radio-Astronomy

Radio-astronomy involves the study of radio waves, whose wavelengths range from a few millimeters to several meters, using special instruments called radio telescopes. These are essentially made of large antennae (single or multiple), which receive the radio waves (rays between 10 and 10,000 MHz) that enter the atmosphere. To surpass these limits, radio astronomy, like optical astronomy, makes use of satellites.

Radio Telescope

The first radio telescope was developed in 1932, quite by chance, by a young American engineer and radio-electrician, of Czech origin, **Karl Jansky**.

Bell Telephone Company gave him the task of locating the origin of the interference (radio "noise") which was hampering radio transmissions along the North Atlantic Coast, and Jansky built a receiver he had invented himself in Holmdel (New Jersey). It was a curious wood and brass scaffolding, 30 meters long and 4 meters high; it rotated on Ford Model T car wheels, and was thus baptized the "carousel."

Using this apparatus, Jansky was the first to receive unknown signals coming from space: a "continuous whistling" which he recorded on 14.6 meter wave length, originating from the Sagitarius constellation, located in the middle of our galaxy, 250 trillion kilometers away. This discovery created a sensation; at the time, it was not yet known that the stars, like other optical sources, are also radio-electric sources (also called radio-sources).

Jansky nonetheless abandoned radio-astronomy in 1938 due to the lack of interest shown in his discovery. However, his name remains associated with the unit of measurement of radioelectric flux of the stars.

Reber radio telescope

The first real radio telescope, as such, was built in 1937 by another American engineer of Dutch origin, Grote Reber. A real radio buff, he used his own savings to build an instrument in his garden in Wheaton, near Chicago. It comprised a parabolic antenna with a diameter of about 3 meters, asimuthaly mobile on its base, like all modern radio telescopes. After two years of patience and unsuccessful experimentation, Reber finally received a signal having a wave length of 1.87 meters.

With this success behind him, he tried, in 1941, to draw the radioelectric map of our galaxy. This document, published in 1944, marked the real beginnings of radio astronomy as a new scientific discipline, officially recognized by astronomers who were eager to participate in it.

Jansky's radio telescope in 1930.

Radioastronomic observatories

The first radio astronomic observatories were created in Cambridge (Great Britain) and Sidney (Australia) after World War II as a result of the progress made during the conflict in the areas of radar and electronics. The International Union of Telecommunications (IUT) contributed valuable aid to radioastronomers by deciding, in 1959, to reserve a frequency band of approximately 1,420 MHz exclusively for the study of signals emitted by cosmic hydrogen over 21 cm of wave length.

One such transmission, which had been announced by the young Dutch astronomer, H.C. van de Hulst, in 1944, was indeed detected, in 1951, by the American physicists, H.I. Ewen and E.M. Purcell, who used a spectroscope.

Giant radio telescopes

Radio telescopes proliferated throughout the world, permitting the detection and location of a multitude of radio sources, some of which were unexpected (like the quasars and the pulsars).

● Jodrell Bank: The first large radio telescope was developed in Jodrell Bank, near Manchester (Great Britain); it went into operation in 1957, the year the first artificial satellite, *Sputnik I*, was launched, and the instrument tracked the Sputnik rocket.

This apparatus, designed by the British physicist, Sir Bernard Lovell, had been built under difficult conditions, especially as regards its financing: Lovell was heavily in debt and even risked going to prison. However, his radio telescope was built; he equipped it with a mobile antenna, 76 meters in diameter, and it was the best instrument in the world for detecting radio waves having a length of more than 50 meters.

● Eifelsberg: The large radio telescope in Green Bank (Virginia), put into operation in 1962 and equipped with an antenna 91 meters in diameter, surpassed that of Jodrell Bank in size. Today the largest mobile-antenna radio telescope in the world is the one in Eifelsberg, near Bonn (Germany). In effect, this instrument is equipped with a parabolic antenna, 100 meters in diameter.

● Ratan 600: The largest Soviet radio telescope, the Ratan 600, went into operation in July 1974 under the aegis of the Causacus Academy of Sciences (USSR); it is located on a plateau, 1,000 meters high, near Zelentchouk, only 40 kilometers from the largest optical telescope in the world. The Ratan 600 is a radio telescope with a diameter of 600 meters, made of approximately 900 rotatable reflectors arranged in rings.

● Arecibo and Nancay: The Arecibo radio telescope, installed by the United States in Puerto Rico, is an apparatus which has a stationary antenna, 300 meters in diameter, and a reflector, made of 38,778 aluminum panels which cover the bottom of a deep, narrow valley.

The Nancay radio telescope, in Sologne (France) is a semi-mobile instrument: a rotatable, 200 x 40-meter reflector surface, sends the signals back to another stationary 300 x 35-meter reflector.

Radiointerferometer

The interferometer, with aperture synthesis, was developed in 1960 by **Sir Martin Ryle**, a British radioastronomer, who received the Nobel Prize in physics along with his colleague, Anthony Hewish in 1974.

Interferometry with aperture synthesis is a method that combines two or several antennae, which simultaneously receive signals from a common source, and these signals are then transmitted to a single receiver. Calculation then permits a real map of the observed source to be made, with a resolution analogous to one which might have been obtained by using a very large instrument (one whose size was equivalent to the distance separating the antennae of the interferometer: and this would not always be possible).

The first large radiointerferometer was built by Sir Martin Ryle, in 1964, in Cambridge (Great Britain), using 3 telescopes, separated by a distance of 1.6 km.

VLA

The largest radiointerferometer in the world is the Very Large Array (VLA), built in the desert near Soccoro (New Mexico). Completed, the VLA comprises 27 metallic parabolas, with a diameter of 25 meters, distributed on three bases 25 km long, arranged in Y-shape. An initial series of six antennae went into operation in 1977.

Radar-astronomy

Radar-astronomy is a particular technique of radio-astronomy, in which radiotelescopes are no longer used simply as passive receivers but like radar, that is, in an active manner, first like transmitters, then like receivers.

Since 1946, radar echoes with rather weak transmission power have been reflected off the moon, as well as off other heavenly bodies, in order to measure their distances and their movements. This same technique was recently used with the Arecibo radio telescope to establish the first topographical map of the surface of the planet Venus (constantly surrounded by a thick layer of clouds).

Telescopes

Galileo's telescope

The invention of the telescope is attributed to the illustrious Italian physicist and astronomer, **Galileo Galilei** (1564-1642), the author of the principle of inertia and of the laws of constant acceleration.

Galileo built his first telescope in 1609, when he was 45. In fact (as he himself realized), Galileo was perfecting an already existing instrument, probably invented about 1590 by Dutch craftsmen. However, their telescopes were equipped with primitive optical systems and were therefore practically useless.

Galileo, on the other hand, built a remarkable instrument using a concave lens, and with it he discovered the true nature of the Milky Way, a multitude of stars, the contours of the moon, the phases of Venus, the satellites of Jupiter, sun spots and the rotation of the sun, all of which contributed more new knowledge than had the combined work of all his predecessors.

Galileo later lost his sight and lived out the final years of his life in solitude, no longer able even to use his marvelous telescope. Galileo's telescope is today preserved in Florence.

Newton's reflecting telescope

The first reflecting telescope was built in 1672 by Isaac Newton, the celebrated English physicist, mathematician and astronomer who, at the age of 24, announced the famous laws of universal gravitation.

A member of the prestigious Royal Society, of which he became president, Newton built the first reflecting telescope; it had a focal distance of 16 meters and was equipped with a spherical metallic lens, 25 cm. in diameter. The instrument was perfected in the 18th and 19th centuries, notably by the Englishman, William Herschel, and the German, Johannes Hevelius, the mayor of Danzig. In 1842, another Englishman, William Parsons, third Count of Rosse, built the first giant telescope on the grounds of his country estate.

Giant telescopes

• Mount Palomar: The huge telescope on Mount Palomar, in California, is equipped with a lens having a 5-meter aperture. It was put into service on June 3, 1948. It was built at the urging of George Hellery Hale, an astronomer, and the inventor of the spectroheliograph.

He received assistance in the form of a 6 million dollar donation from the Rockefeller Foundation. This was not a commonplace undertaking: the most eminent American astronomer had convinced the richest family in the United States to finance the most expensive scientific instrument in history.

It took twenty years to build the giant telescope (twice the size of the large telescopes of the time), and for a quarter of a century it would be the largest optical telescope in the world. It was also the first to be equipped with a Pyrex lens (weighing 14.5 tons); until then, Pyrex had only been used to make crockery.

For some years, however, Hale's giant telescope has been relegated to a less important position, and it has become difficult to use it due to the increasing pollution which comes from the neighboring city of Los Angeles.

• Zelentchouk: The Zelentchouk observatory telescope, located at an altitude of 2,050 meters in the Caucasus Mountains (USSR), is today the largest in the world. The telescope has a 5-meter aperture, and was designed by Dr. Icannissiani. It went into operation in 1974, after extreme difficulties in manufacturing its enormous, 42-ton lens made of borosilicated glass which was ground by the Lomo center in Leningrad. In addition, the lens had to be rebuilt three times; it is one-and-a-half million times more powerful than the human eye.

This telescope has allowed the first observation of 23rd magnitude stars. However, it has a major defect; unlike all observatories in the Western world, it is closed to foreigners.

The Zelentchouk telescope is most certainly the last of the "giants" in optical astronomy. Indeed, the construction of such lenses has now reached the limits of know-how and financial possibilities (even for a powerful State); further, optical performance is not guaranteed.

All the new telescopes built since have more modest dimensions (3 to 4 meters) than the monsters of Zelentchouk and Mount Palomar. Astronomers have decided that it is more efficient to select very good sites and improve the sensors. The Las Palmas telescope, in the Canary Islands (Spain), has a lens 4.2 meters in diameter and is the third most powerful telescope in the world.

Infra-red telescope

The largest infra-red telescope in the world, equipped with a lens having a 3.8-meter aperture, was installed in 1977 by Great Britain at an altitude of 4,200 meters at the top of Mount Mauna Kea, an extinct volcano on the island of Hawaii. Infra-red telescopes can still use wide lenses, as they are easier to manufacture than lenses used in "visible" optics.

Multiple-lens telescopes

The world's first multi-lens telescope was recently installed in the Mount Hopkins observatory (Arizona), by the Smithsonian Astrophysical Observatory (SAO) and the University of Arizona. It comprises six lenses with a 1.85-meter aperture, and is the equivalent of a single-lens telescope with a 4.5-meter aperture.

This new kind of optical telescope was devised by astronomers in order to reduce the size as well as the cost of the instruments, without sacrificing performance. The multiple-lens telescope operates by means of synthesis of aperture: several average-sized lenses are combined (and aligned by laser) inside a single pointing device to provide images of the same object which are then focused at a single point.

Space Telescope

The American *Space Telescope* will be the largest space telescope. In 1985, a satellite, weighing 10.4 tons, and 13-m in length, will be placed in a circular earth orbit at an altitude of 500 km, by the American space shuttle, Columbia. The telescope will have a 2.4-meter aperture. Its performance will be superior to that of current earth-bound instruments, such as the Mount Palomar or the Zelentchouk telescopes: in effect, the Space Telescope will be placed outside of earth's atmosphere and its disturbances.

The Space Telescope will be able to detect objects which are 50 times smaller than those detected by the largest telescopes on earth. Further, it will be able to observe a volume of space which is 350 times greater, and objects as far away as the outer limits of the universe, about 15 billion light years away.

The construction of this unique instrument cost Lockheed, the main contractor, and Perkin Elmer, in charge of the optics, about one billion dollars and ten years of effort. Several European firms also participated in the construction: British Aerospace (G.B.), Dornier (Germany) and Matra (France).

In the opinion of the specialists, the Space Telescope will revolutionize astronomy.

Airborne Telescopes and Space Astronomy

The first astronomical observations carried out by airplane date back to June 29, 1927, when a twin-engine British Imperial Airways plane was used to photograph a total eclipse of the sun above the London fog. This observation technique has since been successfully used during the last twenty years in the United States and in France, notably with the Caravelle planes and even periodically with the supersonic Concorde.

Kuiper Observatory

Since 1965, NASA has been equipped with Learjet and Convair 990 jet airplane observatories, which have allowed, in particular, the discovery of the infrared emissions originating from the center of our galaxy.

Since 1975, NASA has also used a giant C141 Starlifter four-engine jet, especially equipped for astronomy, and in particular for the study of infrared rays. This apparatus, named the Kuiper Observatory after the famous Dutch astronomer, can be used more than 200 nights per year. It is capable of flying for 3 hours with 9 tons of observation instruments, at an altitude of more than 14,000 meters, that is, above 85% of the atmosphere and 99% of the water vapor which surrounds our planet. The Kuiper Observatory is equipped with a telescope having a 91-cm aperture and it has already made many important discoveries, one of which was the observation of the rings of Uranus, in 1977.

2. Meteorology

Meteorology is thought to have begun following the violent storm on November 14, 1854, in the Black Sea during which 3 vessels and 38 ships of the English and French navies anchored at Sebastopol (Crimea) sank and cost the lives of 400 sailors.

Napoleon III's War Minister then asked the famous French astronomer, **Urbain Le Verrier** (1811—1877) if this phenomenon might have been foreseen. Le Verrier noted that the storm had crossed all of Europe, in three days, before it hit the Anglo-French fleet, and he concluded that the catastrophe could have been forecast.

Thus, some consider Le Verrier to be the father of modern meteorology, a science to which numerous other astronomers have also contributed.

Artificial Rain

In 1946, **Vincent J. Schaefer** and **Irving Langmuir** of the General Electric Co. observed that rain could be produced by seeding a cloud with dry ice (solid carbon dioxide).

On November 13, 1946, they flew over the Mount Greylock area (Massachusetts) and sprayed dry ice into clouds (over a distance of 6 km, at a height of 4,500 meters). It is believed that rain did fall, but that it evaporated about 2000 meters from the ground because of the dryness of the atmosphere.

Based on this result, further experiments were carried out in several countries. In 1948, in France, 90,000 tons of water fell on the Paris area after 15 kg of dry ice had been sprayed into clouds.

In 1950, it was discovered that silver iodine crystals had identical properties to those of dry ice.

After having been in fashion for some years, the method was abandoned since the appropriation of a natural resource as essential as water by one area, or by one country, involved too many risks.

Hygrometer

This apparatus measures the degree of humidity in the air and was developed by **Francesco Folli**, in 1664.

It was replaced a century later, in 1781, by the hair hygrometer developed by Horace Benedict, from Saussure (1740—1799): the longer the hair, the more humid the air. This was an absorption hygrometer. Condensation hygrometers subsequently appeared: the quantity of condensed water vapor is measured in a metallic vessel.

Mercury Barometer

Torricelli Barometer

The mercury barometer is the invention of the Italian physicist and geometer, Evanisto **Torricelli** (1608—1647), a student of Galileo.

Using this apparatus, Torricelli succeeded in demonstrating that air had its own weight, which varied depending on the circumstances, and that one could measure the weight variations (atmospheric pressure) by studying variations in the height of the mercury column in the tube.

Wheel Barometer

The first wheel barometer was built by the English astronomer, **Robert Hooke** (1635—1703), in 1665. It was considerably perfected by the French engineer, J. Fortin (1750—1831), whose name it still bears.

Recording Barometer

The oldest system of a recording barometer was invented by the Englishman **Moreland**, in 1670.

Measurement of Altitude

It was necessary to await the invention of the balloon for the barometer to be used to measure altitude. A barometer was carried to an altitude of 3,467 meters on December 1, 1783, aboard a balloon; the French physicist, Jacques Charles (1746—1823), the first to use hydrogen to inflate balloons, was the inventor of the technique.

This performance was subsequently improved in 1804 by the famous French physicist and chemist, Louis-Joseph Gay-Lussac (1778—1850), who reached a height of 6,977 meters, a record at the time.

A barometer can constitute an altimeter. In effect, atmospheric pressure decreases in inverse proportion to an increase in altitude.

Aneroid barometers

The aneroid barometer was invented by the Frenchman L. Vidie, who applied for the patent in 1844.

The mercury barometer provided

On August 20, 1804, Gay-Lussac and Biot went aloft to study temperature differences at various altitudes.

precise and simple measurements, but it was difficult to handle and was limited to a rather narrow range of measurements. Vidie designed an apparatus without liquid; this is the reason for the term aneroid (which means "not wet"). His non-mercury barometer had the advantage of being very sturdy and could function in any position whatsoever.

The modern aneroid barometers are precision instruments which, when the various corrections are carried out, provide pressure values with a mercury precision of 0.5 mm to 0.025 mm.

Rain Gauge

This is a very old apparatus which allows the measurement of the quantity of rain which has fallen in a given spot and at a given time, and was being used in India four centuries before Christ. It was then re-discovered in Korea, in 1442, under the reign of King Sejo. In the West, the use of the rain gauge in meteorology developed beginning in 1639, due to the work of the Italian, Benedetto Castelli.

Wind Gauge

The blade wind gauge was produced in 1644 by the English mathematician and astronomer, **Robert Hooke** (1635—1703).

The principal of this apparatus, which measures wind speed, had been established in 1450 by the Italian, Alberti.

The wind pressure gauge was invented in 1775 by the Scottish physicist, James Lind.

The cupel wind gauge, still used today, was developed in 1846 by the Irish astronomer, Thomas R. Robinson.

Sodar

Sodar (Sound Detection and Ranging) is a kind of radar which uses acoustical waves transmitted vertically and retro-broadcast by temperature turbulence; these allow measurement of the three components of wind.

The principle of this still experimental apparatus was described in 1968—1969 by the Americans **L. G. McAllister** and **C. G. Little**.

Doppler sodar

In 1979, the French company, Bertin, patented the Doppler sodar, as a result of the research carried out by M. Fage.

Sodar was developed with a view to improving the safety of air traffic. It is designed for the continuous measurement of wind by means of telede-

A weather balloon in 1937.

In 1906, Alexander Bell invented a meteorological kite.

tection. The thermal structure of the atmosphere at high altitudes, that is, its degree of heat, is also measured. This indicates wind shear near airports, and can also predict the movements of atmospheric pollutants.

Thermometer

The first thermometer was invented in 1593, by the Italian, **Galileo Galilei** (1564—1642). It was a gas thermometer and it indicated temperature variances, without being able to measure them. Some years later, in 1612, the Italian Santorio developed a water thermometer (See medical thermometer, p. 198).

Alcohol thermometer

The first alcohol thermometer was produced in 1654 by the Italian enameller, **Mariani**, based on information given him by the Arch-Duke of Tuscany, **Ferdinand II** (1621—1670).

Regardless of the liquid which fills the thermometer (alcohol, mercury, etc.) the principle is the same: the liquid expands or contracts under the effect of temperature variations and rises or descends in a closed tube. The two problems to be resolved concerned reliability (the expansion must always be the same) and the gradation of the instrument. To calibrate a thermometer, two fixed points must be chosen and the height of the liquid in the tube at the fixed points must be precisely determined. Subdivisions can then be made.

Fahrenheit thermometer

Daniel Gabriel Fahrenheit (1686—1736) developed his mercury thermometer in 1715. A native of Danzig, he spent his life between Holland and England. After having learned to blow glass, he decided to build thermometers. He chose mercury, which freezes at -38.8°C and boils at 357°C, and therefore has widespread possibilities. Fahrenheit chose two original fixed points: the fusion of a refrigerating mixture (crushed ice and ammoniac salt) and the normal temperature of the human body. Then he divided the interval into 96 degrees: 0°C corresponds to 32° Fahrenheit and 100°C to 212°F. These thermometers were so reliable that they had a great commercial success. They were quickly adopted in England, which still uses the Fahrenheit scale, as well as in the United States and Australia.

The Celsius thermometer

The Swedish physicist, **Anders Celsius** (1701—1744) built a mercury thermometer in 1741; he set 0° as the boiling point of water, and 100° as the point at which water freezes. He divided the interval into 100 degrees.

At the same time, the Frenchman, Jean-Pierre Christin, from Lyon, presented a thermometer in France which used the same reference points, but on an ascending scale. Revolutionary France adopted the metric system on April 7, 1795, and defined the thermometric degree as "the hundredth part of the distance between the factor of ice and that of boiling water." This is the origin of the term "centigrade degree." The world Conference on Weights and Measures accorded it the name of Celsius degree (symbol: C) in October, 1948.

Today, the Kelvin scale, which does not begin with the temperature at which water freezes but with absolute zero (-273.16C), is one of the 6 basic units in the international system.

Weather Balloons

The first weather balloons, designed to explore the atmosphere automatically at a given altitude, lifted off in 1898. It was at that time that **Leon Teisserenc de Bort** (1855—1913), a French meteorologist, launched his first weather balloons in Trappes (France), and discovered the stratosphere.

Radio Probe

The first radio probe, designed and produced by the two French meteorologists, **Robert Bureau** (1892—1965) and **Pierre Idrac** (1885—1935), flew for the first time on March 3, 1927, under a balloon launched in Trappes.

This apparatus allowed the first measurements of temperature, humidity and air pressure at different altitudes; it provided immediate results due to automatic transmission via radio.

Weather Map

The first known weather map was drawn up in 1686 by the famous English astronomer, **Edmund Halley** upon his return from a trip to the southern hemisphere. He put forth the first explanation of regular winds, monsoons and tradewinds.

The first isobar maps, credited to the astronomer, Marie-Davy were published in the meteorological review of the French Imperial Observatory in 1863. The isobar map graphically represents the pressure differences at the surface of an area. Lines connect the points where the atmospheric pressure is the same, just as the level lines indicate relief.

Water current map

The first maps of water currents, prepared by the American Matthew Fontaine Maury (1807—1873), met with immediate success: 210,000 maps were sold from 1848 to 1858. According to estimates prepared at the time, they saved 50 million dollars for maritime shipping companies.

Weather Radar

In 1924, the Englishmen E.V. Appleton and M.A.F. Barnett demonstrated the existence of the ionosphere (the area of the atmosphere in which the components of air are highly ionized), using the reflection of continuous waves. However, the first campaign to use radar for weather study was carried out in 1949, in the United States, for the Thunderstorm Project, under the direction of **M. H. Byers** and **R. R. Braham**.

Weather radar allows tracking of weather balloons equipped with reflectors, detection of large storm clouds and rain, and the study of the internal structure of some cloud masses. Since 1962, it has been widely used in many countries.

Lidar

Lidar is an optical radar, with a laser as transmitter and a telescope as receiver, backed up by a photoelectric detector. It would appear that G. Fiocco and L.D. Smullin were the first to use the Lidar for meteorological purposes in 1976, to detect aerosols at altitudes of up to 140 km.

Weather Satellites

The first weather satellite was launched by the United States on April 1, 1960. The *Tiros I* satellite, like most American weather satellites, was built by RCA.

It was the United States which also launched the first geo-stationary weather satellite, *SMS I*, put into orbit in May, 1974. This satellite was built by Ford Aerospace, a subsidiary of Ford Motor Company.

Space meteorology has essentially made use of two categories of satellites for the last 20 years:

• flying satellites, placed in a circular orbit at an altitude of 1,000 to 1,500 km, and quasi-polar, slanted from 90 to 100 toward the equator.

• geo-stationary satellites which gravitate in a circular orbit around the equator at an altitude of 36,000 km. Since they travel at a speed which is equal to the speed of the earth's rotation, they appear stationary to an observer on the earth.

50 YEARS OF METEOROLOGY	
Nov. 11, 1936	First weather map shown on English television.
1939	First telephonic weather report service installed in New York.
1942	Development of meteorological radar.
May 4, 1944	Due to the opinion and weather forecast of the meteorologist, Stagg, General Eisenhower delayed the Normandy invasion until June 6th; it had originally been scheduled for June 5th.
March 1946	Establishment of 13 floating meteorological stations in the North Atlantic.
1946	Vincent J. Schaefer and Irving Langmuir performed the first successful experiment on cloud seeding in Schenectady (New York, USA) (see "artificial rain").
1950	Possible dating of past climactic events, as far back as 30,000 years (work of Willard Libby using carbon 14). Harold Urey developed a method for measuring the past temperatures of the seas.
1954	Official classification of clouds adopted by the OMM.
1954	First numerical forecasts performed daily (USA and Sweden).
1959	Network of weather rockets established in the USA. The rockets reached an altitude of 55 km.
April 1, 1960	Launching of the Tiros 1 satellite (122 kg.). This launching marked an essential stage in the history of weather satellites.
August 20, 1964	The Nimbus 1 satellite took the first high resolution photographs at night.
December 6, 1966	Launching of the first peostationary weather satellite over the Pacific.
1969	The first GARP project was launched: BOME (Barbados Oceanic and Meteorological Experiment).
May-June 1979	The Executive Commitee of the OMM adopted the new international meteorological code, which became applicable on January 1st, 1982.
1980	Formulation of the world-wide climatological program (OMM-Geneva).

3. The Conquest of Space

Rockets

25 years of Rockets

Although the origin of space rockets goes back some fifty years, notably to the famous German V2, it was not until 1957 that a rocket escaped earth's gravity and put into orbit the well-known Sputnik.

Since 1957, more than 2,500 rockets have been launched. The USSR holds the absolute record, with more than 1,400 rockets, while the United States has launched a few more than 800. The USSR currently launches almost 100 rockets per year; six times as many launched by the United States. But, the United States utilizes more advanced technology and its launchings are ensured of better performance.

First liquid-propelled rocket

Americans willingly attribute the first liquid-propelled rocket to **Robert H. Goddard** (1882—1945), who, on March 16, 1926, launched a small rocket (almost a toy!) a distance of 12.5 meters. The flight lasted 2.5 seconds and was not mentioned by Goddard until 10 years later.

A physics professor at Worcester College (Massachusetts), Goddard had published a work describing "a method for reaching extreme altitudes" as early as 1919. Goddard applied for more than 200 patents for liquid-propelled rockets, most of which used liquid oxygen for fuel; he even built a liquid-propelled engine having a thrust of 300 kg. However, his work was practically ignored by his contemporaries.

The first European to launch a liquid-propelled rocket was the German Johannes Winkler, whose small HWI rocket (using liquid oxygen and methane) was launched on February 21, 1931 in Breslau (Germany).

In fact, the exact date of the first liquid-propelled rockets is not known, since at the time numerous researchers were at work in their various countries. Some think the invention dates back to 1895, when the Peruvian engineer, Pedro E. Paulet, built and patented an engine, 10 cm in diameter, in which a mixture of nitrogen peroxide and gas, fired by an igniter, was injected.

V2 Rocket

The most powerful of the rockets used during World War II was the German A4, alias V2, which inspired all postwar rocket builders.

The V2 was developed under the direction of General Walter Dornberger, assisted by the engineers, Werner von Braun and Thiel, who worked first in Kummersdorf, near Berlin, and then in the famous test center of Peenemunde, on the Baltic Sea. The first test launching took place on October 3, 1942. Nearly 4,300 V2's were launched over London in 1944—45. However, due to the rocket's well-known lack of precision, their use did not change the outcome of the war.

Soviet rockets

The USSR has always been very circumspect about its rockets, whose characteristics and construction have remained secret for a long time. It has nonetheless been possible to identify at least five types of Soviet rockets, categorized by an American code name, since their real Soviet official designation is not known. All these rockets (A,B,C,D,F) are derived from Soviet ballistic missiles, except for the giant rocket (G) which no one in the West has yet seen and for which the tests have failed.

The principal craftsmen of the Soviet rocket program are the academician, Mtislav Keldych, the principal builders **Sergei Korolev**, Yanguel and Vladimir Chelomei, the directors of the engineering departments in charge of the construction of the various rockets, as well as the engineer, Valentin Glouchko, director of the Gas Dynamics Laboratory (GDL) in Leningrad which produced most of the Soviet liquid-propelled engines.

A V2 rocket on the launching site at Kalbschafen in 1945.

The Korolev rocket

The best known of the Soviet rockets is that produced by **Sergei Korolev** (1906–1966); this was the rocket which put the first satellite, Sputnik I, into orbit on October 4, 1957. This rocket was not shown to the public until ten years later at the 1967 Bourget Air Show.

The rocket's construction was original (a series of stages comprising numerous engines, each having a weak unit thrust); this rocket was approximately 30 meters in length and weighed 300 tons; it developed a thrust of 514 tons at liftoff.

The rocket burned kerosene and liquid oxygen, a mixture known ever since Goddard carried out his investigations in the United States. The rocket's high performance was the result of its size.

Nearly 850 rockets of this type (A) have been launched over the past 25 years in the USSR. They are used to put Soviet cosmonauts into orbit. This is the most widely used rocket in the world.

Yanguel rocket

The Yanguel rocket has been used to launch some of the 1,500 Cosmos satellites that have been put into orbit since 1962. The rocket is 20 meters in length and was derived from the Soviet ballistic missile, Sandal (NATO code name) which provoked the Cuban missile crisis in 1962. It was not until 1981 that the name of the principal constructor of the rocket, used for some 20 years, was learned.

The engineer was **Mikhail Yanguel**, the son of a peasant family and today a Soviet academician. In fact, the USSR celebrated Mr. Yanguel's 70th birthday by revealing that since 1962, after having worked with Korolev, he had directed one of the most important organizations involved in the construction of Soviet rockets.

Chelomei rocket

The most powerful Soviet rocket currently in use is the type D rocket, a 1,000 ton machine. Used since 1965 to place the large Soviet scientific and geo-stationary satellites into orbit, it is also used for the Salyut space stations (20 tons).

This rocket, which uses the stage architecture designed for the Korolev rocket, was built under the direction of the engineer **Vladimir Chelomei**. It is also known under the name of "Proton-launch," after the large scientific satellites (12 tons) which it put into orbit when it was first created.

American rockets

The United States was caught unawares by the 1957 Soviet exploit which put the first Sputnik I satellite into orbit. In response, it put together a satellite launcher using the Redstone missile, alias Jupiter A, developed in only a few months by **Wernher von Braun** and his team.

A whole fleet of rockets was subsequently built for civil use by NASA (National Aeronautics and Space Administration), created in 1958 to compete with the Soviet program, and for military use by the U.S. Air Force.

The Atlas rocket

The most widely used American rocket was the Atlas rocket, produced by General Dynamics. Its first launch was successfully carried out on December 17, 1957.

No fewer than 464 Atlas rockets were launched during 25 years of service, placing civil and military satellites into orbit. The Atlas was chosen to send the first American astronaut into space: John Glenn (born in Cambridge, Ohio, in 1921) made his orbital flight on February 20, 1962.

The Atlas is derived from the long range MX 774 missile, which was developed by Karel Bossart, an American engineer of Belgian origin, who was then working at the Jet Propulsion Laboratory (JPL) at Caltech, in Pasadena (California). This rocket, 40 meters in length, has a narrow structure and is made of stainless steel; it weighs 147 tons and is propelled by a kerosene and liquid oxygen engine with a thrust of approximately 200 tons. For the upper stage, it used, in particular, the Agena rocket produced by Lockheed; this was the first rocket stage which was able to be re-ignited in flight. Since 1962, and to date, this rocket has been associated with the General Dynamics Centaur stage, which is the first cryogenic stage (liquid hydrogen and oxygen) to be used by the United States and throughout the world. The development of this 6.8-ton stage cost NASA 700 million dollars. Fifty-seven Atlas-Centaur rockets have already been launched, and these rockets will continue to be used through 1985.

Titan rocket

The Titan rocket, developed by **Martin Marietta**, is one of the most powerful rockets in the American arsenal. The first version, Titan I, which flew in February 1959, was 27 meters in length, weighed 100 tons and was equipped with two kerosene and liquid oxygen engines, with a thrust of 136 tons at liftoff. It gave birth to an intercontinental missile, the Titan 2, of which 62 models are still in service today in silos located throughout the United States.

In 1962, it was replaced by the Titan 3 version, which comprised a Titan 2 rocket and two large powder propulsive engines having a 1000-ton thrust. It was manufactured by United Technologies.

The most powerful version was the Titan 3E-Centaur, which used the General Dynamics cryogenic upper stage. This rocket, capable of placing 17 tons into a low orbit around the earth or of sending 4 tons into interplanetary space, was used to launch the Viking probe ships to Mars. A new, improved version, the Titan 34 D, using the Boeing IUS powder upper stage, went into service in 1982.

Delta rocket

The McDonnell Douglas Delta rocket, which flew for the first time on May 13, 1960, has been the NASA work horse during recent years in placing geo-stationary satellites into orbit (circular orbit around the equator at an altitude of 36,000 km).

More than 160 Delta rockets have been launched over the past 20 years, with a constant improvement in performance; the payload has in fact increased from 100 kg to 1,100 kg. This rocket weighs more than 130 tons and is propelled by Rocketdyne kerosene and liquid oxygen engines with a thrust of 90 tons, backed-up by nine Thiokol powder propulsive engines with a unit thrust of 23 tons. A new, improved version, Delta 3920-PAM, can place 1.4 tons into geosynchronous transfer orbit; it went into operation in 1982.

Scout rocket

The smallest of the American launchers, the Scout, was developed by the Vought Company (LTV Group), and was also one of the most widely used. More than 100 Scouts have been launched since the rocket went into operation in February 1961. The Scout is also the first all-powder American rocket (4 stages).

This rocket is 22 meters in length, weighs 22 tons at liftoff, and its performance has improved continuously; in 20 years, the payload has increased from 45 to 200 kg.

The Scout has been used in particular to launch numerous foreign satellites, especially European ones. This is the only American rocket to have been launched outside of the United States. Italy launched it from a sea platform located off the coast of Kenya in April, 1967.

Saturn 5 rocket

The largest rocket to date is the gigantic American Saturn 5 rocket developed by NASA and built under the direction of Wernher von Braun, based on the Juno 5 project, alias Horizon, of the U.S. Army.

This monster, 110 meters in length, weighs 2,900 tons and develops a thrust of 3,400 tons at liftoff. The first stage is, in effect, made up of 5 FI engines which burn 13,500 liters of liquid propellant per second (liquid kerosene and oxygen). These are still the largest liquid-burning engines made to date.

The two upper stages use high performance engines which burn a higher energy mixture (liquid hydrogen and oxygen). The Saturn 5 was thus capable of placing almost 120 tons into a low orbit around the earth, or 50 tons to the moon.

The rocket's first trial was carried out in November, 1967, at Cape Canaveral (Florida). Saturn 5 was subsequently used for all the American moon flights between 1969 and 1972, and in 1973 this rocket launched the first American space station.

Space Shuttle

The NASA Space Shuttle is the first rocket in the history of astronautics able to be recovered and used again. It is a real aerospace vehicle; weighing more than 2000 tons, it lifts off like a rocket (vertically); the main part (Orbiter) is a sort of delta winged aircraft which weighs 100 tons and is placed into orbit around the earth at low altitude (160 to 1,100 km) for one to four weeks. Orbiter then reenters the atmosphere at a glide and lands on a runway like a plane (horizontally).

The space shuttle can carry a 30-ton payload and a crew of 4 to 7 astronauts, two of whom are the pilots. Due to its jointed arms, it can place all sorts of satellites into orbit. This revolutionary rocket prefigures the space ships of the future, which, like all other means of transport, will be reusable.

Decided upon by President Richard Nixon in 1972, the shuttle was principally constructed by Rockwell International. The United States foresees spending 15 billion dollars to build a fleet of 4 shuttles.

The first, Columbia, had its maiden flight on April 12, 1981, with John Young and Robert Crippen at the controls. Columbia lifted off from Cape Canaveral (Florida) and landed 54 hours later at Edwards A.F. Base (California). The Columbia shuttle has already carried out five successful flights. The second shuttle, Challenger, had its first flight in April 1983; the third, Discovery, had its first flight in August, 1984, and the fourth, Atlantis, is scheduled for launching in 1986.

Other rockets

Ariane rocket

Ariane is the first rocket which Europe has succeeded in building. Its first launching successfully took place on December 24, 1979, in Kourou (French Guinea).

In 1973, ten western European countries (France, Germany, Great Britain, Belgium, Italy, the Netherlands, Spain, Denmark, Sweden and Switzerland), members of the European Space Agency (ESA), decided to build this new European launcher to replace the Europa rocket for which final development failed.

Ariane is a 200-ton rocket, built according to standard design. It has three stages. The first two stages burn a mixture of nitrogen peroxide and dissymetric dimethlhydrazine (UDMH mixture), while the third stage uses cryogenic propellants (liquid hydrogen and oxygen).

Ariane was constructed under the direction of the French National Center for Space Study (NCSS), and by (about) 60 firms from the 10 participating countries.

Indian rocket

The most modest of all currently used rockets throughout the world is the SLV 3, an Indian rocket, which, in July 1980, succeeded in placing a small satellite, Rohini, into orbit; it weighed 35 kg, and was launched from Sriharikota Island (India).

This small, three-stage powder rocket weighs 17 tons and is 25 meters in length. It was constructed by the Indian Space Research Organisation (ISRO), under the direction of Professor Satish Dhawan and the chief engineer, Abdul Kalam.

India is presently developing two other, more powerful rockets, the ASLV and the PSLV, which will be tested during the mid-1980s.

Private rockets

During the last few years, some private initiatives in the field of rocket development, until now exclusively controlled by governments, have been undertaken.

Billigrakete

The best known private rockets are without a doubt those developed by the German engineer, Lutz Kayser, whose OTRAG Company (Orbital Transport und Raketen Aktien Gesellschaft), founded during the 1970's, built very rudimentary but very economical rockets used to launch observation satellites for the third world countries. However, following several modest and unsuccessful tests from a launching site in Zaire, and an aborted attempt at a Libyan installation, the adventurous promoter of the Billigrakete has abandoned these activities. The project has since been taken over by his associate, Frank Wukasch, and the OTRAG Company is more reasonably turning its attention to the manufacture of economical rocket probes.

Percheron

Several American initiatives must also be pointed out, in particular that of Space Sciences Inc. (SSI), created by a Texas businessman to launch observation satellites. The first Percheron rocket to be manufactured by SSI exploded on its launch pad on August 5, 1981, on Matagorda Island (Texas). Construction of this missile, 17 meters in length, cost 1.7 million dollars.

Volksrocket

Another American, Robert Truax, the founder of the Private Enterprise Project, has begun the small-scale manufacture of a small, 7-meter rocket which uses liquid propellants. His ambition is to send a man into space (to an altitude of 80 km). The cost would be 10,000 dollars to any daring apprentice-astronauts willing to risk their lives using the Volksrocket (the rocket of the people).

Satellites

Sputnik I, the first artificial earth satellite

The first artificial earth satellite, Sputnik I (the Russian word means "companion"), was launched on October 4, 1957, by an A1 rocket, built by Korolev. This simple sphere, made of polished steel, with a diameter of 58 cm, and weighing 84.5 kg would astound the entire world with its incessantly repeated "beep, beep."

The satellite was produced under the aegis of the USSR Academy of Sciences, directed by Professor Mtislav Keldych. The operation, decided upon by Nikita Khrushchev in order to establish Soviet space supremacy, was prepared by a committee, one of whose participants was Leonid Brezhnev.

In effect, it was during the international geophysical year (1957—1958) that the conquest of space had its true beginnings with the launching of the first artificial earth satellites, soon to be followed by moon and interplanetary probes allowing exploration of the solar system.

Explorer I satellite

The first Explorer I was launched on January 1, 1958 by a Jupiter C rocket, alias Juno I, a product of the American Army. This small satellite (14 kg), built under the direction of Professor James Van Allen (born in 1914), of the Jet Propulsion Laboratory (JPL) at the famous California technical university, Caltech, led to the discovery of the radiation belts which surround the earth: the Van Allen belts. Solar protons and electrons captured by earth's gravity are concentrated in these belts.

The rockets used to launch Explorer I was a U.S. Army Redstone missile, based on the V2 developed by Wernher von Braun (1912—1977), who immigrated to the United States in 1945, along with 120 of the best German specialists from Peenemunde.

Cosmos satellites

Since March 16, 1962, the USSR has placed more than 1,500 satellites into orbit, all of which have gone under the simple and unique name of Cosmos, followed by an order number.

In fact, such apparent banality hides the wide variety of missions (at least 15 different kinds) involved in both civil (science, weather, observation of the earth, etc.) and military uses (the latter accounting for more than 70% of the satellites). The Cosmos series is used by the USSR to camouflage its military space missions: reconnaissance, electronic spying, telecommunications, navigation and even the eventual interception of enemy satellites.

Up until now, the USSR is the only country to have carried out broad scale development and testing in space of a system of killer satellites, which are capable of attacking another satellite in orbit and destroying it.

These data would indicate that the USSR has a lead over the United States. This is not the case. The Soviets are behind in the fields of electronics and computer technology, as well as in miniaturization. This means that they must place far heavier payloads in orbit; overall, their performance is low when compared with the American payloads.

Space Craft (Manned)

To date, only two countries, the United States and the USSR, have launched space craft with astronauts or cosmonauts (Soviets) on board. There have been flights around the earth, moon flights, and the setting up of orbital space stations used to shelter space crews on long-duration flights.

Since the historic flight of the Soviet, Yuri Gagarin, on April 12, 1961, astronauts and cosmonauts from 12 countries have been sent into space, including three women, Sally Ride (June 18, 1983), Valentina Tereschkova (June 16, 1963), and Svetlana Savitskaya (August 19, 1982)

Soviet manned space craft

The USSR has used four kinds of manned space craft to send men into space. First the one-man Vostoks which were the pioneers in human astronautism, then the Voskhod, a two-man craft, which was followed by the Soyuz, first a three-man and then a two-man craft, and finally the Soyuz T, a three-man craft. All Soviet manned space crafts have been launched from the Tyuratam-Baikonour cosmodrome in the Kazakhstan (USSR).

Vostok

The first Soviet manned space craft, and also the first artificial satellite to have a man on board, was the Vostok. This vessel was initially tested with dogs as passengers; five trials were run between May 1960 and March 1961, under the name of Korabl.

On April 12, 1961, Vostok 1 carrried the young cosmonaut, Yuri Alexerievitch Gagarin (born in 1934, died in 1968, in an airplane accident). Gagarin orbited the earth at an altitude of 327 km in 108 minutes. Six Vostoks were subsequently launched, until 1963. The last carried the first space woman, Valentina Tereschkova.

The Vostoks were 4.7-ton vessels, and the inhabitable cabin (2.4 tons) was a simple sphere measuring 2.3 meters in diameter. The cosmonaut was installed in the cabin, in an ejectable seat. Indeed, for the return, the cosmonaut left the cabin (which fell like a shell) and parachuted from an altitude of 7,000 meters.

Voskhod

The Voskhod was the successor of the Vostok; it was a 5.7- ton vessel, capable of carrying two cosmonauts, as a result of a rearrangement of the cabin interior. There were only two manned space flights of this type, un-

der this name. The flights continued under the name of Cosmos, with animals on board (bio-satellites).

The last, Voskhod 2, allowed Alexi Leonov (born in 1934) to carry out the first space walk, which lasted for 20 minutes, on March 18, 1965. A tunnel hooked up to the vessel's exit was used, and Leonov had a great deal of difficulty reentering it, since his space suit had expanded in the space vacuum.

Today, Leonov is the assistant commander of the City of Stars, the famous training center for Soviet cosmonauts, located near Moscow.

Soyuz

The Soyuz (a Russian word meaning "union") was a 6.7-ton vessel mainly designed for link-ups with the new Salyut orbital space stations. More complete than its predecessors, Soyuz comprised three parts: a propulsion module (2.7 tons) for orbital maneuvers, a pressurized cabin (1.2 tons) for work in space, and a reentry module (2.8 tons), equally pressurized and used for crew reentry. This reentry capsule, equipped with parachutes and a braking system, permitted a soft landing, and the crews were recovered either on land or at sea.

The Soyuz was the most widely used manned space craft over a period of 15 years. About 40 Soyuz of this kind were launched between 1967 and 1981, with Soviet cosmonauts and 9 cosmonauts from socialist countries who were invited by the USSR to participate. The first Soyuz flight, with a single cosmonaut on board, tragically ended in the death of Vladimir Komarov, who had been a member of the first Voskhod 1 crew. The braking parachutes snaked during descent, and unfortunately the cosmonaut crashed with his space vessel.

The flights following were carried out with three-man crews, for which the vessel had been designed. However, the flight of Soyuz 2 also ended in catastrophe, due to an accidental depressurization in the cabin; the crew, comprising Vladislav Volkov, Victor Patsaiev and Georgi Dobrovolski, travelling without spacesuits, were asphyxiated during reentry. Subsequent to this tragic accident, the Soviets decided that from then on all Soyuz flights would be carried out with crews equipped with spacesuits; this only allowed for two men on board.

Soyuz T

A new, improved version, Soyuz T (three-man craft) was put into service in 1980 by the USSR, in order to trans-

Astronauts training in the Gulf of Mexico prior to the Gemini flight in 1965.

port three cosmonauts equipped with space suits to the new orbital Salyut space stations. The first Soyuz T underwent a dry-run test. The subsequent flights were carried out with crews of two, then three, cosmonauts, in June and November 1980. A fourth Soyuz T was launched in 1981, with a three-man crew.

American manned space craft

The United States has used four kinds of space craft for their manned space program. These vessels were the Mercury cabin, a one-man craft, the two-man Gemini, and the three- man Apollo, linked up with a LEM module. Since 1981, the United States has used the Space Shuttle, which has replaced the standard rockets and non-reusable manned vessels. The Shuttle can carry up to seven astronauts, and can, if necessary, take on another 10 astronauts sent from earth on a rescue mission. All the American manned vessels placed into orbit to date, as well as the first space shuttles, have been launched from Cape Canaveral (Florida).

Mercury

The first American manned space craft, Mercury, was built by McDon-

nell Douglas for NASA; these were small, simple capsules (1.2 tons; 2.9 meters in length) and they could carry one suited astronaut for a short-duration flight. The capsules were launched by Atlas rockets.

The United States carried out six flights using Mercury cabins with astronauts aboard in 1961 and 1962. It was on board the Mercury cabin, renamed Freedom 7, launched on May 5, 1961, that the first American was launched into space for a sub-orbital flight lasting 15 minutes. The astronaut was Alan B. Shepard, and he reached an altitude of 187 km before splashing down in the Atlantic Ocean. This flight was followed, on February 20, 1962, by that of John Glenn, today a Senator from Ohio. He circled the earth for 4 hours 55 minutes on board the Mercury vessel, renamed Friendship 7.

Gemini

The Gemini cabins (3.7 tons), also constructed by McDonnell Douglas, were the direct descendants of the preceding cabins, but they had a rearranged interior which could accommodate two astronauts and which comprised a propulsive compartment allowing, in particular, a space rendezvous with an Agena rocket, previously sent into orbit.

The Gemini cabins were launched with Titan 2 rockets. After only two dry-run test flights, NASA carried out 10 flights with two-man (Gemini) cabin crews. The first orbital flight of a Gemini cabin (Gemini 3 flight) took place on March 23, 1965, with Virgil Grissom on board (it was to be his only flight) and John Young, who was to hold the world's record for astronauts, with 5 space flights (including the first Space Shuttle flight).

Using these manned vessels, the United States carried out the first orbit transfers (Gemini 3, in March 1965) and the first space rendezvous (Gemini 6 and 7, in December 1965), as well as the first docking of two vessels placed in earth's orbit (Gemini 6 with Agena, in March, 1966).

Apollo

The Apollo vessel, built by North American Rockwell, was a manned satellite with an entirely new concept, designed specifically for moon flights. The Apollo cabin was linked up with the Lunar Module (LEM), in order to form the "lunar train" launched by the giant Saturn 5 rockets to the moon. The Apollo vessel weighed about 30 tons on liftoff, with its service module comprising fuel and the means for survival. The space cabin, conical in shape, was occupied by three astronauts and weighed no more than 5.7 tons on reentry into earth's atmosphere. For the reentry phase, it was protected by a thick thermal buckle. The round-trip earth-moon voyage on board this small space cabin (6 m³) lasted about 7 days.

The first attempt to launch an Apollo cabin ended in a tragic catastrophe on January 27, 1967. During a count-down rehearsal with Virgil Grissom, Edward White and Roger Chaffee, a fire abruptly started in the cabin, and the crew burned to death. This accident caused a year's delay in the Apollo program. NASA launched 11 Apollo cabins, nine of which were for voyages to the moon. The Apollo program put a total of 12 Americans on lunar soil between July 1969 and December 1972.

Lunar Exploration Module

The Lunar Module, called the space spider, because of its appearance, was developed by the American company, Grumman, to ensure what would inevitably be the most critical phases of the lunar flight: landing and liftoff. This "space climber," weighing 15 tons, made of aluminum and covered with gold leaf, allowed the team of two astronauts to descend (in 12 minutes) to the moon and to leave it (in 7 minutes) for hook-up with the Apollo cabin which was circling the moon at an altitude of about 100 km, with the third astronaut on board.

First moon walk

The LEM, subsequently renamed "Eagle," used by Neil Armstrong (born in Wapakoneto, Ohio, in 1930) and Edwin E. Aldrin, Jr., was the first manned space craft to land on a body other than earth. During the course of the Apollo 11 flight, on July 20, 1969, two Americans were the first men to walk on the moon, thus making the dream of Cyrano de Bergerac, Jules Verne and so many others come true.

The astronauts Armstrong and Aldrin remained on the moon for 21 hours 36 minutes, and each carried out a walk of a little more than 2 hours in the Sea of Tranquility. This was an event in the history of astronautics that will be passed along to posterity. It was also an event that millions of television viewers watched at the moment it took place.

The LEM functioned without any failure during all the Apollo flights. It even allowed the crew of the Apollo 13 mission to be saved on April 11, 1970, after the in-flight explosion of an oxygen tank in the cabin. The crew (James Lovell, John Swigart and Fred

Haise) were able to return safe and sound, thanks to the LEM oxygen tanks.

Pressure Suits

All the American astronauts and the Soviet cosmonauts (with the exception of some Soyuz crews, one of whom paid for the lack with his life), have worn tight pressure suits during their space flights, at least on board transport vessels. Voyages on board orbital space stations and the American Space Shuttle will henceforth take place without the use of pressure suits.

The first pressure suits

The pressure suits used for the first astronauts were derived from those previously developed for jet pilots. In 1934, the pneumatic tire company, B.F. Goodrich, at the request of the American aviator Wiley Post, built the first pressurized suit that could be worn in the cockpit of an airplane.

The pressure suits used during the American Mercury, Gemini, and Apollo space flights were developed, respectively, by the American companies, B.F. Goodrich, David Clark and ILC. They were supple, made of plastic, and the joints also provided sealed seams. These suits could withstand pressures of only 2 to $2.7.10^3$ pascals, or 0.2 to 0.3 kg/cm². Their function was to protect the astronauts in case of inopportune depressurization in the cabin. During the Gemini and Apollo flights, these pressure suits were also used for extra-vehicular activities, that is, space walks.

Pressure suits for the pilots of the American Space Shuttle

The pressure suits for the astronauts of the Shuttle, developed by the firm, Hamilton Standard (United Technologies group), derive from previous designs. The suits are supple and stand pressures of 0.3 kg/cm² suit. However, a new "hybrid" suit (supple fabric and metallic joints) is in the process of being developed by the Acurex Corporation for future flights. It can withstand pressures of 0.56 kg/cm².

Nonetheless, during the first 4 test flights of the Space Shuttle, the crews temporarily wore a different pressure suit supplied by the U.S. Air Force. This was an adapted version of the suits used by American pilots of high performance YF 12A jet planes, flying at Mach 3, at an altitude of 30,000 meters. These special pressure suits were specially designed for ejection at very high altitude.

Space Probes—
The Explorers of the Cosmos

The idea of sending probes to the planets goes back to the origins of astronautics, that is, to the 1920's. The German Hermann J. Oberth published his first book, "Die Rakete zu den Planetenraumen" (*The Rocket in Interplanetary Space*) in 1923; the book described the broad principles of interplanetary space flight. It was followed by another work in 1929, "Wege zur Raumschiffahrt" (*The Paths of Space Flight*), which completed the description of cosmic flights; the latter did not, however, begin until 30 years later.

At the same time, another German, Walter Hohmann, was the first to calculate the conditions for such space flights (flight time, mass of propellants, etc.), as well as the best orbits for reaching the planets. Some orbits are still used today, and they are known as the Hohmann orbits, in honor of their inventor who described them in 1925 in "Die Erreichburkeit der Himmelskorper" (*The Possibilities of Reaching Heavenly Bodies*). This work was not translated in the United States until 1960!

The first probes were sent to other bodies in the solar system in 1966. On February 3, 1966, the Soviet Luna 9 was the first artificial satellite to make a soft landing on the moon. The United States in turn carried out this exploit on June 2, 1966, with the landing of the automatic probe, Surveyor I, which sent back 11,500 photos of the surrounding site (Sea of Tranquility), followed in 1967 by the Surveyor 5 probe which performed the first analyses of the lunar soil. The USSR then sent the Luna 16 probe, in September 1970; it carried out the first bore-sampling of lunar soil (to 2 meters in depth). These samples were brought back to earth. Then, with Luna 17 and Luna 21, in November 1970 and January 1973, the USSR placed the first telecommanded mobile robots on the moon (launched from earth); these covered respectively 9 and 37 km on the surface of the moon. (As regards the first steps taken by man on the moon, see Apollo, p. 348).

The USSR also launched the Venera 7 probe, which landed for the first time on the planet Venus on December 15, 1970, and then Venera 9 and 10, which landed on Venus on October 22 and 25, 1970, and transmitted pictures of the surface of Venus. The United States sent up the Mariner 10 probe; in March 1964, it passed as close as 1,000 km to Mercury, the planet nearest to the sun. The United States also launched the Viking 1 and

2 probes which were the first to land on Mars, respectively on July 20 and September 3, 1976. One of these probes is still in operation.

More recently, the United States successfully launched probes that will fly over Saturn and Jupiter, the two largest planets in the solar system. First Pioneer 10 and 11 flew over the two planets, in December 1973 and December 1974; then the Voyager 1 and 2 probes carried out flights nearer Jupiter and Saturn, as well as their numerous satellites, before continuing their flight to Uranus and Neptune. These two giant planets, farther still, should be approached in January 1986 and August 1989, respectively.

Space Stations

The United States and the USSR are the only countries to have launched long-duration manned space stations around the earth. The Soviets successively placed six Salyut missiles into orbit, while the Americans launched only one Skylab after the Apollo program. However, both in the United States and in the USSR, as well as in Europe, those in charge of space programs are ever aware of the fact that the future of space astronautics depends upon the use of orbital stations, which will be permanently manned by astronaut-worker and astronaut-researcher teams.

Salyut

The USSR was the first country to put a manned space station into operation. Since April, 1971, the Soviets have placed 7 space stations in orbit. The last to date, Salyut 7, was launched on April 19, 1982. It took on board 7 Soviet cosmonauts, one of whom was the second woman in space, as well as the first Frenchman sent into space. The preceding Salyut 6 station, launched on September 29, 1977, returned to the atmosphere on July 29, 1982. For nearly 5 years (58 months), this station linked up with 16 Soyuz and 4 Soyuz T vessels, which brought 27 cosmonauts, 19 of whom were Soviet and 8 of whom were cosmonauts from the Intercosmos member countries, into space; it also received 12 Progress cargo-vessels, carrying 22 tons of freight.

Salyut is a large satellite, weighing about 20 tons; it is cylindrical in shape and has a habitable volume of 91 m³. Although much smaller than the American space station, Skylab, this Soviet space station can accommodate a crew of 4 cosmonauts for one week. In general, however, the crew for one of these long-duration flights (several months) comprises only two cosmonauts.

Skylab

On May 14, 1973, the United States launched the Skylab orbital space station, the largest built to date. Eleven days later, a crew of three astronauts, transported by an Apollo cabin, joined the station. Built by McDonnell Douglas and Martin Marietta, using elements recovered from the Apollo space program, the Skylab space station (an abbreviation of "sky laboratory") weighed 75 tons.

Placed in the S-IVB stage of a Saturn 5 rocket, Skylab had an inhabitable volume of 350 m³, with unprecedented comfort for a manned craft (kitchen, beds, shower, etc.). Skylab was put into a circular orbit around the earth, at an altitude of approximately 500 km; it carried considerable scientific equipment, and in particular, an astronomic observatory. The station was successively occupied by three crews, each comprising 3 astronauts.

These crews were respectively launched on May 25, 1973, July 28, 1973 and November 16, 1973, for flights of 29, 58 and 84 days (an American record). Skylab permitted more than 175,000 photographs of the sun and more than 46,000 photos of the earth to be shot.

The station was abandoned in February 1974 and it disintegrated above Australia on July 11, 1979. The largest part of the debris found on the ground measured 2 meters by 1 meter, and weighed 1/2 ton.

Spacelab

Europe, too, built a manned space laboratory, the Spacelab, currently designed to be carried in the storage compartment of the American Space Shuttle; however, it will later give rise to an independent orbital station. Spacelab (an abbreviation for "space laboratory") was built by the German company, ERNO, and an industrial consortium of 10 European countries (Germany, France, Italy, Belgium, Denmark, Sweden, Switzerland, the Netherlands, Spain and Great Britain). It comprises a pressurized mod-

Preparing Spacelab in November, 1983.

ule and can take up to four specialized astronauts, as well as instrument-bearing probes which can be directly exposed to the vacuum of space. Overall maximal weight is 11.3 tons, of which 5.5 tons to 9.1 tons comprise the payload (scientific and technical instruments).

The first Spacelab flight is scheduled for late 1988, with the first European astronaut. Three astronaut-researchers have been trained for the mission by the European Space Agency (ESA): the German Ulf Merbold, the Dutch Wubbo Ockels, and the Swiss Claude Nicollier.

Extra-Terrestrial Vehicles

Lunar Rover

The first moon vehicle, the Lunar Rover, was manufactured by the famous airplane builder Boeing. It was used during the last three American moon flights (Apollo 15, 16, 17) in 1971-72.

The Lunar Rover allowed the American astronauts to cover, respectively, 24 km, 27 km and 35 km around the landing sites of the three LEMs, in order to explore the moon, to plant instruments and to collect rocks. This two-seater vehicle, with an aluminum chassis and 4-wheel drive, weighed 218 kg when empty and 672 kg when two astronauts and their equipment (181 kg each) and tools (92 kg) were added on.

The Lunar Rover measured 3 meters in length and 2 meters in width when in operating condition, but it could be completely folded for transport in the Lunar Module. The four wheels were made of a fine, supple wire netting and each was activated by an electric engine placed in the hub and supplied by a 36 volt battery. The vehicle had a cruising range of 60 km and a maximal speed of 14 km/h.

The Lunar Rover was piloted like an ordinary car (with, however, a broom handle for a steering wheel). It was also equipped with a color television camera, which transmitted images of the landscapes, and an antenna which allowed the crew to remain in constant contact with the earth.

Lunakhod

The first mobile lunar robots, named Lunakhod, were designed by Soviet specialists to carry out moon exploration from earth, since they have not yet put a man on the moon.

Two Lunakhods were used by the USSR. The first, carried by the Luna 17 probe, explored the area of the Mare Imbrium, from November 1970 to October 1971, covering a distance

The Boeing-built Lunar Rover in 1970.

of 9.6 km in short phases. The second, placed by the Luna 21 probe in the Sea of Tranquility, covered 37 km between January and July 1973.

The Lunakhod looked like a big kettle; it was pressurized to protect the electronic equipment and was mounted on eight wheels activated by electric engines located in the hubs. The craft, 2.2 meters long and 2.2 meters wide, and 1.35 meters high, weighed 840 kg.

Two television cameras located in the front of the vehicle permitted the operator on earth to "see" the landscape, in order to telecommand, in real time, the robot's movements according to the configuration of the terrain. The Lunakhod was also equipped with various scientific instruments (spectrometer, telescope, etc.), and a panel of reflector-maser prisms permitting the earth-moon distance to be precisely measured.

Space Design

Desirous of improving comfort in the Skylab space station, NASA called upon the talents of the most famous industrial design specialist, Raymond Loewy, a Frenchman who had settled in the United States in 1919. This former lieutenant in the French Army designed logos for numerous firms and also left his mark on soda bottles, cigarette packages, cars, trains, as well as space cabins.

Raymond Loewy became involved in the interior arrangement of the Skylab living quarters, for the layout of the cabins, the choice of colors, etc. It was also Loewy who recommended a porthole through which the crew could observe earth when they so desired, and this was one of the elements responsible for the mission's success.

Some of his ideas were also used in the design of the Shuttle's living quarters. All Raymond Loewy's archives, comprising 3,500 projects undertaken for NASA, were sold at auction by Sotheby's of London in July 1981, on the occasion of the designer's 88th birthday.

4. Addendum

The Moon, Earth's Laboratory

Abandoned since the last Apollo flight in 1972, the moon may well constitute an "astronomers' paradise" during the next decades. Soviet and American plans contain dozens of projects: the establishment of a conventional telescope on the earth's surface; the construction of giant radiotelescopes on the hidden side of the moon, in order to listen to messages coming from the cosmos; the construction of metallurgy and electronics factories, etc . . .

Photograph of the moon by Lunik 3, a satellite.

The First Artificial Comet (1984)

On December 27, 1984, the German satellite, IRM, launched barium 110,000 km from the earth; the barium sublimated and, due to the action of solar wind, formed a "tail" 10,000 km long. The comet is especially visible at night, in the Pacific. It is observed by an English satellite and is photographed continually by four observatories on earth and two airplanes. It is hoped that the functioning of solar wind, the flux of particles emitted without interruption by the sun, may be better understood.

Space Telescope (1986)

The American Space Telescope will be the largest space telescope. In 1985, a satellite, weighing 10.4 tons, and 13 meters in length, will be placed in a circular earth orbit at an altitude of 500 km, by the American space shuttle, Columbia. The telescope will have a 2.4-meter aperture. Its performance will be superior to that of current earth-bound instruments, such as the Mount Palomar or the Zelentchouk telescopes: in effect, the Space Telescope will be placed outside of earth's atmosphere and its disturbances.

The Space Telescope will be able to detect objects which are 50 times smaller than those detected by the largest telescopes on earth. Further, it will be able to observe a volume of space which is 350 times greater, and objects as far away as the outer limits of the universe, about 15 billion light years away.

The construction of this unique instrument cost Lockheed, the main contractor, and Perkin Elmer, in charge of the optics, about one billion dollars and ten years of effort. Several European firms also participated in the construction: British Aerospace (G.B.), Dornier (Germany), and Matra (France).

In the opinion of the specialists, the Space Telescope will revolutionize astronomy. Free from the impurities of the earth's atmosphere, this telescope will transmit extremely sharp images back to earth.

Two cameras, two spectrometers and one photometer will permit it to function in the visible, infrared and ultraviolet zones. Hundreds of experiments are already planned for the first fifteen years of operation. Defective parts will be changed by the Shuttle crew or brought back to earth. The estimated cost is one billion dollars.

An Encounter with Halley's Comet (1986)

The passage of comets in the sky has always fascinated man, but the comet to which the famous astronomer, Edmund Halley (1656–1742), gave his name is certainly the most celebrated of them all.

It was noted for the first time in 467 B.C. Chroniclers of very different epoques and countries have spoken of it, using terms which are always very imaginative: dragon of fire, curved sabre, long-tailed dragon…. Queen Mathilde embroidered it onto the Bayeux tapestry; in effect, the comet was present at the Battle of Hastings in 1066.

Aided by Newton, Halley managed to calculate the comet's cycle: approximately 76 years. He was able to forecast its future passages. Its last visit was in 1910, when it caused numerous floods, but was also responsible for a wine of very good quality: the "comet wine."

Halley's comet will return to the neighborhood of Earth in 1986. For scientists, this will be an unequalled opportunity to study the composition of the comet's tail. Comets are often vestiges of gas and dust clouds which formed the solar system, and they may go as far back as the very origin of life.

Vega, Giotto, and Planet A

Four scientific projects have been set up to better understand Halley's comet: two Soviet probes, Vega 1 and Vega 2, will first study the planet Venus, before heading towards the comet. They will be joined by the Giotto European probe, and by the Japanese Planet A, in the beginning of March 1986.

These probes, the weights of which vary between 15 and 120 kg, will pass at approximately 10,000 km from the nucleus, and their approach speed will be close to 80 km/s. Rendezvous will last only a few minutes, and one can easily imagine the total effort required for the success of such an undertaking. The data will subsequently be transmitted back to Earth.

The comet will be visible to the naked eye beginning in December 1985. For inhabitants of the southern hemisphere, it will shine in all its brilliance. Inhabitants of the northern hemisphere will be able to see it under optimal conditions towards the end of April 1986; but it will then be less brilliant, and its tail will seem rather shorter.

Let us enjoy the spectacle and not be too fussy; the next passage of Halley's comet will not take place until 2051!

Keck Observatory (1992)

A sum of 70 million dollars was granted to the California Institute of Technology by the Keck Foundation for the creation of the largest telescope in the world. The diameter of the lens will be 10 meters. Construction of the Keck Observatory is scheduled to begin in 1986, in Hawaii. The telescope will be operational in 1992.

INDEX

A

Aaron, Charles, 85
Abacus, 294
Abt, 5
Accordion, 91
Achenwall, Gottfried, 236
Aciclovir, 218
Ackert and Keller, 327
Ackroyd-Stuart, Herbert, 324, 325
Acoustics, 221-222
Acupuncture, 225
Adams, T., 75
Ader, Clement, 19, 109
Adhesive tape, 176
    bandage, 164
Adler, Harold A., 187
Adrenalin, 208
Adrian, Edgard, 200
Ads, classified, 87
Aebisher, N., 290
Aerosol spray, 164
Aerospatiale Co., 21
Afterburner, 328
Agfa Co., 104
Agfacolor, 99, 100
Agricola, 164
Agriculture, 59-78
Agri Electronics, 73
AIDS, diagnostic test for, 228
Aiken, Herrick and John, 283
Air, 242-243, 245
Aircraft. See Air transportation
Air hostesses, 23
Air mattress, 164-165
Airplane. See Air transportation
Air-raid shelter, 188
Air transportation, 17, 23-25, 84
    aids, 23-24
    airplane, 19-22, 66-67, 75, 183
    balloon, 17-19
    engine, 326-328
    future, 191-193
    helicopter, 22-23
    military, 53-56
Aitkins, Dr., 164
Alarm clocks, 277
Albert, A.C., 286
Albone, Dan, 64
Alexeyev, 26
Algae, cultivaion of, 266
Algebra, 234-237
Algorithm, 298
Alizarine, 258
Alka-Seltzer, 220
Alkea, 211
Al-Khwarizmi, Mohammad ibn-Musa, 232-234
Allis-Chalmers, 63
Allison, Judge, 187
Alphabet, 81-83
Alquerque, 125, 126
Alternator, 319, 321
Altitude, measurement of, 341
Aluminum, 27, 148, 256-257
Amalgams, 222-223
Amati, N., 89
Ambulance, 197
Amdahl, Gene, 298, 299
America's Cup, 144
Ammunition, 36, 40, 44, 45
Ampere, 118
Ampere, Andre-Marie, 250
Ampex Corp., 113, 115
Amyan, Claudius, 205
Anaglyphic principle, 105
Anaxagoras, 251
Anaximander, 275, 338
Anchor, 28

Anderson, Charles E., 115
Andre, H., 320
Andrews, Ed, 169
Andrews, William A., 144
Anesthesia, 208, 210
Angroplasty, transluminal, 211
Angle, 224
Animal husbandry, 71-74
Animation
    cartoon, 102, 105-106
    computer, 106, 303
Anschutz, Ottomar, 101
Answering car, 13
Antenna, radio, 110
Anthelin, Ludwig, 155
Anthrax sera, 202
Antibiotics, 215-216
Antibodies, 218
Antisepsis, 211
Antoine, Tasiglio, 213
Aortic valve transplant, 208
Appendectomy, 205-206
Appert, Leon, 274
Appert, Nicolas, 77-78
Appia, Luis, 197
Appleby, John F., 61
Apples, Granny Smith and Golden, 69
Appleton, E.V., 343
Appolonius of Perga, 232
Aquaculture, 265
Aqua fortis, 288
Aquaspace, 187
Arabic Writing, 82
Arago, Francois, 250
Archer, Henry, 118
Archimedes, 89, 91, 232, 234, 238, 239, 313
Arc lamp, 169
Arctic Energies, 192
Argand, Aime, 28, 169
Argand, Jean Robert, 236
Aristarcus, 233, 240
Aristotle, 232-233, 260
Arithmetic, 231
Arkwright, Richard, 281
Armaments, 35-56
Armati, Salvino degli, 220-221
Armendaud brothers, 325, 326
Armor, 42-43, 181, 185
Arms, 35-36
Armstrong, Edwin, 111
Aron-Rosa, Daniele, 222, 228
Arsenic, 70
Art and computers, 303-304
Artillery, 39-42
Arts
    music, 89-96
    photography, 96-106
    visual communication, 81-89
Artus, 204
Asepsis, 210-211
Ash-tray, magic, 186
Asimov, Isaac, 308
Aspdin, J., 272
Aspirin, 215
Assembly-line, 13
Astley-Cooper, 210
Aston, William Francis, 328
Astrolabe, 275, 337
Astronomy, 337-340
Atari, 135
AT & T, 110, 111, 120-122
Athotis, 210
Atmospheric pressure, 244
Atom, 251, 252
Atomic bomb, 47
Atomic clock, 278-279
Atomic physics, 251-254
Atomic pile, 329

Auenbruger, L., 198
Audio. See Sound reproduction
Auscultation, 197-198
Autogyro, 22, 23
Automobiles, 7, 13, 42, 183, 184, 188
    body and accessories, 16-17
    mechanics, 13-16
Autopsy, 205
Auto racing, 135-137
Avalon Hill Society, 135
Avco Research Laboratories, 320
Averroes, 210
Avery, O.T., 263
Aviation. See Air transportation
Avicenne, 210
Avogadro, Count, 251, 252
Azzaro, Loria, 161

B

Babbage, Charles, 293, 300-302
Babcock and Wilcox, 317
Baby carriage, 165
Backlund, Erik Olaf, 208
Backus, John, 299
Bacon, 320
Bacteriology, 202-203, 227, 261-262
Baekeland, Hendrick, 257
Bailey, Timothy, 283
Bailey, Jeremiah, 62
Bain, Alexander, 111, 112
Bainier, C., 290
Baird, John L., 112-114, 117
Bakelite, 257
Balan, Jean-Aime, 272
Balance scale, 239
Ball, 224
Ball
    bearings, 13-14
    games, 125, 136
    point pen, 84
Ballistivet Inc., 227
Balloons, 17-19, 53, 185, 343
"Ballule," 185
Bally Co., 171
Balmain and Rochas, 157
Bandages, 164, 210
Banister, J., 93
Banknotes, 174
    sensor, 1, 193
Bank of America, 175
Barbed wire, 73
Barberi, 91
Barbie doll, 127
Barber, John, 325
Bardeen, John, 111, 306, 307
Barnard, Christian, 208
Barnay, Antoine, 121
Barnet, C., 303
Barnett, M.A.F., 343
Barnett, William, 321-322
Barometer, 30, 341
Barraquer, 221
Barraud, Robert, 186
Barrios, Alfred, 201
Barris, George, 188
Barski, G., 264
Bar stool, 184
Barthélemy, René, 113
Baseball, 136
BASF, 95, 116
Basketball, 136
Bassoon, 91
Bathtub, 165
Bathyscaphe, 32
Baton, conductor's, 93
Battery, 14, 48, 118, 317-320

Battleship, 50-52
Baudot, Emile, 299
Baudry, Stanislas, 8
Bauer, A., 85
Baulieu, Etinne-Emile, 215
Baumann, E., 165
Bautray, Jean-Pierre, 214
Baxter, S., 28
Bayard, Hippolyte, 97
Bayer Co., 215, 260
Bayonet, 35
Beach, Chester A., 151
Beart, Robert, 270
Beater, 151
Beaufort, Sir Francis, 30, 83
Beaufoy, Colonel, 30
Beauty inventions, 159-164
Beaux, Ernest, 162
Becoulet and Bellet, 274
Becquerel, Antoine, 318
Becquerel, Edmond, 98, 100
Becquerel, Henri, 225-253
Beebe, William, 265
Beekeeping, 72
Begout, Henri, 68
Beidler, G.C., 89
Beiersdorf, Paul, 159, 164
Belgian National War Armaments Factory, 11
Bell, 222
Bell, Alexander Graham, 26, 120-122
Bellet, Dominique, 227
Belling, C.R., 166
Bell Laboratories, 96, 109, 113, 114, 121, 122, 190, 293, 300, 301, 303, 306, 318
Bell, Patrick, 60, 61
Bell, Stuart L., 175
Bell System, 112-114, 121, 122
Bell Telephone Co., 120, 289, 294, 302, 338
Bell, training, 188
Belt, safety, 17
Benedict, Horace, 341
Benedictus, Edouard, 274, 275
Bennet, Harper, 98
Bentall, E., 143
Bentz, Mrs. Melitta, 151
Benz Co., 8, 325
Benz, Karl, 14, 15, 324
Berger, Hans, 200, 204
Berges, Aristide, 319
Berge, Wolfgang, 167
Beringer and Flobert, 36
Berliner, Emile, 94
Beroverio, 274
Berry, Captain, 23
Berson, Salomon, 200
Berthier, Pierre, 37, 256
Berthollet, Claude-Louis, 256
Berthon, 104
Bertillon, Alphonse, 177
Bertin Co., 222, 332, 341
Berzelius, 112
Bessemer, Henry, 256
Bethlehem Steel Co., 41
Bevan, Edward J., 258
Bevon, Ernest, 158
Beverages, 76-77
B.F. Goodrich Co., 159, 349
Bhaskara, 233, 234
Bicchi, A., 326
Bic Co., 163
Bich, Baron, 163
Bickford, Willim, 44-45
Bic Marine, 148
Bicross, 137
Bicycle, 9-10, 12-13, 181-182, 184, 186, 187, 189

cycling, 137-138
Bienvenu, 22
Biesenberger, H., 206
Bifocals, 221
Bikini, 155
Billigrakete, 346
Billingsly, N., 309
Bill of exchange, 174
Biodynamic agriculture, 65-66
Biogas, 333
Biological control, 70
Biomass, 333
Biotechnology, 264
Bircher-Brenner, Dr., 164
Birdseye, Clarence, 78
Birkinshaw, John, 5
Biro, H., 84
Bischop, Alexis, 16
Bissel Carpet Sweeper Co., 165
Bissel Melville R., 165
Bit, 300
Bizarre inventions, 181-189
Blaaum, M., 298
Black powder, 44
Blackton, J. Stuart, 106
Blain, Alexander, 277
Blancart, Pierre-Louis, 69
Blanchart, F., 146
Blenkinsop locomotive, 4
Bleriot, Louis, 20
Blickensdorfer, George C., 88
Bloch, Felix, 250
Blondel, Andre, 31
Blood, 203-205
    letting, 216-217
    pressure, 198
Bloomer, Amelia, 155
Bloomers, 155
Blues music, 93
Blumlein, Alan Dower, 94, 109
BMW Co., 11, 326
Boat. See Maritime transportation and Naval vessels
Bobeck, A.H., 303
Bocage, Andre, 199
Bocci, 136
Boehm, Theobald, 90
Boeing Co., 13, 20-21, 24, 190, 351
Bohemian glass, 274
Bohr, Niels, 252
Bohuon, Claude, 227
Boiler, 25, 315-317
Bollèe, Amedee, 13, 313
Bolyai, Janos, 237
Bombelli, Raffaele, 234
Bomber, aircraft, 54, 56
Bombs, 45-47
Bondstones, 272
Bone transplant, 207
Bonnemain, 166
Bonnet, A., 305
Books, 84-87
Boole, George, 237, 293
Booth, A.D., 304
Booth, H.C., 173
Booth, Oliver, 139
Bordas, F., 78
Borden, Gail, 77
Boreau, Therese, 204
Borelli, 19
Borie, Henri-Jules and Paul, 272
Borsch, J.L., 221
Bossart, Karel, 345
Bottger, Johan Friedrich, 254
Bottles, 77, 153, 212-213, 274
Bouchardat, G., 257
Bouchon, B., 281
Boudou, Gaston, 160
Boules, 136

Bourdarias, J.P., 211
Bourdin, Claude, 313, 325
Boutan, Louis, 265
Bow, 89-90
Bowling, 137
Box
    cold, 218
    mail, 117-118
Boxers, P., 262
Boxing, 137
Boyle, David, 285
Boyle, R., 217
Boyle, Robert, 49, 153, 244, 334
Braconnot, Henri, 258
Brahama, Joseph, 169
Braham, R.R., 343
Braille, Louis, 83
Brain cell transplant, 208
Brakes, 6, 40
Brahmagupta, 231, 234
Bramah, Joseph, 30, 279, 285, 286
Branca, Giovanni, 325
Branderberger, J.E., 258
Brandt, Edgar, 42
Branly, Edouard, 109, 110, 119
Braswell, Ty, 332
Brattain, Walter, 111, 306, 307
Braun Electric Co., 277
Braun, Ferdinand, 250
Brayton engine, 322
Breeder reactor, 329
Brendel, Walter, 211
Brequet, Louis, 14, 277, 322
Brett, Jacob, 120
Brick, hollow, 272
Bridge, 125, 135
Bridges, dentistry, 224-225
Briggs, Henry, 235
Brinster, R.L, 264
Bristol Siddley Co., 327
British Aerospace Co., 21, 340, 352
British National Physical Laboratory, 46
Broadcasts
    radio, 110
    television, 113
Bronze, 254, 255
Brossard, Y., 204
Brother Co., 88, 89
Broughton, Jack, 137
Brown, Harold P., 177
Browning, John Moses, 37, 39
Browning, William, 76
Brown, Jazzbo, 93
Brown, Robert, 261
Brown, Samuel, 15, 321
Bruce, D., 203
Bruch, Walter, 114
Brunel, Isambard K., 30
Brunelleschi, 276
Brunel, Marc, 30, 282
Brunet, Carmela, 187
Bruno Court Co., 162
Brunot, James, 132
Bubble memory, 303
Buchanan, B.G. 305
Buckley, S.B., 73
Budding and Ferrabee, 69
Buehler, William, 290
Buhot, Yves, 204
Building materials, 271-273
Bulbs, bow, 30-31
Bull, Lucien, 104
Bulpitt, W.H., 152
Bumper, 17
Bunsen, Robert Wilhem, 318
Bureau brothers, 40
Burger tractor, 64
Burglar trap, 186
Burial in space, 188
Burkhoder, P., 216
Burnet, William, 69
Burroughs, W.S., 302
Burstyn, Gunter, 42
Burton, Charles, 165

Burton, John, 317
Burton, Luke, 283
Bus, 8
Busch, Hans, 198
Buschnel, Noland, 135, 309
Bushnell, David, 49
Business, 174-177
Butler, Edward, 10
Butter, powdered, 76
Button, 155
Butts, Alfred M., 132
Byers, M.H., 343
Byers, W.Q., 28
Byte, 300

C

Cabbage Patch dolls, 127
Cabin, ejectable, 24
Cable
    telegraph, 119-120
    telephone, 122
Cabriol, Alexandre, 32, 272
Caesarian section, 211-212
Caesar, Julius, 337
Cage, 90
Cahill, Thaddeus, 88, 91
Cailler, Francois-Louis, 75
Calculator, 293-296; see also Computer
Calculus, 236
Calendar, 337-338
Calhamer, Alan B., 127
Callot, Jacques, 288
Calotype, 98
Calvin, Melvin, 265
Cam distributor, 14
Camera, 96-97
    movie, 98, 102-103
    video, 112, 116-117
Camera obscura, 96
Cameron press, 86
Camm, Sir Sidney, 56
Camp, Walter, 138
Canal lock, 30, 320
Canned food, 78
Cannon, 39-41
Cantor, Georg, 237
Capek, Karel, 308
Cappiello, Leonetto, 86
Cap, priming, 36
Capsule, ejectable, 24
Caquot, Albert, 265
Car. See Automobile
Carafe, chilled, 77
Carbine. See Rifle
Carbonated mineral water, 76
Carbon copy, 87-88
Carbon-14 dating, 254
Carburetor, 14-15, 324
Cardano, Geronimo, 14, 83
Cardano, Girolamo, 233
Card games, 125, 131-135
Cards
    credit, 175
    perforated, 302, 303
    smart, 303
    sorter, 301
Carlier, Francois, 168
Carlson, Chester, 89
Carnot, Sadi, 246
Carolus, Johann, 87
Caron, P.A., 277
Carothers, Wallace Hume, 158, 259, 260
Carpet
    ball, 136
    sweeper, 165
Carre, Edmond, 77, 284
Carre, Ferdinand, 284
Carrel, Alexis, 262
Carriage
    artillery, 40
    baby, 165

Carrier, aircraft, 50-52
Carrier, Willis, 164
Carrol, James, 9
Carrol, Tom, 64
Carson, W.L., 95
Cartier, Louis, 277, 279
Cartoon, animated, 102, 105-106
Cartridge, 36
Carts, supermarket, 176
Cartwright, Alexander, 136
Cartwright, Edmund, 282
Caryotype, 264
Caselli, Abbe Giovanni, 112
Cash register, 174, 175
Casino games, 125-126
Casio Co., 89
Cassettes, 95, 116
Castelli, Benedetto, 341
Catalytic reformation, 271
Catamaran, 142
Cataract surgery, 221
Caterpillar Co., 64-65
Cathode ray, 250-251
Cave, F., 287
Cavendish, Henry, 17, 243, 248
Cavities, 222-223
Cavitron, 218
Cayley, Arthur, 237
Cayley, Sir George, 19, 22
Caxton, W., 86, 87
CBS, 114
Celerifer, 9
Celestis Group, 188
Cell, human, 261
Cellophane, 258
Celluloid, 98, 258
Cellulose, 258
Celsius, Anders, 342
Cement, 28, 271-272
Center board, 143
Central heating, 165-166
Ceramics, 256
Cesalpino, Andrea, 204
Chaillou, A., 202
Chain, B., 216
Chain
    bicycle, 9, 35
    tire, 17
Chair
    dentist's, 223-224
    electric, 177-178
    psychiatric, 186
Chakrabarty, M., 264
Chalais-Meudon Research Center, 46
Challis, J., 19
Chalmers, James, 118
Chamberlen, P., 213
Chambers, Arthur, 137
Champagne, 76
Chanel, Coco, 162
Channel, input-output, 307
Chaplin, D.M., 318
Chapman, Frederick H., 30
Chariot, 3, 42
Charles, Daniel, 143
Charles, Jacques, 17, 18
Charlton, John P., 118
Charron, Girardot and Voight, 42
Chaussy, Christian, 211
Checkers, 125, 126
Chelomei, Vladimir, 344, 345
Chemistry, 254-260
Cheret, Jules, 86
Chichester, Francis, 144
Chilvers, Peter, 147
Chimera, 71
Chinese writing, 81, 82

Chiropractic, 225
Chlorine, 256
Chloroform, 208
Chlorophyllic function 261
Chocolate bars, 75
Cholera sera, 202
Chretien, Henri Jacques, 104
Christian Dior Perfume Co., 162
Christiansen, Ole kirk, 129
Christin, Jean-Pierre, 342
Chromosomes, 262
Chronometer, 277
Chronophone, 95
Chronophotography, 101
Chrysanthemum, 69
Chrysler Co., 13
Chuquet, Nicols, 234, 235
Church, A., 293
Cigarette, 166, 167
C.I.I. Honeywell Bull, 175
Cinema. See Motion pictures
CinemaScope, 104
Cinematographe, 102-103
Circulation, blood, 203-204
Clapeyron, Emile, 244
Clarinet, 90, 91
Clarke, Arthur C., 122
Clarke, C.A., 204
Clark, Graeme, 222
Clark, Marione, 9
Classified ads, 87
Claude Georges, 170, 245, 249, 333
Clausius, Rudolf, 246
Clement, 11, 286
Clement, Father, 70
Clementine, 70
Clepsydra, 275
Clerk, Dugald, 323
Cleve, Per Theodor, 243
Clocks and watches, 275-279
Cloning, 264
Clothes dryer, 167
Clothing and accessories, 155-159, 186
Cloth, painted, 84
Coalbrookdale ironworks, 4
Cobalt bomb, 218
Coca-Cola, 77
Cocaine, 208
Cockerell, Christopher, 27
Cocoa powder, 75
Coconut, 69
Coding, 82-83
Codpiece, 156
Coffee, 76
    pot, filters, and grinder, 151
    instant, 77
Cogwell, J., 146
Cog-wheel, 3
Cohen, Samuel, 47
Coheror, radio, 110
Coin
    games, 128-129
    money, 175-176
Coke, 256
Colburn, Irving, 274
Cold cream, 159
Coleco, 127
Colmerauer, Alain, 304
Colonne, Edouard, 93
Color film, 98-100, 104-105, 112-116
Colposcopy, 213-214
Coltman, Professor, 199
Colt pistols, 36-38
Colt, Samuel, 36
Combine harvester, 63-64
Combing iron, 160-161
Combustion chamber, 324, 325
Comet, artificial, 352
Comic book, 86-87
Comic strip, 86
Compact disc, 94, 117
Compass, 28-29
Complex numbers, 234

Composite picture, 177
Composition, musical, 93
Compton, Arthur H., 47
Computer, 73, 106, 295-307
    games, 130-131, 134, 135
Concert, music, 93
Concrete, reinforced, 273
Condenser, 248, 315-316
Condom, 215
Conducting, orchestral, 93
Conics, 232
Constantine, 82
Contact lens, 184-185, 221
Conte, Jacques-Nicolas, 53, 84
Contraception, 214-215, 277
Control, automatic, 307
Control Data Corp., 298
Conversational graphics console, 301
Conway, John Horton, 129
Cookbook, 87
Cook, David, 147
Cooke, William F., 118-119
Cooking, 151-155
Cook, Thomas, 177
Cooley, Denton A., 208
Cooley, James P., 176
Coolidge, William David, 170
Cooling system, engine, 15-16, 328; see also Air Conditioning and Refrigerating
Cooper-Hewitt, 171
Cooper, J.S., 207
Cooper Lasersonics, 218
Coordinates, algebraic, 235-236
Copernicus, Nicolaus, 240-241
Coquille, Emile, 136
Corbato, F., 299
Corbiere, Jacque, 259
Corelli, Arcangello, 89
Cordonnier, Leon, 10
Core sampling, 270
Cork, champagne, 76
Corkscrew, 152
Corlies, John B., 175
Corliss, John B., 266
Cornealize, 221
Cornea transplant, 207
Corn flakes, 75
Corning Glass Co., 221, 274
Cornu, Paul, 22
Corvisart, Baron Jean, 198
Cosmetics, 159-164
Cosmetic surgery, 206
Coster, S., 276
Cotton, William, 283
Coulomb's Law, 248
Coupez, Fernand, 213-214
Coupleux and Givelet, 91-92
Courreges, 157
Courtel, Emile, 106
Cousteau, Jacques-Yves, 265
Cow
    bag, 73
    phoney, 182
Cozzi, Joseph A., 188
Cracking, 271
Crampton, Thomas R., 4-5
Crane, 282
Crank and connecting rod system, 315
Cray, Seymour, 298
Cream separator, 73
Credit card, 175
Cregut, R., 204
Crenner, Francis, 200
Crescent roll, 74
Creusot-Loire Co., 271, 273, 332
Crick, F.M., 263
Crime, 177-178
Crinoline, 155
Criss-cross, 132
Cristofori, B., 90
Crompton, R.E. Bell, 166
Crompton, Samuel, 281
Croissant, 74
Crookes, William, 171, 250

Crop duster, 66-67
Cross, Charles F., 258
Crosset, Charles, 158
Crossword puzzle, 126
Crowder, Norman, 304
Crowns, dental, 224-225, 228
Cryogel, 227
Cryolite, 70
Cryosurgery, 206-207
Cryptography, 82-83
Crystal, 273-274
   set, 111
Ctesibios, 91
Cugnot, Nicolas Joseph, 13
Culbertson, Ely, 125
Cullen, William, 284
Cuneiform, 81
Cunnings, Alexander, 169
Currents
   electric, 248-249
Curie, Marie, 217, 253
Curie, Pierre, 217, 253, 278
Curie, Paul-Jacques, 278
Curler, hair, 160
Curley, Benjamin, 295
Currency, 175-176
Curtis, J., 75
Cushing, Harvey, 206
Cushion, animal, 73
Cutte, H., 277
Cybernetics, 300
Cycling, 137-138; see also Bicycle
Cyrillic alphabet, 82
Cyrus the Great, 117

D

D'Aalandes, Marquis, 17
Daquerre, Jacques, 98
Daquerrotype, 98
Daimler, Gottlieb, 14, 15, 17, 323, 324
Daimler Motorem Gesellschaft, 16
Dalferth, Robert, 165
Dallmeyer, Thomas R., 96, 97
Dalton, John, 251
Dam, 317-320
D'Amecourt, Panton, 22
Dana, Merril A., 188
Dandy, Walter, 199, 206
D'Arsonval, Arsene, 78
Dandy horse, 9
Dangon, Claude, 62
Daniell, John F., 318
Darby, Abraham, 5, 255-256
Darby, Newman, 147
Darcon, General, 48
D'Arezzo, Guido, 93
Darrow, Charles, 131
Darton, William, 132
Darwin, Charles, 261
Data processing, 270, 293-298, 300, 302-303
D'Aurillac, Gerbert, 276, 298
Dausset, Jean, 207, 264
Davenport, F.S., 60
David, Georges, 204
Daviel, J., 221
Da Vinci, Leonardo, 3, 9, 19, 22, 23, 30, 35, 139, 164, 221
Davis, K.H., 306
Davis, M.R., 302
Davis, R., 305
Davy, Sir Humphry, 169, 208, 256
Dy, Benjamin, 87
DDT, 70, 71
De Beers, 266
De Bort, Leon Teisserenc, 342
De Busbecq, Augier, 70
De Chemant, Dubois, 224
De Couloumb, Charles, 248
De Crenan, Marquis, 7
De Dion and Bouton, 324
De Forest, Lee, 110, 111

Dederick, P.K., 63
Deere, John, 60, 64
De Fermat, Pierre, 231, 235-236
De Forges, Sazerac, 46
De France, Henri, 113, 114
De Fredy, Pierre, 141
De Garengest, J.C., 223
De Gebelin, Court, 134
De Giardin, Emile, 87
De Girard, Philippe, 281
Dehydration freezing, 78
De Jouffroy, Marquis, 25
De Keiser, Aelgret, 254
De la Cierva, Juan, 22
De la Condamine, Charles-Marie, 257
Delage, Guy, 143
De la Grange, Joseph-Louis, 231
Delamare-Debouteville, 323
Delambre, Jean, 238
De Laplace, Pierre S., 236
De la Quintinie, Jean, 70
De Larderel, F., 331
Delarothiere,   Joseph-Auguste, 283
De Laval, Charles-Gustave, 73, 325, 326
De Lavinde, Gabriel, 83
De Lavoisier, Antoine L., 204, 242-244, 261, 332
Delco Co., 14, 325
De l'Ecluse, Charles, 70
Deleuil and Archereau, 170
Del Gallo brothers, 171
Della Porta, Giambattista, 96
De Lome, Dupuy, 18, 48
Deloraine, Maurice, 46
De Martel, Thierry, 206
De Mestral, Georges, 159
Demian, C., 91
Democritus, 251
De Moleville, Bertrand, 68
De Monet, Jean-Bapiste, 261
De Montgolfier brothers, 17, 313
Demotic writing, 82
Denayrousse, 32
Denis, J., 204
Denisyuk, Yuri, 290
Denner, J.C., 91
Dentistry, 222-225, 227
Dentures, 223, 224
Denys the Small, 338
De Oliveira Barros, A.J.C., 288
Department store, 175
Deprez, Marcel, 321
Dereux, J.A., 92
De Rivaz, Isaac, 323
De Roberval, Gilles Personne, 239
Derom, Fitz, 208
De Rothschild, Mayer Amschel, 166
De Rivaz, Isaac, 14
De Sarran, Aguilhon, 223
Descartes, Rene, 96, 221, 235, 236
Descroisilles, 151
Design, computer assisted, 304
De Sivrac, Count, 9
De Sourches, Marquis, 7
Despland, A.R., 69
Destructive time keeping, 275
De Torres, A., 90
Deutsche, Erdl, 271
Deutsch Gramophone, 94
De Vaucanson, Jacques, 285, 286
Deverell, R.S., 331
Deville, Sainte-Claire, 256
De Vries, Hugo, 262
De Worde, Wynkin, 87
Diamonds, 266
Diapason, 277
Diaphragm, lens, 97
Dice, J.R., 73
Dickson, W.K.L., 101
Didier, Leon, 104
Didot, Francois, 85

Diesbach, 258
Diesel engine, 5, 326
Diesel, Rudolf, 324-325
Dieting, 164
Differential calculus, 236
Digital Co., 304
Digital Equipment Corp., 296
Di Luzzi, Mondino, 205
Dining car, 6, 7
Diode, 111
Diophantus, 231, 234
Dior, Christian, 157, 162
Diplomacy, 126-127
Diptheria sera, 202
Directories, city, 118
Dirigible, 18, 19
Disc, video, 116, 117
Disk
   laser, 289
   magnetic, 302-303
Dishwasher, 151, 152
Disney, Walt, 130
Distributor, 14, 325
District heating, 332-333
Diving, 31-32, 277
DKW, 10
DNA, 260-264
Dodd, George, 309
Dog
   four-eyed, 183-184
   hygiene belt, 185
   toy barking, 187
Doi, Toshi Tada, 94
Dolby noise reduction, 95
Dolby, Ray, 95, 115
Doll, Henri, 270
Dolls, 127
Domingo, Alain, 187
Dominoes, 127
Donald, J., 200
Donkin, Bryan, 78
Door, automobile, 17
Doppler sodar, 341-342
Dornberger, Walter, 344
Dorn, Ernst, 243
Dotter, Charles, 211
Doubleday, Abner, 136
Dowsing, J.H., 166
Drake, Alfred, 322
Drake, Edwin L., 269, 330
Drake, Jim, 147
Drawing, 84
Drebbel, Cornelio, 49
Dressings, 210
Drew, Dick, 176
Drill, dental, 224
Drilling, 270-271, 285, 330
Drinkea, Philip, 205
Drinking trough, 73
Drinks, 76-77
Drum, magnetic, 302
Drums. See Percussion instruments
Dry cell, 319-320
Dry cleaning, 155
Dryer
   clothes, 167
   hair, 159, 161
Dubois, A., 212
Du Bois-Reymond, 217
Dubus-Bonnel, 274
Duchateau, A., 224
Ducos du Hauron, Louis, 98-100
Ducretet, Eugene, 119
Duda, R.O., 305
Du Fay, Francois, 248
Dufour, Guillaume-Henri, 197
Duggar, H., 216
Duhamel, Henri, 145
Dummy, store, 156
Du Mont, Allen, 112
Dumont, Frank and Ben, 141
Dunant, Henri, 197
Duncan and Suberbie, 10
Dungeons and Dragons, 132

Dunlop, Alfred Norman, 173
Dunlop Co., 16, 65
Dunlop, John Boyde, 16
Dunwoody and Pickard, 111
Duplication, 87-89
Du Pont de Nemours, 158, 258-260
Du Pre, Jean, 86
Dupuytren, G., 212
Durand, Georges, 136
Durand, Pierre, 78
Duret, Francois, 228
Dwelling, underwater, 265
Duinelle, W.H., 225
Dyes, 160, 258-259
Dynamic loudspeaker, 96
Dynamite, 45
Dynamo, 319

E

Ear Surgery, 207, 222
Earthenware, 254
Eastman, George, 98, 99
Eastman Kodak Co., 98-100
Eau de Cologne, 161
Ebersolt, Gilles, 185
Echography, 200
Eckert, J.P., 294, 296
Eckert, W., 296, 297
EDF, 331
Edison Co., 177
Edison, Thomas A., 8, 88, 93-94, 98, 101-103, 109, 127, 170, 320
Edmond, John M., 266
Edoux, Leon, 167
Edwards, Robert, 214
Egberts, E., 283
Egg sheller, 185
Egyptian writing, 81, 82
Ehrlich, J., 216
Eimert, Herbert, 92
Einstein, Albert, 47, 241
Einthoven, William, 199
Ejection seat, 24, 25
Ektachem DT, 60, 227
Elbaz, Pierre, 228
Electricity, 247-251, 317-320
Electrocardiography, 198-199
Electrodynamics, 248-249
Electroencephalogram, 200, 206
Electromagnetism, 249-250
Electronics, 250-251
Electrophoresis, 218
Electrostatics, 247-248
Electrotechnology, 320-321
Electrotherapy, 217
Elevator, 167-168
Elf-Aquitaine, 271
Ellis, T.D., 302
Ellis, William Webb, 138
Elster, Julius, 289
Ely, Eugene, 52
Embryonic transplant, 227, 264
Emery, H.L., 63
EMI, 94, 199
Emitter, radio, 109
Empire S.A., 279
Emulsion technology, 98
Encoding, 299
Endocrine gland transplant, 207
Energy, 313-334
   pack, 218, 220
Engines, 321-325
   aircraft and rocket, 54-55, 326-328
   automobile, 14-16
   locomotive, 4-5
   outboard, 30
   steam, 25, 315-317
Engraving, 288
Envelope, 118
Equal sign, 235
Equations, 233
Erard, S., 89, 90
Eraser, 84

Ericsson, John, 30, 48, 322
Erlangen Program, 237
Erman, L.D., 305
Escape chute, 168, 169
Escapement, 275-276
Eschapasse, 211
Espitallier, Georges, 273
Ethanol, 333
Ether, 208
Etherington, John, 159
Eton, William, 210
Euclid, 231, 232
Euter, Leonhard, 231, 233, 234, 236
European Space Agency, 346, 351
Evans, Bob, 307
Evans, Oliver, 283
Everyday life, 151-178
Evinrude, Ole, 30
Evrard, J.M., 223
Ewen, H.I., 339
Ewing, Maurice, 265
Exchange, telephone, 121
Expert system, computer, 304-305
Explosives, 44-45
Extra-terestrials, 190
Eye glasses, 220-221, 227

F

Fabre, Henri, 20
Fage, M., 341
Fahlberg, Constantin, 75
Fahrenheit, Daniel G., 342
Falcon, H., 281
Falling bodies, theory of, 239-240
Fallopia, Gabriele, 214
Fallopian tube transplant, 208
Fallout shelter, 183
Fan, 155
Fantasmagoria, 101, 102
Faraday, Michael, 118, 217, 245, 250, 272
Farcot, Jean-Christian, 26, 211
Farina family, 161
Farman, Henri, 19
Farming. See Agriculture
Farnsworth, Philo, 112
Faroux, Charles, 136
Fashion, 155-159
Fashioning, automatic, 283
Fateman, R.J., 305
Faucet, foot-touch, 165
Fauchard, Pierre, 224, 225
Fauvelle, J.B., 270
Favre, A., 91
Fecundation, 262
Feigenbaum, E.A., 305
Fem, Luis, 302
Fencing, 138
Feodorov, Sviatoslav, 222
Ferchault de Reamur, Rene, 71
Ferdinand II, 324
Ferguson, Harry, 65
Fermi, Enrico, 47, 329
Ferrari, Ludovico, 233
Ferraris, Galileo, 321
Fertilization, artificial, 74
Fertilizer, 66
Fetal cell transplant, 208
Fiat Co., 13
Fiberglass, 273, 274
Fibonacci, Leonardo, 232
Field, Cyrus, 120
Figg, James, 137
Fighter, aircraft, 53, 55
Filing machine, 287
Fillings, tooth, 222-223
Film-disc, Kodak, 98
Film, photographic, 97-99, 104
Filter, coffee, 151
Findlay, G., 216
Fingernail transplant, 208
Fingerprint identification, 178
Finiguerra, Masa, 288

Finlay, C., 203
Finne, Orance, 338
Fiocco, G., 343
Fire
 extinguisher, 168
 resistant boat, 28
 ship, 48
Firestone Co., 65
Firing simulator, 181
Fischer, E., 208
Fischer, Hans, 99, 100
Fish
 breeding, 73-74
 electronic, 167
Fisher, Alva J., 173
Fishing rod, 181
Fitch, John, 25
Flame thrower, 46
Flavoring, artificial, 75
Flax, spinning, 281
Fleming, J.A., 111, 119
Fleming, Sir Alexander, 216
Flight, first human, 17
Flintlock, 35
Florey, H., 216
Florsdorff, E.W., 78
Fluid mechanics, 240
Fluorescent tube, 171
Fluoride, 222
Fluosol, 204, 205
Flushing system, 169
Flute, 90
Fly, 156
Flying saucers, 190
FM radio, 111
Focq, N., 285
Fokker, Anthony, 53
Folberth, W.M., 17
Folli, F., 204, 341
Fontaine, Hippolyte, 321
Fontana, Niccolo, 233
Food processor, 152
Food preserving, 77-78
Foods, 74-76
Football, 138
Forbes, Walter A., 175
Forceps, 212, 213
Ford Aerospace Co., 343
Ford, Henry, 13, 65
Ford Motor Co., 17
Forester, Jay, 297, 307
Forest, Fernand, 14, 15, 323
Fork, 152, 186
Forlanini, Enrico, 22, 26, 30
Forsyth, Alexander, 36
Fortin, J., 341
Forton, L., 86
Foster, John, 338
Foucault, Leon, 250
Foucher, Guy, 208
Foude, William, 30
Fountain pen, 84
Fourcault, Emile, 274
Fourcin, Andrian J., 228
Fox, Uffa, 143
Fractures, 210
Frafenberg, Ernst, 215
Francis, James, 314
Franklin, Benjamin, 91, 171, 177, 221
Freeman, Alan, 187, 211
Freeze-dried food, 78
Freezing, industrial, 78
Freight car, 6
Frequency modulation (FM), 111
Fresneau, Francois, 157
Fresnel, Augustin, 28
Freud, Sigmund, 217
Friis, Harold, 122
Frisbee, 127-128
Froelich, John M., 64
Froment, Gustav, 321
Frost, George, 111
Froude, William, 240
Frounson, Alan, 227
Frozen food, 75-76
Fruit diet, 164

Frying pan, 154
Fuel cell, 320
Fuel, nuclear, 329
Fujitsu Co., 193, 298, 309
Fuller, C.S., 318
Fulton, Robert, 26, 48-50
Funakubo, Hiroyasu, 308
Functions, 236
Furnace, 255-256
Furniture, 185
Fuse, 44-45, 249
Fusion, controlled, 330
Fust, Johann, 85
Future, 190-194

G

Gabor, D., 290
Gadget, 152-153
Gaggia, 151
Gagolite, 82
Gaillard, Peter, 62
Galilei, Galileo, 198, 239, 241, 276, 339, 342
Gallagher, James, 24
Gallet, Charles, 161
Galley, 48
Galton, Sir Francis, 178
Galvani, Luigi, 217, 318
Games and toys, 125-135
Gametes, 260
Gamgee, John 140
Garand rifle, 37
Garbage disposal, 169
Garcin, J., 141
Gardening, 67-70
Garnerin, Jacques, 23
Garner, Philip, 186
Gas chamber, 178
Gases, 242-245
 chemical warefare, 46, 70
Gasoline, 13, 330
Gaspar, Bella, 99, 100
Gatling gun, 39
Gatling, Richard J., 39
Gauchot, Paul, 24
Gaulard, Lucien, 321
Gaumont, 95, 103
Gauss, Karl F., 231, 234, 237
Gauthier, Georges, 10
Gay and Silver, 287
Gay-Lussac, Louis, 244, 251, 341
G.D. Searle, Inc., 215
Geiger counter, 289
Geiger, Hans, 251, 289
Geigy, 71
Geiser, Jean-Marc, 147
Geissler, Heinrich, 250
Geitel, Hans F., 289
Gelatino bromide process, 98
Genentech, 228
General Dynamics Co., 24, 47, 345
General Electric Co., 96, 133, 154, 170, 199, 341
General Motors Co., 13, 301
Generation of electricity, 317-320
Genetics, 260-264
Genetic System, 228
Gensberg, Samuel, 171
Gensfleich, Johann, 85
Gentex, 221
Gentile, Benedetto, 129
Geoffroy-Dechaume, Vincent, 185
Geometry, 231-234, 236-237
Geophysical Analysis Group, 270
Geothermal energy, 266, 331
Gerhardt, Charles, 215
Germanus, Donnus Nicolaus, 338
Gestetner, D., 88
Gevacolor, 99, 100
Giffard, Henry, 18
Gigax, Gary, 132
Gilardoni, Joseph and Xavier, 272

Gilbert, William, 247
Gilboy, Bernard, 144
Gillette, King Camp, 162-163
Gillette Safety Razor Co., 163
Gilliland, Ezra, 121
Ginsberg, Charles P., 115
Giova, Flavio, 29
Gladstone, 60
Glas, Louis, 169
Glass, 273-275
 bells, 67
 blowing, 273, 274
 fiber plaster, 210
 lens, 96
Glider, 19, 139
Glidden, Joseph, 73
Glomar Challenger, 266
Glouchko, Valentin, 344
Glow discharge, 249
Gnomon, 275
Go, 128, 131
Goddard, Robert, 327, 344, 345
Godowsky, Leopold, 99
Goldenmark, Peter, 94
Gold, underwater, 265
Golf, 138, 139
Goldstine, H.H., 293, 294
Gonder, M.J., 207
Goodhue, L.D., 164
Goodyear Co., 16, 182, 223, 257
Gordon-Bennett, J., 136
Gordon, J., 289
Gorrie, John, 283, 284
Gottlieb, D., 216
Gougeon, Jon and Joe, 143
Gould, R. Gordon, 289
Graebe, Charles, 258
Graham, G., 276
Gramophone, 94
Graphic tablet, 302
Gravity, 238-240, 319
Gray, Elisha, 120, 121
Gray, Stephen, 248
Gray, William, 120
Greek alphabet, 82
Greenhouse, 67-68
Gregoire, Marc, 154
Gregorian calendar, 337-338
Gregory XIII, 337
Gresham, Thomas, 176
GR-5, 257
Griffith, A.A., 326
Grinder, coffee, 151
Grinder machine, 287
Gringmuth, Axel, 167
Gromnos, W., 211
Groom and Boone, 198
Ground transportation, 4-17
Groupomatic, 204
Grove, Sir William R., 318
Growth hormone, 218
Grumman Co., 348
Grundig, 95, 116
Gruntzig, Andrea, 211
Guerlain, Pierre and Jacques, 162
Guillotin, Dr. Joseph I., 178
Guillotine, 178
Guitar, 90, 188
Guldner, Hugo, 325
Guns. See Armaments
Gunzberg brothers, 105
Gurney, Goldsworthy, 317
Gutenberg, 84-85, 287
Gutta-percha, 272
Gutte, 263
Gutters, 172
Gynecology, 211-215

H

Haarmon, Dr. Wilhelm, 75
Haber, Fritz, 265
Haemmerle, Bernard, 184
Haensel, V., 271

Hahnemann, C.S., 225
Hair
 styling, 159-161
 transplant, 207
Haldane, T.G.N., 333
Hale, D.A.S., 211
Hale, George H., 340
Hall, Charles M., 257
Halle, J., 217
Halley, Edmund, 31-32, 343, 352
Halley's Comet, 352
Hall, John, 78
Halsted, 211
Hamilton Beach Co., 161
Hamilton, Franck, 297
Hamilton, L., 151
Hammer, steam, 287
Hammond, Laurens, 92
Hamov, Jacques, 214
Hancock, T., 257
Hancock, Walter, 8
Handball, 139
Handlebars, 9
Handy, W.C., 93
Hang glider, 139
Hannoversche Maschinenbau A.G., 323
Hannoyer, M., 182
Hansen, Malling, 88
Han Sin, 129
Hargreaves, James, 280
Harmonica, 91
HAPP, solar, 191
Hardy, James D., 208
Harp, 89, 90
Harpsichord, 89
Harrington, 224
Harrington, John, 169
Harrison, James, 283
Harrison, John, 277
Hart and Pan's tractor, 64
Hart, C.W., 64
Hartmann, Louis, 154
Hart, P.E., 305
Harvester, 60-64
Harvey, William, 203
Harwood, J., 277
Hasler, Blondie, 144
Hass, Earl, 213
Hassenfratz, J., 204
Hatpin, 157
Hat, top, 159
Haussman, Jean-Michel, 258
Haute couture, 156
Hauy, Valentin, 83
Havilland Co., 20
Hawaii Surf Co., 148
Hawker Siddley Harrier, 56
Hay, 62-63
H-bomb, 47
Headache glasses, 227
Heads or tails, 128-129
Health, 164
Heard, John F., 182
Hearing aid, 221
Heart transplant, 208
Heater, 332-333
Heating, 165-166, 172
Heatley, N., 216
Heathkit Co., 171, 309
Heavy water, 257
Hedge-trimming machine, 69
Hein, Piet, 129
Heisenberg, Werner, 252
Heister, L., 223
Helicoptor, 22-23
Heliogeothermics, 332
Heliogravure, 85
Heliophysics, 332
Heli-Stat, 18, 19
Helium, 245
Hemanometer, 198
Hemolytic disease, 204
Hennebique, Francois, 273
Henriod, Paul, 187
Henry, B. Tyler, 37
Henry, Edward, 178

Herbarium, 68
Herd management, 73
Heredity, 260-261
Hering, C., 225
Heroult, Paul, 27, 257
Herreshof, Nathanael G., 27, 142, 143
Herriott, D.R., 211
Herrman, Emmanuel, 118
Herschel, John, 221
Hershaw, Samuel, 152
Hershel, William, 339
Hertwig, Oscar, 262
Hertz, Heinrich, 109, 110, 250
Hevelius, Johannes, 339
Hewlet-Packard, 296
Hex, 129
Hickman, Ronald P., 171
Hieractic writing, 82
Hierglyphics, 81, 82
Hildebrand and Wolfmuller, 10, 16
Hiller, L., 93
Hiller, M., 303
Him, Gustav, 317
Hindret, Jean, 282
Hipparchus, 233, 275, 337
Hippocrates, 197, 225, 260
Hippodemie, 141
Hirst, J.H., 181
Hitachi, 116, 117, 194, 298
Hittorf, Johann, 250
HLA system, 264
Hobby horse, 9
Hodgson, Peter, 133
Hoffman, Felix, 215
Hoff, Marcian E., 296
Hohmann, Walter, 249
Holden, Colonel, 11
Holden, Daniel, 72-73
Holden, Richard and George, 29
Hollerith, Herman, 293, 298, 301, 302
Hollister, Donald, 171
Hologram, 290
Holt, Benjamin, 64-65
Holzworth, 326
Home knitting machine, 283
Homeopathy, 225
Honda Co., 12, 141
Honold, Gottlieb, 14
Hood, David, 28
Hooke, Robert, 261, 277, 341
Hoop games, 129, 130
Hopper, Grace, 299
Horie, Kenichi, 188
Hormones, 214, 216, 218
Horn, 90-91
Horner, Matthias, 91
Hornsby, R., 69
Horse racing, 139
Horses, corrective lenses for, 184, 185
Horslay, Sir Victor, 206
Hose, garden, 69
Hosegawa, Goro, 131
Hosiery, knitting, 282
Hospital, 197
Hotchkiss, 37
Hotplate, 153
Houdart, Raymond, 206
Houdry, Emile, 271
Houghton and Tompion, 276
Hounsfield, G., 199
Hourglass, 275
Hovercraft, 27
Howard, A., 66
Howe, Elias, 172
Howe, John I., 157
Howitzer, 40
Hughes Aircraft Co., 289
Huguet-Huard, 60
Hula-hoop, 129
Hull, Dr., 164
Hulls, Jonathan, 25
Hunley, Captain, 50
Hunt, Seth, 157

Hunt, Walter, 157, 172
Hurd, Earl, 106
Hussey, Obad, 61, 63, 64
Hutchinson, Miller R., 221
Hutchinson, William, 18
Huygens, Christian, 236, 241, 276, 277, 315
Hyatt brothers, 258
Hybridomas, 264
Hydraulics, 313-315
Hydroelectric power, 317-320
Hydrofoil, 26-27
Hydrogen, 245
Hydrogen peroxide, 256
Hydroponics, 67
Hydrostatics, 239-240
Hygienic napkins, 213
Hygrometer, 341

I

IBM, 88, 290, 293, 294, 296-304, 306, 307
Icannissiani, Dr., 340
Ice breaker, 27
Ice cream, 75
Ice sailing, 139-140
Ice skating, 140, 142
Ichbiah, Jean, 300
ICM Co., 304
Ideal Toys, 132, 134
Identification, criminal, 177-178
Ideograms, 81, 82
Idrac, Pierre, 343
IFREMER, 266
IG Farben, 221, 257, 259
Ignition device, 14
I Hing, 275
Illustrated poster, 86
Image Cascade, 193
Immune system, 263
Immunization. See Vaccination
Imperial Chemical Industries, 259, 260
Incandescent lamp, 170
Incendiary devices, 45-46
Incubator, 213
"Indecipherable Square," 83
Indigutin, 259
Induction, 250
Induction coil, 14, 322
Industry, 269-290
Infarctus, 211
Infinity, 232-233
Information systems, 293-310
Infra-red
photography, 99
telescope, 340
Ingenhousz, G., 261
Ink, 83, 84
INS, 193
Insecticide, 70-71
Insemination, artificial, 71-72, 213
INSERM, 200
Instruction, computer assisted, 304
Instruments, music, 89-92
Instrument table, dental, 224
Insulation, 120
Integral calculus, 236
Integrated circuit, 295-196
Intel Corp., 295
Intelligence, artificial, 304
Intercom, 280
Interferon, 218
Internal combustion engine, 321
International Harvester, 64, 65
International Sleeping Car Co., 6
International Union of Telecommunications, 339
Inui, Takeo, 30
In vitro processes, 262-264
Ionescu, 208
IRCAM, 303
Iris diaphragm, 97

Iron
cores, 307
electric combing, 160-161
lung, 205
Isaacson, L., 93
Isaacson, M., 303
Ishiko, Masahisa, 188
Isotopic separation, 328
Isselbacher, Kurt J., 227
ITT, 46
Ivanovsky, D., 203
Iverson, Ken, 299

J

Jablochkoff, Paul N., 169-170
Jackhammer, 287
Jacobi, Ludovic, 71, 74
Jacobs, Aletta, 215
Jacoby, Walter, 207
Jacquard, Joseph-Marie, 282, 302
Jacquin, Claude, 185, 282
Jammet, Henri, 207
Jansky, Karl, 204, 338
Jansen, J., 198
Janssen, Jules, 101, 243
January 1st, 338
Jarre, Jean-Michael, 92
Jars, preserving, 77-78
Jaum, A., 289
Javel, water, 256
Jazz, 93
Jeans, 156
Jenner, Edward, 201, 203
Jet
aircraft, 20, 21, 54-55
car, 14
Jif-Waterman, 84
Joachim, Frederic-Guillaume, 32
John Shearer and Sons, 60
Johnson, Alfred, 144
Johnson, Robin-Knox, 144
Johnston brothers, 224
Joliot-Curie, Irene, 217, 253
Joliot-Curie, Frederic, 217, 253
Jolly, J.B., 155
Jones, William, 233
Jousse, Pierre, 188
Judet, Jean and Robert, 207
Judo, 140
Judson, Whitecomb, 159
Jukebox, 169
Jumbotron, 193
Junkers, 325
Jurgens Co., 88
JVC, 116, 117
JVX, 190

K

Kahn, Julian S., 164
Kall, Robert, 187
Kalmus, Herbert, 104
Kammerer, Frederick, 153
Kano, Jigoro, 140
Kapany, Narinder S., 290
Kaplan, Viktor, 315
Kappafloat, 275
Kastler, Alfred, 279
Kato, Ichino, 194
Kawasaki Co., 12
Kay, John, 280
Kayser, Lutz, 346
Keel, boat, 28, 143
Keldych, Mtislav, 344, 346
Keller, A.J., 207
Keller-Dorian, 104
Kellog, Edward W., 96
Kellogg, William, 75
Kellybuilt, 73
Kelly, William D., 207
Kelman, 221, 221
Kelsall, Derek, 143

Kelvin, Lord, 110, 333
Kelvin scale, 238
Kemeny, John, 300
Kempf, Martine, 13, 228
Kennerly, A.E., 177
Kenyon, B.C., 169
Keratotomy, radial, 222
Kerton, P.S., 73
Keystone, 63
Key, tin can, 78
Kidney stones, 211
Kidney transplant, 207
Kier, Samuel, 269
Kilburn, R.M., 297
Kilby, Jack S., 295-296
Killbourne, Dave, 139
Killer, 132
Kimberley-Clark Co., 171, 213
Kimberly, Gary, 147
Kinetograph, 101-102
Kinetoscope, 102
King-Smith, Eric, 218
Kinley, K.T., 271
Kircher, P., 101
Kitasato, S., 202
Kitchen, 151-155
Kites, 129
Kitsen, Arthur, 28
Klatte and Corbiere, 158, 259
Kleenex, 171
Klein, Felix, 237
Klietsch, Karl, 85
Knerr, Richard P., 129
Knitting, 282-283
Knoll, Max, 198
Koch, R., 202-203
Kodak, 98-100, 104
Koebek, A., 70
Koening, F., 85
Kohler, George, 218, 264
Kolb, Homer, 147
Koller, K., 208
Korolev, Sergei, 344-345
Krafft, Charles, 205
Krebs, Arthur, 18
Kremer, Gerhard, 338
Kroc, Ray, 175
Kyeser, Conrad, 315

L

Labitte, H., 10
Labaraque, Antoine, 210
Lacassagne and Thiers, 170
Lace, machine-made, 281
Laemmle, C., 106
Laennec, Rene T.H., 198
Lailletet, Louis-Paul, 245
Laisner, Harry, 325
Lallemant, Pierre, 9
Lamarck, 261
Lambert, Alain, 200
Lambert, Jean-Henri, 236
Lambot, J.L., 272
Lamoureux, Charles, 93
Laminated glass, 274
Lamps, 169-171, 187
Lanchester, William, 324
Land, Edwin H., 99, 100, 105, 213, 221
Landing gear, 24, 25
Land sailing, 140
Landscaping, 67
Landsteiner, K., 204
Land transportation, 3-17
Lane, John, 60, 63
Langen, Eugen, 322
Lang, Franz, 325
Langmuir, Irving, 199, 341, 343
Languages, computer, 299-300, 304
Lanston, Tolbert, 85
Laplace, P., 204
Laposky, Ben F., 303

Large, A.L., 152
Larrey, Baron Dominique J., 197
Lassers, 100-101, 211, 222, 227, 289-290
Las Vegas, 125-126
Latarget, Raymond, 207
Lathe, slide, 286
Latin alphabet, 82
Laurent, Alexander, 168
Laurin-Klement, 11
Lauterbur, P.C., 199, 200
Lauter, Richard, 207
Laveran, A., 203
Lavigne, Alexis-Marie, 156, 158
Lawlor, Si, 144
Lawn mower, 68, 69
Lawrence, (camera) 97
Lawrence, (refrigeration) 72
Lawson, Harry, 9
Leclabart, J.M., 160
Leblanc, Nicolas, 163
Lebouder autoplane, 183, 184
Le Bris, Jean-Marie, 19
Le Chatelier, Henri, 318
Leclanche, Georges, 319
Lederberg, J., 305
Le Duc, Alain, 211
Leduc, Rene, 328
Leech cup, 213
Lee, Ivy L., 175
Leeuwenhoek, Antonie Van, 198, 261
Lee, W., 158
Lee, William, 282
Lefaucheux, Casimir, 36
Legendre, Adrien-Marie, 231, 236-237
Lego bricks, 129
Leibniz, Gottfried W., 236, 293
Leisure, 164-173
Leith, E.N., 290
Lelievre, Rene, 160
Le Lorier, Guy, 208
Lemaire, 66
Lemale, Charles, 325
Le Mans rally, 136
Lemiere, Denise, 183
Lemoine, Louis, 287
Lemoine, Roger, 160
Lenoir, Etienne, 14, 16, 321, 322
Lenormand, Sebastien, 23
Lenses
corrective, 220-221, 227, 228
photographic, 96-97
Lentheric, 160
Lenticular system, 104
Lepaute and Caron, 275
Leroux, Henri, 215
Lescure brothers, 154
Lesser, V.R., 305
Leucippus, 251
Leveque, Donnet, 20
Le Verrier, Urbain, 340-341
Levi, Renato, 143
Levret, A., 213
Levy, Lucien, 111
Lewis, Gilbert Newton, 257
Lewis, John, 224
Lewis, J.P., 280
Lewis, Samuel, 271
Lexcen, Ben, 28
Leyden jar, 248
Libby, William F., 254, 278
Libois, Joseph, 121-122
Library, 87
Lie detector, 186-187
Life (game), 129
Light bulb, 109, 171
Lighthouse, 28
Lighting, 169-171
Lightning rod, 171
Light-pen, 302
Lightwave communications, 122
Liley, A.W., 204
Lilienthal, Otto, 19, 139
Lillehei, Richard, 207

Lilly, John, 218
Limb transplant, 207
Limoge, Aime, 208
Limonaire brothers, 91
Limousin, 215
Line, high-tension, 321
Lindenmann, 218
Lind, James, 341
Link, Edwin A., 265
Linogravure, 288
Linoleum, 171
Linotype, 85
Liou Shu-lien, 165
Lipman, Harry L., 118
Lippmann, Gabriel, 198
Lipstick, 161
Liquid
crystal, 129
gases, 245
List, computer, 302
Lister, J., 210, 211
Liston, 210
"Literary piano," 88
Lithography, 85, 86
Little, C.G., 341
Little, Dave, 172
Littre, 202
Liver cancer, test for, 227
Liver transplant, 208
Lloyd and Garren, Drs., 164
Lobatchevski, Nicolai, 237
Lockheed Co., 21, 191, 192, 340, 345, 352
Lock, canal, 320
Locks, 17, 279-280
Locomotives, 4-5
Lodge, Raymond, 351
Logarithms, 235, 236
Logic, 237
Log, mechanical, 30
Longbow, 35
Long, C.W., 208
Longrave, William, 327
Looms, 281-283
L'Orange, Prosper, 325
L'Oreal, 160, 161
Lottery, 129
Loubat, Emile, 7
Loudspeaker, 95-96
Louis, Dr. Antoine, 178
Love, Judge Jack, 187
Lovell, Sir Bernard, 339
Lowes, R., 204
Lucretius, 251
Lully, Jean Baptiste, 93
Lumiere August, 100, 102-103
Lumiere, Louis, 99, 100, 102-103
Luminous pantomimes, 102, 103
Lunar exploration. See Moon exploration
Lund, C.F., 74
Lundstrom, Carl F., 153
Lung
iron, 205
transplant, 208
Lunn, Arnold, 145
Luria, S., 262
Luther, Martin, 137
Lwoff, Andre, 218, 262

M

Macadam, 272
Macalpine-Downie, Rod, 143
Macdonough and R. Evans, 160
MacGaffey, I.G., 173
Mach, Ernst, 240
Machine gun, 37-40, 53, 54
Machines, agricultural, 59-65
Machine tools, 285-287, 306
Mackintosh, 157-158
Madox, Richard L. 98
Maelzel, J.N., 93
Maerowitz, Mordecai, 130
Mager, J., 92
Magic lantern, 101, 102

Magic Wand, 129, 130
Magneto, 14, 318
Magnetoelectric motor, 320-321
Magnetohydrodynamic
    generator, 320
Mah-jong, 130
Mahler, Ernest, 213
Maigret, P., 204
Mail, 117-118
Maiman, H., 211, 289
Malka, Pascal, 148
Mallandin, Charles, 323
Malpighi, Marcello, 204
Manchly, J., 296, 297
Manned space crafts, 247-351
Mannequin, 156-157
Mannes, Leopold, 99
Manning, William, 63
Man-of-war, 48
Manson, P., 203
Maps, 338, 343
Marathon, 141
Marbles, 130
Marcan, L., 288
Marcel Dassault, 204
Marc, Helene, 206
Marchoux, E.E., 202
Marconi, Guglielmo, 119
Marconi rigging, 143-144
Marey, Jiles, 101, 104
Margarine, 75
Marie-Davy, 343
Mariotte, Edme, 244
Maritime transportation, 120,
    182, 183, 188, 271
    boats, 25-28
    diving equipment, 31-32
    navigational aids, 28-31
    See also Naval vessels
Marquet, Jean, 207
Marsch, Sylvester, 5
Marsh, Albert, 166
Martin, C. and A., 28
Martin, Constant, 92
Martin, H., 84
Martin, J., 161
Martin, L., 202
Martin, Marietta, 345, 350
Martin, Pierre, 256
Martin, W.A., 305
Martin-Zucharelli, 262
Marty, Christian, 148
Maury, Matthew Fontaine, 343
Mary Stuart (Queen of Scotland),
    138
Maser, 289
Mason and Ito, Drs., 164
Massey, Edward, 30
Massey-Ferguson, 65
Massey-Harris, 62, 64, 65
Massicot, Guillaume, 287
Masson, Antoine, 14, 322
Masson, Michel, 279
Mastermind, 130
Matches, 153
Matchlock, 35
Mathe, Georges, 204, 207
Mathematics, 231-237
Mathijsen, Antonius, 210
Matrices, 237
Matsushita, 116, 117, 193, 194
Mattel, 127, 135
Mattress, air, 164-165
Maudslay, H., 286
Maunoir, Theodore, 197
Mavica, 100
Maxim, Hiram, 39
Maxwell, James Clark, 98, 100,
    109, 110, 250
May and Smith, 112
Maybach, Wilhelm, 10, 13, 15, 16,
    323, 324
Mayer, L.B., 106
Mayo diet, 164
Mayow, John, 242
May, Wesley, 23
Mazzitello, Joe, 115

MCA, 117
McAdam, John, 272, 287
McAllister, L.G., 341
McCarthy, J., 300
McCormick, Cyrus, 61
McCready, Paul, 22, 184
McDonald, Maurice, 175
McDonald, Richard, 175
McDonalds, 175
McDonnell-Douglas, 47, 192,
    345, 348, 350
McKhann, J., 207
Measures, 237-238
Mechain, Pierre, 238
Mechanics, 240-242
    automobile, 13-16
Media and communication, 109-
    122
Medicaments, 219-220
Medicine, 197-228
Mege-Mouries, Hippolyte, 75
Meikle, Andrew, 61, 62
Meikle, James, 62
Meilland, Antoine, 70
Meiosis, 261
Meissner and Armstrong, 92
Melies, Georges, 104, 105
Melvin, Arthur K. "Spud," 129
Memory
    card, 175-176
    computer, 302
Mendeleyev, Dimitri, 252
Mendel, Gregor, 103, 260, 261
Menzies, Michael, 61
Merchandising, electronic, 175
Merlin, Joseph, 141
Merrifels, 263, 264
Merrill, J.P., 207
Merryman, J.D., 296
Mesotherapy, 225
Messier, Georges, 24
Metallurgy, 254
Meteorology, 340-343
Meter (measurement), 238
Methane, 271, 333
Methanol, 334
Metric system, 238
Metronome, 93
Metropolitan Vickers, 328
Michaux, Ernest and Pierre, 9
Michel, Charles, 218
Michelin, Andre and Edouard, 16
Michelin Co., 3, 16, 17
Mickey Mouse, 130
Microbes, 203
Microbiology, 202, 261
Microscope, 198, 222, 228
Microwave
    oven, 153
    radio, 172
Middleton, Empson Edward, 144
Mikhail Mil Co., 22
Miles Laboratories, 220
Military. See Armaments
Milk
    concentrated, 77
    powdered, 76-77
    refrigeration, 72
Milking machines, 72
Miller, Carl, 89
Miller, Patrick, 142
Miller, Stanley, 265
Millet, Felix, 10
Mill, Henry, 88
Milling machine, 287
Milstein, Cesar, 218, 264
Mincom, 115
Mineral chemistry, 254-257
Mines, 52-53
Miniskirt, 157
Minolta, 100
Minthrop, D.L., 270
Minus sign, 235
Mirror, 171
Missiles, 46-47, 53. See also Roc-
    kets

Mitchell, M., 305
Mitosis, 261
Mitsubishi, 298
Mitz, Vladimir, 206
Mizumo, Shumeihi, 309
Mock, Elmar, 279
Models, 156-157
Model T Ford, 13
Moderator, 329
Mohr, Linda, 227
Moitessier, Bernard, 145
Moloney, Richard T., 171
Molotov cocktail, 46
Malt, Donald A., 207
Morel, G., 262
Money, 174-175
Monier, Joseph, 272
Moniz, Egaz, 199, 206
Monoclonal antibodies, 218
Monod, Lucien, 152
Monod, Robert, 210
Monopoly, 131
Monotype, 85
Montagu, John, 74-75
Monte Carlo Rally, 136
Monte, Del, 137
Montgolfier balloon, 17-18
Moody, John, 147
Moon exploration, 349, 351
Moore and Hascall, 63
Moore, Wendall F., 23
Morestin, H., 206
Moreno, Gilbert, 186
Moreno, Roland, 175, 303
Morey and Johnson, 172
Morey and Son, 76
Morgan, H., 262
Morgan, William G., 147
Mori, Kei, 193
Morris and Betts, 282
Morrison, Alexander, 224
Morrison, Fred, 127
Morrison, J.B., 224
Morrison, Warren Alvin, 278
Morse, Samuel, 118, 119
Mortar, 41-42
Morton, George Thomas, 205
Morton, William, 205, 208
Moslem prayer watch, 279
Mosquito
    pill, 218
    swatter, 167
Motion pictures, 101-106
Motorbike, 10
Motorcycle, 9-12, 16
Motorcycling, 141
Motors, 320-321. See also En-
    gines
Mouchot, 332
Moviemaking
    camera, 98, 101-103
    computers and, 303-304
    projector, 98, 102
Mowing machine, 62-63
Moyes, Bill, 139
Moynier, Gustave, 197
Mudge, T., 276
Mulder and Ryke, 28
Mule jenny, 281
Muller, 221
Muller, H., 66
Muller, P., 71
Muller, Jacques, 279
Muller, Johan, 234
Mullins, Anderson H., 69
Munters, C., 154
Munz, Emile, 196
Murchland, William, 72
Murder Party, 131, 132
Murphy, John Benjamin, 205-
    206
Music, 187
    composition and performance,
    93
    computers and, 303
    instruments, 89-92
    sound reproduction, 93-96

Music box, 91
Mustard gas, 46
Mutations, 262
Muting, 218
Muybridge, Eadweard, 101, 103
Myler, 211

N

Nagelmackers, Georges, 16
Naismith, James, 136
Naito, Ryochi, 204, 205
Namco, 135
Napier, John, 294
Napkin, hygienic, 213
Napalm bomb, 45
Nartov, A.C., 306
NASA, 13, 122, 192, 200, 331,
    332, 337, 340, 345-346, 348,
    349
Nash, John F., 129
Nasmyth, James, 287
Natta, Giulio, 260
Naval vessels, 48-53
    sailing, 142-145
Navigational aids, 28-31, 290
NEC, 298
Needle, 157
Negative numbers, 234
Negatives, 97-98
Negro spirituals, 93
Nelson, C.K., 75
Neon tubes, 170-171, 249
Neoprene, 258
Nepier, John, 235, 236
Nerst, Walter, 246
Nerve gas, 46
Nerve transplant, 207
Nestle, 77, 160
Neuman, M.H.A., 296
Neumatic system, 93
Neurosurgery, 206
Neutron bomb, 47
Newcomen, Thomas, 25, 246, 315
Newcome, Thomas, 87
Newell, A., 304
Newick, Dick, 143
New Look, 157
Newspaper, 87
Newton, Sir Isaac, 236, 240-242,
    339
Nicholson, Charles E., 144
Nidetch, Jean, 164
Niepce de Saint-Victor, Abel, 98
Niepce, Nicephore, 9, 98, 288
Nightingale, Florence, 197
Nipkow, Paul, 112, 114
Nishi, 262
Nissan Co., 13
Nissen, George, 147
Nitrogen, 227, 242-243
Nitroglycerin, 45
Nobel, Alfred, Emil, and Louis,
    45
Nobili, Leopoldo, 318
Noble, Silas, 176
Noel, Suzanne, 206
Nogues, Henri, 104
Nordheim, Sondve, 145
Nordwestdeutsche Kraftwerke,
    334
Northrup Co., 192
Northrup, J.H., 282
Nose transplant, 207
Notation, music, 93
Novocaine, 208
Nova Founders, 185
NTSC procedure, 113, 114, 117
Nuclear energy, 290, 328-330
Nuclear Magnetic Resonance
    (NMR), 200
Nucleic acids, 263
Nudity, athletic, 141
Numa Pompilius, 211
Numbers, 234
Numerals, 232

Nursing, 197
Nylon stockings, 158
Nyu Wa, 91

O

Oars, steamboat with, 25
Oberth, Hermann J., 327, 349
Oboe, 91
Observation plane, 183, 184
Oceanography, 264-266
Odolek, 24
Oersted, Hans Christian, 249,
    256
Office equiment, 87-89
Offset printing, 86
Offshore drilling, 271
Ohm, Simon, 249
Oil
    drilling, 330
    engine, 324
    fire, 271
    lamps, 169
    painting, 84
    tanker, 27
Oki, 298
Oliphant, 218
Olivetti Co., 89
Olympics, 141
Olympus Co., 95
Omniboat, 186
Omnibus, 8
Onnes, Heike Kamerlingh, 245
Oparine, Alexander, 265
Oppenheimer, Robert, 47
Optical soundtrack, 104
Optical fibers, 290
Optical videodisc, 117
Optics, 220-222
Orange juice concentrate, frozen,
    77
Orata, Sergius, 74
Orchestral conducting, 93
Orenteich, N., 207
Organic chemistry, 257-260
Organico, 259
Organs, 91-92
Organs, artificial, 209
Orsippus of Megare, 141
Orthodontics, 224
Oscars (film awards), 106
Oscillator, spiral spring, 276-277
Oscillograph, cathode-ray, 250-
    251
Oshawa, 262
Osius, Fred, 151
Osteotherapy, 225
Osterhoudt, J., 78
Othello, 131
Otis Co., 168
Otis, Elisha Graves, 167
Otospongiosis surgery, 222
Otto, Nicolas, 14, 322, 323
Outboard engines, 30
Outcauld, R.F., 86
Outerbridge, Mary Ewing, 146
Outran, John, 7
Oven, microwave, 153
Owens, Michael J., 274
Oxygen, 242-243, 245
Oxygen tent, 218
Oyster farming, 74

P

Pacemaker, 211
Packaging, plastic, 78
Pac-Man, 135
Paddlewheel, 25-27, 313
Pages, 204
Paget, Arthur, 283
Pain, Gerald, 228
Paint, 84, 173
Painting, 84
Palissy, Bernard, 254

Palmer, Daniel D., 225
PAL procedure, 114, 117
Pan American Airways, 20
Pan, C.H., 64
Panhard and Levassor, 197
Pantelegraph, 112
Paper, 83-84, 287
Paperback book, 87
Paper clip, 89
Papin, Denis, 25, 153, 315
Papyrus, 83
Parachute, 23, 24
Parade, fashion, 156
Parasites, 203
Parchment, 83
Pare, Ambroise, 165, 210
Paris, J.A., 101
Paris, Mathew, 338
Parker Brothers Inc., 129, 131
Parmentier, 76
Parsons, Sir Charles A., 325
Parsons, William, 339
Pascal, Blaise, 6, 232, 239, 244, 299
Paschen, Friedrich, 249
Passenger car, 6
Pasteur, Louis, 201-203, 210-211, 215, 260-263
Pathe, Charles, 103, 104
Pathe-Nathan, 104
Paulet, Pedro E., 344
Pearson, G.L., 318
Pechelharing, 204
Pechiney-Ugine-Kuhlman, 259
Pederson, Aren B., 160
Pelican beak, artificial, 186
Pelouze, Theophile-Jules, 258
Pelton, Lester A., 314, 315
Pemberton, John, 77
Pen, 84
Penaud, Alphonse, 19, 24
Pencil, 84
Pendulum clock, 276
Pendulum, fixed image, 111-112
Penicillin, 215, 216
Penstock, 319
Pepper mill, 186
Percussion, 198
  caps, 36
  instruments, 91, 92
Percy, Pierre-Francois, 210
Peres-Conde, Jose, 185
Perforated card,
Perforated stamp, 118
Performance, musical, 93
Perfume, 161-162
Perier brothers, 25
Perignon, Dom Pierre, 76
Perikatuma, 125
Periodic table, 252
Perkins, Jacob, 283
Perkin, William H., 258
Perky, Henry D., 75
Perlon, 259
Permanent, hair, 159, 160
Pernot plant, 332, 333
Peraud, M., 84
Perrault, Charles, 69
Perrelet, A.L., 277
Perrin, Jean, 250, 251
Perron, L., 276
Pescara, Raul de Pasteras, 325
Petoit, Roger, 76
Petrie, W., 169
Petroleum, 269-271
Petty, William, 88, 142
Petzval, J., 96
Pfeiffer, H., 65-66
Pfleumer, Fritz, 95
Phenakistoscope, 101
Philippe, A., 277
Philips Co., 321
Philips Data Systems, 289
Philips Laboratory, 171
Phillips Co., 94, 95, 116, 117, 135
Phoenician alphabet, 82

Photoelectric cells, 289
Phonograph, 93-94, 103-104
Photocopy, 89
Photographic recording of sounds, 95
Photography
  motion pictures, 101-106
  still, 96-101
  underwater, 264-265
Photogravure, 288
Photosensitive mediums, 97-99
Photovaltaic cell, 318
Physics, 237-254
Phytobiology, 65-67
Phytotherapy, 225
Pi, 233
Piano, 90, 91
Piasecki, Frank N., 18
Piccard, Auguste, 32
Pictet, Raoul-Pierre, 245, 284
Picture, composite
Pilatre de Rozier, Jean-Francois, 17
Pilkington, Alistair, 274-275
Pills, 215, 218
Pilot, automatic, 23
Pin, 157
Pinball machine, 171
Pinchon, Dom, 74
Pincus, Gregory, 215
Pinet, Andre, 122
Pinon, F., 204
Pipe, 171
Pisciculture, 264
Pistols, 36-39
Piston, 315, 317, 323
Pitt, Hiram and John, 62
Pixii, Hippolyte, 318
Pizarro, 74
Place, Dwight and Francis, 254
Plague sera, 202
Planck, Max, 252
Planetary probes, 349-350
Planing machine, 285-286
Plante, Gaston, 318, 320
Plant, Shibuara, 95
Plasma, 204
Plasters, 210
Plastic
  bottle, 153
  explosive, 45
  packaging, 78
  safety shield, 17
  wood, 273
Plat, 264
Plate, 153
Plate glass, 274-275
Plateau, Joseph, 101
Platenius, 78
Plates, photogrphic, 99
Platform tennis, 146-147
Playbach, W., 323
Player piano, 91
Plenciz, M.A., 203
Plimpton, James L., 141
Plotter, 300-301
Plow, 59-60
Plunkett, Roy J., 260
Plus signs, 235
Pocket book, 87
Poiseuille, Louis, 198
Poiteau, Pierre A., 68
Poker, 131-132
Polaroid Corp. 99, 100, 105
Polche, 264
Polyacrylonitrile, 260
Polyamides, 259
Polyester, 260
  bicycle, 12-13
  boat, 143
Polyethylene, 259-260
Polymers, 259
Polymetallic nodules, 265
Polystyrene, 259
Polytetrafluorethylene, 260
Polyvinylchloride, 259

Poncelet, Jean V., 313
Poniatoff, Alexander, 115
Pong, 135
Pool, fish, 73-74
Popcorn, 74
Pope, Franklin, 6
Popert, Symour, 304
Popov, Alexander, 110
Porcelain, 254
Portland cement, 272
Portrait lens, 96
Post, 117-118
Postcard, 118
Poster, 86
Potain, Francois, 198
Potatoes, 74
Pot, coffee, 151
Potter's wheel, 3
Poulet, Pedro P., 327
Poulsen, Valdemar, 115
Pouligny, Bernard, 184
Poultry farming, 71
Powder
  black, 44
  smokeless, 45
Powdered milk, 76-77
Power plants, 191-192, 329, 332-334
Prau, 143
Pravaz, C., 217
Praxinoscope, 101
Prechamber, 325
Prefabricated panel, 273
Premagnetization, 95
Preparations, medical, 215
Prepared piano, 90
Preserving, 77-78
Press
  hydraulic, 286
  printing, 84-86
Pressure cooker, 153-154
Pressure suit, 349
Pressurization, 24
Priestley, Joseph, 242-244, 248, 261
Printer, computer, 301
Printing, 84-86, 287-289
Prison, 187
Probability theory, 236
Probes, space, 349-350, 352
Product Systems Ltd., 12
Progin, Xavier, 88
Protein, 263
Protheses, 209, 221-222
Programming, computer, 298-299
Projector, movie, 102
Propeller, 15, 30
Prussian blue, 258
Prosthesis, auditive, 221-222
Pruning shears, 68
Psychoanalysis, 217
Ptolemy, Claudius, 233, 234, 240, 338
Ptolemy II, 28
Public lighting, 170
Public relations, 175
Public telephone, 120
Public transportation, 6-9
Publishing, 86-87
Pulley, 3
Pullman, George M., 6
Pullman, sleeping car, 6, 7
Pulse code modulation, 95, 96
Pulse-jet engine, 328
Punction balloon, 211
Puppets, 132
Purcell, E.M., 339
Puzzles, 132
Pyrethrin, 70
Pyrethroids, 71
Pyrex, 274
Pythagoras, 231, 240, 251

Q

Quant, Mary, 157
Quantum theory, 241-242

Quartz crystal watch, 277-278
Quevedo, Torres Y., 300

R

Racing
  auto, 135-137
  boat, 141
  horse, 139
Radar, 46, 343, 339
Radial-ply tire, 17
Radiator, 15, 16, 166
Radiglois, Jean-Marc, 184
Radio, 109-111
  beacon, 31
  microwave, 122
  probe, 343
Radioactivity, 252-253
Radio-astronomy, 337-339
Radiocarbon dating, 254
Radioconductor, 110
Radio-immunology, 200
Radiointerferometer, 339
Radiology, 199-200
Radiotelescope, 337-339
Radiotherapy, 217
Railroads, 4-6
Rails, 5
Railway, elevated, 8
Rain, artificial, 341
Raincoat, 157-158
Rainey, Paul M., 96
Rallies, auto, 136, 137
Rambaud, 160
Rambert, Phillippe, 211
Ram, hudraulic, 313
Ramon, Gaston, 202
Ramsay, Sir William, 243
Ramsden, Jesse, 248
Rand Corp., 300, 302
Ransome, Robert, 59
Rateau, Auguste, 326
Ratier and Guibal, 158
Ray, J., 261
Rayleigh, Lord John W., 343
Raynard, Emile, 101
Rayon, 158, 258
Raytheon Co., 153
Razors, 162, 163
RCA, 110, 112, 115, 117, 297, 300, 343
Reaction painting, 188
Reactor
  nuclear, 329-330
  wave energy, 266
Reaming machine, 286
Reaping machine, 60-61
Reard, Louis, 155
Reber, Grote, 338
Reboh, R., 305
Receiver, 111
Recoilless rifle, 42
Recorde, Robert, 235
Recorder, video, 113, 115-116
Recording, first public tape, 95
Recording tape, 95
Records, phonograph, 94
Red Cross, 197
Reddy, Raj, 305
Redier, Antoine, 277
Reed instruments, 90
Reeves, A.H., 96
Refining, 271
Reflection coating, anti, 97
Refrigerating, 283-285
Refrigerator, 154
Refueling, in-flight, 23-24
Register, 306
Registration, vehicle, 6-7
Regnault, Paul-Albert, 158
Reimer, Karl, 75
Rein, 158
Reinforcement brace, 273
Remington Rand Co., 301
Remington Small Arms Co., 88
Remote processing, 300

Renard, Charles, 18
Renaudot, Theophraste, 87
Renault Co., 13, 14, 17
Renault, Louis, 7, 14
Renouard de Villayer, Jean-Jacques, 117, 118
Resin plasters, 210
Respiration, 203-205
Reverdin, Jacques-Louis, 207
Revolver, 36-37
Reynaud, Emile, 102, 103, 105
Reynolds, Osborne, 240
Reynolds, Richard, 5
Rhaeticus, Georg, 234
Rhazes, 210
Rhesus factor, 204
Rhind, Henry, 234
Ricardo, Henry, 325
Ricci, Robert, 161
Rice, Chester W., 96
Rickover, Hyman G., 50
Ridley, Harold, 221
Riemann, Bernhard, 237
Riley, C.V., 66, 70
Ring throstle, 281
Rink, skating, 140
Rip-stop knitting, 282
Ritchell, G.E., 18
Ritter, Johann W., 318
Ritty, James J., 174
Ritzsch, Thimothee, 87
Riva-Rocci, 198
RNA, 262, 263
Road roller, 287
Robert brothers, 17, 18
Robert, Gaspard, 101
Robert, Nicolas, 287
Roberts (planing machine), 285-286
Roberts (tank), 42
Roberts, Charles, 135
Roberts, R., 281, 282
Roberval balance, 239
Robinson, Frank, 71, 77
Robinson, Thomas, R., 341
Robotics, 307-310
Robot, Dr., 76
Robots, 171, 173, 193-194, 197, 307-310
Rocket belt, 23
Rockets, 46-47, 344-346
  engines, 55, 327-328
Rock, John, 215
Rockwell International, 346, 348
Rodent weapon, 167
Rodet, Xavier, 303
Rodiaceta, 260
Roe, 206
Rogallo, Francis M., 139
Roger, Armand, 161
Roger, Ernest, 119
Rogers, Moses, 26
Role games, 131, 132
Rolex Co., 277
Roller, 65
Roller skates, 141, 142
Rolls Royce Co., 328
Roman balance, 239
Rontgen, Wilhelm K., 199, 252
Rooks, Robert, 186
Roos, Michael J., 171
Root, C.S., 77
Rosenblatt, Frank, 295, 304
Rose of Peace, 70
Ross, D.T., 299
Ross, R., 203
Rostard, Jean, 213
Rotor, tail, 22
Rotary dial telephone, 121
Rotary printing, 288-289
Rotenone, 70
Rotheim, Erik, 164
Rouge, 163
Rougerie, Jacques, 187
Roulette, 126

Roulis 2000, 187, 188
Rouquayrol, 32
Roux brothers, 104
Roux E., 202
Rowland, Thomas F., 271
Rowing races, 141
Royer de la Bastie, Francois, 274
Rubber, 257
Rubik, Erno, (Rubik's cube), 132
Rudolff, Cristoff, 235
Rugby, 138
Ruhmkorff, Heinrich D., 14, 319, 322
Rumsey, James, 25
Rush, H.P., 66
Ruska, Ernst, 198
Russell, Bertrand, 237
Russell, John S., 30
Rutgers, 215
Rutherford, Daniel, 243
Rutherford, Lord Ernest, 251-252, 289
Ryle, Sir Martin, 339
Rynd, 217
Rythmostat, 200

S

SAAB, 326
Saccharin, 75
Saccheri, Gerolamo, 232
Sailing, 142-145
  ice, 139-140
  land, 140-141
  ships, 27-28
Salimbeni, A., 202
Salmon, Robert, 60, 62
Sanders, John, 145
Sandwich, 74-75
Sane, 48
Sanger, E., 328
Sanger, L., 263
Sarnoff, David, 114
Satellites, 122, 337, 346-347
Sater, Mel, 115
Sato, T., 222
Sauna, 172
Sauria, Charles, 153
Sauvage, Nicolas, 6
Sauve, Louis, 207
Savery, Thomas, 315
Saw, 286-287
Sawyer, Philip, 290
Sax, Adolphe, 91
Saxophone, 91
Scales, 239
Scali, Francois, 187
Scanner, 199-200, 290
Scarsdale diet, 164
Schaefer, Vincent J., 341, 343
Schalk, C., 259
Schalow, A.L., 289
Schanks, William, 233
Scheaffer, J.J., 218
Scheele, Carl W., 242, 256
Scheider, Ralph, 175
Schick, 163
Schickard, Wilhelm, 294
Schmidt, Egberg, 211
Schmidt, Paul, 328
Schnorr, Baron, 254
Schonbein, Christian F., 258
Schott, Coctor, 96
Schrodinger, Erwin, 252
Schudlam, Molyneux, 143
Schueller, Eugene, 160
Schutzenberger, Paul, 258
Schutz, H.H., 164
Schwarzmann, E., 206
Schweitzer, Hoyle, 147
Science, 231-266
Sclaro, A., 202
Scooter, 10, 11, 185
Scotch tape, 176
Scott, Alfred A., 10
Scrabble, 132

Screw, hydraulic, 313, 314
  threads, 286
Seat-belt, 17, 185-186
Seaplane, 20
Sears, Roebuck & Co., 161
SECAM procedure, 113, 114, 117
Secret writing, 82-83
Securit glass, 275
Security devices, 279-280
Seebeck, Thomas, 318
Seeburg, Justus P., 169
Seelig, Anton, 153
Seiger, Aka, 208
Seiko Co., 173
Seismograph, 269-270
Selenyi, Paul, 89
Selenium, 112
Semiconductors, 250
Semmelweis, I., 211
Semple, William F., 75
Senefelder, Aloys, 85
Senet, 132-133
Senning, Ake, 211
Sense isolation chamber, 218
Seol, Mon Te, 9
Sequin, Louis-Laurent, 323
Sequin, Marc, 4, 317
Sera, 202
Servet, Marcel, 204
Set theory, 237
Sewing machine, 172
Shampoo, 161
Shank, 143
Shape-memory alloys, 290
Sharp, S.H., 288
Shaver, electric, 163
Shaw, D., 304
Sheckley, Robert, 132
Sheet metal, 272-273
Sherman, J.Q., 164
Shibata, Dr., 164
Shillibeer, George, 8
Ships. See Maritime transportation and Naval vessels
Shock-absorber, 12
Shockley, W., 111, 306, 307
Shoe, 158, 182-183, 185, 186, 188
Sholes, Christopher L., 88
Sholto, John, 137
Shore, J., 93
Shortliffe, E., 305
Shuttle, flying, 280
Sickle, 68
Sidecar, 11
Siemens and Malske, 8, 168
Siemens, Wilhelm, 256
Signals, railroad, 5-6
Signature, secret, 176
Sikorsky Co., 20
Sikorski, Igor, 22
Silbermann, G., 90
Sillam, Bernard, 211
Silliman, B., 269
Silly Putty, 133
Silver, Sherman, 207
Silverstone, Peter, 208
Simms, F.R., 11, 17
Simon and Schuster, 87
Simon, F., 304
Simpson, J., 208
Simulation, 300
Simulation games, 133
Singer, Isaac, 172
Sinjou, Joop, 94
Skating, 140, 142
Skiing, 145
Skinner, B.F., 304
Skin transplant, 207
Skoda, 198
Slagle, J.R., 305
Slaughterhouse, refrigerated, 72-73
Slaviter, Henry A., 205
Sleeping cars, 6, 7
Slide, photographic, 100-101
Sloan, John, 171
Slocum, Joshua, 144

Slow motion, 104
Small arms, 35-39
Smeaton, John, 28, 32, 271
Smith and Rennie, 30
Smith, Miss Granny, 69
Smith, Oberlin, 94, 95
Smith, William H., 177
Smith's disc harvester, 60
Smokeless powder, 45
Smullin, L.D., 343
Snail shells, edible, 76
Snap-fastener, 158
Snell, T.B., 73
Snippe, Walter, 185
Snow skiing, 145
Soanes, W.A., 207
Soap, 163-164
Soccer, 145
Sodar, 341-342
Soddy, Sir Frederick, 253
Software, computer, 298
Soil, farming without, 67
Solar energy, 172, 331-332
  aircraft, 22
  calendar, 337
  bicycle, 187, 189
  lenses, 228
Solaro, Ascanio, 45
Solitaire, 133
Solvay, 164
Sommelier, German, 287
Sommerfeld, Arnold, 252
Sony Co., 94, 95, 100, 114, 116, 117, 193
Sosigeny, Alexander, 337
Sound reproduction, 93-96, 103-104
Sowing games, 133
Space, 332, 351-352
  astronomy, 337-340
  conquest of, 190-192, 344-351
  meteorology, 340-343
Space Sciences Inc., 346
Spade, 67
Spallanzani, Lazzaro, 71
Spark plug, 14, 323
Spawning, 73, 74
Speak and Spell, 134
Speakman, 160
Special effects, movie, 105
Species, definition of, 261
Speculum, 212
Speech, 301, 306
Spencer, Percy Le Baron, 153
Sperry, Elmer, 23
Sperry, Lawrence, 23
Spilbury, John, 132
Spindle whorl, 283
Spinning, 280-281
Spinning jenny, 280
Splangler, J.M., 173
Spleen transplant, 208
Spohr, L., 93
Spontaneous generation, 260
Sports, 135-148
Spray
  aerosol, 164
  hair, 161
Sprayer, 66
Sprengel, H., 170
Sprinkling plane, 66-67
Sprague, Frank J., 8
Square root, 235
Stadnick, George O., 188
Stahl, Georg E., 242
Staite, W.E., 169
Staff, music, 93
Stalling, unconfined, 73
Stampfer, Von, 101
Stamps, 118
Stanhope, C., 85
Stanley Aviation Corp., 24
Stanley, Wendell M., 262
Starlz, Thomas, 208
Starter, 12
Statics, 238-242
Statistics, 236

Staufinger, Hermann, 259
Steam
  boat, 25-27
  car, 13
  engine, 315-317, 325-326
  heating, 166
  locomotive, 4-5
  tractor, 64
Steel, 256
Steers, George, 144
Stein, J.A., 90
Stencil, 88
Stencil process, 104
Stenography, 82
Stenuit, Robert, 265
Stephenson, George, 4, 5, 317
Stephenson, John, 17
Stephenson, Robert, 4, 5
Steptoe, Patrick, 214
Stereophonics, 94, 109
Sterikit, 228
Stethoscope, 198
Stevens, John, 30
Stevens, John C., 142, 144
Stevens, Robert L., 28, 30
Stevens, William P., 143
Stevin, Simon, 140, 238, 239
Stewardess, air, 23
Stholer, Franz, 160
Stibitz, G., 293, 294, 302
Stieff Co., 134
Stifel, Michael, 234, 235
Still, Andrew T., 225
Still photography, 96-101
Stimson, S.A., 23
Stirling, Robert, 321
Stitches, 210
Stock exchange, 176
Stockhausen, Karlheinz, 92
Stockings, 158, 282
Stockman Brothers, 156
Stockman, Fred, 156
Stone, Marvin C., 77
Stoney, George, 250
Storage
  battery, 318-320
  blood, 204
  department, 175
  subterreanean, 271
Stores, department, 175
Stradivarius, N., 89
Strasburger, E., 261
Stratojet, 328
Straus, Oscar L., 156
Straw baler, 63
Straws, drinking, 77
Stress card, 201
Stringed instruments, 89-90
Stroboscope, 101
Stroke, C.W., 290
Strout, Jeremiah, 282
Strowger, Almon B., 121
Style, 155-159
Submachine gun, 39
Submarine, 43, 49-51, 192
Subterreanean storage, 271
Subway, 8
Suetonius, 83
Sugar, 74
Suit, lady's, 156
Sulfur, 70
Sullivan, Jeremiah S., 185
Sullivan, W.N., 164
Sulzer, 5, 221, 334
Sumero-Akkadian writing, 81
Sumitomo Co., 71, 194
Sunbeam Co., 151
Sundial, 275
Sunglasses, 221
Super computers, 298
Supercross motorcycling, 141
Superficial musculo-aponevrotic system of the face, 206
Superheater, 317
Superheterodyne, 111
Supermarket, 176
Supersonic plane, 21-22
Surgery, 205-211
Suriray, Jules Pierre, 14

Survis, Dave, 173
Suspenders, 158
Sutherland, I.E., 302, 304
Sutter, Eugene, 160
Sutton, Thomas, 97
Suwa Seikosha, 173
Suzuki Co., 12
Suenska Maskinverken, 325
Swaby, Donald, 167
Swain brothers, 143
Swan, Joseph, 170
Swatch, 279
Swimming, 145-146
Swings, 133, 134
Symbols, 234-235, 237
Syme, 210
Synchronization, movie sound, 103-104
Synchrotron radiation, 253
Synthetic fibers, 158
Synthetic rubber, 257
Synthesizer, 92
Syntony of circuits, 110
Syringe, 217
Szilard, Leo, 47

T

Tablets, 215
Tachi, Susumu, 310
Tachyscope, 101
Talbot, Henry Fox, 97, 98
Talking doll, 127
Tampax Company, 213
Tampons, 213
Tankers, 27, 192, 330
Tank, diving, 32
Tanks, 40-44, 49
Tanks, test, 30
Tape
  magnetic, 302
  measure, 158
  perforated, 302, 303
  recorder, 94-95
  video, 115
Tarnier, S., 213
Tarnower, Herman, 164
Tarot, 134
Tartini, Giuseppe, 89
TAV, 192
Taylor (reaping machine), 60
Taylor (turbulent chamber), 325
Taylor, Bruce, 309
Taylor, Daniel F., 157
Taylor, Mike, 184
Taylor, Frederick W., 13
Taylor, G.I., 28
Taylor, H.D., 96, 97
Taylor shorthand, 82
Taxi, 6-7
Tchechett, Victor, 143
Tea, 77
Technicolor, 104-105
Teddy bear, 134
Teeth
  extraction, 223
  false, 224
  transplant, 207
Tefal Co., 154
Teflon, 260
Telecommunications, 117-122
Telegraph, 118-120
Telephone, 120-122
Telephoto lense, 96-97
Telescope, 339-340, 352
  radio, 337-339
TV, 112
Television, 111-115, 173, 193; see also Video
Telford, Thomas, 287
Teller, Edward, 47
Temperature, 246-247
Tempered glass, 274
Tempering, 255
Tenia, laser, 211
Tennis, 146-147

Terillon, 210
Terminal, computer, 301
Tesla, Nicholas, 110, 321
Testicle transplant, 207
Test-tube baby, 214
Tetanus sera, 202
Texas Instruments, 129, 130, 134, 296
Thales, 232, 247
Thaumatrope, 101
Thenard, Louis-Jacques, 256
Theremin, L., 92
Thermal energy, 165-166, 333
Thermocouple, 318, 320
Thermodynamics, 245-247
Thermo-Fax, 89
Thermometer, 187, 198, 342
Thermos, 154
Thiel, W., 327
Thimonnier, Barthelemy, 172
Thomaseus, 223
Thomas, P., 304
Thompson, William, 264
Thomson and Simon, 95
Thomson Co., 199, 254, 290
Thomson, Elihu, 285
Thomson, Joseph, 250, 251
Thomson, R.G., 88
Thomson, Robert W., 16
Thomson, Sir William, 245, 246
  3D, 105
3M Co., 89, 115, 176, 210, 218, 222, 227, 228
Threshing machine, 61-62, 64
Thorpe, J., 281
Throttle valve, 324
Tie, 159
Tie break, tennis, 146
Tide power, 265-266, 330, 331
Tide tables, 29
Tiemann, Ferdinand, 75
Tile, 272
Tilsby, John, 157
Time clock, 277
Time, standard unit, 238
Timiriadzer, 67
Tin cans, 78
Tires, 3, 16-17
Tissot, 119
Toaster, 154
Tobacco, 70
Tomas, John, 223
Tomato, 70, 193
Tomes, Sir John, 224
Toothpaste, 222
Toothpick, 176
Top hat, 159
Topiary art, 67
Topographic images, 228
Torino, 76
Topedo, 50, 52
  boat, 51-52
Torque converter, 14
Torque rotor, anti, 22
Torricelli, Evangelista, 30, 244, 341
Toshiba Co, 116, 193, 298, 310
Toulouse-Lautrec, 86
Touraine, Jean-Louis, 208
Tourte, F., 89-90
Townes, C.H., 289
Townsend, Sir John, 249
Toys and games, 125-135, 187
TPA enzyme, 228
Track, 134-135
Track and field, 147
Tractor, 59, 64-65
Trains. See Railroads
Trampoline, 147
Tramway, 7-8
Transformer, 321
Transfusion, blood, 204
Transistor, 111, 306, 307
Translation, automatic, 193, 304
Transluminal angioplasty, 211
Transmission, 9, 12, 14
Transplant, 207-208

Transportation
  air, 13-25
  land, 3-13
  maritime, 25-31
Trautonium, 92
Trautwein, F., 92
Travel agency, 177
Traveler's check, 176, 177
Travers, Morris W., 243
Treads, aquatic, 188, 189
Treatments, medical, 215-220
Trees, ornamental, 67
Tressaquet, Pierre, 287
Trevithick, Richard, 4, 316
Trick clock, 279
Trick photography, 105
Tricycle, 173, 186
Triebert, Frederic, 91
Trigonometry, 233-234
Trimaran, 142-143
Trinitron color tube, 114
Triplex, 274, 275
Tritheme, Abbe, 83
Trolleybus, 8-9
Troop carrier, 42
Trough, drinking, 73
TRS Hobbies, 132
Truax, Robert, 346
Trucks, 326
Trumball, Douglas, 193
Trumpet, 91
Truong, Trong, 296
Tsiolkowski, Konstantin, 327
Tsukuba, 192-193
Tube
  color TV, 114
  electrostatic, 306
  lighting, 170-171
  radio, 109, 111
Tukey, John, 300
Tulip, 70
Tuna net, 74
Tuning fork, 93
Tungsten filament, 170
Tunnel, underwater, 266
Tupper, Earl W., 155
Tupperware, 155
Turbines, 13, 313-315, 325-327
Turbocharged engine, 326
Turbodrilling, 271
Turbo aircraft, 20, 54-55
Turbo machines, 325-328
Turing, Alan, 293, 296, 304
Turner, D.A., 178
Turrets, 48-49
Turri, Pellegrini, 88
Turthmosis III, 68
Two-wheelers, 9-13
Tyndall, John, 290
Typography, 287-288
Typewriter, 88

U

UFO's, 190
ULM flyers, 147
Ultrsound surgery, 222
Underwater, 264-266
  cable, 119-120
  tank, 43, 49
Union Carbide, 207
United Airlines, 20
Universal joint, 14
Upnatjeks, J.V., 290
Ur, 135
Uranium enrichment, 290
Urey, Harold, C., 257, 329
U.S. Army, 37
U.S. Navy, 17-18, 50

V

Vaccinations, 201-203
Vacuum, 244
Vacuum cleaner, 173

Vaginal sponge, 71-72
Vail, Alfred L., 119
Valve, butterfly, 15
Van Allan, James, 347
Van Beneden, Edouard, 261
Van de Bursen, 176
Van de Hulst, H.C., 339
Van de Velde, 215
Van Eyck brothers, 84
Van Helmont, Jan B., 242
Van Houten, Conrad, 75
Van Kampelen, Baron, 301
Van Musschenbroeck, Petrus, 248
Van Tassel, J.H., 296
Variolation, 203
Vectors, 236
Velcro, 159
Velocifer, 9
Velocipede, 9, 13-14
Venetian glass, 274
Ventilation, artificial, 164
Veraart, Claude, 222
Verdun, Pierre, 152
Verhoeven, Abraham, 87
Verne, Jules, 333
Vertical takeoff and landing aircraft, 55-56
Veseler, George, 87
Vessels. See Maritime transportation and Nval vessels
Viau, Robert, 187
Vickers-Armstron, 20
Viete, Francois, 234
Video, 115-117
  games, 134, 315
Video Answer Service, 193
Videocassette recorder, 116
Videodisc, 116, 117
Video-endoscopy, 228
Vidie, L., 341
Vieille, Paul, 45
Vihlen, Hugo, 145
Vincent, Clovis, 206
Violin, 89
Virag, Ronald, 200
Virology, 262
Viruses, 203
Visual communication, 81-89
Vitamins, 226-227
Voice
  coding, 228
  synthesizer, 303
Voiscope, 228
Voisin, Gabriel, 19, 20
Vollyball, 147
Volta, Alessandro, 14, 118, 238, 318, 319, 323, 333
Volvo Co., 17, 326
Von Anhalt, Prince Jurgen, 188
Von Baeyer, Adolf, 259
Von Behring, E., 202
Von Bekesy, Georg, 222
Von Bergmann, E., 211
Von Braunmuhl, J., 95
Von Braun, Wernher, 327-328, 345-347
Von Gerstner, Franz A., 6
Von Guericke, Otto, 244, 247-248, 315
Von Jacobi, Moritz H., 320
Von Kleist, J., 248
Von Lave, Max, 252
Von Linde, Karl, 245, 285
Von Mayer, Julius R., 246
Von Neumann, J., 293-294
Von Ohain, Hans, 327
Von Platen, B., 154
Von Schirnahaus, 254
Von Sauerbronn, Baron Carl von Drais, 9
Von Shertel, Hans, 28
Von Sohne, Julius A., 87
Von Steinhill, Carl A., 118
Von Welsbach, Auer, 170
Von Zeppelin, Count Ferdinand, 18

Voogt, Henry, 144
Voronoff, Serge, 207
Vuiturinus, Georg, 135

W

Waaler, Johann, 89
Waksman, S., 216
Wales, Nathaniel, 154
Walker, Charles D., 218
Walker, William H., 98
Walkman, 95
Waller, Augustus D., 198
Wallis, John, 233
Walter, Brooks, 16
Walter, Helmuth, 328
Walton, Fred, 171
Wands, Jack R., 227
Ward, Nathaniel B., 68
Wargames, 135
Warren, J., 208
Washing machine, 173
Waste converter, 73
Watches and clocks, 275-279
Water
  composition of, 243-244
  current map, 343
  energy, 172, 331-332
  frame, 281
  heavy, 328, 329
  lock, 275
  walker, 181
Waterhouse diaphragm, 97
Waterman, L.E., 84
Watermark, 177
Waterproof watch, 277
Water skiing, 145
Waterwheel, 313, 314
Watson, George L., 28
Watson, J.D., 263
Watson-Watt, Sir Robert, 46
Watt, James, 87, 166, 307, 315-316
Watt, Thomas, 25
Wave energy, 252, 265, 266
Wave oscillator, 109-110
Wave, permanent, 160
Waves, radio, 109, 110
Weaponry, laser, 289
Weapons. See Armaments
Weather. See Meterorology
Weaver, W., 304
Webb, Mathew, 145
Weber, W., 95
Wedgewood, R., 88
Weed, Harry D., 17
Weight-driven clock, 276
Weighted keel, 28
Weights, 237-238
Weight Watchers, 164
Wessel, Caspar, 236
Widman, Johann, 235
Wiener, Norbert, 307
Welding, 285
Well-logging, 270
Wells, H., 208
Wermer, Ernst, 96
Werner, Eugene and Michel, 10
Western Electric Co., 109, 110, 122
Westinghouse Electric Corp., 6, 110, 117
Westinghouse, George, 6
Wham-O Manufacturing Co., 128, 129
Wheatstone, Charles, 119, 277
Wheel
  balance, 276-277
  evolution of the, 13
  fifth, 6
  vertical, 313
Wheel lock, 35
Whist, 135
Whitehead, Robert, 52
White, Phillip, 262
Whitney, Eli, 287

Whitten, Glenne, 227
Widman, Johann, 235
Wiener, Norbert, 300, 307
Wilcox, Howard A., 266
Wilkes, H.V., 296, 299
Wilkes, M., 299
Wilkonson, J., 286
Williams, F.C., 306
Williams, Greville, 257
Williams, Robin, 228
Williams, W.H., 64
Willis, J., 82
Wilsdorf, Hans, 277
Winchester carbine, 37
Winchester Co., 37
Windmill, 330
Wind gauge, 341
Winding clocks, 277
Wind instruments, 90-91
Wind power, 27-28, 330-331, 334
Windshield wiper, 17
Wind surfing, 147-148
Wingfield, Walter C., 146
Winkler, Johannes, 344
Winower, 62
Wire, barbed, 73
Wired glass, 274
Wirth, N., 300
Withworth, J., 285, 286, 287
Witzell, S.A., 73
Woerpel, Rick, 186
Wohler, Friedrich, 256
Wood, 217
Wood, Kenneth, 152
Woodrow, J.C., 204
Wood, Walter, 61
Woolf, Arthur, 316
Work-heat equivalent, 246
Workshop, portable, 171
Worth, Charles-Frederick, 156
Worth, Henry Shuttle, 157
Wren, C., 217
Wright, Samuel, 157
Wright brothers, 19, 26
Wrist TV, 173
Wristwatches, 177
Writing, 81-83
  materials, 83-84
Wynne, Arthur, 126

X

Xerography, 89
Xerox Co., 89
X-rays, 199, 252

Y

Yachting, 142
Yale, Linus, 279-280
Yalow, Rosalyn, 200
Yanguel, Mikhail, 344, 345
Yao, 128
Yersin, A., 202
Yeu Nan Jin, 187
Yin, Lo I., 200
Yogurt, 74
Yoo Byung Eun, 186
Young, Thomas, 221
Yperite, 46

Z

Zeigler, H., 289
Zeiss, Karl, 97, 274
Zeltner-Genas, 158
Zephinie, Gerard, 168
Zeppelin, 18, 19
Zero, 231
Ziegenspeck, Fritz, 30
Ziegler, Karl, 260
Zipper, 159
Zukauskas, Raymond, 172
Zukor, A., 106
Zuse, Konrad, 296
Zworykin, Vladimir Kosma, 112, 116, 198

CALLING ALL INVENTORS

Here's your chance to be included among the ranks of Thomas Alva Edison, Alexander Graham Bell, and other famous inventors whose creations have changed the course of history.

Your invention may just be the one that makes it into THE WORLD ALMANAC BOOK OF INVENTIONS" next year—the only book to bring together over 2,000 inventions that have shaped our world. With this prize comes a lifetime subscription to THE WORLD ALMANAC" and a handsome, personalized commendation suitable for framing.

It's all part of the exciting new contest being sponsored by World Almanac Publications, and there are only a few, simple rules to follow.

OFFICIAL RULES

In conjunction with the publication of the first edition of THE WORLD ALMANAC BOOK OF INVENTIONS, World Almanac Publications is holding a contest to determine the fourteen most interesting or important inventions of the recent past for which a patent is issued between January 1, 1983 and February 28, 1986. The following rules apply to all entries:

1. For the invention that is being entered in the contest, a patent must have been issued by the United States Patent Office between the dates stated above.

2. The invention must fall into one of the following fourteen categories: Transportation; Weapons; Agriculture; The Arts; Media and Communication; Games, Toys and Sports; Everyday Life; The Bizarre and The Future; Medicine; The Sciences; Industry; Information Systems; Energy; or Space.

3. An official entry blank will be included in THE WORLD ALMANAC BOOK OF INVENTIONS or can be obtained by writing to World Almanac Publications, 200 Park Avenue, New York, NY 10166. Only these entry blanks or those that have been made available at the 9th Annual International Inventors Exposition '85 will be considered valid.

4. The entry must be accompanied by a photograph or drawing of the invention, a photocopy of the front page of the "patent copy" (which includes the patent number and illustration), and a complete description of its intended use and the manner in which it operates. These documents will not be returned to the entrant unless a stamped, self-addressed return envelope is included with them. Please include on the envelope in which you mail the documents to World Almanac

Publications the category—Transportation, etc.—to which you are applying for consideration.

5. Entries must be received by midnight, March 7, 1986.

6. Five inventions per category will be selected by World Almanac Publications staff for submission to a panel of authorities in each of the subject fields, the panel to be chaired by Jane Flatt, Publisher of THE WORLD ALMANAC. These experts will choose the winners in each category on the basis of a simple majority of votes.

7. The fourteen winners will be notified by mail on or about June 1, 1986 and asked to send a short biography and photograph of the inventor and invention to World Almanac Publications for inclusion in the next edition of THE WORLD ALMANAC BOOK OF INVENTIONS. A list of winners and a short description of their inventions will be made available to any entrant who requests one and accompanies such request with a stamped, self-addressed return envelope.

8. Families and employees of Scripps Howard, Random House, Inc., their divisions and subsidiaries, Baker & Hostetler, Zachary & Front, Inc., and Sarris Bookmarketing Service are not eligible to enter this contest.

THE WORLD ALMANAC BOOK OF INVENTIONS
——————————— Contest Entry Form ———————————

PLEASE PRINT:

_____
Name

_____
Street Address

_____
City   State   ZIP Code

_____
Signature   Date

Description of Invention (attach separate sheet if necessary) :

_____

_____

_____

Please complete the information requested above and send it along with a photograph/illustration of your invention and proof of issuance of patent between January 1, 1983 and February 28, 1986 to:

THE WORLD ALMANAC BOOK OF INVENTIONS
World Almanac Publications
200 Park Avenue
New York, NY 10166

Please don't forget to write the category of your invention on the envelope.